Springer Hydrogeology

The Springer Hydrogeology series serves as a comprehensive platform for advancing knowledge and fostering collaboration in the diverse and multidisciplinary field of hydrogeology. This book series seeks to cater to a wide audience, including researchers, educators, practitioners, policymakers, and students who are passionate about understanding and solving groundwater-related challenges. It aims to provide high-quality, peer-reviewed monographs, edited volumes, textbooks, and conference proceedings to disseminate cutting-edge research, practical insights, and innovative approaches in hydrogeology and related disciplines.

The series covers the entire spectrum of hydrogeological topics, including but not limited to: groundwater science and management, isotope hydrogeology, hydrogeological modeling, groundwater resources and systems, hydrogeochemistry, ecohydrology and groundwater-energy-food nexus, cryohydrogeology, urban and agricultural hydrogeology, emerging technologies in hydrogeology, climate change impacts on groundwater and groundwater-dependent ecosystems, and geotechnical and environmental aspects.

The Springer Hydrogeology series is aligned with the United Nations Sustainable Development Goals (SDGs), particularly SDG 6 (Clean Water and Sanitation) and SDG 13 (Climate Action). By publishing works that address sustainable groundwater use, pollution mitigation, climate resilience, and integrated groundwater resource management, the series supports efforts to achieve equitable and sustainable access to groundwater resources for current and future generations. Contributions that explore the intersection of hydrogeology with other SDGs—such as those on biodiversity, sustainable cities, and food security—are also strongly encouraged.

Peiyue Li · Xiaodong He · Jianhua Wu ·
Vetrimurugan Elumalai
Editors

Sustainable Groundwater and Environment: Challenges and Solutions

Springer

Editors
Peiyue Li
School of Water and Environment
Chang'an University
Xi'an, China

Jianhua Wu
School of Water and Environment
Chang'an University
Xi'an, China

Xiaodong He
School of Water and Environment
Chang'an University
Xi'an, China

Vetrimurugan Elumalai
Department of Hydrology
University of Zululand
Durban, South Africa

ISSN 2364-6454 ISSN 2364-6462 (electronic)
Springer Hydrogeology
ISBN 978-3-031-82193-6 ISBN 978-3-031-82194-3 (eBook)
https://doi.org/10.1007/978-3-031-82194-3

This Springer imprint is published by the registered company Springer Nature Switzerland AG
The registered company address is: Gewerbestrasse 11, 6330 Cham, Switzerland

Foreword

"A great journey starts with a small step." This proverb, as timeless as it is profound, perfectly encapsulates the spirit of the work presented in this book, Sustainable Groundwater and Environment: Challenges and Solutions. As we stand at the crossroads of environmental sustainability and the urgent need to safeguard our precious groundwater resources, I am honored to write the Foreword for this book. This book is a testament to the power of international collaboration and the importance of scientific research in addressing the pressing issues of our time.

Groundwater is the inevitable lifeline beneath our feet, plays an indispensable role in sustaining life, promoting economic development, and maintaining an ecological balance. However, our groundwater resources are under increasing pressure due to human activities and climate change. In South Africa, for example, over 60% of our municipalities are in water-stressed regions, with groundwater over-extraction and contamination posing significant challenges. Similarly, in China, rapid urbanization and industrialization have led to severe groundwater pollution and overexploitation, affecting over 400 million people. Safeguarding these resources while ensuring sustainable development is a complex challenge that requires a deep understanding of the intricated interplay between hydrogeological processes and human-induced impacts.

The collaboration between Chang'an University in China and the University of Zululand in South Africa, as detailed in this book, is an exemplar of how international cooperation can lead to significant advancements in our understanding of these complex issues. The partnership, which has been generously supported by the National Natural Science Foundation of China (NFSC) and the National Research Foundation (NRF) of South Africa, began with a small step but has now become a great journey, yielding fruitful results over the past eight years.

Under the leadership of Prof. Peiyue Li and Prof. Vetrimurugan Elumalai, the joint research project has led to more than 20 journal articles, a book, five co-organized international conferences, and the co-supervision of more than ten master and Ph.D. students. These achievements are a testament to the power of collaborative research and the commitment of both universities to advancing knowledge in this critical field.

On September 13, 2024, a Memorandum of Understanding (MOU) was signed between the two universities, marking a significant milestone in their ongoing collaboration. This is a clear demonstration of the commitment of both institutions in fostering international collaboration and advancing research on sustainable groundwater and environmental challenges.

Looking forward, it is my hope that this book will serve as a catalyst for further research and collaboration in this field. The challenges we face are significant, but the potential rewards are immense. By working together, we can deepen our understanding of the complex issues surrounding groundwater and environmental sustainability, and develop effective strategies to protect these precious resources for future generations.

In closing, I would like to express my deepest appreciation to the governments of China and South Africa, the NFSC, the NRF, and the universities of Chang'an and Zululand for their unwavering support of this important work. I am confident that this collaboration will continue to yield significant insights and contribute to the global effort to ensure sustainable groundwater and environmental management.

Prof. Xoliswa Mtose
Vice Chancellor
University of Zululand
Durban, South Africa

Preface

In our present era, the significance of sustainable groundwater management is increasingly recognized. Groundwater is a critical resource that sustains not only our agricultural and industrial activities but also supports the health and vitality of ecosystems. However, the sustainability of this precious resource is threatened by a host of factors including climate change, over-extraction, pollution, and land-use changes. Thus, the task of ensuring sustainable groundwater management, while balancing the needs of the environment, presents a complex yet urgent challenge.

In 2017, a joint initiative was launched by the National Natural Science Foundation of China (NSFC) and the National Research Foundation (NRF) of South Africa to sponsor ten research projects to foster collaboration between Chinese and South African researchers, addressing environmental and resource issues of mutual interest. Our project, which forms the basis of this book, focused on the critical issues concerning the evolution of aquifer permeability and groundwater quality in the essential critical zones of the Maputland aquifers in South Africa and the loess aquifers in China. Through this collaboration, we investigated various facets of these critical zones, thereby enhancing our understanding of the complexities inherent in the nature of these aquifers. This book is a product of this collaborative effort. It comprises 16 chapters contributed by researchers from an array of universities and institutes in China, South Africa, Egypt, Kuwait, Malaysia, Mexico, Ghana, and Nigeria. The topics of the chapters range from fundamental hydrological circulation of groundwater and techniques for hydrogeological research to groundwater pollution and remediation and groundwater management in different regions. This book, while originating from a collaboration between China and South Africa, has been enriched by contributions from a broader research network encompassing additional Global South countries. Given the global nature of groundwater issues, it is our belief that their resolution necessitates a collaborative approach that transcends national boundaries. This inclusivity not only broadens the scope of our research but also fosters a comprehensive and multinational perspective on groundwater challenges.

The first chapter "Hydrogeology and the Global Significance of Groundwater" emphasizes the critical role of groundwater as a vital resource for domestic use, agriculture, industry, and ecosystems. It highlights key issues in agricultural hydrogeology, particularly the impact of irrigation on groundwater resources, and discusses challenges such as over-extraction, contamination, and climate-induced changes in water availability. The chapter proposes strategies to advance hydrogeology, including enhanced research, improved data sharing, citizen science, international collaboration, and government support. By implementing these strategies, the hydrogeological community can contribute to the sustainable management and protection of groundwater resources globally.

The second chapter "Towards Sustainable Groundwater Management in Kuwait: Strategic Directions" discusses the critical challenges and solutions related to groundwater sustainability in Kuwait, particularly within the context of achieving the United Nations Sustainable Development Goal (SDG) 6, which focuses on clean water and sanitation for all. It provides a comprehensive roadmap for addressing Kuwait's water challenges, integrating modern technology, policy reforms, and regional cooperation. It emphasizes the importance of sustainable groundwater management to ensure long-term water security in arid regions like Kuwait. The proposed strategic directions, such as innovative recharge techniques, IoT systems, and desalination advancements, offer valuable insights for achieving the UN's SDG 6 in water-scarce environments.

The third chapter "Impacts of Artisanal Small-Scale Mining and Climate Change on Groundwater Security: The Case of the Global South" examines the combined impacts of artisanal small-scale mining (ASM) and climate change on groundwater resources in the Global South. ASM, widespread in regions such as Africa, Latin America, and Asia, is often unregulated, leading to heavy metal contamination (e.g., mercury, cyanide) and physical degradation of groundwater systems. Coupled with this, climate change exacerbates groundwater depletion through altered precipitation patterns and prolonged droughts, reducing aquifer recharge. Case studies highlight that these dual forces threaten water quality and availability, ultimately jeopardizing livelihoods, agriculture, and health. The chapter calls for integrated management strategies, stricter regulation, and technological advancements like managed aquifer recharge (MAR) to mitigate these issues. It also emphasizes the need for international cooperation, particularly in capacity building and policy development, to safeguard groundwater security in vulnerable regions.

The fourth chapter "Groundwater and Public Health: Understanding the Links and Mitigating Risks" focuses on the critical relationship between groundwater quality and public health, emphasizing the risks posed by geogenic and anthropogenic contamination. Naturally occurring contaminants, such as arsenic and fluoride, affect millions globally, leading to diseases like arsenicosis and fluorosis. Human activities, including industrial discharges, agriculture, and sewage, introduce heavy metals, nitrates, and emerging pollutants like microplastics and PFAS into groundwater, resulting in neurological, developmental, and immune system disorders. The chapter underscores the importance of health risk assessments, pollution source identification, and remediation strategies. It advocates for comprehensive groundwater management, including improving water quality and remediating contaminated sites. Case studies, particularly from China, illustrate the success of interventions like water quality improvement projects in mitigating arsenicosis and fluorosis. The chapter highlights the need for interdisciplinary approaches to safeguard groundwater and public health.

The fifth chapter "Effect of Groundwater-Lake Interaction on Water and Salinity Balance for Lakes in Arid Regions" provides a comprehensive review of the literature on groundwater's role in the water and salinity balance of lakes in arid regions. It emphasizes the importance of groundwater in maintaining the water environment and ecosystem health of lakes. The chapter systematically examines the influence of groundwater on lake hydrology and salinity using quantitative methods, focusing on both direct and indirect techniques such as geophysical methods, remote sensing, and isotopic and hydrochemical fingerprinting. The review highlights the challenges and future research directions in this field, contributing to the sustainable management and development of lake ecosystems in arid regions.

The sixth chapter "Groundwater in Arid and Semi-arid Regions: Issues of Scarcity and Quality" examines groundwater conditions in the Delingha area of the Qaidam Basin on the Tibetan Plateau, focusing on the impact of climate change. It uses the Standardized Precipitation Evapotranspiration Index (SPEI) to assess drought conditions and their correlation with groundwater levels. The research finds that while drought conditions have generally improved, certain periods experience severe drought. Groundwater levels in inhabited areas show a weak correlation with SPEI due to human activity. The study identifies the northwestern part of the area as optimal for sustainable groundwater development.

The seventh chapter "Machine Learning Techniques in Hydrogeological Research" provides a systematic review of machine learning applications in hydrogeological research. It introduces common machine learning methods like naïve Bayes, artificial neural networks, support vector machines, decision trees, and random forests. The chapter summarizes how these methods have been used to predict groundwater variables, map hydrogeological variables, identify groundwater contamination sources, and assess contamination risk. It also discusses the challenges and future prospects of using machine learning in hydrogeology, emphasizing the need for enhanced data sharing and interdisciplinary cooperation.

The eighth chapter "Groundwater Salinization in Coastal Regions and the Control Mechanisms: Insights for Sustainable Groundwater Development and Management" presents a comprehensive analysis of groundwater salinization in coastal regions, exploring its concepts, mechanisms, impacts, and management strategies. It examines case studies from diverse geographical contexts, including the Mediterranean coastal areas, South Asian coastal aquifers, and the Gulf Coast of the United States. The chapter emphasizes the interaction between natural hydrogeological processes and anthropogenic influences that drive salinization, and it assesses the environmental, socio-economic, and health impacts. It also discusses emerging technologies and innovative approaches for managing coastal aquifers, providing recommendations for future research and practice.

The ninth chapter "Characteristics of Groundwater Level Dynamics in Xi'an (2010–2020) and Groundwater Level Prediction for 2030 Using Regression Analysis and BP Neural Network: Implication to Groundwater Resources Management" focuses on the characteristics of groundwater level dynamics in Xi'an between 2010 and 2020, and it predicts groundwater levels for 2030 using regression analysis and a BP neural network. The study analyzes groundwater level changes and their contributing factors in the Guanzhong Basin of northwest China. It presents data on groundwater level variations and their relationship with various factors such as precipitation and human activities. The chapter contributes to the understanding of groundwater dynamics in arid regions and provides a basis for effective groundwater management and planning in Xi'an.

The tenth chapter "Evaluating Uranium Toxicity in Groundwater and Associated Health Risks in a Hard Rock Aquifer, South India" evaluates the levels of uranium toxicity in groundwater and its associated health risks in a hard rock aquifer in South India. The study collected 54 groundwater samples and found that 20.3% exceeded the WHO's permissible limit for uranium. The research used geospatial analysis and health risk assessment methods to identify high-risk groups, particularly children and infants, and linked the contamination primarily to geological formations and human activities. The findings stress the need for regulatory controls and remediation strategies.

The eleventh chapter "Groundwater Resources Availability in the Shangzhou Basin Area, China: A Modeling Research" presents a modeling study on the availability of groundwater resources in the Shangzhou Basin area of China. Using the MODFLOW model, the research assessed the current groundwater extraction rates and predicted future changes under different scenarios. The study found that the maximum allowable extraction volume without causing significant environmental impact is around 16 million m^3/a. The research provides scientific support for sustainable groundwater management in the region and contributes to the South-to-North Water Diversion Project.

The twelfth chapter "Utilization of Green Chemistry Nano-Emulsion in Remediation of Real Contaminated Coastal Sediments by HOCs and Pb to Protect Shallow Aquifer" discusses the use of green chemistry nano-emulsion for the remediation of coastal sediments contaminated with hydrophobic organic compounds (HOCs) and lead (Pb). The study presents a method for synthesizing a nano-emulsion that effectively removes HOCs and Pb from sediments, protecting the shallow aquifer. The research was conducted on a contaminated site in Egypt, where the nano-emulsion was found to significantly reduce TPHs and Pb levels, suggesting its potential as an efficient and environmentally friendly remediation technique.

The thirteenth chapter "Magnetic Induction on pH, Oxidation-Reduction Potential, Conductivity and Dissolved Oxygen Level of Water and Its Properties: A State-of-the-Art Review" reviews the effects of magnetic induction on water quality parameters such as pH, oxidation-reduction potential (ORP), conductivity, and dissolved oxygen (DO) concentration. It discusses the principles of magnetic water treatment (MWT) and its applications in improving water parameters and properties. The study highlights that MWT can be a cost-effective and eco-friendly alternative to conventional water treatment methods, with potential contributions to contamination control, microbial inhibition, and enhancement of fish and plant growth, as well as healthcare.

The fourteenth chapter "The Deficit Libres-Oriental Aquifer (Balsas Region) of Eastern-Central Mexico Under Global Warming: Hydrochemistry, Moisture Recycling and Rock-Water Interaction" assesses the hydrochemical characteristics and stable isotope compositions of groundwater in the Libres-Oriental aquifer of east-central Mexico across four seasons, considering the context of global warming and recurrent droughts. The groundwater exhibited homogeneity with Ca-Mg-HCO_3 facies, indicating decadal-scale mixing of moisture sources. Autumn samples showed higher variations in isotope fractions, suggesting additional recycling processes during wetter months. The study highlights the importance of sustainable groundwater management in the face of climate change, as the aquifer faces overextraction and depleted precipitation.

The fifteenth chapter "Radionuclides as Environmental Contaminants of Concern: Threats to Public Health Through Soil and Groundwater" examines the distribution, behavior, and health impacts of radionuclides in Nigeria, with a global perspective. It reveals significant spatial variability in radionuclide concentrations, with areas like the Jos Plateau showing higher natural radioactivity. Groundwater investigations also showed variations, with some areas having elevated radon levels. The study underscores the need for standardized measurement techniques, expanded geographical coverage, and integrated research approaches linking environmental data with epidemiological studies. It emphasizes the potential long-term health risks posed by localized radionuclide hotspots and contributes to the global discourse on radionuclide contamination management.

The sixteenth chapter "Future Research Imperatives in Hydrogeology" focuses on emerging research themes in hydrogeology, emphasizing the critical role of groundwater in sustaining global water supplies amid increasing pressures such as climate change, population growth, and contamination. Key areas explored include the impact of climate change on groundwater recharge, water quality, and the interaction between surface and groundwater systems. It highlights technological advancements like remote sensing and isotopic analysis to monitor and manage groundwater resources effectively. The chapter also discusses the potential of Managed Aquifer Recharge (MAR) as a sustainable water management strategy and the growing importance of machine learning in hydrogeochemical modeling for improved groundwater management.

This book covers a broad range of topics in groundwater and environmental sciences, offering valuable insights for researchers, students, practitioners, and policymakers. Its contents will contribute significantly to advancing knowledge and addressing critical challenges in these fields.

Xi'an, China — Prof. Peiyue Li
Xi'an, China — Dr. Xiaodong He
Xi'an, China — Prof. Jianhua Wu
Durban, South Africa — Prof. Vetrimurugan Elumalai

Acknowledgments

We are happy to see the publication of this edited book. It is the joint efforts from publisher, editors, contributors and reviewers that make the publication of this book possible. Therefore, at this moment, we would like to acknowledge the help received from Qiao Shu, our editorial contact. She is always there for giving help whenever we have questions or queries about the book publication. Of course, without the contributions from all the contributors, this book would have never been possible. We thank them for their interests in this book and their cooperation in timely submission of the chapters. We are also grateful for the critical comments on the early versions of the chapters rendered by reviewers. Without their comments, the book would have never become so attractive. Their comments have been constructive and insightful and have helped the contributors to further improve the quality of their chapters.

The editors of this book would like to express their gratitude for the support provided by the governments of China and South Africa in recent years, especially the National Natural Science Foundation of China (NFSC) and the National Research Foundation (NRF) of South Africa. Without their support, we would not have been able to collaborate closely and achieve fruitful results over the past eight years. We would also like to thank the universities, Chang'an University in China and University of Zululand in South Africa, who have provided us with many conveniences in carrying out international collaborative research. On September 13, 2024, the two universities signed the Memorandum of Understanding (MOU), marking another leap forward in international cooperation between the two sides.

The year 2024 is special for the leading editor of this book, Prof. Peiyue Li. He turns 40 this year. He spent his first 20 years in his home province, Hebei Province, China, and the next 20 years in Shaanxi Province, China where he has been staying for study and work since college. At this special time, he would like to express his gratitude to every teacher, classmate, colleague, friend, and family member he has met over these 40 years. Their ongoing support and encouragement have made Peiyue Li who he is today.

Finally, we acknowledge the research grants from the National Natural Science Foundation of China (42072286, 42472316 and 41761144059), the National Key Research and Development Program of China (2023YFC3706901), the State

Top Foreign Experts Recruitment Program (G2023171005L, G2023171004L, G2022171007L, G2021171005L), the Qinchuangyuan "Scientist + Engineer" Team Development Program of the Shaanxi Provincial Department of Science and Technology (2022KXJ-005), the National Ten Thousand Talents Program (W03070125) and the Fok Ying Tong Education Foundation (161098). These research grants have been helpful in enabling us to conduct international research collaborations.

Contents

Editors and Contributors

About the Editors

Prof. Peiyue Li is a full professor in the field of hydrogeology and environmental sciences at Chang'an University, China. He has wide experience in groundwater quality assessment, hydrogeochemistry, and groundwater modeling. He has edited and co-edited ten textbooks and academic monographs, and published over 200 articles in refereed journals on topics that range from groundwater quality assessment and groundwater hydrochemistry to groundwater pumping tests and in situ tracer tests. He at present serves as Associate Editors for Exposure and Health, Mine Water and the Environment, Archives of Environmental Contamination and Toxicology, Environmental Monitoring and Assessment, Human and Ecological Risk Assessment, Applied Water Science, Water Practice and Technology, and Discover Water. He has been awarded one of the most highly cited researchers by Clarivate, Elsevier, and other organizations.

Dr. Xiaodong He is currently working as a young research scholar in the field of hydrogeology and environmental sciences at Chang'an University, China. He holds a Ph.D. in hydraulic engineering from Chang'an University, and has wide experience in hydrogeochemistry, water-rock interaction, hydro-ecology and water security. He has published more than 20 scientific papers in international journals on topics that range from groundwater quality to groundwater pollution, of which 12 papers have been selected as ESI top 1% highly cited papers. He had been selected as top 2% highly cited scientists worldwide (2022). He at present serves as guest editor of Human and Ecological Risk Assessment. He has led and participated in more than 10 research projects funded by the National Natural Science Foundation of China (NSFC) and other national and provincial agencies.

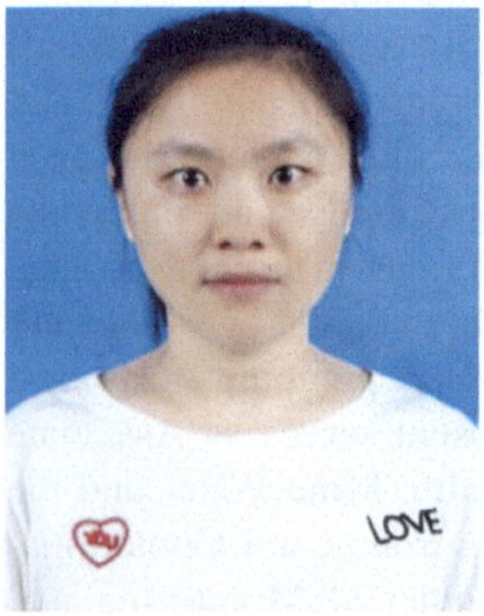

Dr. Jianhua Wu is a full professor at Chang'an University. Her research mainly focuses on hydrology and water resources in arid areas, loess seepage and its disaster mechanisms, and groundwater environmental protection. She has led and participated in more than 10 research projects funded by the National Natural Science Foundation of China (NSFC). She has published more than 80 academic papers in international journals. She has 16 papers selected as highly cited papers and hot papers by ESI (Essential Science Indicators). She currently serves as an associate editor for the international journal Exposure and Health and is a reviewer for various international journals. She has been awarded one of the most highly cited researchers by Clarivate, Elsevier, and other organizations.

Prof. Vetrimurugan Elumalai is an established Hydrogeologist expertise in contaminant hydrology and groundwater modelling. He is excelling as an established researcher and mentor and Head of the department of Hydrology, University of Zululand, South Africa. He is South African National Chair of Hydrology (SARChI-Tier 1). His project experience covers groundwater resource investigation and development, groundwater characterization, contaminant hydrogeology and urban development, surface water groundwater interactions, remote sensing, and GIS application in Water Resource Management. He has published about 80 research papers in reputed journals. He is registered scientist with SACNASP. He is a life member of International Association of Hydrological Science, International Association of Hydrogeologists, International Association of Hydrological Sciences and Geological Survey of South Africa.

Contributors

Sani I. Abba Department of Civil Engineering, Prince Mohammad Bin Fahd University, Al Khobar, Saudi Arabia

Ashraf R. Abouelmagd Egyptian Petroleum Sector, Nasr City, Cairo, Egypt

Mahamuda Abu Department of Geological Engineering, University for Development Studies, Tamale, Ghana

Johnson C. Agbasi Department of Geology, Chukwuemeka Odumegwu Ojukwu University, Uli, Nigeria

Osei Akoto Department of Chemistry, Kwame Nkrumah University of Science and Technology, Kumasi, Ghana

Nader Al-Awadhi National Liason Officer for IAEA, Kuwait Institute for Scientific Research, Kuwait City, Kuwait

Amjad Al-Rashidi Water Research Center, Kuwait Institute for Scientific Research, Kuwait City, Kuwait

Bedour Alsabti Water Research Center, Kuwait Institute for Scientific Research, Kuwait City, Kuwait

Chiedozie C. Aralu Department of Pure and Industrial Chemistry, Nnamdi Azikiwe University, Awka, Anambra State, Nigeria

Huijuan Chen Key Lab of Geo-Environment of Qinghai Province, Xi'ning, China;
Bureau of Qinghai Environmental Geological Prospecting, Xi'ning, China

Jie Chen Key Laboratory of Subsurface Hydrology and Ecological Effect in Arid Area of Ministry of Education, Chang'an University, Xi'an, Shaanxi, China;
School of Water and Environment, Chang'an University, Xi'an, Shaanxi, China;
Key Laboratory of Eco-Hydrology and Water Security in Arid and Semi-Arid Regions of Ministry of Water Resources, Chang'an University, Xi'an, Shaanxi, China

Godfred Darko Department of Chemistry, Kwame Nkrumah University of Science and Technology, Kumasi, Ghana

Johnbosco C. Egbueri Department of Geology, Chukwuemeka Odumegwu Ojukwu University, Uli, Nigeria;
Research Management Office (RMO), Chukwuemeka Odumegwu Ojukwu University, Uli, Anambra State, Nigeria

Esam El-Sayed Faculty of Science, Geology Department, Minia University, El-Minia, Egypt

Vetrimurugan Elumalai Department of Hydrology, University of Zululand, Richards Bay, Durban, South Africa

Rakesh Roshan Gantayat Department of Hydrology, University of Zululand, Durban, South Africa

Oscar Agesandro García-Arriola Posgrado en Ciencias del Mar y Limnología, Universidad Nacional Autónoma de México, Ciudad de México, CP, Mexico

P. Gopinathan Resource Quality Assessment Division, CSIR-Central Institute of Mining and Fuel Research, Dhanbad, Jharkhand, India

Zilong Guan PowerChina Northwest Engineering Corporation Limited, Xi'an, Shaanxi, China

Opoku Gyamfi Department of Chemistry Education, Akenten Appiah-Menka University of Skills Training and Entrepreneurial Development, Asante-Mampong, Ashanti, Mampong, Ghana

Khaled Hadi Water Research Center, Kuwait Institute for Scientific Research, Kuwait City, Kuwait

Song He PowerChina Northwest Engineering Corporation Limited, Xi'an, Shaanxi, China;
Key Laboratory of Water and Soil Resources Conservation and Restoration in the Middle and Lower Reaches of Yellow River Basin, MNR, Zhengzhou, China;
School of Water and Environment, Chang'an University, Xi'an, Shaanxi, China

Xiaodong He School of Water and Environment, Chang'an University, Xi'an, Shaanxi, China;
Key Laboratory of Subsurface Hydrology and Ecological Effects in Arid Region of the Ministry of Education, Chang'an University, Xi'an, Shaanxi, China;
Key Laboratory of Eco-Hydrology and Water Security in Arid and Semi-Arid Regions of the Ministry of Water Resources, Chang'an University, Xi'an, Shaanxi, China

Wenxu Hu Faculty of Geosciences and Engineering, Southwest Jiaotong University, Chengdu, China

Ibrahim Ishag Key Laboratory of Subsurface Hydrology and Ecological Effect in Arid Area of Ministry of Education, Chang'an University, Xi'an, Shaanxi, China;
School of Water and Environment, Chang'an University, Xi'an, Shaanxi, China;
Key Laboratory of Eco-Hydrology and Water Security in Arid and Semi-Arid Regions of Ministry of Water Resources, Chang'an University, Xi'an, Shaanxi, China

Yujie Ji Shaanxi Weihe Ecological Zone Protection Center (Shaanxi Weihe River Basin Control and Protection Center), Xi'an, Shaanxi, China

Xiaomei Kou PowerChina Northwest Engineering Corporation Limited, Xi'an, China

Hwang Sheng Lee Department of Mechanical and Material Engineering, Lee Kong Chian Faculty of Engineering and Science, Universiti Tunku Abdul Rahman, Cheras, Kajang, Selangor, Malaysia;
Centre for Advanced and Sustainable Materials Research, Universiti Tunku Abdul Rahman, Cheras, Kajang, Selangor, Malaysia

Peiyue Li School of Water and Environment, Chang'an University, Xi'an, Shaanxi, China;
Key Laboratory of Subsurface Hydrology and Ecological Effects in Arid Region of the Ministry of Education, Chang'an University, Xi'an, Shaanxi, China;
Key Laboratory of Eco-Hydrology and Water Security in Arid and Semi-Arid Regions of the Ministry of Water Resources, Chang'an University, Xi'an, Shaanxi, China

Weichao Liu School of Water and Environment, Chang'an University, Xi'an, Shaanxi, China;
Key Laboratory of Subsurface Hydrology and Ecological Effects in Arid Region of the Ministry of Education, Chang'an University, Xi'an, Shaanxi, China;
Key Laboratory of Eco-Hydrology and Water Security in Arid and Semi-Arid Regions of the Ministry of Water Resources, Chang'an University, Xi'an, Shaanxi, China

Yuan Liu PowerChina Northwest Engineering Corporation Limited, Xi'an, Shaanxi, China

Joo Ling Loo Department of Mechatronics and Biomedical Engineering, Lee Kong Chian Faculty of Engineering and Science, Universiti Tunku Abdul Rahman, Cheras, Kajang, Selangor, Malaysia

Han Lu School of Water and Environment, Chang'an University, Xi'an, Shaanxi, China;
Key Laboratory of Subsurface Hydrology and Ecological Effects in Arid Region of the Ministry of Education, Chang'an University, Xi'an, Shaanxi, China;
Key Laboratory of Eco-Hydrology and Water Security in Arid and Semi-Arid Regions of the Ministry of Water Resources, Chang'an University, Xi'an, Shaanxi, China

Mohamed Bakry Masod Analysis and Evaluation Department, Egyptian Petroleum Research Institute, Cairo, Egypt

Seth Obiri-Yeboah Department of Pharmaceutical Sciences, Kumasi Technical University, Kumasi, Ghana;
Department of Chemistry, Kwame Nkrumah University of Science and Technology, Kumasi, Ghana

Chaitanya B. Pande Indian Institute of Tropical Meteorology, Pune, Maharashtra, India

B. Preethi Department of Geology, College of Engineering Guindy (CEG), Anna University, Chennai, India

Guangxiong Qin Key Lab of Geo-Environment of Qinghai Province, Xi'ning, China;
Bureau of Qinghai Environmental Geological Prospecting, Xi'ning, China

Ahmed M. Rashad Analysis and Evaluation Department, Egyptian Petroleum Research Institute, Cairo, Egypt

Priyadarsi D. Roy Instituto de Geología, Universidad Nacional Autónoma de México, Ciudad de México, CP, Mexico

Chidambaram Sabarathinam Water Research Center, Kuwait Institute for Scientific Research, Kuwait City, Kuwait

Dhanu Radha Samayamanthula Water Research Center, Kuwait Institute for Scientific Research, Kuwait City, Kuwait

José Luis Sánchez-Zavala Instituto de Geología, Universidad Nacional Autónoma de México, Ciudad de México, CP, Mexico

R. Saravannan Department of Geology, College of Engineering Guindy (CEG), Anna University, Chennai, India

Sekar Selvam Department of Geology, V.O. Chidambaram College, Tuticorin, Tamil Nadu, India

Abeer El Shahawy Department of Civil Engineering, Faculty of Engineering, Suez Canal University, Ismailia, Egypt

Moustafa Gamal Snousy Egyptian Petroleum Sector, Petrotrade Co, Cairo, Egypt

T. Subramani Department of Geology, College of Engineering Guindy (CEG), Anna University, Chennai, India

Yan Tian PowerChina Sinohydro Bureau 3 Co.,LTD, Xi'an, Shaanxi, China

Henry C. Uwajingba Department of Surveying and Geoinformatics, Federal Polytechnic, Ugep, Cross River State, Nigeria

Jie Wang Faculty of Geosciences and Engineering, Southwest Jiaotong University, Chengdu, China;
Sichuan Province Engineering Technology Research Center of Ecological Mitigation of Geohazards in Tibet Plateau Transportation Corridors, Chengdu, China

Liwei Wang Faculty of Geosciences and Engineering, Southwest Jiaotong University, Chengdu, China;
Sichuan Province Engineering Technology Research Center of Ecological Mitigation of Geohazards in Tibet Plateau Transportation Corridors, Chengdu, China

Jianhua Wu School of Water and Environment, Chang'an University, Xi'an, Shaanxi, China;
Key Laboratory of Subsurface Hydrology and Ecological Effects in Arid Region of the Ministry of Education, Chang'an University, Xi'an, Shaanxi, China;
Key Laboratory of Eco-Hydrology and Water Security in Arid and Semi-Arid Regions of the Ministry of Water Resources, Chang'an University, Xi'an, Shaanxi, China

Yong Xiao Faculty of Geosciences and Engineering, Southwest Jiaotong University, Chengdu, China;
MOE Key Laboratory of Groundwater Circulation and Environmental Evolution, China University of Geosciences (Beijing), Beijing, P. R. China;
Fujian Provincial Key Laboratory of Water Cycling and Eco-Geological Processes, Xiamen, China

Fei Xu School of Water and Environment, Chang'an University, Xi'an, Shaanxi, China;
Key Laboratory of Subsurface Hydrology and Ecological Effects in Arid Region of the Ministry of Education, Chang'an University, Xi'an, Shaanxi, China;
Key Laboratory of Eco-Hydrology and Water Security in Arid and Semi-Arid Regions of the Ministry of Water Resources, Chang'an University, Xi'an, Shaanxi, China

Zibo Xu School of Water and Environment, Chang'an University, Xi'an, Shaanxi, China;
Key Laboratory of Eco-Hydrology and Water Security in Arid and Semi-Arid Regions of Ministry of Water Resources, Chang'an University, Xi'an, Shaanxi, China

Hongjie Yang Faculty of Geosciences and Engineering, Southwest Jiaotong University, Chengdu, China;
Sichuan Province Engineering Technology Research Center of Ecological Mitigation of Geohazards in Tibet Plateau Transportation Corridors, Chengdu, China

Junzhu Yang Key Laboratory of Subsurface Hydrology and Ecological Effect in Arid Area of Ministry of Education, Chang'an University, Xi'an, Shaanxi, China;
School of Water and Environment, Chang'an University, Xi'an, Shaanxi, China;
Key Laboratory of Eco-Hydrology and Water Security in Arid and Semi-Arid Regions of Ministry of Water Resources, Chang'an University, Xi'an, Shaanxi, China

Junyan Yang School of Water and Environment, Chang'an University, Xi'an, Shaanxi, China;
Key Laboratory of Subsurface Hydrology and Ecological Effects in Arid Region of the Ministry of Education, Chang'an University, Xi'an, Shaanxi, China;
Key Laboratory of Eco-Hydrology and Water Security in Arid and Semi-Arid Regions of the Ministry of Water Resources, Chang'an University, Xi'an, Shaanxi, China

Junzhu Yang Key Laboratory of Subsurface Hydrology and Ecological Effect in Arid Area of Ministry of Education, Chang'an University, Xi'an, Shaanxi, China;
School of Water and Environment, Chang'an University, Xi'an, Shaanxi, China;
Key Laboratory of Eco-Hydrology and Water Security in Arid and Semi-Arid Regions of Ministry of Water Resources, Chang'an University, Xi'an, Shaanxi, China

Shaokang Yang Key Lab of Geo-Environment of Qinghai Province, Xi'ning, China;
Bureau of Qinghai Environmental Geological Prospecting, Xi'ning, China

Augustine Chung Wei Yap Department of Mechanical and Material Engineering, Lee Kong Chian Faculty of Engineering and Science, Universiti Tunku Abdul Rahman, Cheras, Kajang, Selangor, Malaysia

Mahmoud F. Zawrah Refractory, Ceramics and Building Materials Department, National Research Center, Dokki, Cairo, Egypt

Pengbin Zhang School of Water and Environment, Chang'an University, Xi'an, Shaanxi, China;
Key Laboratory of Subsurface Hydrology and Ecological Effects in Arid Region of the Ministry of Education, Chang'an University, Xi'an, Shaanxi, China;
Key Laboratory of Eco-Hydrology and Water Security in Arid and Semi-Arid Regions of the Ministry of Water Resources, Chang'an University, Xi'an, Shaanxi, China

Yuqing Zhang Faculty of Geosciences and Engineering, Southwest Jiaotong University, Chengdu, China

Bin Zhao Key Laboratory of Subsurface Hydrology and Ecological Effect in Arid Area of Ministry of Education, Chang'an University, Xi'an, Shaanxi, China;
School of Water and Environment, Chang'an University, Xi'an, Shaanxi, China;
Key Laboratory of Eco-Hydrology and Water Security in Arid and Semi-Arid Regions of Ministry of Water Resources, Chang'an University, Xi'an, Shaanxi, China

Dongyang Zhao Key Lab of Geo-Environment of Qinghai Province, Xi'ning, China;
Bureau of Qinghai Environmental Geological Prospecting, Xi'ning, China

Xiaoguang Zhao PowerChina Northwest Engineering Corporation Limited, Xi'an, Shaanxi, China

Zhen Zhao Key Lab of Geo-Environment of Qinghai Province, Xi'ning, China;
Bureau of Qinghai Environmental Geological Prospecting, Xi'ning, China

Zhiwen Zheng School of Water and Environment, Chang'an University, Xi'an, Shaanxi, China;
Key Laboratory of Subsurface Hydrology and Ecological Effects in Arid Region of the Ministry of Education, Chang'an University, Xi'an, Shaanxi, China;
Key Laboratory of Eco-Hydrology and Water Security in Arid and Semi-Arid Regions of the Ministry of Water Resources, Chang'an University, Xi'an, Shaanxi, China

Xiaoping Zhou PowerChina Northwest Engineering Corporation Limited, Xi'an, Shaanxi, China

Hydrogeology and the Global Significance of Groundwater

Peiyue Li and Vetrimurugan Elumalai

Abstract This chapter underscores the global significance of groundwater as a vital resource for domestic use, agriculture, industry, and ecosystems. Particularly, the chapter highlights key scientific issues in agricultural hydrogeology, emphasizing the implications of agricultural irrigation on groundwater resources. Additionally, it discusses the challenges faced in domestic, industrial, and ecological contexts, including over-extraction, contamination, and climate-induced changes in water availability. To address these challenges, the chapter proposes strategies for advancing hydrogeology in the new era, including enhanced research and monitoring, improved data sharing, incorporating citizen science, enhanced international collaboration, planned collaboration among academics, industry, research institutes, and governments, and continued funding and support from governments. By embracing these strategies, the hydrogeological community can continue to make significant contributions to the sustainable management and protection of groundwater resources worldwide.

Keywords Groundwater resources · Groundwater quality · Groundwater pollution · Modern hydrogeology · Hydrogeochemistry · Groundwater management · Groundwater development

---

P. Li (✉)
School of Water and Environment, Chang'an University, No. 126 Yanta Road, Xi'an 710054, Shaanxi, China
e-mail: lipy2@163.com; peiyueli@chd.edu.cn

Key Laboratory of Subsurface Hydrology and Ecological Effects in Arid Region of the Ministry of Education, Chang'an University, No. 126 Yanta Road, Xi'an 710054, Shaanxi, China

Key Laboratory of Eco-Hydrology and Water Security in Arid and Semi-Arid Regions of the Ministry of Water Resources, Chang'an University, No. 126 Yanta Road, Xi'an 710054, Shaanxi, China

V. Elumalai
Department of Hydrology, University of Zululand, Kwa-Dlangezwa, Richards Bay, Durban 3886, South Africa

P. Li et al. (eds.), *Sustainable Groundwater and Environment: Challenges and Solutions*, Springer Hydrogeology, https://doi.org/10.1007/978-3-031-82194-3_1

1 Introduction

Groundwater is a key part of the Earth's ecosystem, playing a crucial role in sustaining human life and the sustainability of ecosystems. As the world's largest freshwater reservoir, groundwater not only nourishes most of the world's lakes and rivers, maintaining the health of wetlands and other ecosystems, but also provides a vast amount of freshwater resources for agricultural irrigation, industrial production, and daily human life (Shiklomanov 2000; Li and Qian 2018). The sustainable development and excellent quality of groundwater play a key role in the healthy development of human society. However, factors such as over-exploitation, human pollution, and climate change are threatening the sustainability of groundwater resources, triggering a series of groundwater environmental issues (Li et al. 2014, 2023a; He et al. 2022a; Wu et al. 2020a). Currently, water resources, water disasters, and water environmental issues have become severe problems restricting human survival and development.

Hydrogeology is a sub-discipline of geology that studies the distribution, movement, and properties of water in the Earth's crust, particularly in aquifers (Hölting and Coldewey 2019). This field is crucial for understanding the complex interplay between water, soil, rock, and living organisms, and it plays a vital role in various societal and environmental contexts. Water is a fundamental resource for life and societal development. It supports ecosystems, serves as a medium for transportation, and is a critical component in industrial processes. Moreover, water is an essential element in agriculture, which is responsible for producing the food that sustains human life (Li et al. 2022). Therefore, understanding the dynamics of water in the environment, which is the primary focus of hydrogeology, is crucial.

Particularly in agriculture, the importance of hydrogeology cannot be overstated. Irrigation is a practice that relies heavily on our understanding of water movement and availability in the subsurface. Hydrogeological knowledge helps in the design of efficient irrigation systems, ensuring that water is delivered to crops in the right amounts and at the right times. This not only maximizes crop yield but also minimizes water wastage, a critical consideration in regions where water resources are scarce. Furthermore, hydrogeology plays a significant role in managing agricultural pollution. Fertilizers and pesticides, commonly used in farming, can leach into the groundwater, causing contamination that can harm human health and the environment (Wang et al. 2023, 2024; Alam et al. 2024). Hydrogeologists study how these substances move through the soil and into groundwater systems, providing necessary information for creating strategies to prevent or mitigate such contamination. Another critical aspect of hydrogeology in agriculture relates to climate change. As global temperatures rise, patterns of rainfall and evaporation are changing, which can affect the availability of water for irrigation (Chandio et al. 2020; Hatfield et al. 2020). Hydrogeologists can model these changes and their impacts on groundwater resources, providing valuable information for agricultural planning and water resource management.

This chapter underscores the significance of groundwater across various sectors, including domestic use, industry, and agriculture. We place particular emphasis on the implications of agricultural irrigation on groundwater, using this as a basis to summarize the key scientific issues that need to be addressed in agricultural hydrogeology. Given the rapid development of hydrogeology, we also outline the primary research topics of contemporary hydrogeology and delineate the challenges that this field currently faces. It is our hope that this chapter will offer readers fresh perspectives on the evolution of modern hydrogeology and the ongoing research into groundwater.

2 Significance of Groundwater

Groundwater plays an indispensable role in our global ecosystem and economy, serving as a crucial resource for numerous sectors. It is the primary source of potable water for approximately half of the world's population and accounts for over 40% of water used for irrigation (United Nations 2022), highlighting its importance in both domestic and agricultural contexts. In industrial sectors, groundwater is utilized for cooling, processing, and cleaning purposes, underscoring its versatility. Beyond its tangible uses, groundwater also contributes significantly to maintaining the health of our ecosystems. It nourishes wetlands and sustains the baseflow of rivers and streams, playing a vital role in preserving biodiversity.

2.1 Groundwater for Domestic Uses

Groundwater is a key source for domestic water supply, providing drinking water for approximately 50% of the global population. It is a vital resource in both urban and rural settings, with rural areas often relying entirely on groundwater, especially in regions where surface water is scarce. On a global scale, about 22% of the total freshwater used is for domestic purposes, with a large proportion coming from groundwater sources (United Nations 2022). For instance, in India, around 85% of rural domestic water needs are met through groundwater, while in the United States, about 44% of the population relies on groundwater for their drinking water (World Bank 2023). The ability of groundwater to be easily accessed through wells and boreholes makes it indispensable in arid and semi-arid regions, where surface water bodies are often absent. In Sub-Saharan Africa, an estimated 75% of people depend on groundwater for their daily water needs (United Nations 2022). These statistics underscore the significance of groundwater in ensuring water security for billions of people across the globe.

Despite its importance, the security of domestic groundwater supply faces numerous challenges, including over-extraction, contamination, and climate-induced changes in water availability. In countries like Mexico and China, aquifer depletion due to over-pumping for agricultural, industrial, and domestic use has led to significant drops in water tables, threatening the sustainability of groundwater as a source for household use (Cruz-Ayala and Megdal 2020; Du et al. 2024). In Mexico City, for example, groundwater extraction accounts for 70% of the city's water supply, yet the aquifer is being drained faster than it can be replenished, causing land subsidence and increasing the cost of water extraction (Cruz-Ayala and Megdal 2020). Pollution also poses a serious risk, with agricultural runoff, industrial chemicals, and inadequate wastewater treatment contaminating groundwater supplies in many regions. In Bangladesh, for example, arsenic contamination affects millions of people who rely on shallow groundwater for their drinking water (Ivy et al. 2023; Loewenberg 2017). Climate change further exacerbates these challenges, as changing precipitation patterns reduce aquifer recharge rates, particularly in already water-stressed regions such as the Middle East and North Africa (World Bank 2018). Sustainable groundwater management, including the implementation of efficient water use practices and pollution control measures, is critical for securing long-term domestic water supplies in these vulnerable areas.

2.2 *Groundwater for Agriculture*

Groundwater is an essential resource for global agriculture, providing a stable and reliable supply of water for irrigation, especially in regions where surface water resources are scarce or seasonal. Globally, agriculture is the largest consumer of freshwater, accounting for approximately 70% of all freshwater withdrawals, with groundwater contributing around 43% of the total water used for irrigation (FAO 2020). In countries such as India, the United States, and China, groundwater plays a particularly critical role in supporting food production. For instance, in India, groundwater accounts for nearly 60% of all irrigation water, helping sustain the livelihoods of millions of smallholder farmers (Shah 2009). The ability of groundwater to provide a decentralized and often self-regulated water supply makes it indispensable for farming in arid and semi-arid regions, where surface water may be unreliable due to seasonal variability or climate change impacts. This widespread reliance on groundwater ensures food security and supports rural economies, particularly in developing countries.

However, the over-extraction of groundwater for agricultural use is leading to significant challenges that threaten the sustainability of this vital resource. In many regions, groundwater is being withdrawn at rates that exceed natural recharge, leading to declining water tables and the depletion of aquifers. For example, in parts of northern India and the North China Plain, groundwater levels are dropping by more than one meter per year due to over-pumping for irrigation (Rodell et al. 2009). This over-extraction not only reduces the availability of water for future agricultural use but also leads to land subsidence, reduced water quality, and increased energy costs for pumping water from greater depths. Additionally, the contamination of groundwater by agricultural chemicals such as nitrates and pesticides poses a serious risk to both human health and the environment (WWAP 2018). Addressing these challenges requires the implementation of more sustainable water management practices, including the promotion of efficient irrigation techniques, improved groundwater governance, and the integration of surface and groundwater management to ensure the long-term viability of agricultural water supplies.

2.3 *Groundwater for Industry*

Groundwater plays a crucial role in supporting industrial activities worldwide, providing a reliable and often high-quality water source for various manufacturing processes, cooling, and energy production. Globally, industrial water use accounts for approximately 19% of total freshwater withdrawals, with groundwater contributing a significant portion of this supply (WWAP 2014). In some industrialized nations, such as the United States and Germany, groundwater is a critical resource for industries like food processing, pharmaceuticals, textiles, and electronics manufacturing. For example, in the United States, about 20% of industrial water use comes from groundwater sources, with industries in arid regions particularly reliant on it due to limited surface water availability (USGS 2015). Groundwater is often preferred for industrial use because it tends to be of higher quality than surface water, requiring less treatment for many industrial applications. Moreover, its stable availability makes it a reliable source for industries that require continuous water supply, such as power generation and chemical manufacturing, where interruptions in water access can lead to significant economic losses.

However, the industrial use of groundwater also poses several challenges, particularly with respect to sustainability and environmental impacts. In many regions, over-extraction of groundwater for industrial purposes is contributing to the depletion of aquifers, particularly in water-stressed areas. For example, in parts of northern China, where industries rely heavily on groundwater, over-pumping has led to significant declines in groundwater levels, exacerbating water scarcity issues (Liu et al. 2011). Additionally, industrial activities can lead to groundwater contamination, with pollutants such as heavy metals, organic chemicals, and solvents seeping into aquifers and posing risks to both human health and ecosystems. This is particularly problematic in regions with inadequate regulation or enforcement of industrial wastewater discharge. The contamination of groundwater not only affects its availability for industrial use but also compromises its use for drinking water and agriculture. To address these issues, industries must adopt more sustainable water management practices, including water recycling and reuse, as well as stricter controls on groundwater extraction and pollution. Furthermore, integrated water resource management approaches that consider both surface and groundwater resources are essential to ensuring the long-term sustainability of groundwater for industrial use.

2.4 *Groundwater in Ecosystems*

Groundwater is an essential element of numerous ecosystems, significantly influencing the health and functioning of both terrestrial and aquatic environments. Various ecosystems, including wetlands, rivers, lakes, and forests, rely on a consistent supply of groundwater to maintain their water levels, especially during dry spells when surface water may be limited. Groundwater contributes to baseflow in rivers, which is crucial for sustaining aquatic habitats and supporting biodiversity during periods of low rainfall (Barlow and Leake 2012). In arid and semi-arid regions, groundwater-dependent ecosystems (GDEs) such as oases, springs, and riparian zones depend almost entirely on groundwater for their survival. For instance, the Okavango Delta in Botswana, one of the largest inland deltas globally, relies on groundwater inflows, which support a rich diversity of wildlife, including several endangered species (McCarthy et al. 2003; Cachazo et al. 2023). Additionally, groundwater supports plant communities with deep-rooted vegetation that can tap into aquifers for water during dry seasons. These ecosystems provide critical services, including carbon sequestration, water purification, and wildlife habitat, underscoring groundwater's vital role in maintaining ecological balance (Costanza et al. 2014).

However, the over-extraction of groundwater for agricultural, industrial, and urban purposes threatens the sustainability of groundwater-dependent ecosystems worldwide. In many regions, declining groundwater levels lead to wetland degradation, spring depletion, and reduced river flows, adversely impacting biodiversity and ecosystem services. For example, in the Western United States, excessive groundwater pumping has resulted in wetland loss and diminished streamflows, endangering fish species like salmon and steelhead that rely on cool, groundwater-fed streams for spawning (Murray and Marmorek 2003). Furthermore, groundwater contamination from pollutants such as nitrates, pesticides, and industrial chemicals exacerbates the issue by degrading water quality and harming aquatic ecosystems (Li et al. 2024a). To address these challenges, a more integrated approach to water resource management is necessary, taking into account the needs of both human activities and ecosystems. Protecting groundwater-dependent ecosystems will require improved groundwater governance, the implementation of sustainable extraction practices, and the restoration of degraded habitats to ensure these ecosystems continue to provide their invaluable ecological services (Nelson 2022).

3 Impacts of Agricultural Irrigation on Groundwater

Agriculture is an important sector for sustainable society. The growing global population and increasing living standards have driven the demand for agricultural products. It is estimated that the world's food demand will increase by about 60% by 2050 (FAO 2013), and irrigation has significantly contributed to the growth of agricultural yields, serving as a key factor in sustainable agricultural development (Fernández et al. 2020). However, irrigation also has numerous negative impacts on the environment, such as the depletion of surface and groundwater resources, water pollution, loss of biodiversity, and soil salinization (Hu et al. 2016). Influenced by various factors such as topography, groundwater level depth, and irrigation quotas, different irrigation methods and water sources have varied impacts on the groundwater environment. Conducting research on the impact of agricultural irrigation on the groundwater environment is crucial for groundwater resource management, groundwater pollution prevention, and soil salinization control in irrigation areas.

In arid and semi-arid regions, agricultural irrigation often exhibits seasonality, leading to periodic fluctuations in groundwater levels in the irrigation areas. This not only repeatedly leaches the soil in the groundwater fluctuation zone, but also affects the biogeochemical environment of the groundwater, thereby influencing the hydrogeochemsitry and the migration and transformation of pollutants (Garamhegyi et al. 2018; Wei et al. 2024). Firstly, after groundwater is replenished by agricultural irrigation, the water level rises. When the groundwater level depth is within the range of 0–3 m, the content of soluble salts in the root zone soil will increase under the effects of evaporation concentration and plant transpiration. Since soluble salt ions have surface aggregation properties, under the action of capillary forces, the salts in the groundwater will also migrate upwards due to capillary action, eventually accumulating in the 0–50 cm depth of the cultivated layer of the soil, causing soil salinization (Ishikura et al. 2017, Fig. 1). However, when irrigation water infiltrates to raise water level to the salt accumulation layer, soluble salts in the soil will be largely dissolved into the groundwater, leading to an increase in groundwater salinity. Fluctuations in groundwater levels caused by irrigation, along with heavy fertilization, can also lead to the accumulation of nitrate and toxic substances in soil and groundwater (Yin et al. 2021; Liu et al. 2016).

Moreover, fluctuations in groundwater levels affect the biogeochemical environment of the groundwater, thereby affecting the migration and transformation processes of pollutants in the soil and groundwater. For instance, in the groundwater fluctuation zone, changes in gas saturation lead to the disruption of the interfacial tension environment, and changes in microbial degradation and adsorption–desorption of pollutants, thereby affecting the distribution of pollutants. For example, for petroleum hydrocarbon pollutants, a decrease in water level accelerates pollutant migration, and an increase in water level leads to an increase in pollutant dissolution. For cadmium convection caused by changes in groundwater level fluctuations alters the migration of cadmium ions (Wu et al. 2020b). For nitrate, an increase in groundwater level leads to a decrease in nitrate content in the soil, while a decrease in groundwater level leads to an increase in soil nitrate content. For ammonia nitrogen, an increase in groundwater level leads to an increase in soil ammonia nitrogen content, while a decrease in groundwater level leads to a decrease in soil ammonia nitrogen content (Zhang et al. 2020). Therefore, under the influence of irrigation, the groundwater dynamic, hydrogeochemistry and biological processes exhibit synergistic evolution (Fig. 1). However, most current research focuses on summarizing the evolutionary patterns of groundwater under the influence of single factors, and

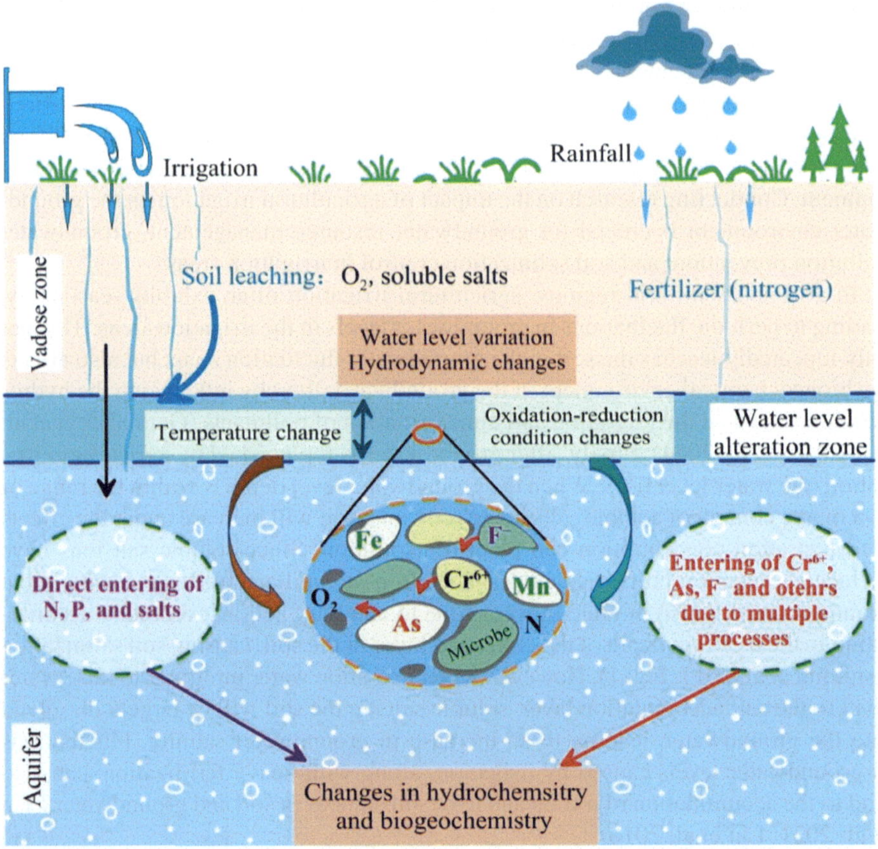

Fig. 1 Schematic diagram showing the biogeochemical changes in groundwater induced by irrigation and rainfall

the inherent mechanism of the coordinated evolution of groundwater under multiple factors is not yet clear.

3.1 Impact of Agricultural Irrigation on Groundwater Levels

Agricultural irrigation is the largest user of water resources, consuming over 70% of global water resources (FAO 2013). In China, agricultural water use accounts for more than 60% of the total water consumption, with irrigation water accounting for over 90% of agricultural water use (Lu et al. 2021). The use of different water sources for irrigation has varying degrees of impact on groundwater levels. About 28% of the irrigation areas in China are irrigated by rivers, such as the Hetao Irrigation District in

Inner Mongolia Autonomous Region, the Qingtongxia Irrigation District in Ningxia Hui Autonomous Region, and the Baoji Gorge Irrigation District, Jinghui Canal Irrigation District, and Luo Hui Irrigation District in Shaanxi Province, which mostly use water from the Yellow River and its tributaries for irrigation. Irrigation with surface water can cause a rise in groundwater levels, which can easily lead to soil salinization problems. About 20% of the irrigation areas in China use groundwater for irrigation. Extracting groundwater for agricultural irrigation has multiple potential impacts on the quantity and quality of groundwater resources. For example, over-extraction of groundwater can lead to a drop in groundwater levels, causing problems such as a water deficit, nitrate pollution, and saltwater intrusion in the groundwater system (Tweed et al. 2018; Tulip et al. 2022). Therefore, studying the impact of irrigation on the groundwater level is of great significance for the sustainable management of groundwater resources (Tulip et al. 2022).

In recent years, scholars have used various methods and technical means to analyze and explore the impact of agricultural irrigation on groundwater levels. Yang et al. (2021) established an in situ observation site for vadose zone moisture migration in the Yongning Irrigation District of the Yinchuan Plain, and found that irrigation infiltration recharge accounted for 68% of the total groundwater recharge. Zhai et al. (2021) analyzed the time series data of groundwater depth in the Qingtongxia Irrigation District from 1998 to 2017, and found that leakage from the irrigation system accounted for over 80% of groundwater recharge, thus considering that irrigation is the main reason for groundwater level changes. In addition, Hu et al. (2016) studied the spatiotemporal relationship between agricultural irrigation and groundwater depletion, further revealing the negative impact of irrigation on the groundwater system. Li (2018) used Hydrus-1D and Visual MODFLOW to establish a numerical model of vadose zone moisture and groundwater migration and transformation in the plain area of the Manas River Basin, and quantitatively analyzed the fluctuation of groundwater levels and changes in groundwater volume under different irrigation modes. Malki et al. (2017) studied the changes in groundwater levels in the intensively irrigated Chetouka-Massa area of Morocco over the past decades, and found that excessive extraction of groundwater for irrigation has led to a sharp drop in groundwater levels in the area. Tulip et al. (2022) used the WetSpass water balance model to assess the impact of irrigation on groundwater recharge in the northwestern region of Bangladesh, and found that irrigation return water is the main source of groundwater recharge in the dry season. All these studies demonstrate the significant impact of agricultural irrigation on groundwater levels, but the impact of irrigation by extracted groundwater and irrigation by surface water on groundwater level fluctuations is not the same. Irrigation by surface water can have a much serious impacts than irrigation using groundwater extracted from the same farmland.

3.2 Impact of Agricultural Irrigation on Groundwater Quality

Agricultural irrigation also greatly affects groundwater quality. The infiltration of irrigation water not only leaches soluble salts in the unsaturated zone, changing the content of various hydrochemical ions in groundwater, but also introduce various fertilizers applied to aquifers, directly affecting the quality of groundwater. Therefore, scholars have carried out a lot of work on the issue of groundwater pollution caused by agricultural irrigation. Liu et al. (2016) found that autumn irrigation washes salts into the groundwater, leading to an increase in groundwater mineralization and deterioration of water quality. Khasanov et al. (2022) evaluated the multi-year spatiotemporal changes of groundwater level, mineralization, and soil salt content in the arid irrigation area of Syrdarya Province, Uzbekistan, and found that the increase in temperature would correspondingly change the salt content of the soil, resulting in higher mineralization of shallow groundwater. In addition to the leaching effect on soluble salts, the infiltration of irrigation water also promotes the occurrence of water-rock interactions, causing significant changes in the solute contents in groundwater (Yu et al. 2022). Gao et al. (2022) specifically conducted research on this topic, and found that the concentrations of almost all ions during the irrigation period were greater than those during the non-irrigation period, indicating that long-term irrigation has an impact on groundwater salinization.

In fact, the process of the impact of agricultural irrigation on groundwater quality is quite complex. First, the groundwater level fluctuations caused by irrigation have a certain impact on groundwater quality (Garamhegyi et al. 2018). Tweed et al. (2020) proposed a conceptual model of the seasonal scale process affecting the arsenic concentration in the irrigation area of the Mekong Delta in Cambodia, revealing for the first time the seasonal changes in the As concentration in the groundwater of this irrigation area with the fluctuation of groundwater. Zhang et al. (2020) also found that as the irrigation water volume increased, the groundwater level rose, which changed the redox environment of the groundwater, and promoted the reduction of iron oxide minerals. The contact of groundwater with arsenic-bearing sediments caused the increase in arsenic concentration in the groundwater. Secondly, the fertilizers and pesticides applied in agricultural activities will infiltrate into the aquifer with the irrigation water, thereby causing direct pollution of the groundwater. The correlation between agricultural irrigation and groundwater nitrate concentration has been fully proven (Alam et al. 2024; Wang et al. 2024; He et al. 2022b; Li et al. 2024b). Malki et al. (2017) found that the groundwater quality in the intensive irrigation area of Chetouka-Massa, Morocco, has been affected by nitrate leaching pollution caused by fertilizationover the past decade. Tweed et al. (2018) introduced the seasonal impact of irrigation measures on groundwater resources in temperate humid regions, and found that the concentration of groundwater nitrate has increased in recent years. In addition, the quality of irrigation water is also one of the important factors affecting groundwater quality. To alleviate water resource pressure, many areas have adopted sewage irrigation methods, which, while having a positive impact on agricultural activities, also have negative impacts related to environmental pollution (Singh 2021a, b). Jampani et al. (2020) revealed the mixing processes controlling groundwater chemistry in wastewater-irrigated agricultural systems, and the increase in SO_4^{2-} concentration in groundwater is related to the contribution of irrigation wastewater. Abuzaid and Jahin (2021) evaluated the impact of irrigation with three water sources: Nile freshwater, agricultural drainage from drainage ditches, and mixed wastewater, on groundwater quality in the Nile Delta region of Egypt, and found that compared with freshwater irrigation, the quality of groundwater under wastewater irrigation and agricultural drainage backfilling conditions significantly decreased by 39% and 23%, respectively.

In summary, the infiltration of agricultural irrigation water affects both the dynamic changes of groundwater and to some extent the physicochemical characteristics of groundwater. Moreover, the change in the groundwater dynamics often accompanies the change in the biogeochemical environment of the groundwater fluctuation zone, thereby further affecting the changes in the groundwater geochemical processes and the migration and transformation process of pollutants. Most existing studies focus on the single impact of agricultural irrigation on groundwater levels and groundwater quality, often overlooking the changes in the geochemical processes and biological processes brought about by changes in the groundwater dynamics. Therefore, it is an urgent scientific problem to construct a coordinated evolution model between the dynamics, geochemical processes, and biological processes of groundwater under irrigation conditions, reveal the migration and transformation mechanism of pollutants under the coordinated evolution of multiple processes, and deeply analyze the impact of agricultural irrigation on the groundwater environments.

3.3 *Scientific Issues to Be Addressed*

From the above literature review, it can be seen that there are many studies on the evolution of the groundwater environment and the impact of agricultural irrigation on the evolution of the groundwater environment. Although some understanding has been gained on the impact of agricultural irrigation on the groundwater environment, there are still the following scientific issues to be resolved, which is especially important for China.

- In the arid and semi-arid regions, the rainy season and irrigation period often overlap, and the changes in the groundwater environment in the irrigation area are affected by both rainfall and irrigation. For groundwater dynamics, how much impact does the rainfall have? And how much impact does the irrigation have? There is still a lack of quantitative research to quantify their contributions. This is not conducive to a deep understanding of the evolution mechanism of the groundwater environment in the irrigation area and the protection of groundwater quality.

- The fluctuation of groundwater levels not only repeatedly leaches soluble salts in the groundwater fluctuation zone, but also changes the biogeochemical environment of the groundwater, thereby affecting the progress of many biogeochemical processes. So what is the mechanism behind this? What is the result of the influence? These are questions that have not been clearly answered.
- Nitrogen is the most typical pollutant in the irrigation area. Driven by water dynamics, the biogeochemical environment of nitrogen migration and transformation has changed. So how will the changes in the biogeochemical environment feedback on the migration and transformation of pollutants? The reactive migration mechanism of nitrogen under water level variations and its simulation is currently a hot and difficult issue in academic research.

In summary, to truly understand the migration and transformation mechanism of typical agricultural pollutants in the groundwater of large-scale irrigation areas, it is necessary to finely depict the impact of agricultural irrigation on the groundwater dynamics, and promote the research of the coordinated evolution in groundwater dynamics, the hydrogeochemistry and biological processes.

4 Important Research Topics of Modern Hydrogeology

Contemporary groundwater research, as we define it as modern hydrogeology, has very wide scope of research, encompassing the relationship between groundwater and natural elements such as geology, soil, ecosystems, and climate. It also covers the relationship between groundwater and human activities, such as the connection between groundwater and human health, groundwater pollution remediation, and groundwater resource management. This chapter summarizes eight primary areas of groundwater research for the reader's reference (Li et al. 2024c). These eight primary areas of groundwater research have close relationships. Formation and evolution of groundwater is the basic of other research fields, because only after getting adequate knowledge of basic hydrogeological information can researchers conduct research in other topics, while achieving sustainable groundwater resources and environmental management is the ultimate goal in modern hydrogeological research, supported by the other research topics which act like the beams and columns of a house (Fig. 2). Each of these areas represents a critical aspect of modern groundwater research, offering valuable insights into the complex and multifaceted role of groundwater in our natural and human-influenced world.

4.1 Formation and Evolution of Groundwater

The research on the formation and evolution of groundwater is a basic aspect of groundwater research. It involves the basic processes of groundwater formation,

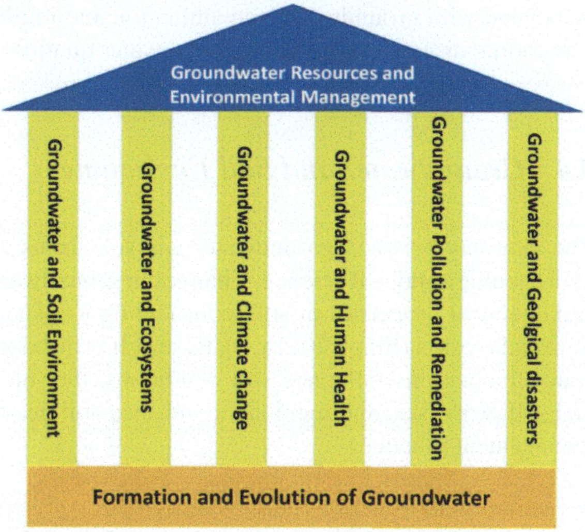

Fig. 2 Relationships among the eight primary areas of groundwater research

distribution, flow, and circulation (Mádl-Szőnyi et al. 2023). The main research topics in this research area includes the hydrogeological conditions of groundwater, changes in groundwater dynamics, physical and chemical processes of groundwater flow, and interactions between groundwater and surface water, atmosphere, organisms, and rocks (Li et al. 2024c).

4.2 *Groundwater Contamination and Mitigation*

Groundwater contamination is a worldwide concern, significantly affecting both human health and environmental balance (Karunanidhi et al. 2021). Key research areas involve identifying contaminant sources, examining the pathways and transformations of pollutants within groundwater systems, assessing pollution risks, and developing effective methods for groundwater cleanup and protection (Li et al. 2024c).

4.3 *Groundwater and Public Health*

As a primary drinking water source, groundwater quality has a direct influence on human health (Guo et al. 2024; Alam et al. 2024; Meng et al. 2023). This area of study focuses on the effects of hazardous substances in groundwater, such as heavy metals and organic pollutants, on public health. It also includes evaluating the health risks associated with groundwater contamination and implementing measures to prevent waterborne diseases through improved water quality management.

4.4 Groundwater and Soil Environment

The interplay between groundwater and soil forms a critical intersection between hydrogeology and soil science. Changes in groundwater levels can alter the physical and chemical properties of soil, contributing to issues like salinization, desertification, and erosion (Singh 2021a, b). Research in this domain investigates how groundwater fluctuations influence soil conditions, the interaction between groundwater and soil processes, and methods to enhance soil health through better groundwater management practices.

4.5 Groundwater and Geohazards

Groundwater plays a crucial role in environmental stability and can contribute to geohazards, such as landslides, subsidence, and ground collapses (Li et al. 2023b; Suganthi and Elango 2020). This research area aims to understand how groundwater dynamics impact geological stability, explore the links between groundwater and geohazards, and develop strategies for mitigating such disasters through effective groundwater management.

4.6 Groundwater-Ecosystem Interactions

The interaction between groundwater and ecosystems, including wetlands, forests, and grasslands, is a pivotal research area in environmental groundwater studies (Li et al. 2024d). This field examines how groundwater influences ecosystem health, the water requirements of different ecosystems, and the interconnected processes that bind groundwater and ecological systems.

4.7 Groundwater and Climate Change

The relationship between groundwater and climate change is a growing research focus. It investigates the effects of climate change on groundwater systems and how groundwater changes feed back into climate dynamics (Amanambu et al. 2020). Key topics include the influence of changing climate patterns on groundwater cycles and resources, as well as the role groundwater may play in moderating or exacerbating climate change effects.

4.8 Sustainable Groundwater Management

Sustainable management of groundwater resources is essential to balance usage with protection (Ahmad and Al-Ghouti 2020). This field includes evaluating and forecasting groundwater supplies, understanding the impacts of extraction, identifying protection zones, and shaping effective groundwater management policies and regulations.

5 Way Ahead of Advancing Hydrogeology in the New Era

The preceding sections have highlighted the significance of groundwater in various sectors, including domestic use, agriculture, industry, and ecosystems. The critical role of groundwater in maintaining global water security and the multifaceted challenges it faces necessitate a concerted effort to advance hydrogeology in the new era. This section outlines key strategies and approaches to ensure the continuous development of hydrogeology in the new era.

5.1 Enhanced Research and Monitoring

Continuous research and monitoring are essential to understand the dynamics of groundwater systems and address emerging challenges. Advances in technology, such as remote sensing, satellite imagery, and data analytics, should be harnessed to improve the accuracy and timeliness of groundwater assessments. This includes monitoring groundwater levels, quality, and recharge rates, as well as assessing the impacts of climate change and human activities on groundwater resources.

5.2 Improved Data Sharing for Hydrogeology

The advancement of hydrogeology in the new era necessitates a multifaceted approach, encompassing various strategies and collaborations. Firstly, data sharing is crucial for the progression of hydrogeological research. By making data accessible and interoperable, researchers can build upon previous findings, validate hypotheses, and uncover new insights. This fosters a collaborative environment that accelerates scientific discovery (Hölting and Coldewey 2019).

5.3 Incorporating Citizen Science in Hydrogeology

Citizen science also plays a pivotal role in advancing hydrogeology. Engaging the public in scientific research not only broadens the scope of data collection, but also enhances public awareness and understanding of groundwater issues (Li et al. 2017; Li and Wu 2019). Such initiatives can lead to more informed decision-making and a stronger sense of community responsibility for water resources (Cruz-Ayala and Megdal 2020).

5.4 Enhanced International Collaboration

International collaboration is essential for addressing the global significance of groundwater. By working together, nations can share knowledge, technology, and resources to tackle common challenges, such as groundwater pollution and overexploitation. This collaboration can lead to the development of innovative solutions that benefit all parties involved (United Nations 2022; World Bank 2023).

5.5 Planned Collaboration Among Academics, Industry, Research Institutes, and Governments

Collaboration among academics, industry, research institutes, and governments is another critical aspect of advancing hydrogeology (Li and Wu 2022). This collaboration can facilitate the transfer of research findings into practical applications, promote the adoption of best practices, and ensure that groundwater resources are managed sustainably. By working in partnership, these stakeholders can create a more resilient and sustainable water future (Ahmad and Al-Ghouti 2020).

5.6 Continued Funding and Support from Governments

Furthermore, the continued funding and support of research programs, such as those funded by the National Natural Science Foundation of China and the National Key Research and Development Program of China, are essential for the advancement of hydrogeology. These programs provide the necessary resources to conduct groundbreaking research, train the next generation of hydrogeologists, and disseminate findings to a wide audience.

Acknowledgements This study was funded by the National Natural Science Foundation of China (42072286, 42472316, and 41761144059), the National Key Research and Development Program of China (2023YFC3706901), and the Qinchuangyuan "Scientist + Engineer" Team Development Program of the Shaanxi Provincial Department of Science and Technology (2022KXJ-005).

References

Abuzaid AS, Jahin HS (2021) Implications of irrigation water quality on shallow groundwater in the Nile Delta of Egypt: a human health risk prospective. Environ Technol Innov 22(2):101383

Ahmad AY, Al-Ghouti MA (2020) Approaches to achieve sustainable use and management of groundwater resources in Qatar: a review. Groundw Sustain Dev 11:100367. https://doi.org/10.1016/j.gsd.2020.100367

Alam SMK, Li P, Fida M (2024) Groundwater nitrate pollution due to excessive use of N-fertilizers in rural areas of Bangladesh: pollution status, health risk, source contribution, and future impacts. Expo Health 16(1):159–182. https://doi.org/10.1007/s12403-023-00545-0

Amanambu AC, Obarein OA, Mossa J, Li L, Ayeni SS, Balogun O, Oyebamiji A, Ochege FU (2020) Groundwater system and climate change: present status and future considerations. J Hydrol 589:125163

Barlow P, Leake SA (2012) Streamflow depletion by wells—Understanding and managing the effects of groundwater pumping on streamflow. USGS Circular 1376. https://pubs.usgs.gov/circ/1376/. Accessed 2 Oct 2024

Cachazo LM, Makati K, Chadwick MA et al (2023) A review of the freshwater diversity in the Okavango Delta and Lake Ngami (Botswana): taxonomic composition, ecology, comparison with similar systems and conservation status. Aquat Sci 85:115. https://doi.org/10.1007/s00027-023-01008-z

Chandio AA, Jiang Y, Rehman A, Rauf A (2020) Short and long-run impacts of climate change on agriculture: an empirical evidence from China. Int J Clim Change Strateg Manag 12(2):201–221

Costanza R, de Groot R, Farber S et al (2014) Changes in the global value of ecosystem services. Glob Environ Chang 26:152–158. https://doi.org/10.1016/j.gloenvcha.2014.04.002

Cruz-Ayala MB, Megdal SB (2020) An overview of managed aquifer recharge in Mexico and its legal framework. Water 12(2):474. https://doi.org/10.3390/w12020474

Du J, Laghari Y, Wei Y-C, Wu L, He A-L, Liu G-Y, Yang H-H, Guo Z-Y, Leghari SJ (2024) Groundwater depletion and degradation in the North China plain: challenges and mitigation options. Water 16(2):354. https://doi.org/10.3390/w16020354

FAO (2020) The State of Food and Agriculture 2020: overcoming water challenges in agriculture. https://www.fao.org

FAO (2013) FAO statistical yearbook, World Food and Agriculture. Food and Agriculture Organization of the United Nations, Rome

Fernández GI, Lecina S, Ruiz-Sánchez MC et al (2020) Trends and challenges in irrigation scheduling in the semi-arid area of Spain. Water 12(3):785

Gao Y, Chen J, Qian H et al (2022) Hydrogeochemical characteristics and processes of groundwater in an over 2260 year irrigation district: a comparison between irrigated and nonirrigated areas. J Hydrol 606:127437

Garamhegyi T, Kovacs J, Pongracz R et al (2018) Investigation of the climate-driven periodicity of shallow groundwater level fluctuations in a Central-Eastern European agricultural region. Hydrogeol J 26(3):677–688

Guo W, Li P, Du Q, Zhou Y, Xu D, Zhang Z (2024) Hydrogeochemical processes regulating the groundwater geochemistry and human health risk of groundwater in the rural areas of the Wei River Basin, China. Expos Health 16(2):291–306. https://doi.org/10.1007/s12403-023-00555-y

Hatfield JL, Antle J, Garrett KA, Izaurralde RC, Mader T, Marshall E, Nearing M, Philip Robertson G, Ziska L (2020) Indicators of climate change in agricultural systems. Clim Change 163:1719–1732. https://doi.org/10.1007/s10584-018-2222-2

He S, Wu J, Wang D, He X (2022a) Predictive modeling of groundwater nitrate pollution and evaluating its main impact factors using random forest. Chemosphere 290:133388. https://doi.org/10.1016/j.chemosphere.2021.133388
He X, Li P, Shi H, Xiao Y, Guo Y, Zhao H (2022b) Identifying strontium sources of flowback fluid and groundwater pollution using 87Sr/86Sr and geochemical model in Sulige gasfield, China. Chemosphere 306:135594. https://doi.org/10.1016/j.chemosphere.2022.135594
Hölting B, Coldewey WG (2019) Hydrogeology. Springer, Heidelberg, Germany
Hu X, Shi L, Zeng J et al (2016) Estimation of actual irrigation amount and its impact on groundwater depletion: a case study in the Hebei Plain, China. J Hydrol 543:433–449
Ishikura K, Yamada H, Toma Y et al (2017) Effect of groundwater level fluctuation on soil respiration rate of tropical peatland in Central Kalimantan, Indonesia. Soil Sci Plant Nutr 63(1):1–13
Ivy N, Mukherjee T, Bhattacharya S et al (2023) Arsenic contamination in groundwater and food chain with mitigation options in Bengal delta with special reference to Bangladesh. Environ Geochem Health 45:1261–1287. https://doi.org/10.1007/s10653-022-01330-9
Jampani M, Liedl R, Hülsmann S et al (2020) Hydrogeochemical and mixing processes controlling groundwater chemistry in a wastewater irrigated agricultural system of India. Chemosphere 239:124741
Karunanidhi D, Subramani T, Roy PD, Li H (2021) Impact of groundwater contamination on human health. Environ Geochem Health 43:643–647
Khasanov S, Li F, Kulmatov R et al (2022) Evaluation of the perennial spatio-temporal changes in the groundwater level and mineralization, and soil salinity in irrigated lands of arid zone: as an example of Syrdarya Province, Uzbekistan. Agric Water Manag 263:107444
Li P, Qian H, Wu J (2014) Origin and assessment of groundwater pollution and associated health risk: a case study in an industrial park, northwest China. Environ Geochem Health 36(4):693–712. https://doi.org/10.1007/s10653-013-9590-3
Li P, Tian R, Xue C, Wu J (2017) Progress, opportunities and key fields for groundwater quality research under the impacts of human activities in China with a special focus on western China. Environ Sci Pollut Res 24(15):13224–13234. https://doi.org/10.1007/s11356-017-8753-7
Li W (2018) Influence prediction of different irrigation modes on groundwater level in plain area of Ma River Basin. Hydro Sci Cold Zone Eng 1(8):20–23 (in Chinese)
Li P, Qian H (2018) Water resources research to support a sustainable China. Int J Water Resour Dev 34(3):327–336. https://doi.org/10.1080/07900627.2018.1452723
Li P, Wu J (2019) Drinking water quality and public health. Expos Health 11(2):73–79. https://doi.org/10.1007/s12403-019-00299-8
Li P, Wu J (2022) Medical geology and medical geochemistry: an editorial introduction. Expos Health 14(2):217–218. https://doi.org/10.1007/s12403-022-00479-z
Li P, Wang D, Li W, Liu L (2022) Sustainable water resources development and management in large river basins: an introduction. Environ Earth Sci 81(6):179. https://doi.org/10.1007/s12665-022-10298-9
Li P, Sabarathinam C, Elumalai V (2023a) Groundwater pollution and its remediation for sustainable water management. Chemosphere 329:138621. https://doi.org/10.1016/j.chemosphere.2023.138621
Li P, Wu J, Zhou W, LaMoreaux JW (2023b) Hazard hydrogeology. Springer, Cham, Switzerland
Li L, Li P, He S, Wang D, Tian Y, Niu L (2024a) Groundwater nitrate pollution source apportionment under varying land use/land cover patterns. Expos Health. https://doi.org/10.1007/s12403-024-00666-0
Li L, Li P, Tian Y, Kou X, He S (2024b) Nitrate sources and transformation in surface water and groundwater in Huazhou District, Shaanxi, China: integrated research using hydrochemistry, isotopes and MixSIAR model. Environ Res. https://doi.org/10.1016/j.envres.2024.120052
Li P, Wu J, Elumalai V (2024c) Recent advances in modern hydrogeology: promoting harmony between nature and humanity. Water 16:1501. https://doi.org/10.3390/w16111501
Li P, Kou X, Wang Y, Niu L (2024d) Building a more sustainable Chinese Loess Plateau. J Earth Sci 35(1):283–287. https://doi.org/10.1007/s12583-024-1970-3

Liu J, Cao G, Zheng C (2011) Sustainability of groundwater resources in the North China Plain. In: Jones J (ed) Sustaining groundwater resources. International Year of Planet Earth. Springer, Dordrecht. https://doi.org/10.1007/978-90-481-3426-7_5
Liu X, Yue W, Ni B et al (2016) Impact of autumn irrigation on groundwater dynamics in Yichang Irrigation Sub-district, Inner Mongolia. J Beijing Normal Univ (Nat Sci) 52(3):380–386
Lu S, Zhao H, Jiang Y, Hao Z, Zhang X, Chen G (2021) Analysis of irrigation water based on multi-source remote sensing data and water balance principle. Shuilixuebao 52(9):1126–1135 (in Chinese)
Loewenberg S (2017) The poisoning of Bangladesh: how arsenic is ravaging a nation. https://undark.org/2017/08/16/bangladesh-arsenic-poisoning-drinking-water/. Accessed 2 Oct 2024
Mádl-Szőnyi J, Batelaan O, Molson J, Verweij H, Jiang X-W, Carrillo-Rivera JJ, Tóth Á (2023) Regional groundwater flow and the future of hydrogeology: evolving concepts and communication. Hydrogeol J 31:23–26
Malki M, Bouchaou L, Hirich A et al (2017) Impact of agricultural practices on groundwater quality in intensive irrigated area of Chtouka-Massa, Morocco. Sci Total Environ 574:760–770
McCarthy TS, Ellery WN, Bloem A (2003) Some observations on the geomorphological impact of hippopotamus (*Hippopotamus amphibius L.*) in the Okavango Delta, Botswana. Afr J Ecol 41(4):373–383. https://doi.org/10.1046/j.1365-2028.1998.89-89089.x
Meng Y, Wu J, Li P, Wang Y (2023) Distribution characteristics, source identification and health risk assessment of trace metals in the coastal groundwater of Taizhou City, China. Environ Res 238:117085. https://doi.org/10.1016/j.envres.2023.117085
Murray C, Marmorek D (2003) Adaptive management and ecological restoration. https://biology.illinoisstate.edu/rcander/BSC337/Reading%20Assignment%20Two/Murray%20and%20Marmorek%20Adaptive%20Management.pdf. Accessed 2 Oct 2024
Nelson RL (2022) Water rights for groundwater environments as an enabling condition for adaptive water governance. Ecol Soc 27(2):28. https://doi.org/10.5751/ES-13123-270228
Rodell M, Velicogna I, Famiglietti JS (2009) Satellite-based estimates of groundwater depletion in India. Nature 460(7258):999–1002. https://doi.org/10.1038/nature08238
Shah T (2009) Taming the anarchy: groundwater governance in South Asia. Routledge, 324pp.
Shiklomanov IA (2000) Appraisal and assessment of world water resources. Water Int 25(1):11–32. https://doi.org/10.1080/02508060008686794
Singh A (2021a) Soil salinization management for sustainable development: a review. J Environ Manag 277:111383
Singh A (2021b) A review of wastewater irrigation: environmental implications. Resour Conserv Recycl 168:105454
Suganthi S, Elango L (2020) Estimation of groundwater abstraction induced land subsidence by SBAS technique. J Earth Syst Sci 129:46
Tulip S, Siddik M, Islam M (2022) The impact of irrigation return flow on seasonal groundwater recharge in northwestern Bangladesh. Agric Water Manag 266:107593
Tweed S, Celle-Jeanton H, Cabot L et al (2018) Impact of irrigated agriculture on groundwater resources in a temperate humid region. Sci Total Environ 613–614:1302–1316
Tweed S, Massuel S, Seidel JL et al (2020) Seasonal influences on groundwater arsenic concentrations in the irrigated region of the Cambodian Mekong Delta. Sci Total Environ 728:138598
United Nations (2022) The United Nations World Water Development Report 2022: groundwater: making the invisible visible. UNESCO, Paria
USGS (United States Geological Survey) (2015) Estimated use of water in the United States in 2015. https://pubs.usgs.gov/circ/1441/circ1441.pdf
Wang D, Li P, Yang N, Yang C, Zhou Y, Li J (2023) Distribution, sources and main controlling factors of nitrate in a typical intensive agricultural region, northwestern China: vertical profile perspectives. Environ Res 237:116911. https://doi.org/10.1016/j.envres.2023.116911
Wang D, Wu J, Li P, Li L, Yang J, Zhang P, He S, Kou X, Wang Y (2024) Seasonal nitrate variations, risks, and sources in groundwater under different land use types in a thousand-year-cultivated

region, northwestern China. Environ Res 251:118699. https://doi.org/10.1016/j.envres.2024.118699

Wei Y, Chen Y, Cao X, Xiang M, Huang Y, Li H (2024) A critical review of groundwater table fluctuation: formation, effects on multifields, and contaminant behaviors in a soil and aquifer system. Environ Sci Technol 58(5):2185–2203. https://doi.org/10.1021/acs.est.3c08543

World Bank (2018) Beyond scarcity: water security in the Middle East and North Africa. MENA Development Report. World Bank, Washington DC. http://hdl.handle.net/10986/27659

World Bank (2023) The hidden Wealth of Nations: groundwater's critical role in a changing climate. https://www.worldbank.org/en/news/feature/2023/06/14/the-hidden-wealth-of-nations-groundwater-s-critical-role-in-a-changing-climate. Accessed 2 Oct 2024

Wu J, Li P, Wang D, Ren X, Wei M (2020a) Statistical and multivariate statistical techniques to trace the sources and affecting factors of groundwater pollution in a rapidly growing city on the Chinese Loess Plateau. Hum Ecol Risk Assess 26(6):1603–1621. https://doi.org/10.1080/10807039.2019.1594156

Wu J, Huang Y, Yin L et al (2020b) Effect of long-term straw returning and groundwater level on cadmium accumulation and availability in soils. J Agro-Environ Sci 39(9):1957–1963

WWAP (World Water Assessment Programme of the United Nations) (2014) Water and energy. The United Nations world water development report 2014. Paris, UNESCO. https://unesdoc.unesco.org/ark:/48223/pf0000225741

WWAP (World Water Assessment Programme of the United Nations) (2018) Nature-based solutions for water. The United Nations World Water Development Report 2018. Paris, UNESCO. https://www.unesco.org/en/wwap/wwdr/2018

Yang L, Xu Y, Ma Y, Song X, Zhao Z, Gong L, Liu H, Song Q (2021) Water movement in the unsaturated zone and groundwater recharge in the Yellow River irrigation area. Res Soil Water Conserv 28(6):168–174 (in Chinese)

Yin Y, Mao Y, Yang L et al (2021) Effects of aerated irrigation on the growth and rhizosphere soil of *Malus hupehensis*. Scientia Silvae Sinicae 57(10):59–70 (in Chinese)

Yu H, Wen X, Wu M et al (2022) Data-based groundwater quality estimation and uncertainty analysis for irrigation agriculture. Agric Water Manag 262:107423

Zhai J, Cao J, Liu K et al (2021) Evolutionary change in groundwater table in Qingtongxia irrigation district and its determinants. J Irrig Drain 40(9):102–110 (in Chinese)

Zhang D, Fan M, Liu H et al (2020) Effects of shallow groundwater table fluctuations on nitrogen in the groundwater and soil profile in the nearshore vegetable fields of Erhai Lake, southwest China. J Soils Sediments 20(1):42–51

Towards Sustainable Groundwater Management in Kuwait: Strategic Directions

Chidambaram Sabarathinam, Amjad Al-Rashidi, Bedour Alsabti, Dhanu Radha Samayamanthula, Khaled Hadi, and Nader Al-Awadhi

Abstract United Nations sustainable developmental goals on water, referred to as UNSDG 6 have been a baseline for developmental activities in many countries. SDG 6 has 6 targets and indicators along with international collaboration, capacity building, knowledge, and local community participation. Kuwait has aligned its 2035 vision to achieve the UNSDG targets and indicators. This chapter summarizes the activities focusing on the targets to achieve clean water and sanitation. Achieving water sustainability in arid and hyperarid regions like Kuwait is a strenuous task depending mainly on the desalinated water, brackish groundwater, and treated wastewater for drinking, domestic, industrial, and agricultural purposes. Kuwait depends on two main transboundary aquifers shared by the neighboring countries for groundwater needs. Groundwater chemistry has been monitored for natural and anthropogenic impacts on the quality and sustainability of the resources. The hotspots of hypersaline groundwater, paleo groundwater, contamination zones, and recharge regions have been identified and regularly examined. Further, utilisation of treated wastewater and desalinated water has led to an increase in water use efficiency. In addition, variation in rainfall patterns, source moisture for recharge, and rainwater chemistry

C. Sabarathinam (✉) · A. Al-Rashidi · B. Alsabti · D. R. Samayamanthula · K. Hadi
Water Research Center, Kuwait Institute for Scientific Research, Safat, 13109 Kuwait City, Kuwait
e-mail: csabarathinam@kisr.edu.kw

A. Al-Rashidi
e-mail: arashidi@kisr.edu.kw

B. Alsabti
e-mail: bsabti@kisr.edu.kw

D. R. Samayamanthula
e-mail: vdhanuradha@kisr.edu.kw

K. Hadi
e-mail: khadi@kisr.edu.kw

N. Al-Awadhi
National Liason Officer for IAEA, Kuwait Institute for Scientific Research, Safat, 13109 Kuwait City, Kuwait
e-mail: nawadi@kisr.edu.kw

P. Li et al. (eds.), *Sustainable Groundwater and Environment: Challenges and Solutions*, Springer Hydrogeology, https://doi.org/10.1007/978-3-031-82194-3_2

for recharging the water resources has been determined to address future challenges. Attempts to increase the groundwater potential of the region through managed aquifer recharge experiments have proved to be successful in the region. The chapter also highlights the challenges in addressing the targets, especially to meet the demands of the increasing population, rapid urban development, and industrialisation with available water resources. Suggestions for water conservation techniques, policy development, and monitoring techniques have been suggested to address the challenges. Incorporating these studies, a cloud based online monitoring system integrating the field real-time data through IoT and interlinking with the end users for the development of strategies to manage the water resources has been proposed in this chapter.

Keywords Open-source platform · Web-based platform · Environmental sustainability · Groundwater management

1 Introduction

Freshwater availability for basic needs is a global challenge that significantly impacts ecosystems, economies, and human societies (Li et al. 2024). The increasing stress on water resources due to climate change has a critical impact, leading to water quantity, quality, availability, and distribution alterations (Li and Wu 2024; Wang et al. 2023; Ren et al. 2024). Kuwait, along with many other countries in the Middle East and North Africa (MENA) region, is dealing with significant constraints on water resources. Climate change exacerbates these challenges, altering water availability through rising temperatures, unpredictable precipitation patterns, and sea level rise. Sustainable groundwater management has become a top priority for ensuring long-term water security in arid countries like Kuwait, where freshwater resources are critically limited. Kuwait's water scarcity issue is compounded by its reliance on non-renewable water resources and groundwater exploitation, leading to the deepening of water levels and the contamination of water resources. Saltwater intrusion is a growing concern, particularly as the country's agricultural sector depends heavily on groundwater for irrigation. Additionally, Kuwait relies on desalination plants to meet its domestic water needs. This process consumes extensive amounts of fossil fuels and increases the country's carbon footprint.

However, water security in Kuwait and the MENA region must be understood through the lens of its unique regional pressures (e.g., rapid population growth, urbanization, and industrial expansion) that intensify demand for water resources. So, to meet their agricultural, industrial, and urban water demands, nations in the MENA region consume over 80% of the water resources accessible, leading to groundwater depletion in Kuwait and other nearby countries (WRI 2021). Despite these significant challenges, Kuwait, being a part of the Gulf Cooperation Council (GCC) countries, has managed to avoid severe food and water shortages through large-scale desalination and global food imports. However, these strategies for environmental and economic costs are unsustainable in the long term.

Climate change has caused water-related disasters, including droughts and floods, which are worse and more frequent (UNESCO 2019). Decreased rainfall and rising surface temperatures have reduced groundwater recharge rates, increasing water shortages in Kuwait and neighboring countries (Khater 2010). The overuse of groundwater and mismanagement of water resources, coupled with inefficient water conservation practices, further aggravate the issue, making Kuwait highly vulnerable to future water crises. Farmers in the Abdally and Wafra areas in Kuwait extract to irrigate their crops, leading to over-extraction. As a result, groundwater levels have declined, water quality has deteriorated, and many farms have been abandoned due to the unsustainable depletion of resources (Al-Rashed et al. 1998). On the other hand, due to excess watering of gardens and sewage and water system leaks, groundwater levels have risen substantially in several residential areas of Kuwait City and its suburbs within the last twenty years (Hamdan 1987; Senay et al. 1987). Sustainable groundwater management is essential to address Kuwait's growing water security concerns. This requires an integrated approach encompassing technological innovation, policy reforms, and public engagement. Groundwater resources must be managed more efficiently to prevent further depletion and contamination by reducing the ecological footprint of water management in Kuwait, possible solutions include renewable energy powering of desalination plants, better agricultural practices, and efficient water recycling. Moreover, aligning Kuwait's water management strategies with the United Nations' Sustainable Development Goal 6 (SDG 6), which focuses on "ensuring access to water and sanitation for all," is critical for long-term water security. SDG 6 emphasizes the significance of sustainable conservation and management of water to meet future water demands, while also supporting broader goals of environmental sustainability, health, and economic development (Romano and Akhmouch 2019; Makarigakis and Jimenez-Cisneros 2019). Achieving this goal will require the integration of water-smart policies, institutional strengthening, and creating awareness among the common public, as well as capacity building and local community participation to conserve water. Figure 1 illustrate the targets of UN SDG 6.

This chapter addresses the problems of water scarcity, over-extraction, and contamination by exploring strategic options for sustainable groundwater management in Kuwait in line with SDG 6 and providing practical solutions. Further, it provides a road map for guaranteeing the long-term sustainability of water resources in Kuwait by positioning the country's specific water issues within the larger framework of regional and global water security.

Fig. 1 The United Nations (UN) sustainable development Goal (SDG) 6 targets

2 UN SDG 6 Targets 6.1 and 6.2

Kuwait's groundwater fields are located in central and northern regions (Fig. 2), and the majority of them are brackish water fields. In contrast, the freshwater fields, indicated in dark blue, are limited to the Rawdhatain and Umm Al-Aish fields in northern Kuwait. These two fields contain freshwater lenses, vital for providing the country with potable water resources. This distribution emphasizes the challenge that Kuwait faces in managing its limited freshwater resources and strategic importance of optimizing the use of its brackish water reserves.

As groundwater demand continues to rise, Kuwait experiences a decline in the groundwater level. The Dammam formation and the Kuwait Group aquifer have been rapidly depleting in Kuwait in the past few years. Studies have projected that, at current withdrawal rates, central Kuwait could see a maximum groundwater decline of 50 m (Mukhopadhyay et al. 1994). However, certain Urban areas face drainage problems during rain events due to the shallow groundwater level, which is compounded by the presence of impermeable gatch layers (Al-Rashed et al. 1998; Riedel and Simon 1973). Over time, the development of groundwater fields in Kuwait

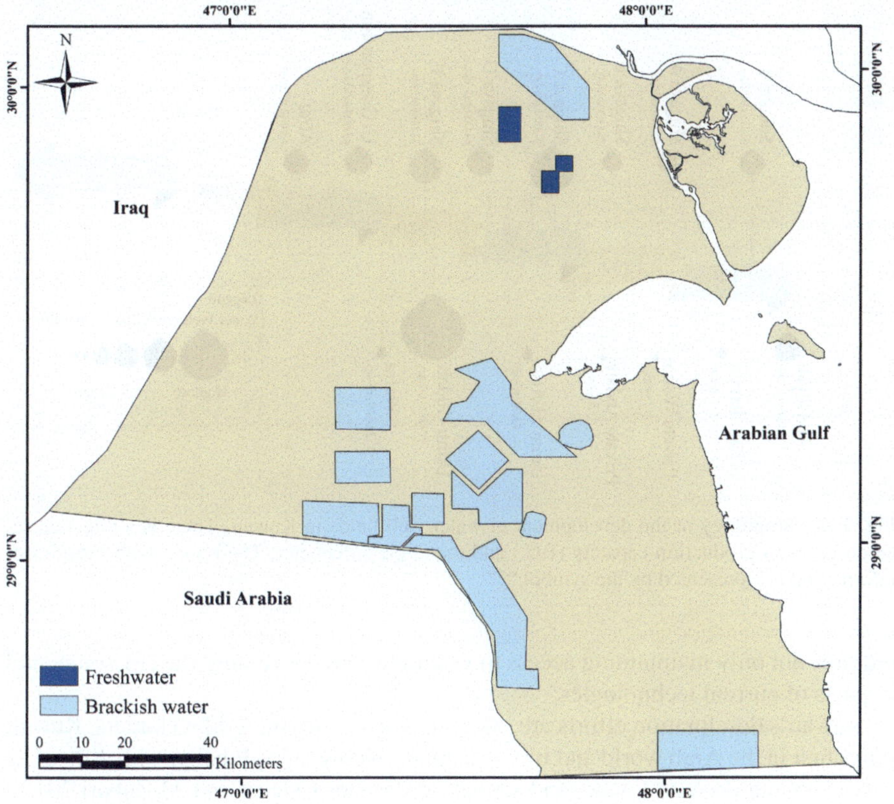

Fig. 2 Groundwater fields in Kuwait, showing limited freshwater fields (dark blue) at Rawdhatain and Umm Al-Aish, with the majority being brackish water fields (light blue)

has increased the groundwater production rate (Fig. 3), due to the progression of well-field construction and expansion over the decades.

Kuwait's freshwater scarcity is aggravated by a growing population, resulting in a pressing need to build desalination facilities and it was initiated in 1953. Since then, the nation has heavily relied on seawater desalination technologies. Multi-Effect Distillation (MED), which uses heat, Multistage Flash Distillation (MSF), and Reverse Osmosis (RO), are the main methods which collectively produce approximately 683.8 Million Imperial Gallons Per Day (MIGPD) of freshwater. Despite being an older technology, MSF still accounts for a substantial portion of Kuwait's freshwater supply due to its capacity to handle large-scale desalination.

Kuwait is making great progress towards achieving water and sanitation for all, in accordance with objectives 6.1 and 6.2. Target 6.2 is on ensuring that everyone has access to proper sanitation and hygiene facilities, especially in light of the increasing demands caused by a growing population, whereas Target 6.1 seeks to ensure that everyone has access to inexpensive, safe drinking water. Achieving these targets

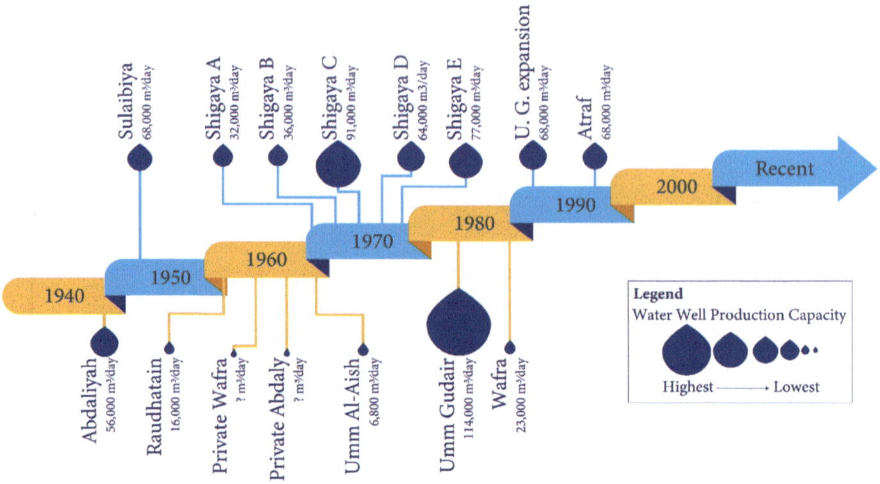

Fig. 3 A chronology of the development of water well fields in Kuwait. There is a direct correlation between production capacity (P.C.) and the circle's diameter. The region with insufficient information is represented by the symbol "?"

requires not only maintaining access to water but also addressing the environmental impacts of current technologies.

Kuwait's desalination efforts are integral to achieving these SDG targets. Kuwait ranks first in the Arab world and fifth in sanitation coverage globally, with 90% of its people having access to water and sanitation services (Aleisa and Al-Zubari 2017). The growth of desalination plants across areas like Doha, Shuwaikh, and Zour has been critical in meeting municipal water demands, with desalination accounting for 84.8% of the water supply (Al-Zubari et al. 2017). This success contributes to SDG target 6.1 because it guarantees that a large number of people have access to water that is safe to drink.

However, while these desalination facilities enhance water accessibility, they present significant environmental challenges. The energy-intensive nature of these processes, largely powered by fossil fuels, contributes to Kuwait's high fuel consumption—54% of the national fuel use is tied to desalination plants—raising concerns about greenhouse gas emissions and overall sustainability. Balancing water security with environmental responsibility is crucial to Kuwait's long-term water strategy. As part of its commitment to SDG 6, Kuwait acknowledges the need to develop more energy-efficient desalination technologies. Research and investment into renewable energy sources for powering desalination plants are essential for minimizing the environmental footprint while ensuring a sustainable freshwater supply.

In addition to desalination, Kuwait has pioneered wastewater management since the 1950s. The Al-Sulaibiya Wastewater Treatment Plant (WWTP), launched in 2005, exemplifies advanced treatment capabilities, using RO technology to convert treated wastewater into potable water. This facility processes about 425,000 cubic meters of municipal wastewater per day, making it a vital contributor to SDG target 6.2 by improving access to sanitation services and facilitating the reuse of water for agricultural and non-potable purposes. Still, difficulties remain. There are growing concerns about the impact on marine ecosystems due to the release of almost 30% of untreated wastewater into the sea. Even though the Sulaibiya WWTP processes 64% of the nation's sewage (Aleisa 2019), only 58% of the cleansed wastewater finds its way back into circulation. Thus, to improve sanitation and environmental health, the discharge of untreated wastewater must be reduced by adopting efficient methods of wastewater recycling.

As the population grows and water demand increases, Kuwait's combined efforts in desalination and wastewater treatment are crucial for long-term sustainability. Advancing these technologies and promoting the reuse of treated wastewater not only addresses immediate water needs but also aligns with the nation's commitment to SDG 6.

3 UN SDG 6 Targets 6.3 and 6.4

Reducing pollution and elimination of harmful substances from treated wastewater would greatly improve water quality, which in turn would boost recycling and safe reuse of water. Sustainable water management techniques and the preservation of aquatic habitats are two of the aims of the United Nations' Sustainable Development Goals (UNSDG) particularly 6.3. Groundwater in Kuwait is predominantly brackish to saline. Seawater intrusion is one of the factors that contribute to the pollution of groundwater, especially in coastal regions (Aliewi et al. 2021). Apart from seawater intrusion, over-extraction of groundwater increased the salinity to more than 50,000 mg/L (Sabarathinam et al. 2020), which is hyper-saline higher than seawater salinity. The saturation states reveal that silicates are less saturated than carbonates along the coastal regions, which are governed by lithology (Sabarathinam et al., 2022). Infiltration of contaminated sewage and denitrifying bacteria influence nitrate concentrations exceeding 50 mg/L in groundwater along Kuwait's Bay, including the process of contaminated bay water intrusion into the adjacent groundwater (Alsabti et al. 2023). Kuwait has two prominent type of aquifers including Kuwait Group and Dammam Formation. The Shagaya water well region was the major well field used for groundwater production. Excessive production caused salinity due to the weathering, dissolution, ion-exchange, and groundwater mixing from the deeper part of the aquifers (Rashid et al. 2022). However, groundwater pollution in Kuwait is primarily caused by oil and gas activities, which introduce contaminants such as hydrocarbons, including PAHs and TPH, into the water (Mukhopadhyay et al. 2017). Groundwater in Kuwait is mostly saline or brackish, with the exception of a small amount of freshwater lenses in the northern depressions of Raudhatain and Umm Al-Aish, which are located in the Dibdibba Formation, an aquifer in the Kuwait group. This freshwater is of fossil origin, having been recharged between the Pleistocene and Recent periods, with occasional replenishment from rainfall or surface runoff infiltrating through several large wadis that flow into the depression (Misak and Hussain 2022). Anthropogenic activities such as oil field operations raised concerns about groundwater quality in several wells at Raudhatain and Umm Al-Aish depressions in north Kuwait (Mukhopadhyay et al. 2017). The severity of pollution in Kuwait after the Iraqi invasion had a notable effect on the groundwater (Mukhopadhyay et al. 2008).

Consequently, Kuwait holds the world's largest wastewater treatment plant in Sulaibiya along with other plants. Treating municipal wastewater using conventional methods and advanced technologies like ultrafiltration and reverse osmosis (RO) improves effluent quality. However, studies reported that treated effluent, if discharged, could also contaminate the groundwater due to the presence of endocrine-disrupting chemicals and persistent organic pollutants, etc. (Saeed et al. 2017; Alshemmari 2021; Abusam et al. 2022). In recent years, it has been noted that leachate from landfills has adversely impacted groundwater with higher salinity and the presence of heavy metals (Al-Rashed 2018). Moreover, Kuwait stands first in developing multi-stage flash desalination plants to produce fresh water. Ultimately, brine discharge into the sea affects the aquatic and marine ecosystem (Darwish et al. 2011). Similarly, RO units utilized in agricultural farms in Kuwait were identified with brine discharged containing anti-scalants, organic salts, and backwash slurries that could infiltrate into groundwater and affect the quality (Bhandary et al. 2024; Bhandary 2023). The development of desalination plants increased globally after the 1990s, where Kuwait once again stands in first place for initiating the largest desalination units through RO technologies. Besides, brine or desalination rejects are a major concern, leading to increased pollution, specifically with metals in the Kuwait Bay region apart from sewage effluents. It was also observed that suspended sediments in the region increase the metal concentration and other inorganic constituents in groundwater by leaching and adsorption (Sabarathinam et al. 2019; Ali and Chidambaram 2021). Correlating the changes in the surficial parameters with time to the groundwater flow and chemistry indicated the interrelationship between the Land surface properties and groundwater (Jayaramu et al. 2024). The recent climate changes were noticed, and the moisture sources for rainfall in Kuwait were identified using the Hysplit backward trajectory model. Western regions are the primary moisture source regions. Winter rainfall events, for example, can travel long distances through the Azores from northeast America. In contrast, November events can originate from the Caspian/Black Sea region in northeastern Kuwait, with relatively shorter travel trajectories. Higher ion concentrations and saturation of aragonite and calcite minerals in some rain samples collected in Northern Kuwait after dust events suggests that these ions may have slowly seeped into groundwater and altered its salinity in certain areas (Svv et al. 2023). Therefore, Kuwait authorities must take into consideration to combat the groundwater pollution through policy makers and develop management strategies. Finally, regular groundwater assessment and the implementation of strategies to enhance water quality are suggested as effective measures to combat pollution, contributing to the achievement of UNSDG 6.3.

Groundwater in Kuwait is extracted for household and domestic utilities, irrigation, blending with desalinated water for potable and industrial purposes etc. Excessive extraction and pumping have depleted the groundwater sources, increased the salinity of aquifers, and reduced the recharge rate (Sadeqi et al. 2024). Therefore, integrated approaches for groundwater management in Kuwait are necessary for the well-being of the eco-hydrology of the aquifers. Exploitation of limited freshwater available in the Raudhatain and Umm Al-Aish depressions of northern Kuwait for bottling companies may deplete the freshwater sources and leave water stress in future that could lead to water scarcity (Hadi 2024; Al-Rashidi et al. 2023). Apart from the conventional methods, nonconventional methods to improve water use efficiency are mainly through Wastewater treatment plants (Al-Shammari et al. 2013), Greywater treatment (Samayamanthula et al. 2019), Desalination (Al-Shammari et al. 2013), Rainwater harvesting (Al-Rashed and Aliewi 2017; Samayamanthula et al. 2021), Artificial groundwater recharge (Mukhopadhyay et al. 2020). The above are a few methods to be adopted that pave the way for the efficient use of water and enhance sustainable management of water to attain the goal UNSDG 6.4. In the journey of water efficiency overtime, Kuwait has initiated and developed several wastewater treatment and desalination plants (Fig. 4).

Around seven wastewater treatment plants have been established since the 1980s due to increased water demand and wastewater generation. It includes several stages of treatments like screening, primary, secondary, tertiary and finally, to remove the microbes, either ultrafiltration or RO were coupled and treated. The different desalination units in Kuwait include conventional types such as Multistage Flash Distillation (MSF), Multi-Effect Distillation (MED) and Membrane-based Reverse Osmosis (RO), with an average of 85% municipal water supplies. Out of several desalination plants initiated in Shuwaikh, Zour (North and South), Shuaiba (North and South), Doha (East and West), and Sibiya, Doha West stands with the highest production capacity of 170.4 MIGD (Million Imperial Gallons per Day) (Hadi 2024). Since the

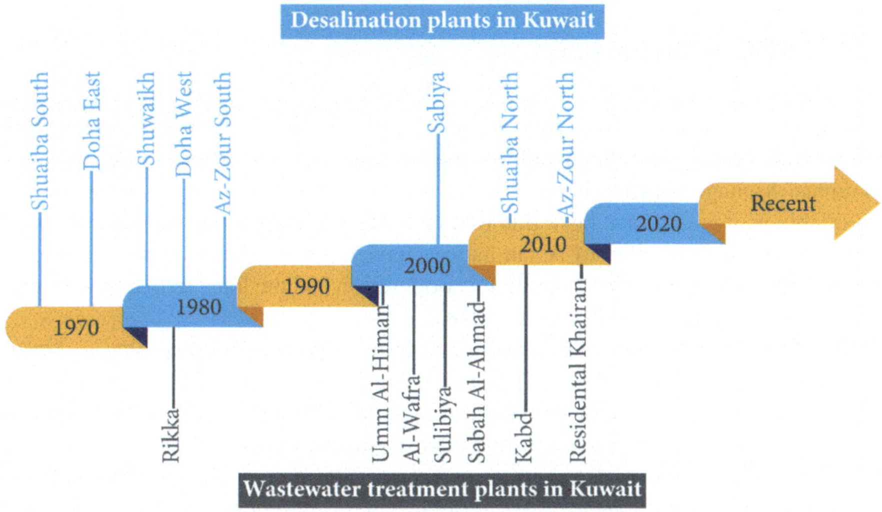

Fig. 4 A chronology of Kuwait's desalination and wastewater treatment facilities

increase in water demand, limited water resources in Kuwait, and dependence on other sources, greywater is one alternate source and economical way for domestic usage. One crucial treatment technique developed to reuse greywater is gravity-governed filtration using simple and economical natural materials like gravel, sand, and active carbon, followed by disinfection. This treatment is highly effective and can be used for landscaping, toilet flushing, and irrigation for non-edible plants. Disinfection types like ultrafiltration or UV treatment can be utilized to artificially recharge aquifers and prevent saltwater intrusion in coastal aquifers (Samayamanthula et al. 2019). One similar study as a pilot plant in school was developed in 2017, which includes screening, filtration, UV and Chlorination, and the effluent was used for toilet flushing and unrestricted irrigation (Alsulaili et al. 2017). The future scope of research in Kuwait concerning grey water is wider as minimal studies were made and not adopted for regular use. Another potential water source is rainwater harvesting, which should be implemented on a small scale or individually despite limited rainfall in Kuwait it serves as a major source of recharge or direct supply without treatment at a lower cost and limits the usage of groundwater sources. The development of watersheds can also benefit Kuwait by utilizing the rainwater for domestic purposes. Artificial recharge of aquifers is one promising approach, but expensive. A minimal study was conducted on a pilot scale by injecting RO-treated water into the aquifers, which helps reduce salinity and limits drying up during scarce rainfall (Mukhopadhyay 2019). A hydrogeological investigation of the Az-Zaqlah depression in north Kuwait was conducted to assess the feasibility of artificial recharge by storing RO-treated municipal wastewater (Mukhopadhyay et al. 2012). One more numerical modelling study for the suitability of artificial recharge was carried out in the Kabd area (Mukhopadhyay et al. 2022).

4 UN SDG 6 Target 6.5

Although transboundary rivers have attracted more research, transboundary aquifers in the Arabian region have been understudied despite their importance as Asian water resources. Management of shared aquifer resources has not been as productive or sustainable as it could have been due to a lack of knowledge, experience, and institutional backing. A number of Asian nations have acknowledged the need of strong regional governance in order to facilitate cooperation and the long-term management of transboundary aquifers (Lee et al. 2018).

The Arabian Peninsula has three major transboundary aquifers: Wajid, Umm Er Radhuma, and Wasia (Mazzoni et al. 2018; Altayyar et al. 2024). Groundwater recharge rates in the Arabian Peninsula transboundary aquifers show high spatio-temporal variability. The recharge rates can be influenced by climate, occasional rainfall events, and irrigated agriculture practices (Sherif et al. 2023). Human activities such as agricultural irrigation and rapid urban development have led to a significant loss of groundwater in the Wajid, Umm Er Radhuma, and Wasia aquifer systems on the Arabian Peninsula (Altayyar et al. 2024). Few transboundary aquifers in the

Arabian Peninsula are noteworthy. In dry and semi-arid regions, such as North Africa and the Middle East, these aquifers play a vital role in supplying water for both human and environmental needs (Eckstein 2021).

In the northeastern part of the Arabian region, over the Paleogene Formations, is the Neogene Aquifer System, which has been extended northward. The Wadi al Batin and Wadi ar Rimah flow towards the Shatt al-Arab and form the boundaries of the Dibdibba Delta basin. The major aquiferous formations that comprise the basin are the Hofuf, Dam, and Hadrukh in Saudi Arabia and in Kuwait and Iraq, it is Dibdibba, Lower Fars, and Ghar. The basin extends over all three nations. Groundwater in Saudi Arabia typically comes from the Lower Hadrukh Formation, composed chiefly of continental sands and gravels, or the Upper Dibdibba Formation, located in northern Kuwait and the southern part of Iraq. The groundwater extraction from the Dibdibba Formation, and thus heavy abstraction, has inverted groundwater flows towards the Paleogene aquifers from the Neogene. These two major factors have affected the region's groundwater regimes. There are about 20 aquifers listed by IGRAC in the Arabian region (Table 1), and more than 10 locations have an overlap as lower and upper aquifers (Fig. 5). It is to be noted that the coastal aquifers of the area are transboundary.

The major aquifer systems of Kuwait are Kuwait (KG) and Dammam Formations (DF). The groundwater of these aquifers is mostly paleo groundwaters. The dating of groundwater reveals that the aquifers reflected recharge in the central and northwestern part of the peninsula in a relatively semi-arid/arid climate under the influence of the Indian Ocean monsoon (IOM). The identification that the aquifer recharges in the southwestern region of Kuwait, Saudi Arabia, is a result of isotopic investigations on rainwater and groundwater (Hadi et al. 2016). During the early and middle Holocene, from approximately 12,000–5,000 years, BP (McClure 1976; Street-Perrott and Perrott 1990), IOM circulation was stronger, and the mean summer position of ITCZ was much farther north than its present position (Demenocal et al. 2000).

In addition, the ITCZ began to recede southward between 5,000 and 6,000 years ago. The retreat of the IOM was reported as the ITCZ moved southwards 5,000 years ago (Neff et al. 2001) resulting in reduction of rainfall (Radha et al. 2023). The enrichment of $\delta^{18}O$ of groundwater samples (Sabarathinam et al. 2020) evidently reflected a warmer period with less amount of rainfall. The lakes surrounding Rub Al-Khali dried up due to a change in monsoonal conditions (McClure 1976). Further enrichment of $\delta^{18}O$ of groundwater samples from the DF aquifer and groundwater samples collected from the KG aquifer dated around 4025 ± 150 years BP probably represented arid conditions that prevailed around 4175 years BP (Fig. 6).

Practically no significant recharge to the DF aquifer was observed after 15,000 years BP till recently, suggesting that the water source of the DF aquifer is finite (Fig. 6). It indicates practically no renewability; therefore, caution should be applied to the use of this groundwater source due to its non-renewability. Given its non-renewability, it is prudent to exercise caution when using groundwater that appears to be older than 10,000 years (Cendon et al. 2019).

Table 1 List of transboundary aquifers of the Arabian Peninsula (IGRAC 2022)

Aquifer code	Aquifer name	Sharing countries	Area (km^2)
AS001	Western Aquifer Basin	Egypt, Israel, Palestinian Territory	12313
AS002	Coastal Aquifer Basin	Egypt, Israel, Palestinian Territory	18945
AS003	Northeastern Aquifer	Israel, Palestinian Territory	1245
AS004	Anti-Lebanon	Lebanon, Syria	3061
AS010	Upper Jezira	Iraq, Syria, Turkey	10294
AS123	Taurus-Zagros	Iran, Iraq, Turkey	207478
AS124	Jezira Tertiary Limestone Aquifer System	Syria, Turkey	130779
AS125	Neogene Aquifer System (North-West): Upper and Lower Fars	Syria, Iraq	66266
AS126	Saq-Ram Aquifer System (West)	Jordan, Saudi Arabia	151718
AS127	Wasia-Biyadh-Aruma Aquifer System (North): Sakaka-Rutba	Saudi Arabia, Iraq	84212
AS128	Neogene Aquifer System (South-East): Dibdibba-Kuwait Group	Iraq, Kuwait, Saudi Arabia	148313
AS129	Tawil Quaternary Aquifer System: Wadi Sirhan Basin	Jordan, Saudi Arabia	45638
AS130	Umm er Radhuma-Dammam Aquifer System (North): Widyan-Salman	Iraq, Kuwait, Saudi Arabia	242966
AS131	Wajid Aquifer System	Saudi Arabia, Yemen	370578
AS139	Wasia-Biyadh-Aruma Aquifer System (South): Tawila-Mahra/Cretaceous Sands	Saudi Arabia, Yemen	157061
AS140	Umm er Radhuma-Dammam Aquifer System (Centre): Gulf	Saudi Arabia, United Arab Emirates, Bahrain, Qatar	291613
AS141	Umm er Radhuma-Dammam Aquifer System (South): Rub'al Khali	Oman, Saudi Arabia, United Arab Emirates, Yemen	686619
AS142	Basalt Aquifer System (West): Yarmouk Basin	Jordan, Syria	6873
AS143	Basalt Aquifer System (South): Azraq-Dhuleil Basin	Jordan, Syria	8457
AS157	Eocene Aquifer	Lebanon, Palestinian Territory	340

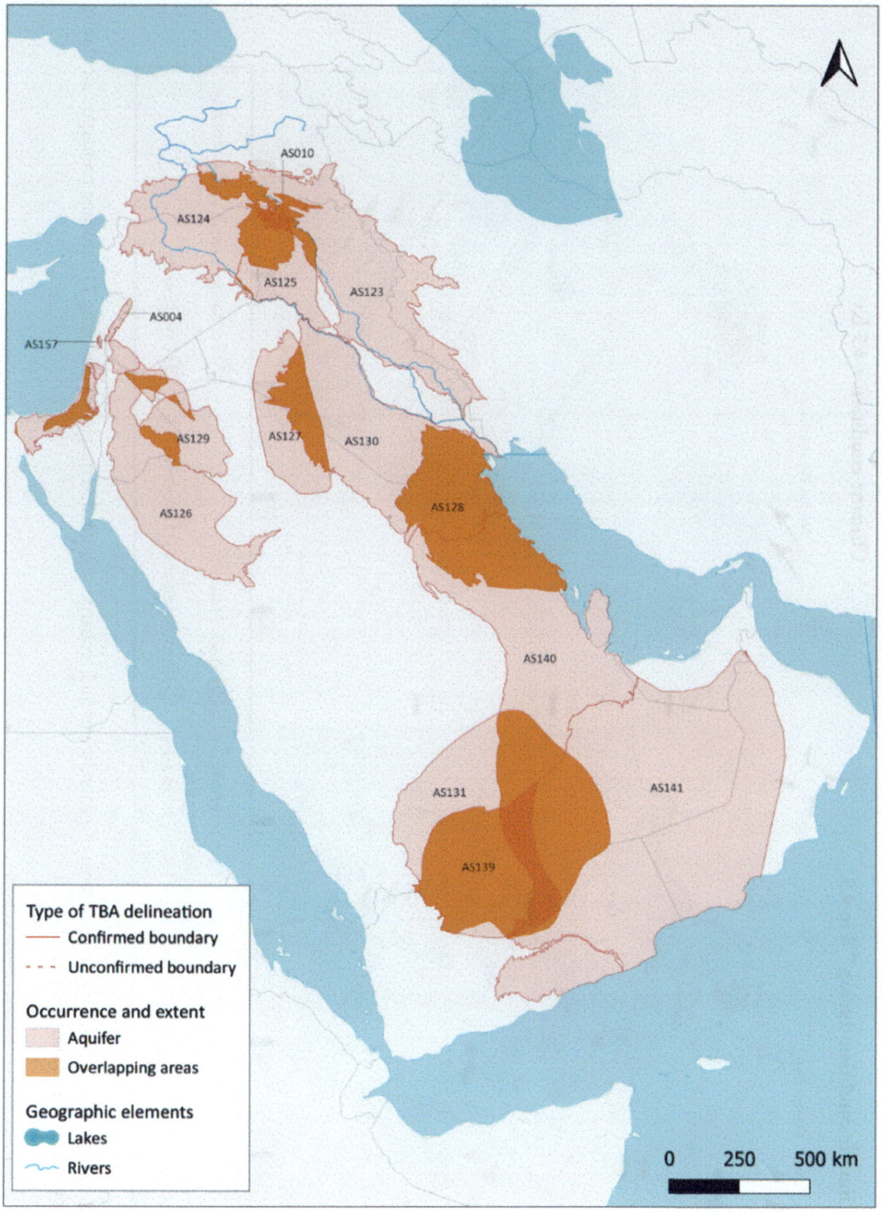

Fig. 5 IGRAC (International Groundwater Resources Assessment Centre), 2022. Transboundary Aquifers of the Middle East [map]. Edition 2022. Delft, Netherlands

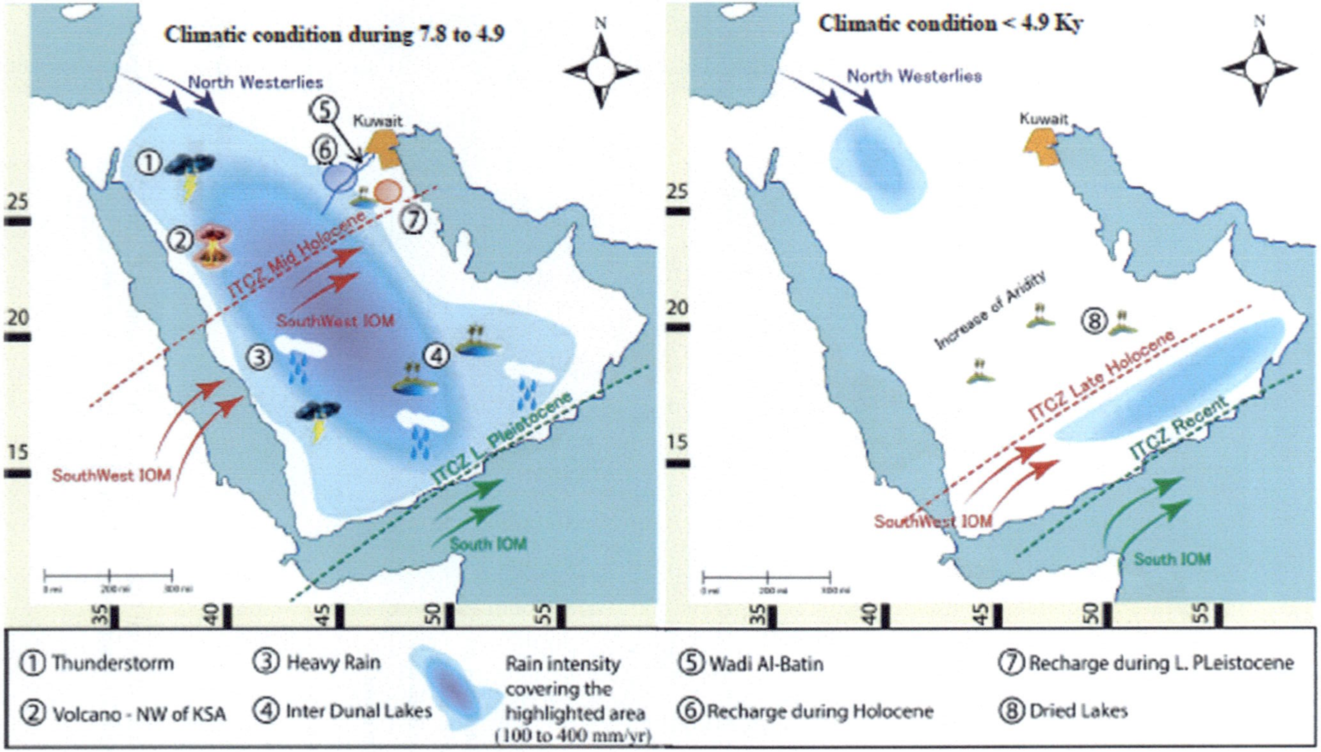

Fig. 6 The shift of ITCZ observed from the isotopic signatures of groundwater and recharge regions determined by adopting Pleistocene recharge calculation of Jasechko (2016). Adopted from Sabarathinam et al. (2020)

Using hydrochemical data from 116 water wells for an extended length of time (1975–2019), the spatio-temporal evolution of the Shagaya well field's hydrochemical characteristics was assessed. The study inferred that, though there was minor variation in the hydrogeochemistry, this variation was mostly attributable to a common recharge source across the territorial region (Rashid et al. 2022). Further, studies on helium in groundwater assisted in identifying that the primary movement was lateral, with some insitu and some ascending vertically across formational flow (Rashid et al. 2023). It is possible that terrigenous helium migrates to shallow waters across fault planes, as evidenced by the higher concentration of this element in groundwater samples close to oil fields.

5 UN SDG 6 Target 6.6 Protect and Restore Water-Related Ecosystems

Kuwait's strategic choice to attain partial food security resulted in three agricultural regions: Abdaly in the north, Sulabiya in the middle, and Wafra in the south. The Wafra Agricultural Area's soil is mainly composed of deep sands, underlain by a gatch bed (impervious layer) at shallow depths, often not surpassing 2 m (Misak et al. 2014). Historically, misuse of irrigation water has led to significant soil salinization in Wafra. Between 1995 and 2006, unchecked groundwater extraction resulted in rising water tables and accumulation of saline water in low-lying areas. This situation deteriorated agricultural productivity and led to the abandonment of many farms (AlAli 2008; Misak and Hussain 2023). The indifference of Kuwait's agricultural sector towards water use has caused considerable overirrigation, resulting in soil salinization and waterlogging (Al-Murad et al. 2017). Recent initiatives by the Ministry of Water and Electricity have sought to regulate groundwater extraction, resulting in less ponding and improved soil conditions (Misak and Hussain 2023).

The excessive rise in groundwater levels has resulted in waterlogging, particularly in low-lying areas. This accumulation creates stagnant water bodies that can persist for extended periods, especially during the hot summer, when evaporation rates are high (approximately 3000 mm/year) (Misak et al. 2014). Al-Murad et al. (2017) found that the connection between inappropriate groundwater use in the Wafra agricultural region and leakage from the Dammam aquifer had largely influenced the total dissolved solids (TDS) concentrations. A badly constructed well in the Dammam aquifer and irrigation return flows from nearby farms situated at a higher elevation caused the problem. Inadequate well sealing has resulted in ongoing leakage because Dammam is a confined aquifer (Al-Murad et al. 2017). Soil salinisation resulted from intensive irrigation using brackish water and high evaporation rates in the region. Starting in 2008, the Public Authority for Agriculture and Fisheries Resources (PAAFR) began an effort to seal off all wells in Wafra that could reach the Dammam aquifer. This led to the closure of 600 wells between June 2008 and October 2009 (Al-Murad et al. 2017).

Due to this issue, the water level rised, resulted in the development of surface water/ponds in the region. The existence of an impervious cohesive layer at 3 mbgl (Al-Sulaimi et al. 1994) confines the irrigation return flow inside this depression and creates a surface water accumulation. Nevertheless, the salinity in these water bodies was very high due to evaporation. Upon complete desiccation, salt encrustations were observed on the surface, making it unsuitable for agricultural use (Al-Murad et al. 2017).

One of the main causes of waterlogging and salinisation is insufficient drainage infrastructure. While some measures have been taken to control flowing wells, comprehensive drainage solutions are still required to mitigate these issues effectively. The formation of saline wetlands and waterlogged areas can disrupt local ecosystems. Changes in soil chemistry and hydrology may affect plant species composition and the overall biodiversity within the agricultural landscape (AlAli 2008). The Kuwaiti government has initiated efforts to manage groundwater extraction and improve soil conditions; however, challenges still need to be addressed due to the scale of the problem and the need for ongoing monitoring and intervention. One of the future solutions for these long-lasting brackish ponds is planting mangroves. Mangroves are uniquely adapted to thrive in saline environments, making them potentialcandidates for use in developed ponds with high salinity levels, such as those in Wafra, Kuwait. Mangroves can tolerate high salinity levels, except for a few salt-tolerant species, which are significantly higher than typical ocean salinity (about 35 ppt) (The Ocean Portal Team 2023). This adaptability makes them suitable for environments where salinity is a concern. Mangroves can enhance nutrient cycling within saline ponds by stabilizing sediments and improving organic matter content. Their root systems help prevent erosion and promote sediment deposition, improving soil quality (Ahmed et al. 2022). The presence of mangroves can also aid in filtering pollutants from water, thereby improving water quality within the ponds. This is crucial in areas where agricultural runoff may contribute to salinity and pollution (Nehemia et al. 2019). In addition, Mangrove ecosystems can sequester significant amounts of carbon dioxide (CO_2) from the atmosphere (Inoue 2019).

6 UN SDG 6 Target 6.A International Collaboration and Capacity Building

In order to provide a steady supply of safe drinking water, Kuwait Institute for Scientific Research (KISR) has collaborated with the International Atomic Energy Agency (IAEA) and the United Nations Development Programme (UNDP) to adopt advanced water management strategies. These collaborations provide essential technical expertise and financial support, assisting Kuwait's water management projects and ensuring alignment with international best practices (IAEA 2018, 2023; The United Nations in Kuwait 2024; GUST 2024). Regarding managing groundwater resources and sharing information, Kuwait has been an essential player in recent years in terms of international collaboration and by assisting in capacity-building to member states of the region. Kuwait spearheaded efforts to address joint groundwater assessment and management issues within the ARASIA Cooperative Agreement for Arab States in Asia for Research, Development, and Training related to Nuclear Science and Technology in conjunction with the International Atomic Energy Agency. The Water Research Centre, KISR, has led two major projects on urban coastal groundwater management and addressing groundwater vulnerability with regard to climate change in the Arab region. Kuwait has coordinated these projects with other neighboring countries (Iraq, Jordan, Lebanon, Oman, Qatar, Saudi Arabia, Syrian Arab Republic, United Arab Emirates, and Yemen). Kuwait has shared the expertise to train the partners on Sea Water Intrusion Model (SEAWAT) and geochemical and isotope data Interpretation. Further, the scientists from KISR have been assigned to various countries like Indonesia, Vietnam, Malaysia, Nepal etc., as experts by the IAEA to impart training and deliver special lectures on groundwater modelling, application of isotope techniques in groundwater studies and to assist other countries in the field of isotope hydrology. Apart from such initiatives, Kuwait also shares knowledge and helps in capacity-building through the agency by conducting training activities for countries on the above-mentioned fields of research.

7 UN SDG 6 Target 6.B Participation of Local Communities

Kuwait recognizes that achieving long-term water sustainability also depends on community participation, as emphasized in SDG 6.B. In collaboration with civil society organizations like the Kuwait Environment Protection Society (KEPS), the Kuwait Municipality plays an active role in engaging local communities in water conservation initiatives. Through public education campaigns and community outreach, these organizations promote responsible water usage and raise awareness about the importance of conserving resources. Educational initiatives targeting schools and universities foster a new generation of environmentally conscious citizens. Apart from KISR institutions like Kuwait University and the Gulf University for Science and Technology (GUST) emphasize water conservation in their curricula, hosting workshops and conferences to prepare future leaders to tackle Kuwait's water challenges with innovative solutions (SDG Portal 2024; Water Action Hub 2024; GUST 2024). When it concerns managing Kuwait's water resources, KISR and the Ministry of Electricity, Water and Renewable Energy (MEWRE) work hand in hand.

Private companies have also contributed significantly to Kuwait's efforts to meet SDG 6 targets. The Kuwait National Petroleum Company (KNPC) has implemented internal measures to recycle water and treat wastewater, setting a benchmark for other industries to follow by reducing water pollution and promoting treated water reuse. Zain Group, a leader in the telecommunications sector, is actively involved in public campaigns promoting water conservation (Zain Group 2023; Kuwait Times

2021). These efforts, often in collaboration with KEPA, aim to raise awareness and engage the community in sustainable water management practices (SDG Portal 2024; Water Action Hub 2024). KISR, in partnership with the Kuwait Environmental Public Authority (KEPA), is at the forefront of improving wastewater treatment and environmental sustainability. Their projects focus on developing innovative wastewater reclamation technologies and optimizing treatment processes. By supporting initiatives for safe wastewater reuse, KISR ensures that Kuwait's wastewater management strategies meet both local and international environmental standards.

8 Current Water Status, Consumption, and Sources

Groundwater, desalinated seawater, and treated wastewater are the three main sources of water in Kuwait. Groundwater, primarily sourced from the Kuwait Group and the Dammam Formation aquifers, is characterized by varying salinity levels and is heavily utilized for irrigation, livestock, landscaping, construction, and non-potable household uses. Desalinated seawater is the principal source of potable water, while treated wastewater is increasingly used for irrigation and industrial purposes.

In 2006, desalination constituted 54.6% of Kuwait's total water use, groundwater production 22.3%, and wastewater reuse 23.1%. With a range of 180–2,018 L per person per day, Kuwaiti residential villas are the fourth most water-consuming in the GCC region. Due to their affordable availability, the higher-income class constitutes the largest consumer category. A shortage of institutional oversight and administration, increased infrastructure development, and the need for freshwater are the three primary obstacles to water resource management in Kuwait.

9 Challenges in Groundwater Management

The primary challenges in managing Kuwait's groundwater resources include:

- **High Per Capita Consumption**: Because of low cost water and relatively high per capita income, Kuwait has one of the highest residential water consumption rates in the Gulf Cooperation Council (GCC) area.
- **Rising Demand**: Rapid population growth and extensive infrastructural development have significantly increased the demand for freshwater.
- **Over-Extraction**: Groundwater levels have been falling and water quality has been worsening due to over-extraction, making water shortage and heading to water stress.
- **Institutional Gaps**: In order to effectively monitor and manage water resources, there is a noticeable absence of thorough institutional frameworks.
- **Environmental Stressors**: Climate change and associated extreme weather events pose additional stress on the already limited water resources.

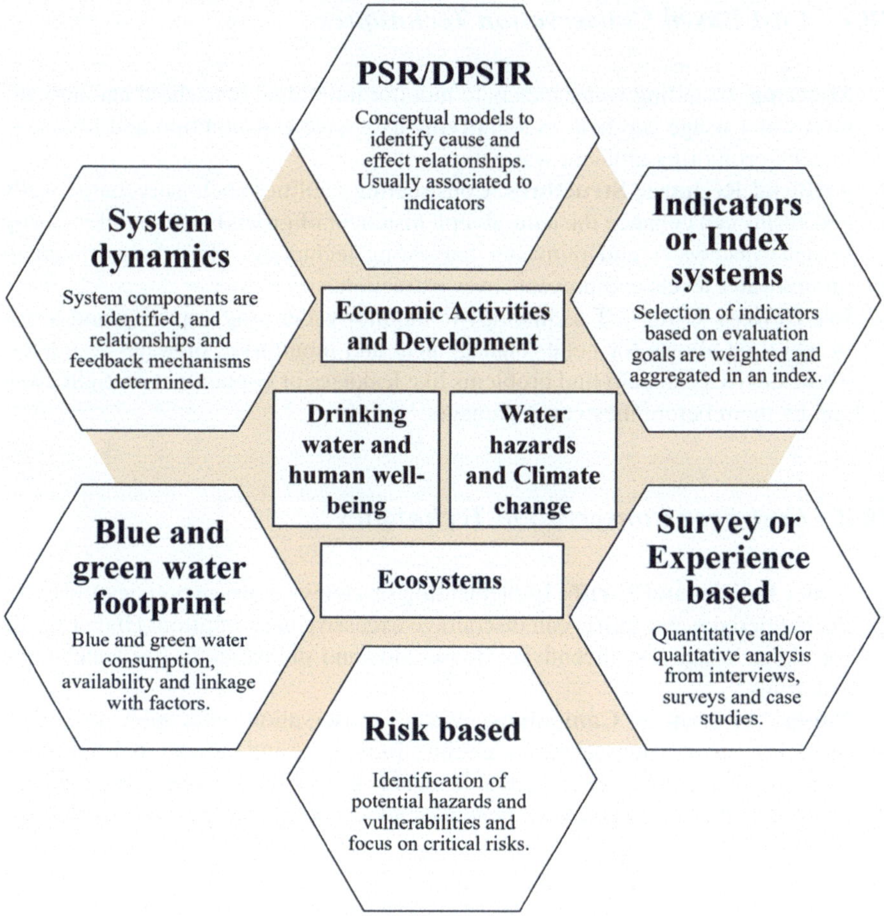

Fig. 7 Water security framework and water security evaluation (modified after Marcal et al. 2021)

- Water security approaches should concentrate on six essential aspects like PSR pressure state response—DPSIR, Diver Pressure State impact response, Indicator systems, System dynamics survey, blue and green water footprint and Risk-based (Fig. 7) considering four dimensions; (a) the well-being of humans with domestic drinking water, (b) the ecosystem, (c) water hazard and climate change and (d) economic activities and development.

10 Strategic Solutions

Addressing these challenges requires a multifaceted approach, incorporating both financial and non-financial strategies for water conservation and management.

10.1 Cost Based Conservation Techniques

1. **Metering**: Installing water meters to monitor individual household and agricultural water usage can help in identifying excessive consumption and leakages, thereby promoting efficient water use.
2. **Artificial Recharge Structures**: Constructing facilities such as recharge wells and basins can enhance the natural replenishment of groundwater aquifers using treated wastewater and rainwater harvesting techniques. This helps maintain groundwater levels and prevents over-extraction.
3. **IoT Devices**: Using IoT technology to monitor water usage, quality, and levels in real time allows for better management and monitoring of water resources. These devices can help find problems like leakages or contamination right away and fix them before they cause damage.

10.2 Cost Free Conservation Techniques

1. **Water Policies and Tariffs**: Implementing stricter water usage policies and introducing tiered water tariffs can discourage excessive consumption. Higher tariffs for higher usage can incentivize households and industries to use water more efficiently.
2. **Public Awareness Campaigns**: Water conservation education initiatives involving media campaigns, community programs, and educational initiatives can foster a culture of sustainable water use. Highlighting conservation's environmental and economic benefits can encourage responsible water use behaviour.

11 Strategic Plans and Initiatives

11.1 Kuwait National Development Plan (New Kuwait 2035)

The New Kuwait 2035 initiative outlines the country's long-term development goals, including strategies for sustainable water management. It emphasizes reducing reliance on fossil fuels, promoting renewable energy, and investing in infrastructure for efficient water use. By aligning water management strategies with national development goals, Kuwait aims to ensure sustainable water resource availability for future generations (MOFA 2024; ESCWA 2023).

11.2 Integrated Water Resources Management (IWRM)

Coordination of land, water, and related resource development and management is the goal of the Integrated Water Resources Management (IWRM) strategy, which Kuwait has adopted. This holistic strategy considers the social, economic, and environmental aspects of water management, ensuring that water use is sustainable and equitable. IWRM promotes stakeholder engagement and cross-sectoral collaboration to achieve comprehensive water management goals.

12 Future Management Strategy

A comprehensive future management strategy for sustainable groundwater management in Kuwait involves developing an integrated, web-based platform that consolidates data from various governmental sectors (Fig. 8). This platform will facilitate effective water resource management, monitoring, and policy development.

The proposed database and platform will be a collaborative effort involving the Kuwait Institute for Scientific Research (KISR) and multiple governmental bodies, including:

- Environment Public Authority

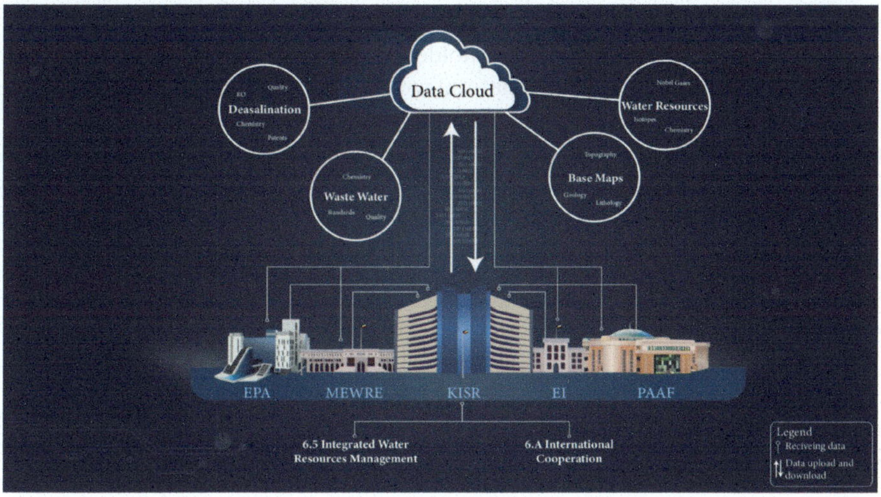

Fig. 8 Conceptual IoT-based water management system for Kuwait, with real-time data stored in the cloud and managed by KISR. The data will be available to different end users EPA (Environment Protection Agency), MEWRE (Ministry of Electricity and Water and Renewable Energy), EI (Educational Institutions) and PAAF (Public Authority Of Agriculture Affairs And Fish Resources), Decision-making, and governance

- Ministry of Electricity & Water and Renewable Energy
- Ministry of Public Works
- Public Authority of Agriculture Affairs & Fish Resources.

13 Conclusion

The State of Kuwait is approaching the UN SDG 2030 goals in line with Vision 2035 of Kuwait concerning water. Though several goal targets are effectively achieved, there are a few goals to be more effectively focused, such as the integrated water management and enhancement of water-related ecosystems and conservation practices from local communities. Effective groundwater management in Kuwait requires a multifaceted approach that combines technological innovations, policy reforms, and inter-sectoral collaboration. By implementing the proposed strategic solutions and future management strategies, Kuwait can address its water challenges and ensure the sustainability of its groundwater resources. By integrating data from different agencies, the platform will provide a holistic view for water resource management, enabling coordinated efforts and informed decision-making. Installing IoT devices across the country will facilitate real-time data collection and sharing. These devices will monitor water levels, quality, and usage patterns, providing a comprehensive, up-to-date database. This real-time information will be crucial for effective water management, allowing for prompt responses to emerging issues and better planning for future water needs. The comprehensive data repository will be hosted on an open-source Geographic Information System (GIS) platform. This will enable easy access and sharing of data among stakeholders, supporting transparency and collaborative decision-making. The GIS platform will also allow for advanced spatial analysis, helping to identify trends and patterns in water usage and resource availability. Integrating real-time monitoring, comprehensive data collection, and collaborative governance will support Kuwait's clean, safe, and sustainable water management plan, securing its water resources for future generations.

Acknowledgements The authors would like to thank the International Atomic Energy Agency for their support through different projects on groundwater management. The authors also extend their thanks to KISR and KFAS for funding studies on groundwater.

References

Abusam A, Saeed T, Al-Jandal N (2022) Fate of estrogens in Kuwaiti municipal wastewater treatment plants. Desalin Water Treat 263:139–144

Ahmed S, Sarker SK, Friess DA, Kamruzzaman M, Jacobs M, Islam MA, Pretzsch H (2022) Salinity reduces site quality and mangrove forest functions. From monitoring to understanding. Sci Total Environ 853:158662

AlAli EH (2008) Groundwater history and trends in Kuwait. WIT Trans Ecol Environ 112:153–164

Aleisa E (2019) Analysis on reclamation and reuse of wastewater in Kuwait. J Eng Res 7(1)

Aleisa E, Al-Zubari W (2017) Wastewater reuse in the countries of the Gulf Cooperation Council (GCC): the lost opportunity. Environ Monit Assess 189:1–15

Ali A, Chidambaram S (2021) Assessment of trace inorganic contaminants in water and sediment to address its impact on common fish varieties along Kuwait Bay. Environ Geochem Health 43(2):855–883

Aliewi A, Bhandary H, Al-Qallaf H, Sabarathinam C, Al-Kandari J (2021) Assessment of the groundwater yield and sustainability of the transboundary Dibdibba aquifer using numerical modelling approach. Groundw Sustain Dev 15:100678. https://doi.org/10.1016/j.gsd.2021.100678

Al-Murad M, Uddin S, Rashid T, Al-Qallaf H, Bushehri A (2017) Waterlogging in arid agriculture areas is due to improper groundwater management, as shown in an example from Kuwait. Sustainability 9(11):2131

Al-Rashed M, Al-Senafy MN, Viswanathan MN, Al-Sumait A (1998) Groundwater utilization in Kuwait: some problems and solutions. Int J Water Resour Dev 14(1):91–105

Al-Rashed M, Aliewi A (2017) Water resources sustainability in Kuwait against United Nations sustainable development goals. In: Sustainability in the Gulf. Routledge, pp 37–56

Al-Rashed AR (2018) The environmental impacts of Kabd landfill on the soil and groundwater in Kuwait: a case study. Int J Geosci 9(5):255–271

Al-Rashidi A, Sabarathinam C, Samayamanthula DR, Alsabti B, Rashid T (2023) Groundwater management for agricultural purposes using fuzzy logic technique in an arid region. Water 15(14):2674

Alsabti B, Sabarathinam C, Svv DR (2023) Identification of high nitrate concentration in shallow groundwater of an arid region: a case study of South Kuwait's Bay. Environ Monit Assess 195(1):143

Al-Shammari SB, Al-Khalaf B, Al-Sharaifi F, Shahalam AM (2013) Quality assessment of treated wastewater in Kuwait and possibility of reuse it to meet growing water demand. Desalin Water Treat 51(22–24):4497–4505

Alshemmari H (2021) An overview of persistent organic pollutants along the coastal environment of Kuwait. Open Chem 19(1):149–156

Alsulaili AD, Hamoda MF, Al-Jarallah R, Alrukaibi D (2017) Treatment and potential reuse of greywater from schools: a pilot study. Water Sci Technol 75(9):2119–2129

Al-Sulaimi J, Viswanthan MN, Szekely F, Naji M (1994) Geohydrological study of Al-Abdally and Al-Wafra farm areas, Vol. IV: Al-Wafra (Report No. KISR 4404). Kuwait Institute for Scientific Research

Altayyar MO, Ali S, Larson A, Boving T, Thiem L, Akanda AS (2024) Quantifying groundwater depletion in Arabian Peninsula transboundary aquifer systems: understanding natural and anthropogenic drivers. Groundw Sustain Dev 101293. https://doi.org/10.1016/j.gsd.2024.101293

Al-Zubari W, Al-Turbak A, Zahid W, Al-Ruwis K, Al-Tkhais A, Al-Muataz I et al (2017) An overview of the GCC unified water strategy (2016–2035). Desalin Water Treat 81(s 5)

Bhandary H (2023) Impact assessment of Inland desalination units on groundwater quality and levels in Kuwait. In: Water resources management and sustainability: solutions for Arid Regions. Springer Nature Switzerland, Cham, pp 361–372

Bhandary H, Sabarathinam C, Akber A, Rashid T, Samayamanthula DR, Jayaramu Y, Alsabti B (2024) Evaluation of environmental geochemical signatures due to RO rejects on arid agricultural farms and tangible solutions. Geosci Front 15(6):101929

Cendon D, Haldorsen S, Chen JY, Hankin S, Nogueira G, Momade F, Achimo M, Muiuane E, Mugabe J, Stigter T (2019) Hydrogeochemical aquifer characterization and its implication for groundwater development in Maputo District, Mozambique. Quat Int. https://doi.org/10.1016/j.quaint.2019.06.024

Darwish MA, Darwish A, Darwish A (2011) Fifty years of MSF desalination in Kuwait and sustainability issues. Desalin Water Treat 29(1–3):343–354

Demenocal P, Ortiz J, Cuilderson T, Adkins J, Sarnthein M, Baker I, Yarusinsky M (2000) Abrupt onset and termination of the African Humid Period. Quatern Sci Rev 19:347–361

Eckstein G (2021) International law for transboundary aquifers: a challenge for our times. AJIL Unbound 115:201–206. https://doi.org/10.1017/aju.2021.18

ESCWA, United Nations Economic and Social Commission for Western Asia (2023) Kuwait Vision 2035. https://nwm.unescwa.org/resources/240

GUST (2024) Gulf University for Science and Technology. SDG 6 Clean Water and Sanitation. https://www.gust.edu.kw (GUST)

GUST, Gulf University for Science and Technology (2024) Water conservation policy. Retrieved from https://www.gust.edu.kw/sdg/water-conservation-policy

Hadi K, Saravana Kumar U, Al-Senafy M, Bhandary H (2016) Environmental isotope systematics of the groundwater system of southern Kuwait. Environ Earth Sci 75(14):1096

Hadi K (2024) Current status, challenges, and future management strategies for water resources of Kuwait. In: Terrestrial environment and ecosystems of Kuwait: assessment and restoration. Springer Nature Switzerland, Cham, pp 141–169

Hamdan L (1987) Study of the subsurface water rise in the residential areas of Kuwait (Vol. 1: Main report) (Report No. KISR 2227). Kuwait Institute for Scientific Research

IGRAC (2022) International groundwater resources assessment centre. Transboundary aquifers of the Middle East map 2022. https://www.un-igrac.org/resource/transboundary-aquifers-middle-east-map-2022

Inoue T (2019) Carbon sequestration in mangroves. In: Blue carbon in shallow coastal ecosystems: carbon dynamics, policy, and implementation, pp 73–99

International Atomic Energy Agency (IAEA) (2018) Water management strategies in arid and semi-arid regions. IAEA Publishing

International Atomic Energy Agency (IAEA) (2023) Climate change and its impact on water security: a scientific review. IAEA Publishing

Jasechko S (2016) Partitioning young and old groundwater with geochemical tracers. Chem Geol 427:35–42

Jayaramu Y, Panda B, Al-Rashidi A, Kumar US, Sabarathinam C, Akbar A, Palanivel P (2024) Unravelling the impact of microclimatic changes on coastal aquifer of multiple land use through integrated techniques: a comparative study using two decadal data. Groundw Sustain Dev 101239

Khater AR (2010) Impacts of climate change on groundwater in the Arab Region

Kuwait Times (2021) Zain launches online campaign to reinforce environmental awareness. Retrieved from https://kuwaittimes.com/zain-launches-online-campaign-to-reinforce-environmental-awareness/

Lee E, Jayakumar R, Shrestha S, Han Z (2018) Assessment of transboundary aquifer resources in Asia: status and progress towards sustainable groundwater management. J Hydrol Reg Stud 20:103–115. https://doi.org/10.1016/j.ejrh.2018.01.004

Li P, Wu J (2024) Water resources and sustainable development. Water 16(1):134. https://doi.org/10.3390/w16010134

Li P, Kou X, Wang Y, Niu L (2024) Building a more sustainable Chinese Loess Plateau. J Earth Sci 35(1):283–287. https://doi.org/10.1007/s12583-024-1970-3

Makarigakis AK, Jimenez-Cisneros BE (2019) UNESCO's contribution to face global water challenges. Water 11(2):388

Marcal J, Antizar-Ladislao B, Hofman J (2021) Addressing water security: an overview. Sustainability 13(24):13702

Mazzoni A, Heggy E, Scabbia G (2018) Forecasting water budget deficits and groundwater depletion in the main fossil aquifer systems in North Africa and the Arabian Peninsula. Glob Environ Chang 53:157–173. https://doi.org/10.1016/j.gloenvcha.2018.09.009

McClure HA (1976) Radiocarbon chronology of late quaternary lakes in the Arabian Desert. Nature 365:755–756

Misak R, Hussain W (2022) Groundwater in Kuwait. The geology of Kuwait. Springer International Publishing, Cham, pp 199–214

Misak R, Hussain W (2023) Groundwater in Kuwait. In: Abd el-Aal AEK, Al-Awadhi JM, Al-Dousari A (eds) The geology of Kuwait. Regional geology reviews. Springer, pp 199–214. https://doi.org/10.1007/978-3-031-16727-0_9

Misak R, El Gamily H, Hussain W (2014) Threats to agriculture lands at Al-Wafra, southern part of Kuwait. J Arid Land Stud 24(1):9–12

MOFA (2024) Ministry of Foreign Affairs, Kuwait. Kuwait Vision 2035. https://www.mofa.gov.kw/en/pages/kuwait-vision-2035

Mukhopadhyay A, Al-Sulaimi J, Barrat JM (1994) Numerical modeling of groundwater resource management options in Kuwait. Groundwater 32(6):917–927

Mukhopadhyay A, Al-Awadi E, Quinn M, Akber A, Al-Senafy M, Rashid T (2008) Ground water contamination in Kuwait resulting from the 1991 Gulf War: a preliminary assessment. Groundw Monitor Remediat 28(2):81–93

Mukhopadhyay A, Al-Haddad A, Kotwicki V, Al-Senafy M, Rashid T, Al-Qallaf H (2012) Feasibility of artificial recharge at the Az-Zaqlah depression in Kuwait. In: Proceedings of the 6th IAHR international groundwater symposium: groundwater modeling and management under uncertainty, pp 19–21

Mukhopadhyay A, Quinn M, Al-Haddad A, Al-Khalid A, Al-Qallaf H, Rashed T et al (2017) Pollution of fresh groundwater from damaged oil wells, North Kuwait. Environ Earth Sci 76:1–16

Mukhopadhyay A (2019) Field assessment of the use of RO-treated wastewater for artificial recharge of aquifers in Kuwait. KISR project (WM037C)

Mukhopadhyay A, Al-Sulaimi J, Al-Sumait AA (2020) Creation of potable water reserve in Kuwait through artificial recharge. In: Artificial recharge of groundwater. CRC Press, pp 175–180

Mukhopadhyay A, Akber A, Al-Saaidi M, Aladwani A (2022) Planning for a multi-well artificial recharge field in Kuwait with the help of numerical modeling. Groundwater 60(5):685–698

Neff U, Burns SJ, Mangini A, Mudelsee M, Fleitmann D, Matter A (2001) Strong coherence between solar variability and the monsoon in Oman between 9 and 6 kyr ago. Nature 411:209–213

Nehemia A, Chen M, Kochzius M, Dehairs F, Brion N (2019) Ecological impact of salt farming in mangroves on the habitat and food sources of Austruca occidentalis and Littoraria subvittata. J Sea Res 146:24–32

Radha SD, Sabarathinam C, Al Otaibi F, Al-Sabti BT (2023). Variation of centennial precipitation patterns in Kuwait and their relation to climate change. Environ Monit Assess 195(1):20

Rashid T, Sabarathinam C, Al-Qallaf H, Bhandary H, Al-Jumaa M, Shishter A, Al-Salman B (2022) Evolution of hydrogeochemistry in groundwater production fields of Kuwait-Inferences from long-term data. Chemosphere 307:135734

Rashid T, Sabarathinam C, Kumar US, Al-Jumaa M, Al Salman B, Naseeb H (2023) Terrigenic helium in brackish groundwaters of Kuwait, probable influences from hydrocarbon resources. Groundw Sustain Dev 23:101048

Ren X, Li P, Wang D, Zhang Q, Ning J (2024) Drivers and characteristics of groundwater drought under human interventions in arid and semiarid areas of China. J Hydrol 631:130839. https://doi.org/10.1016/j.jhydrol.2024.130839

Riedel G, Simon AB (1973) Geotechnical properties of Kuwaiti "gatch" and their improvement. Eng Geol 7(2):155–165

Romano O, Akhmouch A (2019) Water governance in cities: current trends and future challenges. Water 11(3):500

Sabarathinam C, Bhandary H, Al-Khalid A (2019) A geochemical analogy between the metal sources in Kuwait Bay and territorial sea water of Kuwait. Environ Monit Assess 191:1–19

Sabarathinam C, Bhandary H, Al-Khalid A (2020) Tracing the evolution of acidic hypersaline coastal groundwater in Kuwait. Arab J Geosci 13(1146)

Sabarathinam C, Rashed T, Dashti F, Bhandary H (2022) Equilibrium states of groundwater chemistry in coastal region of Kuwait. Desalin Water Treat 263:248–256

Sadeqi D, Aliewi AS, Al-Qallaf H, Rashed T (2024) Groundwater Salinization in Kuwait: a major threat to indigenous ecosystems. In: Terrestrial environment and ecosystems of Kuwait: assessment and restoration. Springer Nature Switzerland, Cham, pp 171–196

Saeed T, Al-Jandal N, Abusam A, Taqi H, Al-Khabbaz A, Zafar J (2017) Sources and levels of endocrine disrupting compounds (EDCs) in Kuwait's coastal areas. Mar Pollut Bull 118(1–2):407–412

Samayamanthula DR, Sabarathinam C, Bhandary H (2019) Treatment and effective utilization of greywater. Appl Water Sci 9(4):90

Samayamanthula DR, Sabarathinam C, Al-Ajeel FK (2021) Hydrochemical attributes and their spatial variation in the rainwater of Kuwait. In: International conference on Mediterranean Geosciences Union. Springer Nature Switzerland, Cham, pp 9–12

SDG Central Statistical Bureau, Kuwait (2024) Goal 6 indicators. SDG Portal. https://sdg.csb.gov.kw/G6_Idicator_EN?id=72

Senay Y, Hamdan L, Yaqubi A (1987) Study of the subsurface water rise in the residential areas of Kuwait (Vol. 3: Hydrogeology). Kuwait Institute for Scientific Research (KISR Report No. 2227)

Sherif M, Sefelnasr A, Al Rashed M, Alshamsi D, Zaidi FK, Alghafli K, Baig F, Alfaifi H, Loni OA, Ahamed MB, Ebraheem AA (2023) A review of managed aquifer recharge potential in the Middle East and North Africa Region with examples from the Kingdom of Saudi Arabia and the United Arab Emirates. Water 15(4):742. https://doi.org/10.3390/w15040742

Street-Perrott FA, Perrott RA (1990) Abrupt climate fluctuations in the tropics: the influence of Atlantic Ocean circulation. Nature 343:607–612

Svv DR, Al-Rashidi A, Sabarathinam C, Alsabti B, Al-Wazzan Y, Kumar US (2023) Temporal and spatial shifts in the chemical composition of urban coastal rainwaters of Kuwait: the role of air mass trajectory and meteorological variables. Sci Total Environ 899:165649

The Ocean Portal Team (2023) Mangroves, Smithsonian Ocean. https://ocean.si.edu/ocean-life/plants-algae/mangroves. Accessed 23 Sep 2024

The United Nations in Kuwait (2024) Sustainable Development Goal 6: clean Water and Sanitation. https://kuwait.un.org/en/sdgs/6 (The United Nations in Kuwait)

UNESCO World Water Assessment Programme (2019) The United Nations World Water Development Report: leaving No One Behind; UNESCO: Paris, France, 186. ISBN 9789231003097

Wang D, Li P, He X, He S (2023) Exploring the response of shallow groundwater to precipitation in the northern piedmont of the Qinling Mountains, China. Urban Clim 47:101379. https://doi.org/10.1016/j.uclim.2022.101379

Water Action Hub (2024) Stakeholder participation (SDG 6.b). https://wateractionhub.org (Water Action Hub)

WRI—World Resources Institute. 17 Countries, home to one-quarter of the World's Population, Face Extremely High Water Stress. Available online: https://www.wri.org/insights/17-countries-home-one-quarter-worlds-population-face-extremelyhigh-water-stress. Accessed 2 Nov 2021

Zain Group (2023) Sustainability agenda. Retrieved from https://zain.com/SR2023/18-sustainability-agenda/

Impacts of Artisanal Small-Scale Mining and Climate Change on Groundwater Security: The Case of the Global South

Seth Obiri-Yeboah, Opoku Gyamfi, Peiyue Li, Osei Akoto, and Godfred Darko

Abstract Groundwater security is threatened by human activities and climate change, especially in developing nations where artisanal and small-scale mining are pervasive. This chapter examines the dual impacts of artisanal small-scale mining and climate change on groundwater resources by focusing on two aspects: quality and availability. Most artisanal small-scale mining is done without considering any regulatory requirements, and results in the contamination of the groundwater. Simultaneously, climate change is increasing the rates of groundwater depletion due to increasing droughts, altering precipitation patterns, and reduced rates of groundwater recharge. Study cases in the Global South illustrate the point of artisanal small-scale mining and climate change: degradation of groundwater is very rapid. This chapter, therefore, underlined the imperative need for integrated management, much firmer regulation, and sustainable mining that could reduce adverse impacts on groundwater security in precarious regions. Future trends have also shown that the activities and impacts related to ASM, as well as those due to climate change, may increase due to unending economic pressures, population growth, and the global

S. Obiri-Yeboah
Department of Pharmaceutical Sciences, Kumasi Technical University, Kumasi, Ghana

O. Gyamfi
Department of Chemistry Education, Akenten Appiah-Menka University of Skills Training and Entrepreneurial Development, Asante-Mampong, Ashanti, Mampong, Ghana

P. Li
School of Water and Environment, Chang'an University, No. 126 Yanta Road, Xi'an 710054, PR China

S. Obiri-Yeboah · O. Akoto · G. Darko (✉)
Department of Chemistry, Kwame Nkrumah University of Science and Technology, Kumasi, Ghana
e-mail: godfreddarko@gmail.com

P. Li
Key Laboratory of Subsurface Hydrology and Ecological Effects in Arid Region of the Ministry of Education, Chang'an University, No. 126 Yanta Road, Xi'an 710054, Shaanxi, China

Key Laboratory of Eco-Hydrology and Water Security in Arid and Semi-Arid Regions of the Ministry of Water Resources, Chang'an University, No. 126 Yanta Road, Xi'an 710054, Shaanxi, China

P. Li et al. (eds.), *Sustainable Groundwater and Environment: Challenges and Solutions*, Springer Hydrogeology, https://doi.org/10.1007/978-3-031-82194-3_3

precious minerals demand, especially in Africa, Latin America, and Asia. Climate models similarly predict more frequent and severe droughts that stress groundwater systems even further. These parallel trends may result in more acute water shortages that undermine livelihoods and intensify conflicts over water resources. The international community and agencies like the United Nations and the World Bank have much to do in the area of capacity building, policy advocacy, and facilitating access to technology that moderates environmental damage to ensure the best sustainable practices that would improve the monitoring and management of groundwater in developing nations for groundwater security future generations.

Keywords Groundwater pollution · Global warming · Environmental degradation · Heavy metal poisoning

1 Background

Groundwater is an important natural resource that has helped sustain livelihoods, especially in places where surface water is either too scarce or inaccessible (Sikka et al. 2021). The Global South (Fig. 1), that is, developing countries with lower economic status in the world, are found throughout Africa, Oceania, Asia, and Latin America. Such nations have a higher dependence on groundwater for drinking water (Panneerselvam et al. 2023), irrigation for agriculture (Gugulothu et al. 2022) and industrial purposes (Hoogesteger 2022). According to some studies, this higher dependence on groundwater is due to the effects of artisanal small-scale mining (ASM), and to others, climate change is the culprit that increasingly threatens the sustainability of groundwater. Any of these h factors have a predisposition to cause not only the degradation of water quality but also the availability and accessibility of groundwater; hence, it is serious in terms of water security (Ofosu and Sarpong 2022; Akbari et al. 2020; Arthur-Holmes et al. 2022).

According to Ofosu and Sarpong (2022), artisanal and small-scale mining (ASM) refers to informal mining activities carried out with very limited technologies and capital investment, usually engaged by individuals or small groups. While ASM is a livelihood provider for millions worldwide, especially in the rural parts of many countries (Yankson and Gough 2019), it is also notorious for considerable environmental degradation (Macháček 2019). Of much concern would be its effect on groundwater quality and quantity.

The mining of gold and gemstones especially uses toxic chemicals such as mercury and cyanide during most ASM operations in the extraction process of the minerals from the ore (Wang et al. 2021). These usually find their way into the soil (Chen et al. 2022) and leach into systems of groundwater (Cacciuttolo et al. 2023) under uncontrolled processes, leading to contamination that could sometimes results in negative irreversible environmental and human health outcomes (Esinam et al. 2024). The long-term accumulation of heavy metals and other pollutants in aquifers not only affects the drinking water supplies for local communities but also impacts

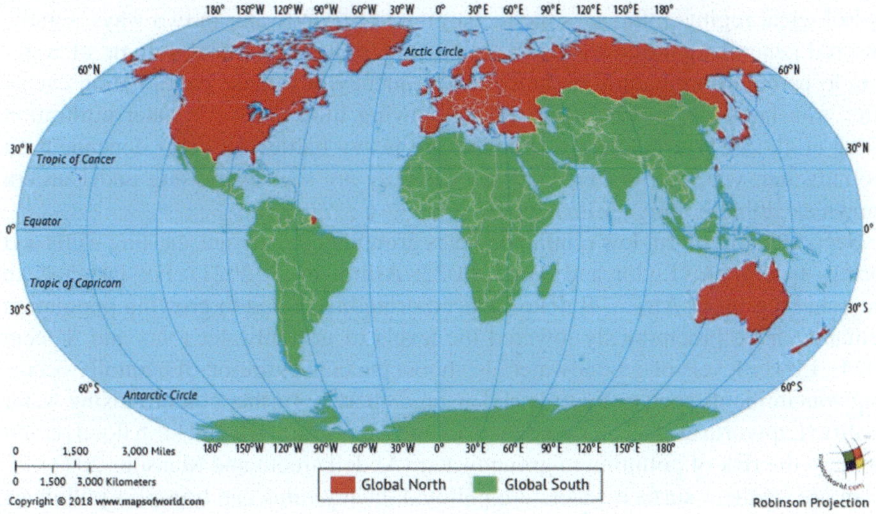

Fig. 1 Socio-economic demarcation of the world's countries into the Global North and Global South

agricultural productivity and ecosystems dependent on groundwater (Hilson 2020). ASM has also been linked to shallow groundwater high in arsenic, lead, and mercury levels, a situation that presents health risks to millions of people (Esinam et al. 2024). Such pollutants lead to fatal health complications when consumed over long periods, including neurological impairment (De Paula Arrifano et al. 2023), kidney damage (Sanders et al. 2019), and even tumours and cancers (Cacciuttolo et al. 2023). Aquifer contamination by ASM activities poses more health risks to women of child-bearing age and children than men (Landrigan et al. 2022; Cossa et al. 2021), because the former are more liable to use untreated groundwater for drinking and culinary purposes.

Climate change is another important contributor to groundwater resource insecurities in terms of their quality and quantity in the Global South (Amanambu et al. 2020; MacAlister and Subramanyam 2018). It is well established that compared to surface waters, groundwater is a far more robust resource in the context of climate variability; it is indeed the impact of climate change on the quantity and quality of the available supplies that is increasingly affecting groundwater. According to several studies, increases in global temperatures, shifts in precipitation patterns, and increases in extremes bring about rapid hydrological cycle changes, marking significant changes in available groundwater supplies (Wang and Liu 2023; Wu et al. 2020; Bano and Arshad 2018). A significant direct impact on groundwater is the disruption of its natural recharge patterns (Meixner et al. 2016).

In most countries of the Global South, rainfall regimes are becoming increasingly unpredictable (Afifi et al. 2016), with longer periods of no rainfall followed by hours or even days of heavy rainfall (Version 2021; Siswanto et al. 2016). This changed hydrological regime influences the recharging of groundwater in two ways. Firstly, lowered general rainfall during the dry seasons reduces the total volume of water able to percolate into aquifers. Increased frequency of intense precipitation events often translates into more surface runoff, leaving little room for water infiltration (Garg et al. 2021). Undeniably, droughts are heavy battles for areas dependent on groundwater, where both frequency and severity are seen increasing under an era known as global boiling (Anozie 2024).

Secondly, long-term low rainfall depletes groundwater storage, causing wells and springs to dry up (Taylor and Aureli 2022; Ashraf et al. 2021). For instance, in the Sahel region of Africa, climate-induced droughts, added to growing population demands, have precipitously lowered the levels of groundwater (Naz and Saleem 2024). Lower levels of groundwater also boost the concentration of naturally occurring contaminants such as fluoride and arsenic, thereby further compromising water quality (Lapworth et al. 2022). At the same time, heavy rainfalls and flash floods could increase the risk of pollution to groundwater (Abdelkareem and Mansour 2023).

In areas where surface waters are polluted, heavy rains can transport pollutants such as pesticides, sewage, and industrial waste into aquifers (Geris et al. 2022). This is especially problematic in regions with shallow groundwater tables, where aquifers are more vulnerable to contamination (Foster and Chilton 2020). In coastal areas, rising sea levels driven by climate change are also contributing to the salinization of freshwater aquifers (Paldor and Michael 2021), further reducing the availability of potable groundwater.

Especially worrying is how ASM and climate change together affect most parts of the Global South, given that within this part of the world, these phenomena often strike ecologically fragile environments (Ofosu et al. 2020). ASM worsens the plight of groundwater systems in addition to climate change, creating a feedback loop that magnifies the impacts of both drivers. For instance, in regions experiencing prolonged droughts due to climate change, local communities may turn to ASM as a means of economic survival, further increasing pressure on already strained groundwater resources (Moyo 2024).

Mining-induced land degradation also reduces the capacity of aquifers to recharge groundwater and even increases flood risk that could introduce contaminants into aquifers (Kausher et al. 2023). Since their activities remain informal and even in some jurisdictions illegal, at present, there is no good wastewater management policy to be adhered to by ASM adherents, and thus their activities remain largely threatening from all perspectives (Ofosu et al. 2024). When there is heavy rainfall, the contaminated water enters the groundwater system and further aggravates the situation. Overall, the impact of ASM and climate change are intricately connected and often mutually reinforcing (Landrigan et al. 2022), requiring an all-hands-on-deck and sometimes adaptive approaches over time. Along with the impacts of ASM, climate change requires a range of different levels of governance in the Global South for groundwater to be safeguarded (Madhav 2020).

In most developing countries, for instance, policies pertinent to water resources and mining and the protection of the environment are either fragmented or poorly enforced (Schoderer and Ott 2022), thus allowing ASM to operate with little oversight. For that matter, strengthening regulatory frameworks and mining operations, coupled with small-scale mining compliance with environmental standards, will go a long way in mitigating the contamination of groundwater (Wuijts et al. 2021). Moreover, to maintain and sustain water security, local communities would require long-term water management strategies that will reduce the impacts of changing climate effects on groundwater quality, quantity and accessibility. Such strategies may include the development and improvement of monitoring systems of the groundwater (Calderwood et al. 2020), the achievement of sustainable mining (Hirons 2020), and investment in technologies for groundwater recharge, such as managed aquifer recharge-MAR (Job 2023). However, the involvement of the local community through educational activities to sensitize the community towards sustainable utilization of the water (Zikargae et al. 2022), especially where it is a dominant economic activity. Groundwater resources throughout the Global South are increasingly coming under the dual blows of artisanal and small-scale mining, along with climate change. Indeed, these two drivers, beyond compromising the quality and quantity of groundwater, further lower the already prevailing socio-economic and environmental vulnerabilities (Kumah et al. 2020).

Despite the evident risks posed by ASM and climate change to groundwater resources, research regarding their joint implications for groundwater security in the Global South remains scanty. The body of existing research has focused mostly on direct ASM environmental impacts (Humphrey et al. 2022; Kaonga 2018; Ireson et al. 2006), or effects of climate change on groundwater (Amanambu et al. 2020; Green and Allen 2011; Holman 2006), or often on both independently.

This chapter, therefore, tries to address such gaps in knowledge by assessing the coupled effects of artisanal and small-scale mining and climate change on groundwater security in the Global South. This study examined the alternative ways in which linked factors affect the quality, accessibility, and quantity of groundwater, the socio-economic implications for the affected communities, and environmental impacts, considering existing water governance frameworks for their potential to mitigate such challenges. This, in turn, contributes to the nascent but growing literature on more holistic approaches to threats facing groundwater security in the Global South.

This chapter reviewed the dual impacts of ASM and climate change on groundwater security, with a regional focus on the Global South. Within the scope of this chapter, it was important to understand, from a multi-disciplinary approach drawing from environmental science, hydrology, and socio-economic factors, how these two phenomena-ASM and climate change-influence the quality and availability of groundwater. It sought to reveal knowledge gaps and give recommendations on policy interventions, as well as sustainable management strategies from both national and international perspectives.

The specific objectives were:

- To estimate the degree of groundwater contamination caused by artisanal small-scale mining in the Global South, using pollutants like heavy metals, mercury, and cyanide.
- To assess the impact of climatic change on the regime and availability of groundwater recharge due to changes in rainfall, drought, and depletion rate in the areas of artisanal mining activities.
- To analyse the interaction between artisanal mining activities and the impact of climate change on their combination in increasing groundwater insecurity.
- To review case studies of selected countries of the Global South, possibly illuminating localized experiences of the challenges surrounding groundwater, as well as the socio-economic impacts of water scarcity faced by mining communities.
- To recommend integrated groundwater management methods that are coherent with policy frameworks and adaptation mechanisms for reducing the adverse impacts of artisanal mining and climate change on groundwater resources.

1.1 Methodology

This chapter sought to collate existing research and findings on the subject at hand, an indication of the state of knowledge on the subject, and hence the identified gaps that would indicate the direction for future research into ASM and climate change combined impacts on groundwater resources, the case of the Global South.

To ensure a thorough review, peer-reviewed journal and general articles, reports from international organizations, and relevant case studies about the topic were considered in detail. Key databases such as Google Scholar, PubMed, Crossref, Web of Science, and Science Direct were used to locate relevant resources. The following are the inclusion criteria.

Most of the reviewed articles were those studies that highlighted ASM and climate change and their impacts on groundwater security in the Global South. Articles that were published within the last 15 years were considered most relevant to the current studies. This chapter was organised around a set of themes identified as important in literature. A series of keywords were used in literature searches. These included environmental impacts of ASM on water security, deforestation and long degradation, chemical pollution, sedimentation, climate change impacts on water security, altered hydrological cycles, extreme weather events, socio-economic impacts, livelihoods and health, and community displacements i.e., refugee situations.

2 Artisanal and Small-Scale Mining and Groundwater Degradation

2.1 Chemicals Leaching and Causing Groundwater Pollution

Key pathways in which ASM affects water quality include chemical leaching, particularly of very dangerous substances such as mercury and cyanide (Malone et al. 2023). In gold processing, for instance, mercury amalgamation is standard amongst small-scale operators. In countries like Ghana, Tanzania, and Peru, tests were done to show that the level of mercury in certain groundwater sources within the immediate vicinities of these mining activities is quite high (Dorleku, Nukpezah and Carboo 2018; Ligate et al. 2022; Custodio et al. 2020). Once released, mercury can seep into the soil and percolate into aquifers, remaining contaminated for extremely long periods (Aleku et al. 2024). Mercury is also bio-accumulative (Kershaw and Hall 2019; Eagles-Smith et al. 2016), hence posing grave risks to the local communities that depend on groundwater for drinking and agriculture.

Similarly, cyanide has also been applied in some ASM practices (Eagles-Smith et al. 2016), leading to cyanide poisoning of water. Though the greatest amount of cyanide is used in somewhat larger operations, its usage in small-scale mining has been reported as well, especially in places like Latin America and Southeast Asia (González-Valoys et al. 2022; Leung and Lu 2016). Scientific studies have shown that cyanide can persist in groundwater systems and cause environmental and health impairment for a long time (Kumar et al. 2017).

2.2 Turbidity and Physical Interference with the Aquifers

Most ASM activities entail large quantities of soil and rock removal, seriously altering the physical form of groundwater systems mostly irreversibly (Asare et al. 2024). For example, mining activities have been known to create gross sedimentation and siltation (Fig. 2) that could block the course and reduce the replenishment capacity of aquifers. There is a possibility of physical interference by ASM in aquifers, altering flow conditions and lowering the water level in parts (Roughton et al. 2016).

Equally, in areas like Southeast Asia and West Africa (Fig. 3), ASM-induced deforestation and degradation have been said to reduce the infiltration of soils through rainfall, causing groundwater storage depletion (Adegbite 2022; Getahun and HAJ 2015).

For instance, studies conducted in countries like Zimbabwe and Peru indicated that digging the bed of these rivers and alluvial deposits results in tremendous amounts of loads of sediment in the water body, raising the turbidity levels and hence affecting the quality of the groundwater (Fielding et al. 2018). At the same time, the sedimentary deposits transport heavy metals, adding to the issue of contamination (Nguyen et al. 2019).

Fig. 2 Artisanal small-scale mining activities and their perceptible impacts on the environment

Fig. 3 Abandoned artisanal small-scale mining site not reclaimed in a forest

2.3 Heavy Metal Contamination

Heavy metal contamination due to ASM can be a more critical factor for groundwater deterioration since heavy metals for base metals like copper, lead, and zinc could be mobilized and leach into groundwaters during its extraction within ASM operations (Nkuna et al. 2022). Above-average metal concentrations in all samples taken from around ASM operations in areas considered for ASM operations include arsenic, lead, and cadmium (Radi et al. 2023; Utermann et al. 2019), highly toxic metals at very low concentration levels (Kiran et al. 2021). There is, hence, a high hazard of chronic diseases for the local population and disorders in the development of children (Bah et al. 2023; Jannetto and Cowl 2023). Some recent studies have used advanced monitoring techniques, GIS, and remote sensing to map the diffusion of heavy metals in groundwater around ASM areas (Wang et al. 2024; Akhtar et al. 2021).

2.4 ASM and Acid Mine Drainage (AMD)

Acid mine drainage has been considered another serious problem that may be associated with ASM activities, especially in areas where sulphide-rich ores are mined (Ofosu and Sarpong 2022). Excavated and exposed to air and water, such ores form sulphuric acid that seeps into the ground to form acidified groundwater (Dey and Majumdar 2024). Once initiated, the process might go on for decades with disastrous implications for local groundwater systems. In particular, ASM, according to research studies carried out in places as South Africa and Latin America, proved to give the biggest contribution to AMD, especially in gold and copper belts, since heavy metals included in surrounding rock can be dissolved in acidified groundwater and thus further aggravate the problem by introducing more contaminants to water sources (Worlanyo and Jiangfeng 2021).

The resulting cocktail is so noxious that even such groundwater cannot be utilized either by humans or in the fields (Raletsena 2021). Remediation of acidified groundwater is often expensive and complex. Therefore, it is a difficult problem for developing countries with a limited budget for environmental management (Al-Hashimi et al. 2021).

3 General Environmental Impacts of ASM

3.1 Land Degradation and Deforestation

The most pronounced impacts of ASM activities relate to land degradation, including soil erosion and deforestation. Most ASM activities require that very large portions of land be cleared to allow access to the mineral deposits, especially in those areas that are most densely forested (Ofosu et al. 2020). This is something that has been well documented, with parts of the vast stretches of their respective forest covers having gone down due to mining-a thing which continues unabated in both the Amazon Basin and the Congo Basin (Álvarez-Berríos et al. 2021; Radley 2022). Recent estimates from various studies place ASM as among the major causes of deforestation across the tropical world, which has disastrous implications for biodiversity and carbon sequestration capacity (Aldridge Mugadza 2022; Kabelong Banoho et al. 2020). As a result, it contributes to reduced trends in biodiversity loss of entire habitats and ineffective ecosystems (Rocha-Santos et al. 2017). ASM fragments wildlife habitats;

the process is now affecting the flora, fauna, and even ecosystem services such as water regulation and soil fertility (Fagariba et al. 2024). Furthermore, the removal of vegetation accelerates soil erosion, resulting in the loss of topsoil which is crucial for plant growth and water infiltration (Gu et al. 2020).

4 Groundwater Regional Case Studies of ASM Impacts

4.1 Ghana (West Africa)

Today, Ghana has become a hotbed of ASM or "galamsey" gold mining, particularly in the Ashanti, Western, and Central regions (Eduful et al. 2020). While a major contributor to the local economies, the environmental impacts of ASM operations are disastrous. Degradation of groundwater in mining communities is emerging as a source of increasing concern due to the high dependence on groundwater for drinking, agriculture, and industry across these areas (Abanyie et al. 2023).

Other recent findings, such as by Owusu Sekyere et al. (2023), also identified significant mercury and lead contamination of in-groundwater sources close to the ASM operations in the Ashanti region. Such metals might thus become used for various amalgamation processes in the course of gold panning and are well known to persist in the environment for years following mining operations (Durkalec et al. 2022). Water samples from the boreholes serving communities up to a radius of 0–20 km from the active mines had elevated values of mercury above the allowable limits by WHO in drinking water (Palumbo-Roe et al. 2020).

Apart from chemical contamination, over-extraction represents another critical form of physical degradation to the groundwater resource within the mining areas of Ghana (Ewusi et al. 2017). In these mining areas, during mineral processing, large volumes of water are used within the ASM operations, normally associated with over-extraction of groundwater from the local aquifers. Affum et al. 2024), affirmed that water table levels in mining areas like Tarkwa and Obuasi have deteriorated within the past couple of years, hence affecting the availability of groundwater for local agriculture. Such depletion during dry seasons worsens the situation of inadequate water, hence straining the communities to rely on these sources for survival.

4.2 Peru (Latin America)

The Peruvian region of Madre de Dios personifies one of the worst cases of environmental degradation because of ASM in the Amazon basin. Large areas are being intoxicated due to this process, and not only surface water but also groundwater is contaminated by it. Most ASM operations in Madre de Dios are unregulated, and the use of mercury in separating the metal from the ore is widely practised (Chirif et al. 2019). A recent investigation by Trinchieri (2023), established high values of mercury in shallow groundwater sources consumed by indigenous communities. In some areas, the concentration of mercury in water was more than two to three times higher than the maximum value considered acceptable by the Peruvian government and in the international drinking water standards.

Apart from mercury, the region of Madre de Dios has suffered serious sedimentation of its groundwater resources due to deforestation and soil erosion resulting from mining processes (Diaz 2021). The removal of vegetation cover during land clearing for mining accelerates the rate at which sediments enter the aquifers. In addition, these sediments lower the quality of the water and slow down the rate of replenishment of the groundwater system, hence its long-term sustainability. Another study cited by Rivera et al. (2019) indicated that sediment loads in aquifers close to ASM areas were higher compared to undisturbed areas and resulted in increased turbidity, reducing the potability of the water for local communities.

4.3 Indonesia (Southeast Asia)

According to Meutia et al. (2023), mercury from ASM gold mining has been identified as one of the leading causes of groundwater degradation in Indonesia, particularly in the islands of Kalimantan and Sulawesi. It would not be an exaggeration to say that almost all ASM gold extraction uses mercury, with an estimated more than 100 metric tonnes released to the environment each year from ASM activities in the country (Aleku et al. 2024). This therefore poses a direct threat to groundwater resources since in tropical settings, mercury is very mobile and easily leaches into the groundwater system (Kyaw and Sakakibara 2022; Meutia et al. 2023).

Putra et al. (2023), commented on Central Kalimantan, as mainly contaminated with mercury through the regional well water prepared for drinking and irrigation. According to them, the amount of mercury in the groundwater samples was 30–50 times higher than the limits set according to environmental safety standards (Arrazay et al. 2023). Besides this, in some areas, cyanide was also extracted because already many miners had started shifting from Mercury to Cyanide-based extraction of gold (Verbrugge et al. 2021). Cyanide, being highly soluble in water, is increasingly polluting the groundwater (Esinam et al. 2024), hence making the site clean-up even more difficult (Roldán et al. 2021).

Another very significant issue relevant to ASM areas in Indonesia is the over-extraction of groundwater (Lo et al. 2022). Gold mining activities require substantial amounts of water, and in the countryside, given the severe limitation of other water sources, this has resulted in a sharp decline in the regional water table. In South Sulawesi, for example, substantial reductions in groundwater levels have resulted from ASM-induced over-extraction, negatively impacting local rice production and drinking water supplies (Rivki et al. 2022). The Indonesian Environmental Forum says that in 2022, for example, some areas must dig deeper wells to get water, which means that more money and energy are spent on getting clean drinking water.

4.4 Zimbabwe (Southern Africa)

ASM gold mining has been increasingly developing during the last twenty years in the Midlands Province of Zimbabwe (Nkomo 2022) while having high impacts on the local water resources including the groundwater (Mulenga et al. 2024). The severe contamination of the groundwater supplies through AMD and toxic metals releases was mainly done by the local ASM operators with generally little or no environmental oversight (Mwamba 2020). Groundwater sources around ASM in the Midlands were acidic (Ouma et al. 2022); the minimum value was recorded at 3.5, while the values remained at a low level from the threshold limit considered to be safe for human consumption. Acidic conditions enhance the leaching of heavy metals like arsenic, lead, and cadmium into groundwater and the deterioration of water quality (Zhang et al. 2023).

Groundwater is equally over-extracted in the ASM areas of Zimbabwe. Because surface water is largely inaccessible, ASM mining has increasingly relied on groundwater extraction for mineral processing. The result has been the localised declines in water table levels further reducing the accessibility of water during the dry season (Manyakaidze et al. 2024; Gona et al. 2022). In other cases, community boreholes that used to serve the entire community have dried up (Arthur-Holmes et al. 2022), compelling the residents to seek water elsewhere and often over long distances.

For instance, many aquifers in this region have had low rates of recharge compared to the rates of extraction, according to Mpala et al. (2020), hence threatening the long-term sustainability of water resources within the Midlands.

4.5 Mongolia (East Asia)

The booming artisanal and small-scale gold mining, especially in the northern parts of the country, contributes much to groundwater degradation in Mongolia (Garcia et al. 2022). Groundwater contamination in places like Zaamar and Bornuur is so critical due to mining without any control, which uses mercury and cyanide in most operations for the separation of gold (Daley et al. 2018).

For instance, very recently, Oyunbat et al. (2021), presented mercury in the groundwater sources of those areas at levels as high as 13 times the allowable limits according to Mongolian and international standards. This contamination thus encompasses important implications for nomadic herding communities dependent on groundwater supplies for their livestock and household purposes (Crul et al. 2024). This has perpetuated into generalized significant groundwater depletion from such arid and semi-arid regions in particular (Brierley et al. 2016). From the perspective of the long-term view, groundwater will remain precarious in the Zaamar and Bornuur regions as a result of continued demand by ASM operations.

4.6 Ethiopia (Horn of Africa)

Most ASM operations in Ethiopia, especially in Oromia and Southern Nations, Nationalities, and Peoples' Region, have degraded the groundwater resource by increasing its heavy metal burden (Rodríguez (2019). Crul et al. (2024) recorded high-level arsenic and lead contamination of the local groundwater resulting from small-scale gold mining activities. It was indicated that the level of arsenic in the wells was more than 1.86 times higher than the maximum level recommended by the WHO for drinking water, hence posing a real hazard to community health (Demissie et al. 2023). Because most of the Ethiopian rural communities depend entirely on groundwater for their drinking and irrigation, this alone has a high socio-economic effect (Mengistu et al. 2021). Long-term exposure to such arsenic-contaminated water has been linked with different chronic diseases including skin lesions and cancers (Rajiv et al. 2022). Over-exploitation of groundwater for ASM operations in arid parts of Ethiopia is also threatening the sustainability of the local water supplies, especially during the dry seasons where recharge rates are normally low (Macháček 2020).

4.7 Democratic Republic of Congo (DRC)—Central Africa

The DRC is an extremely rich mineral country; thus, ASM contributes significantly to the extraction of cobalt, gold, and diamonds (Katz-Lavigne et al. 2023). Meanwhile, the DRC has been classified as one of the countries whose aquifers have been seriously degraded by ASM activities in the provinces of Katanga and Ituri (Bikubanya and Geenen 2023). A study by Kouadio et al. (2023), indicated that ASM activities in these places have polluted ground waters with heavy metals such as cobalt, cadmium, and uranium, among others, as well as toxic substances emanating from mine tailings like sulfuric acid. Groundwater sources near mines in Katanga had up to more than 30 times the acceptable levels of drinking water according to WHO standards (Julien et al. 2022). Contamination on this scale presents a serious health risk to the people living nearby, as locals rely on the groundwater for drinking, washing, and farming. Groundwater abstraction for washing and processing minerals from shallow aquifers has been taken to an exceptionally high level in Ituri Province. This heightened degree of water scarcity, combined with quite erratic rainfall, has led to an increase in water shortage (Yamfwa et al. 2024).

5 Climate Change and Water Resources in the Global South

5.1 Groundwater Recharge and Climate Change

Climate change impacts the earth's hydrological cycle (Fig. 4) (Konapala et al. 2020), leading to increasingly erratic precipitation patterns in most parts of the world (El-Sayed and Kamel 2020). Again, because it depends on regular rainfall, groundwater recharge is very sensitive to such changes (Anurag et al. 2021; Moeck et al. 2020). Some studies in recent years have documented how changes in rainfall patterns reduce the rate at which groundwater is recharged in Africa and parts of Asia. For example, Moeck et al. (2020) cited that with the longer length of the dry season and increasing erratic heavy rainfall events, the risk of surface runoff rather than infiltration would imply the rate of groundwater recharge is in decline across the semi-arid regions of Africa. In Latin America, where heavy storms frequently create a lot more runoff relative to groundwater recharge (Poveda et al. 2020), there are implications for local aquifers, including those in northeastern Brazil (Marengo et al. 2023).

In addition to this, increased global temperatures raise the rates of evaporation in many of these regions, further decreasing the volume of water available for groundwater recharge (Fig. 5) leading many to turn to unconventional sources of water for domestic use (Scanlon et al. 2023).

Greater temperatures have also impacted the snowmelt in mountainous regions (Kumar et al. 2022; Javadinejad et al. 2020), which may provide the most important source of groundwater recharge. For example, in portions of the Andes, glacial melting—resulting from increased temperatures—provides seasonal water that renews groundwater aquifers (Ruiz-Pereira et al. 2023). However, with glacier mass shrinking due to ongoing climate change (Hugonnet et al. 2021), this source

Fig. 4 A summary of the lives of millions in the Global South as reported by the United Nations High Commissioner for Refugees (UNHCR) in 2024

Fig. 5 Impacts extreme droughts lead some to unconventional sources of water for use

of recharge becomes unstable in the long term (Cepeda et al. 2022). In areas where seasonal monsoons are relied upon for recharging groundwater, like South Asia, the variability in the pattern of the monsoon, where rainfall is much heavier but over a shorter period, promotes ineffective groundwater recharging (Jameel et al. 2023). According to a study by Shang et al. (2020), although the intensity of rainfall has increased, South Asia is receiving a shorter monsoon, resulting in less overall recharge, which is worsening the state of groundwater availability for agriculture and domestic uses.

5.2 Drought and Groundwater Depletion

Throughout the Global South, longer and harsher droughts are becoming the new norm (Calverley and Walther 2022), fueled by changes in climate (Satoh et al. 2022). However, in most cases, when drought dries out streams and lakes, groundwater turns out to be the only source of water. Though very helpful in those situations, high dependency on this resource in such times accelerates the deterioration of aquifers (Sarremejane et al. 2022). It has been indicated that the high level of extraction during the drought periods in Maharashtra and Gujarat in India has reduced the level of groundwater, pushing many areas toward a water crisis (Sarremejane et al. 2022; Sengupta 2021).

Similarly, East Africa has also gone through the worst droughts within the past decade lowered its groundwater storage significantly (Palmer et al. 2023; Adhikari et al. 2015). However, over-extraction rates are driving significant drops in groundwater levels in Kenya and Ethiopia, consequently threatening agricultural production and household water supplies according to the report by Merem et al. (2020). Where communities rely on this resource, the threat that has been posed to food security by the depletion of this basic commodity brings about economic instability (Granit 2022).

Groundwater depletion, particularly during drought, is not a question of quantity alone, but also one of quality (Petersen-Perlman et al. 2022). This fall in groundwater levels makes the shallow aquifers more susceptible to pollution by contaminants (Jarsjö et al. 2020), while the deeper aquifers become more expensive to extract and are not always replenishable (Zwarteveen et al. 2021), thus increasing disparities in access to water among socio-economic groups. Furthermore, in coastal regions, reduction in groundwater levels and rising sea levels may cause saline intrusions and contaminate freshwater aquifers (Basack et al. 2022; Bosserelle et al. 2022). Bangladesh presents a case where the rise in the sea level and over-extraction of groundwater have resulted in widespread salinization of ground-water (Jamei et al. 2022).

5.3 Increasing Trends of ASM and Climate Change in Developing Countries

According to Ofosu et al. (2020), ASM activities in developing countries have largely been brought forth by socio-economic factors such as poverty and unemployment that have pushed many into this occupation. For example, that study affirmed that in sub-Saharan Africa alone, millions are involved in ASM, which are large contributors to the local economies at a very high environmental cost. Hilson and Hu (2022) have said ASM is indeed a source of livelihood, yet it has an overall negative effect on natural resources such as water due to the unregulated manner of its operation. The processing of mainly gold extraction with hazardous chemicals such as mercury leads to leakage into the soil to the groundwater system where water is found.

Besides that, ASM is also common in hydrologically sensitive ecological zones (Tarekegn et al. 2022). The mining acts disrupt the local hydrology and as such reduce the resilience of the ecosystems to natural hydrological cycles. Large areas are cleared off mining, in which all vegetation responsible for water retention and holding the soils together is removed (Worlanyo and Jiangfeng 2021). As argued by Mensah and Tuokuu (2023), the resultant deforestation and soil erosion from ASM in Western Africa have been to a great extent reducing groundwater recharge, therefore depressing the water table and increasing the competition for the already limited supplies.

Simultaneously, climate change is worsening the socio-economic gaps already posed by the effects of unregulated mining regarding water security in developing nations (Edokpayi et al. 2020). According to the Intergovernmental Panel on Climate

Change, developing countries are the most affected by the effects of climate change (Intergovernmental Panel on Climate Change (IPCC) 2011), while specific areas of Asia, Latin America, and Africa are very susceptible to altered rainfall patterns, higher occurrences of droughts, and temperature increases.

5.4 Synergistic Effects on the Safety of Groundwater

This means the compounded risk to groundwater security due to ASM and climate change. That is, ASM compounds environmental vulnerabilities already created by climate change through a reduction in the rate of groundwater recharge and contamination of the aquifers (Rüttinger et al. 2020). The physical disruption of the landscape by mining activities includes, by definition, weakening natural processes allowing water to seep into underground aquifers, due to the heaviness of rainfall. For example, the destruction of vegetation and soil cover reduces infiltration, allowing more surface runoff, while less water penetrates the Earth's crust to improve the groundwater level.

Apart from that, ASM effects are magnified by climate change, whereby more and more surface water, in instances of drought, is not available, shifting people to groundwater use. Already over-extracted for agriculture, domestic use, and pollution from ASM, the risk of its depletion due to such extraction increases with each passing day (Adegbite 2022). In areas like Latin America, the depletion of groundwater resources because of ASM and changes in climate variability has created a vicious cycle; the declining availability of water for agriculture and more conventional works pushes more people towards mining as a source of income, further exacerbating environmental degradation (Ofosu et al. 2020).

5.5 Future Expectations

The future of groundwater security in developing countries will continue to be highly precarious unless the current ASM and climate change processes are put under check. Global demand for minerals, impelled by newer sets of minerals for technologies such as renewable energy and electronics, is bound to continue scaling upwards (Wellmer 2022). According to estimates, demands for minerals like cobalt and lithium are going to spiral, pushing more and more communities across developing countries into ASM mining (CEEW et al. 2023).

Indeed, coupled with that, the projections beyond 2018 reported about others on climate change issued by the IPCC indicate further increases in the frequency and intensity of extremes, including the ones associated with long droughts, thus making groundwater an increasingly reliable resource in areas where surface water is undependable (Chiang et al. 2021; Krishnan et al. 2020).

These synergistic stressors between ASM and climate change will, unless this continues with no kind of mitigation, degrade groundwater resources beyond their local sustainable use. For instance, in southern Africa, where both mining activities and climate change are expected to intensify, the availability of groundwater could decrease by 29.96 and 33.54% by the middle of the twenty-first century (Mohammed et al. 2023). This would have devastating effects on food security, public health, and economic development in these regions.

5.6 The Role of the International Community

The problems have reached such proportions and complexity that it is first and foremost a matter of international cooperation that may help to mitigate the synergetic effects of ASM and climate change on groundwater security (ASEAN Secretariat 2021). International organizations such as the United Nations and the World Bank, and even regional ones like the African Union, have been instrumental in fostering sustainable development. However, increasingly specific efforts are required toward the regulation of ASM, climate change, and groundwater management (Fukuda-Parr and Muchhala 2020). The international community could feel assured of reinforcement regarding the national capacities, both in terms of management of ASM and protection of groundwater resources. The promotion of the formalization of ASM can be one of the basic focal points since mining activity needs to be considered under a legal framework that looks into environmental protection (Arthur-Holmes and Ofosu 2024). This involves the adoption of best practices for mining water management, including the avoidance of toxic chemicals and the rehabilitation of mined lands to their natural states of retaining water (Agboola et al. 2020).

What is more, the international community should support climate adaptation strategies that would particularly focus on how to manage groundwater (Petersen-Perlman et al. 2022). For instance, investment in technology which saves water, including rainwater harvesting and artificial groundwater recharge would help communities in the developing nations to be resilient against climate change and also from mining pressures. International financial institutions including the World Bank and the IMF can provide the required funding and technical assistance to upscale such initiatives with much ease (Henning 2020). Lastly, there is international cooperation about transboundary water resources. Many developing countries share an aquifer with neighbours; hence, jointly managing the groundwater is even more urgent to avoid over-extraction and contamination (Stevanović and Stevanović 2021). While the UN's Water Convention and other international agreements do provide a framework for the joint management of shared water resources, these need further development and adaptation in response to the dual pressures of ASM and climate change (De Chaisemartin 2020).

6 Conclusion

The impacts of ASM and climate change as stand-alone environmental unfriendly activities and events have been widely cited as two of the most complex and important threats to groundwater security across the Global South. ASM is characterised by informal and unregulated methods that have rapidly been spreading across such regions as Sub-Saharan Africa, South East Asia and Latin America due to the high global demand for precious minerals such as gold, cobalt, and diamond. As stated by Hilson (2020), mining activities have an immense impact on the groundwater systems via the use of dangerous chemicals such as mercury and cyanide that contaminate both surface water and underground water sources.

The stress on the groundwater resources is heightened by an unpredictable climate, events of droughts, and unusual rainfall patterns. The changing climate has enhanced the intensity of the environmental degradation caused by ASM, mostly due to disruptions in hydrological cycles and reductions in groundwater recharge rates. As temperatures continue to rise globally, in an era defined as "global boiling", much more frequent and longer droughts will continue to stress groundwater resources, a fact that was measured in the work of Scanlon et al. (2023).

Groundwater is a main source of freshwater supply for domestic and agricultural needs in most regions of the developing world. Various studies have estimated that over-abstraction rates, increased by mining activities, are unsustainable and lead to the over-exploiting and eventual depletion of the resource. It is safe to say such groundwater depletion, apart from contamination resulting from mining activities, is a dual threat to water security, especially within rural and disadvantaged communities. Also worthy of note is the connection of ASM with climate change showing several disturbing trends. ASM is set to rise persistently because of increased demand for their products. This may push mining to more sensitive ecosystems hence making more threats to the groundwater.

Climate models also predict, concurrently, increased variability in rainfall and an increase in extreme events, which collectively reduce the reliability of groundwater recharge (Davamani et al. 2024). In the absence of significant interventions, these combined stresses are likely to lead to heightened water scarcity, reduced agricultural productivity, and increased conflict over water resources.

These issues are interrelated; put together, they serve an essential function for the global community. Solutions should be aggressively sought both at the local and global levels: international entities such as the United Nations and the World Bank, alongside regional organizations like the African Union and ASEAN, need to work out policies that entrench sustainable mining and enhance resilience to climate variability. This shows the urgent need, at the policy level, for the introduction of superior regulatory regimes that consider the environmental impacts of ASM, besides ensuring proper adaptation to climate change within national and regional water management plans (Grison et al. 2023). The international community can assist developing countries in adopting effective groundwater monitoring systems with appropriate technical capacities and financial support. These technologies are very necessary in monitoring groundwater quality and quantity. This will help in making more informed decisions in the management of water resources.

Establishing cross-border agreements on sharing water resources is important in regions where one water basin spans several countries, thereby not damming them indefinitely, intending to ensure the sustainable and equitable management of such surface waters. Other critical areas which require international collaboration involve capacity-building and enhancing local governance and people's empowerment. Equipping local governments and communities with the means to oversee artisanal and small-scale mining operations and enforce sustainable water management practices might alleviate some environmental damage resulting from mining.

Conception and implementation of policies involve several local stakeholders, they are bound to be both culturally and environmentally appropriate, thus increasing their success rate. Security related to groundwater for developing countries will depend on how both the ASM and climate change issues are engaged. The effort involves international leadership to establish and institute global partnerships toward sustainable development and water security. Issues of unregulated mining within a period of changing climate are very complex and have grown in urgency on a worldwide basis. The protection of groundwater supplies for the next generation with collaboration, forward-thinking policy, and good on-the-ground strategies must be seen as a responsibility.

References

Abanyie SK et al (2023) Sources and factors influencing groundwater quality and associated health implications: a review. Emerg Contam 9(2):100207. https://doi.org/10.1016/j.emcon.2023.100207

Abdelkareem M, Mansour AM (2023) Risk assessment and management of vulnerable areas to flash flood hazards in arid regions using remote sensing and GIS-based knowledge-driven techniques. Nat Hazards 117(3):2269–2295. https://doi.org/10.1007/s11069-023-05942-x

Adegbite A (2022) Practical approach to reclamation of post-mining sites. Doctoral thesis of Fakultät 2 Umwelt und Naturwissenschaften. https://doi.org/10.26127/BTUOpen-5871

Adhikari U, Nejadhashemi AP, Woznicki SA (2015) Climate change and eastern Africa: a review of impact on major crops. Food Energy Secur 4(2):110–132. https://doi.org/10.1002/fes3.61

Affum AO, Kwaansa-Ansah EE, Osae SD (2024) Estimating groundwater geogenic arsenic contamination and the affected population of river basins underlain mostly with crystalline rocks in Ghana. Environ Challeng 15:100898. https://doi.org/10.1016/j.envc.2024.100898

Afifi T et al (2016) Human mobility in response to rainfall variability: opportunities for migration as a successful adaptation strategy in eight case studies. Migr Dev 5(2):254–274. https://doi.org/10.1080/21632324.2015.1022974

Agboola O et al (2020) A review on the impact of mining operation: monitoring, assessment and management. Results Eng 8:100181. https://doi.org/10.1016/j.rineng.2020.100181

Akbari M, Najafi Alamdarlo H, Mosavi SH (2020) The effects of climate change and groundwater salinity on farmers' income risk. Ecol Ind 110:105893. https://doi.org/10.1016/j.ecolind.2019.105893

Akhtar MN et al (2021) Smart sensing with edge computing in precision agriculture for soil assessment and heavy metal monitoring: a review. Agriculture 11(6):1–37. https://doi.org/10.3390/agriculture11060475

Al-Hashimi O et al (2021) Remediation: occurrence, migration and adsorption modelling. Molecules 26:5913

Aldridge Mugadza A (2022) The disastrous effects of deforestation and forest degradation in the climate vulnerability era. CIFILE J Int Law 3(6):2022. https://doi.org/10.30489/CIFJ.2022.324393.1049

Aleku DL, Lazareva O, Pichler T (2024) Mercury in groundwater—Source, transport and remediation. Appl Geochem 170:106060. https://doi.org/10.1016/j.apgeochem.2024.106060

Arthur-Holmes F, Busia KA, Yakovleva N, Vazquez-Brust DA (2022) Artisanal and small-scale mining (ASM) methods and the UN Sustainable Development Goal 6: perceived implications for clean water supply. Environ Sci Policy 137:205–215. https://doi.org/10.1016/j.envsci.2022.08.017

Álvarez-Berríos N, L'Roe J, Naughton-Treves L (2021) Does formalizing artisanal gold mining mitigate environmental impacts? Deforestation evidence from the Peruvian Amazon. Environ Res Lett 16(6). https://doi.org/10.1088/1748-9326/abede9

Amanambu AC et al (2020) Groundwater system and climate change: present status and future considerations. J Hydrol 589:125163. https://doi.org/10.1016/j.jhydrol.2020.125163

Anozie C (2024) Left behind: funding climate action in the global South. Texas A&M Law Rev 11(2):333–363. https://doi.org/10.37419/LR.V11.I2.2

Anurag H et al (2021) Modeling the impact of spatiotemporal vegetation dynamics on groundwater recharge. J Hydrol 601:126584. https://doi.org/10.1016/j.jhydrol.2021.126584

Arrazay S et al (2023) Spatial distribution and human health risks of mercury in the gold mining area of Mandailing Natal District, Indonesia. Environ Monit Contam Res 3:33–42. https://doi.org/10.5985/emcr.20230003

Arthur-Holmes F, Ofosu G (2024) Rethinking state-led formalisation of artisanal and small-scale mining (ASM): towards mining licence categorisation, women empowerment and environmental sustainability. Resour Policy 93:105058. https://doi.org/10.1016/j.resourpol.2024.105058

Asare D et al (2024) Impact of different illegal artisanal small-scale mining techniques on soil properties in a major mining landscape in Ghana. Environ Chall 17:101008. https://doi.org/10.1016/j.envc.2024.101008

ASEAN Secretariat (2021) ASEAN regional plan of action for adaptation to drought 2021–2025. https://asean.org/book/asean-regional-plan-of-action-for-adaptation-to-drought-2021-2025/

Ashraf S, Nazemi A, AghaKouchak A (2021) Anthropogenic drought dominates groundwater depletion in Iran. Sci Rep 11:9135. https://doi.org/10.1038/s41598-021-88522-y

Bah HAF et al (2023) Determinants of exposure to potentially toxic metals in pregnant women of the DSAN-12M cohort in the Recôncavo Baiano, Brazil. Int J Environ Res Public Health 20(4):2949. https://doi.org/10.3390/ijerph20042949

Bano I, Arshad M (2018) Climatic changes impact on water availability. In: Arshad M (ed) Perspectives on water usage for biofuels production. Springer, Cham. https://doi.org/10.1007/978-3-319-66408-8_2

Basack S et al (2022) Saltwater intrusion into coastal aquifers and associated risk management: critical review and research directives. J Coastal Res 38(3):654–672. https://doi.org/10.2112/JCOASTRES-D-21-00116.1

Bikubanya D-L, Geenen S (2023) From adverse incorporation to decent work? Evidence from artisanal and small scale gold mining in DR Congo. University of Antwerp, Centre of Excellence—Working Papers. https://repository.uantwerpen.be/docstore/d:irua:17493

Bosserelle AL, Morgan LK, Hughes MW (2022) Groundwater rise and associated flooding in coastal settlements due to sea-level rise: a review of processes and methods. Earth's Future 10(7):1–26. https://doi.org/10.1029/2021EF002580

Brierley GJ et al (2016) Introduction: landscape and ecosystem diversity in the yellow river source zone. Springer Geogr. https://doi.org/10.1007/978-3-319-30475-5_1

Cacciuttolo C, Cano D, Custodio M (2023) Socio-environmental risks linked with mine tailings chemical composition: promoting responsible and safe mine tailings management considering copper and gold mining experiences from Chile and Peru. Toxics 11(5). https://doi.org/10.3390/toxics11050462

Calderwood AJ et al (2020) Low-cost, open source wireless sensor network for real-time, scalable groundwater monitoring. Water 12(4):1066. https://doi.org/10.3390/W12041066

Calverley CM, Walther SC (2022) Drought, water management, and social equity: analyzing Cape Town, South Africa's water crisis. Front Water 4:910149. https://doi.org/10.3389/frwa.2022.910149

CEEW et al (2023) Addressing vulnerabilities in the supply chain of critical minerals. https://www.ceew.in/sites/default/files/addressing-critical-minerals-supply-chain-vulnerabilities-india.pdf. Accessed 1 Oct 2024

Cepeda Arias E, Cañon Barriga J, Salazar JF (2022) Changes of streamflow regulation in an Andean watershed with shrinking glaciers: implications for water security. Hydrol Sci J 67(11):1755–1770. https://doi.org/10.1080/02626667.2022.2105650

Chen MH, Yu XZ, Feng YX (2022) Tracing the pollution and human risks of potentially toxic elements in agricultural area nearby the cyanide baths from an active private gold mine in Hainan Province, China. Environ Geochem Health 44(10):3279–3296. https://doi.org/10.1007/s10653-021-01038-2

Chiang F, Mazdiyasni O, AghaKouchak A (2021) Evidence of anthropogenic impacts on global drought frequency, duration, and intensity. Nat Commun 12(1):1–10. https://doi.org/10.1038/s41467-021-22314-w

Chirif A et al (2019) Peru: eforestation. Lima

Cossa H et al (2021) Health studies in the context of artisanal and small-scale mining: a scoping review. Int J Environ Res Public Health 18(4):1–24. https://doi.org/10.3390/ijerph18041555

Crul MRJ, Schneider J, Lelie F (2024) VU research portal super-diversité. https://www.elitesproject.eu/publications/books%0Ahttps://research.vu.nl/ws/portalfiles/portal/129674647/W+T+C++bisschop+-+thesis.pdf

Custodio M et al (2020) Human risk from exposure to heavy metals and arsenic in water from rivers with mining influence in the Central Andes of Peru. Water 12(7):1946. https://doi.org/10.3390/w12071946

Daley E et al (2018) Gender, land and mining in Mongolia 1:65–111. http://mokoro.co.uk/wp-content/uploads/Gender_Land_and_Mining_in_Mongolia_WOLTS_Research_Report_No.1_January_2018-1.pdf

Davamani V et al (2024) A critical review of climate change impacts on groundwater resources: a focus on the current status, future possibilities, and role of simulation models. Atmosphere 15(1):122. https://doi.org/10.3390/atmos15010122

De Chaisemartin M (2020) Measuring transboundary water cooperation within the framework of Agenda 2030: a proposal for a revision of SDG Indicator 6.5.2. Water Int 45(1):60–78. https://doi.org/10.1080/02508060.2019.1708659

De Paula Arrifano G et al (2023) Neurotoxicity and the global worst pollutants: astroglial involvement in arsenic, lead, and mercury intoxication. Neurochem Res 48(4):1047–1065. https://doi.org/10.1007/s11064-022-03725-7

Demissie S et al (2023) Prevalence of arsenic-induced skin lesions and associated factors in Ethiopia: community-based study. Toxicol Rep 11:153–161. https://doi.org/10.1016/j.toxrep.2023.07.007

Dey S, Majumdar A (2024) Current status of pollution in Major Rivers and Tributaries of India and protection-restoration strategies. In: Kanhaiya S, Singh S, Dixit A, Singh AK (eds) Rivers of India. Springer Cham. https://doi.org/10.1007/978-3-031-49163-4_4

Diaz J (2021) Ecologies of gold: understanding the social, political, and ecological impacts of mercury use in informal, small-scale gold mining in Madre de Dios, Peru. Ph.D. thesis of the University of California. https://escholarship.org/content/qt1dj406jq/qt1dj406jq.pdf. Accessed 1 Oct 2024

Dorleku MK, Nukpezah D, Carboo D (2018) Effects of small-scale gold mining on heavy metal levels in groundwater in the Lower Pra Basin of Ghana. Appl Water Sci 8(5):1–11. https://doi.org/10.1007/s13201-018-0773-z

Durkalec M et al (2022) Factors influencing lead, mercury and other trace element exposure in birds from metal mining areas'. Environ Res 212:113575. https://doi.org/10.1016/j.envres.2022.113575

Eagles-Smith CA et al (2016) Mercury in western North America: a synthesis of environmental contamination, fluxes, bioaccumulation, and risk to fish and wildlife. Sci Total Environ 568:1213–1226. https://doi.org/10.1016/j.scitotenv.2016.05.094

Edokpayi JN et al (2020) Adaptive management approach, water conservation and wastewater treatment in BRICS Nations. https://doi.org/10.1016/B978-0-12-818339-7/00005-9

Eduful M et al (2020) The Illegal Artisanal and Small-scale mining (Galamsey) "Menace" in Ghana: is military-style approach the answer? Resour Policy 68(May). https://doi.org/10.1016/j.resourpol.2020.101732

El-Sayed A, Kamel M (2020) Climatic changes and their role in emergence and re-emergence of diseases. Environ Sci Pollut Res 27(18):22336–22352. https://doi.org/10.1007/s11356-020-08896-w

Esinam S et al (2024) Sustainable groundwater purification: leveraging local materials for lead, Iron, and Cyanide removal in Ghanaian mining communities 14(9):637–648

Ewusi A et al (2017) Mining and metal pollution: assessment of water quality in the Tarkwa mining area. Ghana Min J 17(2):17–31. https://doi.org/10.4314/gm.v17i2.4

Fagariba CJ, Sumani JBB, Mohammed AS (2024) Artisanal and small-scale gold mining impact on soil and agriculture: evidence from upper Denkyira East Municipality, Ghana. Eur J Environ Earth Sci 5(3):12–20. https://doi.org/10.24018/ejgeo.2024.5.3.443

Fielding CR, Alexander J, Allen JP (2018) The role of discharge variability in the formation and preservation of alluvial sediment bodies. Sediment Geol 365:1–20. https://doi.org/10.1016/j.sedgeo.2017.12.022

Foster S, Chilton J (2020) Sustainable management of groundwater. Oxford Res Encycl Environ Sci. https://doi.org/10.1093/acrefore/9780199389414.013.635

Fukuda-Parr S, Muchhala B (2020) The Southern origins of sustainable development goals: Ideas, actors, aspirations. World Dev 126(September 2015):104706. https://doi.org/10.1016/j.worlddev.2019.104706

Garcia AK et al (2022) Drylands facing change. Drylands Facing Change. https://doi.org/10.4324/9781003174486

Garg KK et al (2021) Impact of rainwater harvesting on hydrological processes in a fragile watershed of South Asia. Groundwater 59(6):839–855. https://doi.org/10.1111/gwat.13099

Geris J et al (2022) Surface water-groundwater interactions and local land use control water quality impacts of extreme rainfall and flooding in a vulnerable semi-arid region of Sub-Saharan Africa. J Hydrol 609(September 2021):127834. https://doi.org/10.1016/j.jhydrol.2022.127834

Getahun YS, Haj VL (2015) Assessing the impacts of land use-cover change on hydrology of Melka Kuntrie Subbasin in Ethiopia, using a conceptual hydrological model. J Waste Water Treat Anal 06(03). https://doi.org/10.4172/2157-7587.1000210

Gona LS et al (2022) Impact of climate change on groundwater potential and recharge in the drought prone Runde catchment of Zimbabwe. Water Supply 22(7):6405–6426. https://doi.org/10.2166/ws.2022.144

González-Valoys AC et al (2022) Environmental challenges related to cyanidation in Central American gold mining; the Remance mine (Panama). J Environ Manag 302. https://doi.org/10.1016/j.jenvman.2021.113979

Granit I (2022) Microgrids through the energy-water-food security Nexus in La Guajira, Colombia: increasing water and food security or jeopardizing groundwater levels? Energy Res Soc Sci 93(September):102814. https://doi.org/10.1016/j.erss.2022.102814

Green TR, Allen DM (2011) DigitalCommons @ University of Nebraska—Lincoln Beneath the surface of global change: impacts of climate change on groundwater

Grison C et al (2023) Integrated water resources management in cities in the world: global challenges. Water Resour Manag 37(6–7):2787–2803. https://doi.org/10.1007/s11269-023-03475-3

Gu C et al (2020) Distinguishing the effects of vegetation restoration on runoff and sediment generation on simulated rainfall on the hillslopes of the loess plateau of China. Plant and Soil 447(1–2):393–412. https://doi.org/10.1007/s11104-019-04392-4
Gugulothu S et al (2022) Geochemical evaluation of groundwater and suitability of groundwater quality for irrigation purpose in an agricultural region of South India. Appl Water Sci 12(6):1–13. https://doi.org/10.1007/s13201-022-01583-w
Henning CR (2020) Regional financial arrangements and the International Monetary Fund: sustaining coherence in global financial governance
Hilson G (2020) The "Zambia Model": a blueprint for formalizing artisanal and small-scale mining in sub-Saharan Africa? Resour Policy 68(April 2018):101765. https://doi.org/10.1016/j.resourpol.2020.101765
Hilson G, Hu Y (2022) Changing priorities, shifting narratives: remapping rural livelihoods in Africa's artisanal and small-scale mining sector. J Rural Stud 92(September 2021):93–108. https://doi.org/10.1016/j.jrurstud.2022.03.010
Hirons M (2020) How the sustainable development goals risk undermining efforts to address environmental and social issues in the small-scale mining sector. Environ Sci Policy 114:321–328. https://doi.org/10.1016/j.envsci.2020.08.022
Holman IP (2006) Climate change impacts on groundwater recharge-uncertainty, shortcomings, and the way forward? Hydrogeol J 14(5):637–647. https://doi.org/10.1007/s10040-005-0467-0
Hoogesteger J (2022) Regulating agricultural groundwater use in arid and semi-arid regions of the Global South: challenges and socio-environmental impacts. Curr Opin Environ Sci Health 27:100341. https://doi.org/10.1016/j.coesh.2022.100341
Hugonnet R et al (2021) Accelerated global glacier mass loss in the early twenty-first century to cite this version: HAL Id: insu-03214987 accelerated global glacier mass loss in the early twenty-first century
Humphrey CE et al (2022) Using automated seepage meters to quantify the spatial variability and net flux of groundwater to a stream. Water Resour Res 58(6):1–18. https://doi.org/10.1029/2021WR030711
Intergovernmental Panel on Climate Change (IPCC) (2011) Summary for policy-makers, climate change, IPCC WG1 fourth assessment report. Cambridge University Press, 19.III.20(IPCC-XVII/Doc. 3d). http://ipcc.ch/meetings/session17/doc3d.pdf%5Cnhttp://link.springer.com/https://doi.org/10.1007/BF02986817
Ireson A, Makropoulos C, Maksimovic C (2006) Water resources modelling under data scarcity: coupling MIKE BASIN and ASM groundwater model. Water Resour Manag 20(4):567–590. https://doi.org/10.1007/s11269-006-3085-2
Jameel Y et al (2023) Shift in groundwater recharge of the Bengal Basin from rainfall to surface water. Commun Earth Environ 4(1). https://doi.org/10.1038/s43247-022-00650-z
Jamei M et al (2022) Computational assessment of groundwater salinity distribution within coastal multi-aquifers of Bangladesh. Sci Rep 12(1):1–28. https://doi.org/10.1038/s41598-022-15104-x
Jannetto PJ, Cowl CT (2023) Elementary overview of heavy metals. Clin Chem 69(4):336–349. https://doi.org/10.1093/clinchem/hvad022
Jarsjö J et al (2020) Projecting impacts of climate change on metal mobilization at contaminated sites: controls by the groundwater level. Sci Total Environ 712. https://doi.org/10.1016/j.scitotenv.2019.135560
Javadinejad S, Dara R, Jafary F (2020) Climate change scenarios and effects on snow-melt runoff. Civil Eng J 6(9):1715–1725. https://doi.org/10.28991/cej-2020-03091577
Job C (2023) Financing managed aquifer recharge projects. Groundw Monit Remediat 43(4):11–17. https://doi.org/10.1111/gwmr.12620
Julien KK et al (2022) Treatment of well water polluted with metallic trace elements by Moringa seeds: case of the Kisanga District in the Democratic Republic of Congo. Int J Multidiscip Res Growth Eval 03(04):356–362

Kabelong Banoho LPR et al (2020) Biodiversity and carbon sequestration potential in two types of tropical rainforest, Cameroon. Acta Oecol 105(March). https://doi.org/10.1016/j.actao.2020.103562

Kaonga V (2018) Αιτηση Εξοδων 1 Page, the Acceptance of Islamic Hotel Concept in Malaysia: a Conceptual Paper

Katz-Lavigne S et al (2023) Driving change in the Democratic Republic of Congo: an initial mapping of participation in mineral regulation and responsible sourcing. In: Working papers/University of Antwerp. Institute of Development Policy and Management; Université d'Anvers. Institut de politique et de gestion du développement.-Antwerp [Preprint].

Kausher R et al (2023) Assessing impacts of mining-induced land use changes on groundwater and surface water quality using isotopic and hydrogeochemical signatures. Sustainability (Switzerland) 15(14). https://doi.org/10.3390/su151411041

Kershaw JL, Hall AJ (2019) Mercury in cetaceans: exposure, bioaccumulation and toxicity. Sci Total Environ 694. https://doi.org/10.1016/j.scitotenv.2019.133683

Kiran, Bharti R, Sharma R (2021) Effect of heavy metals: an overview. Mater Today: Proc 51(xxxx):880–885. https://doi.org/10.1016/j.matpr.2021.06.278

Konapala G et al (2020) Climate change will affect global water availability through compounding changes in seasonal precipitation and evaporation. Nat Commun 11(1):1–10. https://doi.org/10.1038/s41467-020-16757-w

Kouadio LM et al (2023) Assessment of metallic pollution of water and soil from illegal gold mining sites in Kong 2, Hiré and Degbézéré (Ivory Coast) assessment of metallic pollution of water and soil from illegal gold mining sites in Kong. J Mater Environ Sci 2023(1):41–61. http://www.jmaterenvironsci.com

Krishnan R et al (2020) Assessment of climate change over the Indian region: a report of the ministry of earth sciences (MOES), government of India. In: Assessment of climate change over the Indian Region: a report of the Ministry of Earth Sciences (MoES), Government of India. Springer Singapore. https://doi.org/10.1007/978-981-15-4327-2

Kumah C, Hilson G, Quaicoe I (2020) Poverty, adaptation and vulnerability: an assessment of women's work in Ghana's artisanal gold mining sector. Area 52(3):617–625. https://doi.org/10.1111/area.12639

Kumar R et al (2017) Remediation of cyanide-contaminated environments through microbes and plants: a review of current knowledge and future perspectives. Geosyst Eng 20(1):28–40. https://doi.org/10.1080/12269328.2016.1218303

Kumar R et al (2022) Assessment of climate change impact on snowmelt runoff in Himalayan Region. Sustainability (Switzerland) 14(3):1–23. https://doi.org/10.3390/su14031150

Kyaw WT, Sakakibara M (2022) Transdisciplinary communities of practice to resolve health problems in Southeast Asian Artisanal and small-scale gold mining communities. Int J Environ Res Public Health 19(9). https://doi.org/10.3390/ijerph19095422

Landrigan P et al (2022) Reducing disease and death from Artisanal and Small-Scale Mining (ASM)—The urgent need for responsible mining in the context of growing global demand for minerals and metals for climate change mitigation. Environ Health: Glob Access Sci Sour 21(1):1–13. https://doi.org/10.1186/s12940-022-00877-5

Lapworth DJ et al (2022) Groundwater quality: global threats, opportunities and realising the potential of groundwater. Sci Total Environ 811:1–29. https://doi.org/10.1016/j.scitotenv.2021.152471

Leung AMR, Lu JLDP (2016) Environmental health and safety hazards of indigenous small-scale gold mining using cyanidation in the Philippines. Environ Health Insights 10:125–131. https://doi.org/10.4137/EHI.S38459

Ligate F et al (2022) Geogenic contaminants and groundwater quality around Lake Victoria goldfields in northwestern Tanzania. Chemosphere 307(P2):135732. https://doi.org/10.1016/j.chemosphere.2022.135732

Lo W et al (2022) Integration of numerical models and InSAR techniques to assess land subsidence due to excessive groundwater abstraction in the Coastal and Lowland Regions of Semarang City. Water (Switzerland) 14(2). https://doi.org/10.3390/w14020201

MacAlister C, Subramanyam N (2018) Climate change and adaptive water management: innovative solutions from the global South. Water Int 43(2):133–144. https://doi.org/10.1080/02508060.2018.1444307

Macháček J (2019) Typology of environmental impacts of artisanal and small-scale mining in African Great Lakes Region. Sustainability (Switzerland) 11(11). https://doi.org/10.3390/su11113027

Macháček J (2020) Alluvial artisanal and small-scale mining in a river stream-Rutsiro case study (Rwanda). Forests 11(7). https://doi.org/10.3390/F11070762

Madhav R (2020) Environmental Governance of Artisanal and small-scale mining in India. Asian Affairs 51(4):895–912. https://doi.org/10.1080/03068374.2020.1834797

Malone A et al (2023) Transitional dynamics from mercury to cyanide-based processing in artisanal and small-scale gold mining: social, economic, geochemical, and environmental considerations. Sci Total Environ 898(March):165492. https://doi.org/10.1016/j.scitotenv.2023.165492

Manyakaidze P, Musavengane R, Simatele M (2024) Groundwater development and management constraints in drought prone Chiredzi and Zvishavane Districts, Zimbabwe. Clim Serv 34(November 2022):100467. https://doi.org/10.1016/j.cliser.2024.100467

Marengo JA et al (2023) Heavy rainfall associated with floods in southeastern Brazil in November–December 2021. Nat Hazard 116(3):3617–3644. https://doi.org/10.1007/s11069-023-05827-z

Meixner T et al (2016) Implications of projected climate change for groundwater recharge in the western United States. J Hydrol 534:124–138. https://doi.org/10.1016/j.jhydrol.2015.12.027

Mengistu TD et al (2021) Challenges and prospects of advancing groundwater research in Ethiopian aquifers: a review. Sustainability (Switzerland) 13(20):1–15. https://doi.org/10.3390/su132011500

Mensah AK, Tuokuu FXD (2023) Polluting our rivers in search of gold: how sustainable are reforms to stop informal miners from returning to mining sites in Ghana? Front Environ Sci 11(May):1–16. https://doi.org/10.3389/fenvs.2023.1154091

Merem EC et al (2020) Issues in transboundary water use in the River Nile Basin area of Africa. World Environ 10(2):27–44. https://doi.org/10.5923/j.env.20201002.01

Meutia AA, Bachriadi D, Gafur NA (2023) Environment degradation, health threats, and legality at the Artisanal small-scale gold mining sites in Indonesia. Int J Environ Res Publ Health 20(18). https://doi.org/10.3390/ijerph20186774

Moeck C et al (2020) A global-scale dataset of direct natural groundwater recharge rates: a review of variables, processes and relationships. Sci Total Environ 717. https://doi.org/10.1016/j.scitotenv.2020.137042

Mohammed N, Aragaw TT, Gnanachandrasamy G (2023) Evaluation of possible health risks associated with groundwater pollution at Kombolcha, a north-central Ethiopian industrial town in the Awash River basin. Environ Dev Sustain 26:31035–31074. https://doi.org/10.1007/s10668-023-04214-9

Moyo VP (2024) Rural Farmers' adaptive and transformative capacity in response to climate change and Artisanal small-scale mining in …. (March). https://scholar.sun.ac.za/bitstreams/55dbcd15-de37-49a5-81d8-4103b49c149a/download

Mpala SC et al (2020) Modelling the water level of the alluvial aquifer of an ephemeral river in south-western Zimbabwe. Hydrol Sci J 65(8):1399–1415. https://doi.org/10.1080/02626667.2020.1750615

Mulenga M et al (2024) Mining in sub-Saharan Africa: status impacts, and interventions. Water 16:756

Mwamba B (2020) Water resources management in Zambia: a case of cumulative impacts associated with copper mining in the upper Kafue catchment, Copperbelt province, Zambia, pp 1–112

Naz U, Saleem MS (2024) Climate-induced vulnerabilities: conflict and migration patterns in the Sahel Region of Africa * Corresponding Author 8(2)

Nguyen BT et al (2019) J Pre-proof

Nkomo M (2022) The anatomy of a gold-rush: politics, uncertainty and the organisation of artisanal mine work and labour in Zimbabwe. Extr Ind Soc 11(July):101124. https://doi.org/10.1016/j.exis.2022.101124

Nkuna R et al (2022) Accessing metals from low-grade ores and the environmental impact considerations: a review of the perspectives of conventional versus bioleaching strategies. Minerals 12(5). https://doi.org/10.3390/min12050506

Ofosu G et al (2020) Socio-economic and environmental implications of Artisanal and small-scale mining (ASM) on agriculture and livelihoods. Environ Sci Policy 106:210–220. https://doi.org/10.1016/j.envsci.2020.02.005

Ofosu G, Sarpong D (2022) Beyond the doom: sustainable water management practices of small-scale mining operations. Resour Policy 77(March):102649. https://doi.org/10.1016/j.resourpol.2022.102649

Ouma KO, Shane A, Syampungani S (2022) Aquatic ecological risk of heavy-metal pollution associated with degraded mining landscapes of the southern Africa River Basins: a review. Minerals 12(2). https://doi.org/10.3390/min12020225

Ofosu G et al (2024) Ban mining, ban dining? Re(examining) the policy and practice of "militarised conservationism" on ASM operations. Extr Ind Soc 17:101432. https://doi.org/10.1016/j.exis.2024.101432

Owusu Sekyere A et al (2023) Evaluating groundwater quality in the Asante Akyem Central District of Ghana. Environ Health Insights 17. https://doi.org/10.1177/11786302231188270

Oyunbat P et al (2021) Spatial distribution, pollution, and health risk assessment of heavy metal in industrial area soils of Ulaanbaatar, Mongolia. Int Arch Photogramm Remote Sens Spatial Inf Sci ISPRS Arch 43(B4–2021):123–133. https://doi.org/10.5194/isprs-archives-XLIII-B4-2021-123-2021

Paldor A, Michael HA (2021) Storm surges cause simultaneous salinization and freshening of coastal aquifers, exacerbated by climate change. Water Resour Res 57(5):1–14. https://doi.org/10.1029/2020WR029213

Palmer PI et al (2023) Drivers and impacts of Eastern African rainfall variability. Nat Rev Earth Environ 4(4):254–270. https://doi.org/10.1038/s43017-023-00397-x

Palumbo-Roe B, Olaka L, Bell R, Mitchel C, Bide T, Odiwuor C, Barlow T (2020) Reconnaissance study of groundwater quality in the artisanal gold mining districts of Migori County, Kenya. Brittish Geological Survey (Open Report OR/21/012), p 45

Panneerselvam B et al (2023) An integrated approach to explore the suitability of nitrate-contaminated groundwater for drinking purposes in a semiarid region of India. Environ Geochem Health 45(3):647–663. https://doi.org/10.1007/s10653-022-01237-5

Petersen-Perlman JD, Aguilar-Barajas I, Megdal SB (2022) Drought and groundwater management: Interconnections, challenges, and policyresponses. Curr Opin Environ Sci Health 28:100364. https://doi.org/10.1016/j.coesh.2022.100364

Poveda G et al (2020) High impact weather events in the Andes. Front Earth Sci 8(May):1–32. https://doi.org/10.3389/feart.2020.00162

Putra RS et al (2023) Some pollution problems to consider the design for remediation. In: Handbook of environmental chemistry, vol 115, pp 33–78. https://doi.org/10.1007/698_2022_896

Radi N, Hirche A, Boutaleb A (2023) Assessment of soil contamination by heavy metals and arsenic in Tamesguida abandoned copper mine area, Médéa, Algeria. Environ Monit Assess 195(1). https://doi.org/10.1007/s10661-022-10862-7

Radley B (2022) Class formation and capital accumulation in the countryside: artisanal and small-scale gold mining in South Kivu, DR Congo. J Agrar Change 22(2):398–414. https://doi.org/10.1111/joac.12474

Rajiv SV, George M, Nandakumar G (2022) Dermatological manifestations of arsenic exposure. J Skin Sex Transm Dis 5(1):14–21. https://doi.org/10.25259/jsstd_3_2022

Raletsena MV (2021) Elementary composition and metabolomic investigation of potato (Solanum tuberosum L.) irrigated with Fly Ash treated acid mine drainage (AMD) water (December)

Rivera IA et al (2019) Concentration, rainfall variability and groundwater. Water 11(2497):16
Rivki M et al (2022) The anthropology of resource extraction. Routledge Taylor and Francis Group [Preprint], no 112
Rocha-Santos L et al (2017) Functional decay in tree community within tropical fragmented landscapes: effects of landscape-scale forest cover. Plos One 12(4):1–18. https://doi.org/10.1371/journal.pone.0175545
Rodríguez, Velastequí M (2019) Addis Ababa University Addis Ababa Institute of Technology, School of Civil and Environmental Engineering Groundwater, pp 1–23
Roldán MD et al (2021) Bioremediation of cyanide-containing wastes. EMBO Rep 22(11):1–5. https://doi.org/10.15252/embr.202153720
Roughton et al (2016) A review on the impact of River and Inland Sand Mining on Nigerian River Basins. Theories of Sustainable Development, pp 1–201. https://www.researchgate.net/profile/Amir-Abdulazeez-2/publication/343481022_A_Review_on_the_Impact_of_River_and_Inland_Sand_Mining_on_Nigerian_River_Basins/links/5f2c0efc458515b72907316a/A-Review-on-the-Impact-of-River-and-Inland-Sand-Mining-on-Nigerian
Ruiz-Pereira S, Alvarado Peterson V, Trombotto Liaudat D (2023) Mountain cryosphere landscapes in South America: value and protection. Conservation 3(1):232–246. https://doi.org/10.3390/conservation3010017
Rüttinger L et al (2020) Impacts of climate change on mining, related environmental risks and raw material supply, p 80. https://www.umweltbundesamt.de/sites/default/files/medien/479/publikationen/texte_106-2020_impacts_of_climate_change_on_mining_related_environmental_risks_and_raw_material_supply.pdf.
Sanders AP et al (2019) Combined exposure to lead, cadmium, mercury, and arsenic and kidney health in adolescents age 12–19 in NHANES 2009–2014. Environ Int 131(July):104993. https://doi.org/10.1016/j.envint.2019.104993
Sarremejane R, Messager ML, Datry T (2022) Drought in intermittent river and ephemeral stream networks. Ecohydrology 15(5). https://doi.org/10.1002/eco.2390
Satoh Y et al (2022) The timing of unprecedented hydrological drought under climate change. Nat Commun 13(1). https://doi.org/10.1038/s41467-022-30729-2
Scanlon BR et al (2023) Global water resources and the role of groundwater in a resilient water future. Nat Rev Earth Environ 4(2):87–101. https://doi.org/10.1038/s43017-022-00378-6
Schoderer M, Ott M (2022) Contested water- and miningscapes—Explaining the high intensity of water and mining conflicts in a meta-study. World Dev 154:105888. https://doi.org/10.1016/j.worlddev.2022.105888
Sengupta S (2021) Indian droughts, drought policy and climate change. Mcrg.Ac.in [Preprint]. http://www.mcrg.ac.in/7thCSC/7thCSC_Full_Papers/SOHINI_SENGUPTA_TISS_Paper_2022.pdf
Shang W et al (2020) Influence of the South Asian high-intensity variability on the persistent heavy rainfall and heat waves in Asian monsoon regions. Int J Climatol 40(4):2153–2172. https://doi.org/10.1002/joc.6324
Sikka AK, Alam MF, Pavelic P (2021) Managing groundwater for building resilience for sustainable agriculture in South Asia. Irrig Drain 70(3):560–573. https://doi.org/10.1002/ird.2558
Siswanto S et al (2016) Temperature, extreme precipitation, and diurnal rainfall changes in the urbanized Jakarta city during the past 130 years. Int J Climatol 36(9):3207–3225. https://doi.org/10.1002/joc.4548
Stevanović Z, Stevanović AM (2021) Monitoring as the key factor for sustainable use and protection of groundwater in karst environments—An overview. Sustainability (Switzerland) 13(10). https://doi.org/10.3390/su13105468
Tarekegn N et al (2022) Modeling the impact of climate change on the hydrology of Andasa watershed. Model Earth Syst Environ 8(1):103–119. https://doi.org/10.1007/s40808-020-01063-7
Taylor R, Aureli A (2022) Unesco-Ihd. Hydrol Sci Bull 17(1):39–40. https://doi.org/10.1080/02626667209493798

Trinchieri I, Tridello A, ME E (2023) Politecnico di Torino Politecnico di Torino. Thesis (October):87316161

Utermann J et al (2019) Heavy metal stocks and concentrations in forest soils. https://doi.org/10.1007/978-3-030-15734-0_7

Verbrugge B, Lanzano C, Libassi M (2021) The cyanide revolution: efficiency gains and exclusion in artisanal- and small-scale gold mining. Geoforum 126:267–276. https://doi.org/10.1016/j.geoforum.2021.07.030

Version A (2021) Anthropogenic intensification of short-duration rainfall extremes

Wang J et al (2021) A review of biocyanidation as a sustainable route for gold recovery from primary and secondary low-grade resources. J Clean Prod 296. https://doi.org/10.1016/j.jclepro.2021.126457

Wang X, Liu L (2023) The impacts of climate change on the hydrological cycle and water resource management. Water 15(13):2342. https://doi.org/10.3390/w15132342

Wang L et al (2024) Integrating portable X-ray fluorescence site survey and ArcGIS models for rapid risk assessment and mitigation strategies at an abandoned arsenic mine site: a case study. Environ Technol. https://doi.org/10.1080/09593330.2024.2354121

Wellmer FW (2022) What we have learned from the past and how we should look forward, Mineral Economics. Springer, Berlin, Heidelberg. https://doi.org/10.1007/s13563-021-00296-x

Worlanyo AS, Jiangfeng L (2021) Evaluating the environmental and economic impact of mining for post-mined land restoration and land-use: a review. J Environ Manag 279:111623. https://doi.org/10.1016/j.jenvman.2020.111623

Wu WY et al (2020) Divergent effects of climate change on future groundwater availability in key mid-latitude aquifers. Nat Commun 11(1):1–9. https://doi.org/10.1038/s41467-020-17581-y

Wuijts S et al (2021) Protection of drinking water resources from agricultural pressures: effectiveness of EU regulations in the context of local realities. J Environ Manag 287:112270. https://doi.org/10.1016/j.jenvman.2021.112270

Yamfwa O et al (2024) Problem of Artisanal Gold Mining in the Mining Area of Mangi, Tshopo Province in the Democratic Republic of Congo. Int J Prog Sci Technol 43(1):hal-04464741

Yankson PWK, Gough KV (2019) Gold in Ghana: the effects of changes in large-scale mining on artisanal and small-scale mining (ASM). Extr Ind Soc 6(1):120–128. https://doi.org/10.1016/j.exis.2018.09.009

Zhang P et al (2023) Water quality degradation due to heavy metal contamination: health impacts and eco-friendly approaches for heavy metal remediation. Toxics 11(10):828. https://doi.org/10.3390/toxics11100828

Zikargae MH, Woldearegay AG, Skjerdal T (2022) Empowering rural society through non-formal environmental education: an empirical study of environment and forest development community projects in Ethiopia. Heliyon 8(3):e09127. https://doi.org/10.1016/j.heliyon.2022.e09127

Zwarteveen M et al (2021) Transformations to groundwater sustainability: from individuals and pumps to communities and aquifers. Curr Opin Environ Sustain 49:88–97. https://doi.org/10.1016/j.cosust.2021.03.004

Groundwater and Public Health: Understanding the Links and Mitigating Risks

Xiaodong He, Zhiwen Zheng, and Han Lu

Abstract As one of the most valuable natural resources, groundwater is the vital water source in ensuring global water security, supporting agricultural production, and safeguarding human health. However, influenced by climate change and human activities, groundwater quality has increasingly deteriorated, posing significant threats to public health. This paper aims to explore the links between groundwater and public health and propose effective strategies to mitigate the associated risks. Geogenic inferior groundwater, such as high-arsenic and high-fluoride groundwater has long affected human health, leading to arseniasis and fluorosis. Human activities, such as industrial discharges, the excessive use of pesticides and fertilizers, and the discharge of untreated sewage, have resulted in excessive concentrations of heavy metals, organic pollutants, and microorganisms in groundwater, contributing to various health issues such as neurological diseases and cancer. Furthermore, emerging contaminants represented by microplastics and per- and polyfluoroalkyl substances (PFAS) continue to be identified. Safeguarding public health, which depends on groundwater, requires joint efforts in accurate sources identification and control, exposure reduction, rehabilitation, and toxicological assessment. Comprehending the links between groundwater quality and public health is essential for mitigating these risks.

Keywords Groundwater quality · Human health · Groundwater contamination · Waterborne diseases · Health risk

X. He (✉) · Z. Zheng · H. Lu
School of Water and Environment, Chang'an University, No. 126 Yanta Road, Xi'an 710054, Shaanxi, China
e-mail: hexiaod3@163.com

Key Laboratory of Subsurface Hydrology and Ecological Effects in Arid Region of the Ministry of Education, Chang'an University, No. 126 Yanta Road, Xi'an 710054, Shaanxi, China

Key Laboratory of Eco-Hydrology and Water Security in Arid and Semi-Arid Regions of the Ministry of Water Resources, Chang'an University, No. 126 Yanta Road, Xi'an 710054, Shaanxi, China

P. Li et al. (eds.), *Sustainable Groundwater and Environment: Challenges and Solutions*, Springer Hydrogeology, https://doi.org/10.1007/978-3-031-82194-3_4

1 Introduction

Water is one of the essential elements for human survival. Groundwater, serves as the primary source of freshwater for human use, accounting for around 95% of available freshwater (Wang et al. 2021). It plays a key role in maintaining human health, supporting economic development, and sustaining ecosystem, peculiarly in arid/semi-arid areas (Xie et al. 2023). In China, for instance, more than 61% of the 657 cities rely on groundwater as their major water supply, while in the arid/semi-arid regions of northern China, the proportion of groundwater in supporting industrial, agricultural, and domestic use reaches 50%, 33% and 65%, respectively (Wang et al. 2018). However, groundwater quality deterioration due to both geological conditions, climate change, and human activities has consistently posed challenges to its utilization. Long-term consumption of contaminated groundwater has negatively impacted human health.

Geogenic inferior groundwater, characterized by naturally occurring contaminants, is widely distributed in the world (Wang et al. 2021). It refers to inferior groundwater caused by natural geological processes, often linked to the chemical composition of groundwater, geological structures, rock and sediment properties, and groundwater circulation (Fig. 1). The presence of arsenic (As), fluoride (F), manganese (Mn), and iron (Fe), along with other inorganic contaminants in groundwater, is typically related to the chemical compositions of the rocks and sediments that groundwater flows through. These chemical compositions may be released into the groundwater system under specific geological conditions, such as changes in the redox environment. Since groundwater flows through the subsurface, it undergoes various physical, chemical, and biological reactions with surrounding rocks and soils, which lead to informing poor groundwater. As is one of the most common contaminants affecting water suitability for drinking. Long-term exposure to As, mainly through drinking water, can lead to high prevalence of skin lesions, and even induce cancer (Selinus et al. 2005; He et al. 2020). High-As groundwater is the leading source of As exposure, causing significant harm to public health in many regions worldwide (Ravenscroft 2007; Wang et al. 2021). The fight against geogenic inferior groundwater, particularly high-As groundwater, continues.

In addition to natural factors, human activities significantly contribute to groundwater contamination. In this study, groundwater pollution refers to the quality deterioration caused by human activities. Once contaminated, groundwater takes an extended time to recover, even if the pollution source is entirely removed, and complete restoration may be unattainable. The primary sources of groundwater contamination include industrial sources (e.g., the discharge and accumulation of wastewater, waste gas, and solid waste), agricultural sources including the application of pesticides and fertilizers, and domestic sources such as leaks from municipal sewage systems and landfills (Fig. 1). Notably, the excessive use of nitrogen fertilizers and pesticides has significantly increased nitrate levels in groundwater across nearly all agricultural areas worldwide (Gu et al. 2013; Alam et al. 2024).

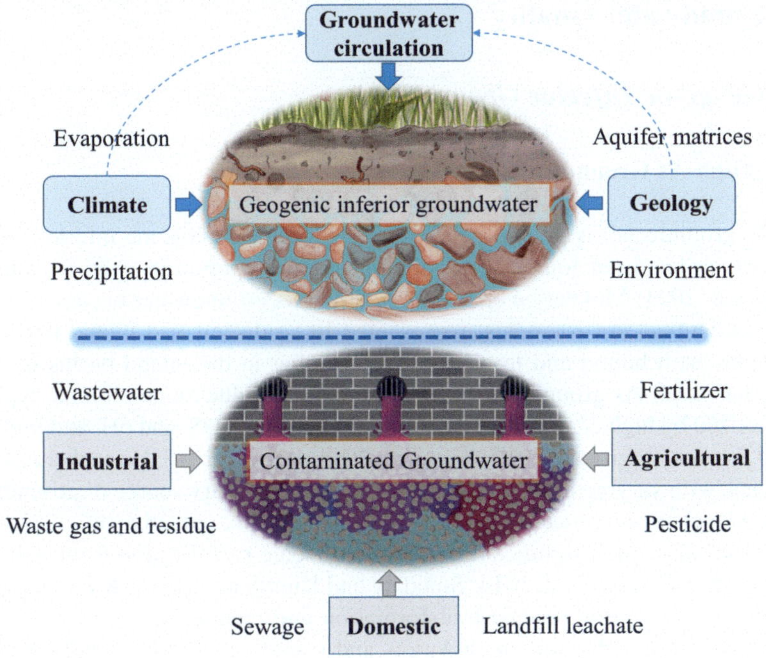

Fig. 1 Types of contaminated groundwater and influencing factors

Organic contaminants, heavy metals, and pathogenic microorganisms from industrial activities and domestic wastewater discharge continue to pose serious threats to groundwater quality. In recent years, emerging contaminants such as antibiotics, endocrine disruptors, microplastics, and per- and polyfluoroalkyl substances (PFAS) have drawn increasing attention. Although the levels of these emerging contaminants in groundwater are usually low, their pollution pathways are often concealed, their types are complex, and their effects are delayed. Therefore, the biotoxicity, environmental persistence, and bioaccumulation of emerging contaminants in groundwater, along with their potential impacts on public health, demand urgent and focused investigation (Xie et al. 2023).

Drinking contaminated water has serious health implications (Karunanidhi et al. 2019). Inferior groundwater has contributed to endemic diseases and exacerbated water scarcity issues. Groundwater contamination from human activities continuously threatens both water security and public health. Therefore, it is vital to better understand the link between groundwater quality and public health, and mitigate health risks during groundwater development and utilization. Effective groundwater management requires a comprehensive understanding of the key contaminants and their exposure routes to humans, along with the development of targeted management tools and policies to safeguard groundwater quality.

2 Groundwater Quality

2.1 Geogenic Inferior Groundwater

2.1.1 High-As Groundwater

High-As groundwater is the most representative form of geogenic inferior groundwater, which has been found on all continents and reported in over 70 countries (Wang et al. 2021; Mukherjee et al. 2024). High-As groundwater occurs in diverse hydrogeochemical environments, represented by arid/semi-arid inland basins and river deltas with humid and tropical climate. Either in the inland basins or in the river deltas, high-As groundwater usually occurs in reducing conditions, typically characterized by high concentrations of Fe, Mn, HCO_3^-, TDS, and pH, and low NO_3^- and SO_4^{2-} levels (Wen et al. 2013; Guo et al. 2014; Jia et al. 2018; He et al. 2020). The sediments, with varying arsenic content, serve as the primary source of groundwater arsenic. Geochemical processes such as sulfide oxidation, reductive dissolution, and competitive adsorption significantly influence arsenic mobilization from sediments into groundwater (Guo et al. 2014; Smedley and Kinniburgh 2002). Microorganisms also play a vital role in arsenic transformation and release, acting as catalysts in arsenite oxidation, reduction, methylation, and volatilization. Microbially mediated Fe reduction and oxidation may control the seasonal mobilization and scavenging of As in groundwater (Wang et al. 2021).

2.1.2 High-F Groundwater

High-F groundwater is also widely distributed in the world. F^- concentrations in groundwater are controlled by various geological, meteorological, and hydrogeological conditions (Li et al. 2019; He et al. 2020). High-F groundwater has been generally found in shallow groundwater, deep groundwater, and geothermal water (Wen et al. 2013; Wang et al. 2018). Groundwater enriches F from water–rock interactions of F-bearing minerals and further accumulate through evapotranspiration occurring in the discharge areas, forming high-F groundwater (Wen et al. 2013). Factors such as arid climate, ion exchange, and slow groundwater flow contribute to F enrichment in shallow groundwater. Deep high-fluoride groundwater is often associated with palaeo-seawater and -brine (Currell et al. 2011; Wen et al. 2013). High-F geothermal water originates from intense water-rock interactions under high temperature. Furthermore, water-rock interactions, evaporation and concentration, mixing of groundwater with geothermal fluids, and ion exchange in aquifers are four primary genetic mechanisms of high-F groundwater (Subba Rao 2017; Jia et al. 2018; He et al. 2021; Wang et al. 2021).

2.1.3 Other Geogenic Inferior Groundwater

High concentrations of Fe and Mn are commonly found together in groundwater due to their similar chemical properties and their co-occurrence in minerals. In China, the high concentration of Fe and Mn is the main cause of groundwater exceeding the standard. The piedmont areas around basins or plains are primary sources of Fe and Mn, where weathering of rocks releases these elements, which are then transported to lower-lying areas, enriching the sediments with Fe and Mn minerals (Jia et al. 2018). Since Fe and Mn is abundant in sediments, pH, redox potential (Eh), dissolved oxygen (DO), dissolved organic carbon (DOC), and microbial activities are main controlling factors (Jia et al. 2018; Xu and Li 2024).

Geogenic high-iodine groundwater has been reported in sedimentary and coastal plains. High-iodine groundwater generally was found in discharge areas with slow groundwater flow (Zhang et al. 2013). High-iodine groundwater exhibited lower SO_4^{2-} levels, indicating that a reducing environment can promote iodine enrichment in the aquifer (He et al. 2021). Additionally, high iodine levels were observed occurred in shallow groundwater with high TDS, suggesting that evaporation significantly contributes to iodine enrichment in these areas. The primary factors controlling iodine mobilization include the dissolution of iodine-containing minerals, shallow groundwater evaporation, and the release of iodine from sediments in environments rich in organic matter reduction (Li et al. 2014). Biogeochemical processes largely regulate iodine mobilization, with organic carbon degradation, reductive dissolution of iron oxides/oxyhydroxides, and microorganism-mediated organic carbon decomposition playing significant roles in iodine enrichment (Amachi et al. 2005; Li et al. 2013).

2.2 *Groundwater Contamination*

2.2.1 Nitrate

Nitrate pollution in groundwater is a global problem. Nitrogen is a common inorganic chemical in the environment, with nitrate being the most mobile form and one of the most prevalent pollutants in aquifers (Gu et al. 2013). Nitrate pollution is particularly problematic in regions with intensive agriculture (Liu et al. 2019). In North and Northeast China, approximately 7400 ha of cropland are contaminated by excessive nitrate levels (Ma et al. 2018). Fertilizers, domestic sewage, livestock manure, and other pollutants enter groundwater through water-driven processes, such as irrigation and precipitation, infiltrating through the vadose zone and increasing nitrate levels in groundwater (Wang et al. 2023). Nitrate contamination is also a growing concern in urban regions with more nitrogen sources than agricultural regions, including solid waste, wastewater, and atmospheric deposition (Tang et al. 2004; Gu et al. 2013). Biogeochemical processes, including nitrification, denitrification, and dissimilatory reduction of nitrate to ammonium (DNRA) control the migration and transformation of nitrate (Wang et al. 2024). Long-term ingestion of nitrate-contaminated groundwater can affect human health. Nitrate can be reduced to nitrite in the intestine, inducing blue baby syndrome (He et al. 2019).

2.2.2 Heavy Metals

Heavy metal contamination is a significant concern in groundwater due to its persistent nature and inability to degrade through microbial processes. Instead, heavy metals can be transformed into more toxic metal–organic compounds through bioaccumulation, thereby posing greater risks to ecosystems and public health. Common heavy metals found in contaminated groundwater include cadmium (Cd), lead (Pb), mercury (Hg), chromium (Cr), and As, which come from both natural and anthropogenic sources. Anthropogenic activities are most main pollution sources include those associated with industrial activities such as mining, smelting, chemical production, and the manufacturing of pigments, pharmaceuticals, textiles, and leather tanning agents (Zamora-Ledezma et al. 2021). These pollutants enter groundwater through wastewater discharge, runoff, and rainfall infiltration.

2.2.3 Persistent Organic Contaminants

Persistent organic contaminants, such as pesticides, pharmaceuticals, and non-aqueous phase liquids (NAPLs), are widespread in agricultural, urban, and industrial settings. A wide range of organic compounds can be released into the environment, contaminating groundwater through their residues or derivatives. Sources of contamination include leaking sewage pipes, underground gas station tanks, and landfill leachate, all of which contribute to significant groundwater pollution. These contaminants may exist in groundwater as pure compounds or mixtures, either dissolved in water or as separate phases (Bhatt et al. 2007). The concentrations and decay rates of many persistent organic contaminants are influenced by processes such as degradation, dilution, and compound-specific adsorption properties. Additionally, some degradation products are more toxic than their parent compounds (Sinclair and Boxall 2003; Schreglmann et al. 2013).

2.2.4 Pathogenic Microorganisms

The primary sources of pathogenic microorganisms in groundwater are human and animal feces (Dong et al. 2024). The main pathogens are Norovirus, Campylobacter, Shigella, Hepatitis A and Giardia. The main factors contributing pathogen intrusion into groundwater included septic system, municipal sewage, heavy rainfall, poor well construction, inadequate treatment, surface-water intrusion, and agricultural activities (Murphy et al. 2017). Pathogenic microorganisms enter groundwater when inadequately protected groundwater sources or wells in aquifers are contaminated by surface water or sewage. This issue is exacerbated by extreme weather events, where heavy rainfall and runoff can cause sewer overflows, basement backups, and localized flooding. Mismanaged agriculture is another significant non-point source of groundwater pathogens. This includes direct fecal deposition on agricultural fields by wild animals and grazing livestock, as well as runoff and leachate from feedlots, animal-rearing facilities, manure storage systems, and livestock enterprises (Dong et al. 2024).

2.2.5 Emerging Contaminants

Emerging contaminants are toxic and harmful chemicals characterized by biotoxicity, environmental persistence, and bioaccumulation, posing significant risks to both the ecological environment and human health. However, they are either not regulated or are inadequately addressed by existing management measures. Emerging contaminants in groundwater include pharmaceuticals, personal care products, endocrine disruptors, food additives, flame retardants, perfluorinated compounds, microplastics, and nanoparticles (Wang et al. 2023). These contaminants are primarily found in unconfined aquifers at concentrations ranging from several to several hundred ng/L. They can be introduced into groundwater from both point and nonpoint sources aboveground or through buried sources. Research on emerging contaminants in groundwater remains limited, and further understanding of their transport, transformation mechanisms, and toxic effects is needed. Taking microplastics as an example, the migration of microplastics in soil and groundwater is influenced by various chemical, physical, and biological factors, with different mechanisms affecting their behavior (Sangkham et al. 2023). In general, the mobility of microplastics decreases with increasing ionic strength, ion concentration, Fe/Al oxide content, cation release on the medium surface, and medium roughness. Conversely, their mobility increases with higher pH levels, concentrations of dissolved organic matter, pore water velocity, medium particle size, and media saturation. Additionally, the growth and development of plants, along with the activity of small animals and microorganisms, also influence the migration of microplastics in soil and groundwater (Dong et al. 2020; Zhao et al. 2023).

3 Link of Groundwater and Public Health

3.1 Exposure Pathways

Groundwater health exposure refers to the process by which pollutants in groundwater enter the human body through various pathways. These pathways include the following: (1) Direct ingestion of contaminated groundwater, (2) Inhalation of airborne pollutants, (3) Skin contact (He et al. 2021). Drinking polluted groundwater is a major route of exposure. Contaminants such as heavy metals can directly enter the human body through consumption, posing significant health risks. Contaminated groundwater can volatilize into the air through soil, rocks or other media, leading to the formation of gaseous pollutants. When individuals are outdoors, they may inhale in these pollutants, potentially affecting their health. In indoor environments, such as homes, schools, or offices, contaminated groundwater used for activities like humidification, bathing, or cleaning can release harmful substances into the air. Prolonged indoor exposure increases the risk of inhaling these pollutants. When contaminated groundwater is used for activities such as bathing, swimming, or washing clothes, harmful substances may be absorbed through the skin (Selinus et al. 2005). Although the skin has limited capacity to absorb pollutants, prolonged exposure can lead to adverse health effects.

3.2 *Geogenic Inferior Groundwater and Health*

3.2.1 As and Arseniasis

Long-term exposure to As is associated with various cancers, including skin, lung, bladder, kidney, and liver. As can cause chromosomal aberrations, alter DNA methylation and repair mechanisms, induce cell proliferation, transform cells, and promote tumorigenesis (Selinus et al. 2005). As-related skin damage is often characterized by raindrop-shaped pigmentation loss, typically accompanied by hyperkeratosis on the palms and soles. Studies have also found a strong association between long-term exposure to As in drinking water and increased risks of hypertension and diabetes (Selinus et al. 2005; He et al. 2020). Asymptomatic microcirculatory disturbances and arterial dysfunction have also been observed in individuals exposed to arsenic. In Thailand, arsenic exposure has been linked to lower IQs in children. In China, Blackfoot disease, caused by As exposure, was reported in Taiwan in the 1950s, with researchers identifying As contaminated artesian well water as the source (Cheng et al. 2016). The signs and symptoms of As toxicity vary widely among individuals, populations, and geographic regions. Long-term consumption groundwater with high As concentrations can lead to arsenicosis. In China, the first case of arsenicosis caused by drinking water was reported in Kuitun, Xinjiang Autonomous Region, in 1980. Severe outbreaks of arsenicosis occurred in the Inner Mongolia Autonomous Region in 1989, and the Chinese government declared it an endemic disease in 1992 (Sun 2004; He et al. 2020). Subsequently, endemic arsenicosis has been identified in other provinces.

3.2.2 F and Fluorosis

F is an essential element in the human diet, and F deficiency associated with an increased risk of dental caries. However, while F is essential for human health, its optimal intake falls within a narrow range. Excessive F intake over time lead to dental fluorosis and, in severe cases, skeletal fluorosis (Li et al. 2019). Fluorosis is categorized into drinking water fluorosis, coal-burning fluorosis, and tea-drinking fluorosis (He et al. 2020). Coal-burning fluorosis is a distinct issue in China, especially in the north regions where residents use high high-fluoride coal for cooking and heating during the winter months (Dai et al. 2007). Tea-drinking fluorosis is common in areas with high concentrations of ethnic minorities, where residents traditionally consume brick tea with high fluoride levels (Selinus et al. 2005). In China, fluorosis has been reported in 29 provinces, with the exception of Shanghai. Long-term drinking high-F groundwater is a primary factor. In Shanghai, where nearly all drinking water supply is sourced from surface water.

3.2.3 Others

Fe, while not considered harmful to human health by drinking water, is regarded as an objectionable impurity in water supplies. In contrast, high Mn concentrations in drinking water have been gained increasing attention in recent years due to its potential neurotoxic effects and its association with impaired intellectual performance in children (Jia et al. 2018). Iodine is essential for the synthesis of thyroid hormones, and its deficiency can lead to iodine deficiency disorders. However, excessive iodine intake can inhibit thyroid functions and contribute to thyroid autoimmunity, such as Hashimoto's thyroiditis (Li et al. 2013). The recommended daily iodine intake for adults is 150 μg (Tang et al. 2013). Consider the health benefits, the presence of key elements like Na, Ca, and Mg in groundwater is also required optimum ranges (Fig. 2). High Na concentrations in groundwater may significantly increase daily Na intake of residents who live in high-salinity regions such as coastal areas, elevating the risk of hypertension and cardiovascular diseases. Excessive intake of Ca and Mg can contribute to the development of coronary artery disease, cerebrovascular diseases, and stone disease (Wang et al. 2020).

3.3 Contaminated Groundwater and Health

3.3.1 Heavy Metals and Health

Metal pollution in groundwater poses a significant environmental threat due to its potential harm to humans, even at low concentrations (Fig. 2, Sarath et al. 2023). Many metals are classified as toxic or carcinogenic due to they can generate reactive oxygen species and induce oxidative stress, resulting in various diseases (Ali et al. 2019). Prolonged or acute exposure increases the production of reactive oxygen species, potentially induce multiple organs (e.g., lungs, kidneys, liver, stomach, and skin) damage (Zamora-Ledezma et al. 2021; Xie et al. 2023). Among the most

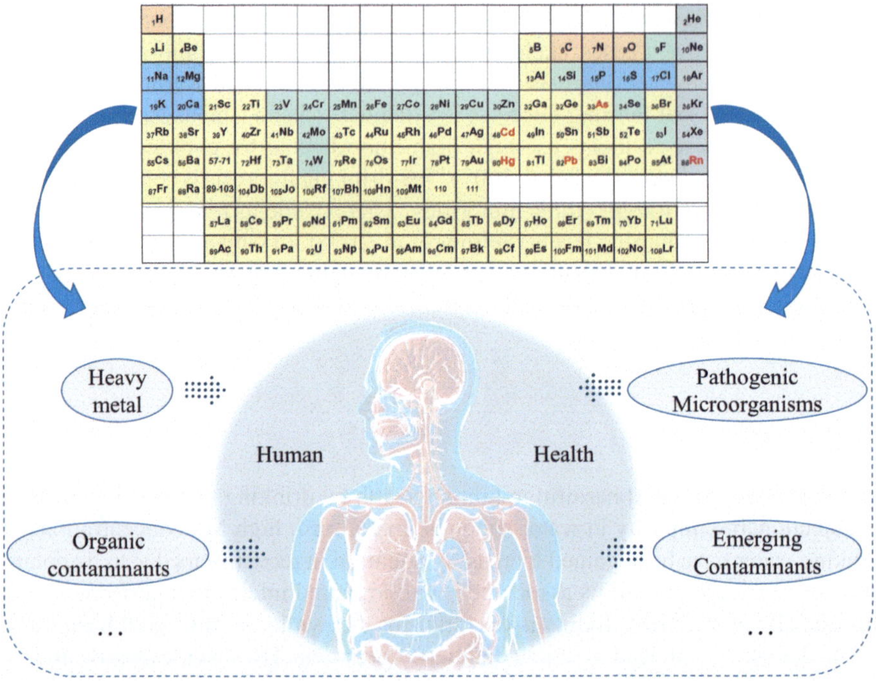

Fig. 2 Health hazards of key contaminants in groundwater

harmful metals found in contaminated groundwater are Cd, Pb, and Hg. High exposure risk to Cd may lead to Cd accumulation in the bones, liver, kidneys, and nervous tissues, increasing cancers risk of the prostate, lung, and liver (Tsutsumi et al. 2014; Wu et al. 2016; Xie et al. 2023). Chronic exposure to Pb such as digestive and respiratory systems elevates the health risk of nervous system, digestive system, hematopoietic system, and kidneys (Koh et al. 2014). Organic Hg exposure can cause mental retardation, deafness, cerebral palsy, and other disorders, particularly in children exposed in utero. Inorganic Hg exposure can lead to renal tubular dysfunction and neuropsychiatric disorders (Wu et al. 2016). Significantly elevated levels of Cr, Ni, Cb, and Pb have been reported in the blood of occupational welders chronically exposed to welding fumes (Mahmood et al. 2015). In areas contaminated with heavy metals, endemic diseases are prevalent. For example, in a town in the Buldana district, India, contaminated with Pb and Cd, 23 deaths from renal failure and 18 cases of associated symptoms, including anorexia and hypertension, were reported (Bawaskar et al. 2010).

3.3.2 Persistent Organic Pollutants and Health

Persistent organic pollutants, such as organochlorine pesticides, polycyclic aromatic hydrocarbons (PAHs), and polychlorinated biphenyls (PCBs), are highly toxic (Alharbi et al. 2018). These pollutants can enter the body indirectly and accumulate through the ingestion of polluted groundwater and food produced using contaminated groundwater. The accumulation and metabolism of these pollutants can induce various biological toxicities, influencing the normal development of children and inducing the nervous system, reproductive system, and even cancer hazards (Mrema et al. 2013). Vafeiadi et al. (2014) included 1117 mothers and their newborns, found that prenatal exposure to persistent organic pollutants (PCBs and hexachlorobenzene) negatively affects fetal growth. Moreover, these toxic substances can cross the blood–brain barrier, leading to nervous system damage, which manifests as reduced learning ability, cognitive impairment, and motor dysfunction. Persistent organic pollutants also weaken the immune system by damaging immune cells, producing reactive oxygen species, and through other mechanisms (Fig. 2).

3.3.3 Pathogenic Microorganisms and Health

The widespread presence of pathogenic microorganisms in groundwater poses a significant threat to public health (Fig. 2). Direct ingestion of untreated groundwater can lead to illness caused by these pathogens. Murphy et al. (2017) approximated that 35.2–59.4 million cases of acute gastroenteritis per year worldwide could be attributable to ingestion of the contaminated groundwater. Historically, groundwater contaminated with bacteria and viruses has accounted for around 50% of all waterborne diseases in the USA (Nwadike et al. 2024). Gastroenteritis is the most prevalent illness induced by bacterial pathogens (Pandey et al. 2014). The most common pathogenic agents found in groundwater include pathogenic bacteria, protozoa, and viruses, which can lead to outbreaks of acute gastrointestinal illnesses as well as sporadic cases (Burch et al. 2022; Xie et al. 2023; Dong et al. 2024). Viral contaminants pose the greatest threat to groundwater quality (Hu et al. 2020). They have been confirmed as the second leading reason of death in children aged under five years (Efstratiou et al. 2017). Compared to bacteria, viruses are smaller, more diverse, more persistent, and more resistant to disinfection methods (Charles et al. 2009). Common viral contaminants in groundwater include rotavirus, enterovirus, hepatitis A virus, and enveloped coronaviruses. Ingesting contaminated groundwater can cause acute gastroenteritis, meningitis, conjunctivitis, hand, foot, and mouth disease, as well as impaired liver function.

3.3.4 *Emerging Contaminants and Health*

Emerging contaminants enter the human body by the natural cycle and pose significant risks to human health (Van et al. 2021). Numerous studies have identified endocrine disruption, neurotoxicity, cardiotoxicity, immunotoxicity, hepatotoxicity, etc., as effects of exposure to these pollutants (Rodriguez-Narvaez et al. 2017; Zhang et al. 2018; Lempart et al. 2020; Xie et al. 2023). For example, perfluorinated alkylated substances (PFAS) in groundwater can cause nail disease, liver damage, various forms of cancer, and impair fetal development, while endocrine-disrupting chemicals (EDCs) can interfere with the endocrine system, leading to birth defects and developmental delays in fetuses (Liu et al. 2022). Animal test results show that after entering the body, microplastics can accumulate in the heart, liver, spleen, lung, kidney and testicles and other organs through the blood circulation, and produce a variety of biological toxic effects, such as inflammation, oxidative stress, immune damage, metabolic disorders (Fig. 2).

4 Mitigating Risks

4.1 Health Risk Assessment

Health risk assessment is a method utilized to evaluate the harmful effects of toxic compositions on human health, primarily using risk levels as the evaluation metric. This approach provides a foundation for decision-making in groundwater environmental management. Health risk assessment process of groundwater is a complex and systematic procedure aimed at assessing the potential impact in groundwater contaminants on human health. It typically includes the following key steps (Fig. 3): (1) Hazard identification: Pollutants are evaluated based on biological and chemical data to determine whether they cause harm, and target pollutants are selected. The scope and affected receptors are analyzed, and the potential adverse effects, harm types, and severity of these pollutants are assessed. (2) Dose-effect analysis: This step assesses the potential health effects of contaminants at specific exposure levels, considering both acute and chronic effects. (3) Exposure assessment: This step involves evaluating the exposure of the population to pollutants, including exposure duration, frequency, pathways, and total exposure calculation. (4) Risk characterization: Based on the consequences of the dose-effect analysis and exposure assessment, the potential health risks from groundwater pollution are calculated, including both carcinogenic and non-carcinogenic risks (USEPA 2011). In the above deterministic health risk assessment process, exposure assessment parameters typically adopt certain values. However, these values should be adjusted based on the specific event and the population's unique characteristics, which introduces some uncertainty. In order to address this uncertainty, Monte Carlo simulation is utilized to validate the risk assessment results (Gao et al. 2022; Du et al. 2024). Groundwater health risk assessment relies on multidisciplinary support from monitoring, chemical analysis, environmental science, toxicology, and epidemiology. This process enables a comprehensive understanding of the potential impact of groundwater contamination on public health, and guides the development of effective risk control and management measures to safeguard drinking water safety.

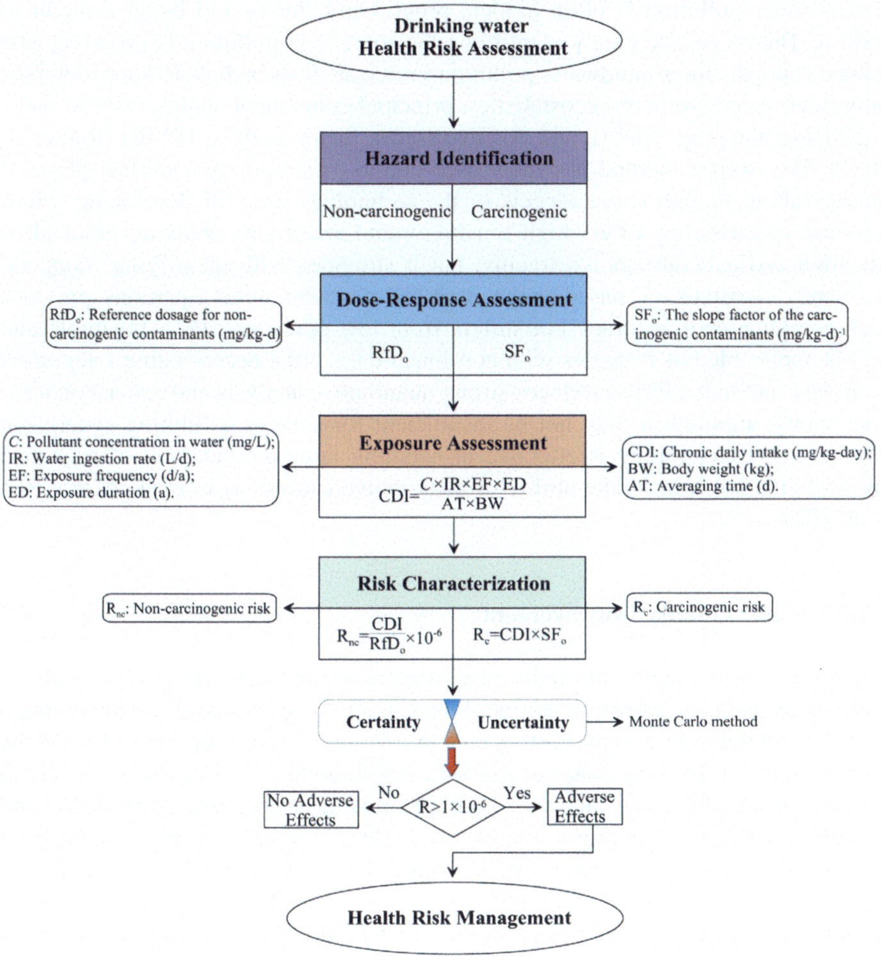

Fig. 3 Procedures of health risk assessment

4.2 *Groundwater Management*

4.2.1 Pollution Source Identification

Groundwater pollution is often hidden, progressing slowly and being difficult to control. Therefore, accurately identifying the sources of pollution is essential. The main techniques for groundwater pollution source analysis include isotope methods, fluorescence spectrometry, geostatistics, principal component analysis (PCA), self-organizing mapping (SOM), and positive matrix factor analysis (PMF) (Ran et al. 2024). The isotope method is widely used due to its simplicity and clear physical interpretation, though some aspects of the technology are still developing. Fluorescence spectroscopy offers high sensitivity and selectivity, enabling rapid identification and accurate source tracing, but it struggles with identifying inorganic pollutants. Geostatistics, based on the inversion of mathematical equations, provides high computational accuracy but suffers from low efficiency. PCA is simple and highly applicable but struggles with non-linear data, often necessitating integration with other methods. PMF produces strong quantitative analysis and can account for uncertainty, although it may not be insufficient for sources exhibiting synergistic effects. SOM technology is effective in handling large datasets and offers broad applicability, but it lacks the ability of quantitative analysis (Liu et al. 2024a; Ran et al. 2024).

4.2.2 Water Quality Improvement

The most effective method to reduce exposure to groundwater contamination is by eliminating the contamination source. Water quality improvement projects remain a fundamental measure to preventing endemic diseases from drinking water. As the identification of drinking water arsenicosis and fluorosis, the Chinese government has prioritized addressing this issue by implementing measures to mitigate As and F^- concentrations in drinking water (He et al. 2020). A study of fluoride's effects on the safe water supply was conducted by Wang et al. (2012), involving 81,786 children who aged 8–12 and 594,698 adults aged over 16 in China. The study revealed a significantly higher link of fluorosis in areas with high-F drinking water compared to areas with safe water supply. Figure 4 depicts the change in the number of residents benefiting from water quality improvement project and the patient number of arsenicosis and fluorosis from 2002 to 2021 in China. The water improvement projects have substantially reduced the patient number of arsenicosis and fluorosis (Fig. 4). With the increase of the number of people benefiting from water improvement projects, the number of arsenicosis (Fig. 4a) and fluorosis (Fig. 4b) patients showed an obvious decreasing trend. While the water supply projects successfully mitigated exposure pathways, the remaining challenge is how to remediate contaminated groundwater.

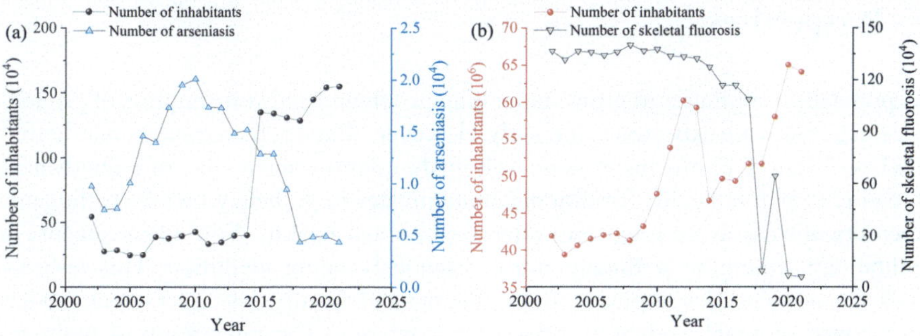

Fig. 4 Number of arsenicosis **a** and skeletal fluorosis, **b** versus inhabitants with water quality improvement

4.2.3 Contamination Remediation

Contamination remediation is a reactive approach to addressing contaminated groundwater. The primary remediation technologies include physical (e.g., pump and treatment, multi-phase extraction, circulating wells), chemical (e.g., in situ chemical oxidation/reduction, permeable reactive barriers), biological (e.g., microbial bioremediation, phytoremediation), and combined remediation technologies (Chen et al. 2024). Pump and treatment method involves pumping contaminated groundwater to the surface, purifying it, and then either injecting back it into the ground or discharging it. This method is suitable for the treatment of dissolved contaminates, such as nitrate and heavy metals. Circulating wells can promote the dissolution and transport of contaminants by creating a three-dimensional circulation pattern, enabling in situ remediation through techniques such as air injection and soil gas extraction. In situ chemical oxidation involves injecting oxidizing agents into groundwater, where pollutants are transformed into non-toxic or less toxic substances via oxidation. Common oxidizing agents include persulfate, hydrogen peroxide, potassium permanganate, and ozone. Contaminants can also be reduced through reactions with reducing agents such as zero-valent iron, polysulfide, and sodium disulfite. Permeable reactive barrier is placed along groundwater flow paths to remove pollutants through adsorption, precipitation, chemical degradation, or biodegradation by reactive materials. Microbial bioremediation typically is used to remove organic pollutants in groundwater, including in-situ bioremediation and bioreactors (Muhammad et al. 2022; Liu et al. 2024a). Phytoremediation technology employs plants to extract, filter, volatilize, or stabilize contaminants within the soil (Zhao et al. 2024). Since single remediation technology often fail to fully address groundwater pollution, combined remediation technologies integrate the strengths of multiple techniques to enhance efficiency and effectiveness. In recent years, novel restorative materials such as nanometer zero-valent iron (nZVI) have been used in the remediation of varied groundwater pollutants, including chlorinated organic contaminates, nitroaromatic compounds, nitrates, hexavalent chromium, arsenic, and heavy metals, due to their effectiveness, low cost, and ease of synthesis (Liu et al. 2024b).

5 Perspectives

As population continues to grow, future human demand and consumption of groundwater resources are expected to intensify. However, climate change and human activities have exerted significant pressure on both the quantity and quality of groundwater, threating to public health. Traditional contaminates (e.g., heavy metals, pathogenic microorganisms), have long-term effects on human health. With advancements in technology, emerging pollutants are continuously being identified. This necessitates the ongoing development and refinement of health risk assessment models to enhance regional evaluation parameter databases. The integration of hydrogeology and medicine is essential to advancing the field of medical geology, particularly in addressing the impacts of emerging pollutants. In the future, identifying the forms, control factors, migration and transformation, and environmental toxicity of emerging pollutants in groundwater system, remains a challenge for groundwater quality.

Acknowledgements This research is supported by the Postdoctoral Fellowship Program (Grade C) of China Postdoctoral Science Foundation (GZC20241452), the China Postdoctoral Science Foundation (2024M752738), and the Fundamental Research Funds for the Central Universities, CHD (300102294101).

References

Amachi S, Mishima Y, Shinoyama H, Muramatsu Y, Fujii T (2005) Active transport and accumulation of iodide by newly isolated marine bacteria. Appl Environ Microbiol 71:741–745. https://doi.org/10.1128/AEM.71.2.741-745.2005

Alharbi OML, Basheer AA, Khattab RA, Ali I (2018) Health and environmental effects of persistent organic pollutants. J Mol Liq 263:442–453. https://doi.org/10.1016/j.molliq.2018.05.029

Ali H, Khan E, Ilahi I (2019) Environmental chemistry and ecotoxicology of hazardous heavy metals: environmental persistence, toxicity, and bioaccumulation. J Chem-NY 1–14. https://doi.org/10.1155/2019/6730305

Alam SMK, Li P,·Fida M (2024) Groundwater nitrate pollution due to excessive use of N-fertilizers in rural areas of Bangladesh: pollution status, health risk, source contribution, and future impacts. Expos Health 16:159–182. https://doi.org/10.1007/s12403-023-00545-0

Bawaskar HS, Himmatrao Bawaskar P, Himmatrao Bawaskar P (2010) Chronic renal failure associated with heavy metal contamination of drinking water: a clinical report from a small village in Maharashtra. Clin Toxicol 48:768–768. https://doi.org/10.3109/15563650.2010.497763

Bhatt P, Kumar MS, Mudliar S, Chakrabarti T (2007) Biodegradation of chlorinated compounds—A Review. Crit Rev Env Sci Tec 37:165–198. https://doi.org/10.1080/10643380600776130

Burch TR, Stokdyk JP, Rice N, Anderson AC, Walsh JF, Spencer SK, Firnstahl AD, Borchardt MA (2022) Statewide quantitative microbial risk assessment for waterborne viruses, bacteria, and protozoa in public water supply wells in Minnesota. Environ Sci Technol 56:6315–6324. https://doi.org/10.1021/acs.est.1c06472

Charles KJ, Shore J, Sellwood J, Laverick M, Hart A, Pedley S (2009) Assessment of the stability of human viruses and coliphage in groundwater by PCR and infectivity methods. J Appl Microbiol 106:1827–1837. https://doi.org/10.1111/j.1365-2672.2009.04150.x

Chen Y, Zhao X, Li Y, Zhang C, Dang Q, Xi B (2024) Environmental behavior and remediation progress of emerging contaminates in polluted sites in China. Environ Eng 42(1):166–176 (in Chinese with English abstract)

Cheng PS, Weng SF, Chiang CH, Lai FJ (2016) Relationship between arsenic-containing drinking water and skin cancers in the arseniasis endemic areas in Taiwan. J Dermatol 43:181–186. https://doi.org/10.1111/1346-8138.13058

Currell M, Cartwright I, Raveggi M, Han D (2011) Controls on elevated fluoride and arsenic concentrations in groundwater from the Yuncheng Basin, China. Appl Geochem 26:540–552. https://doi.org/10.1016/j.apgeochem.2011.01.012

Dai S, Li W, Tang Y, Zhang Y, Feng P (2007) The sources, pathway, and preventive measures for fuorosis in Zhijin County, Guizhou, China. Appl Geochem 22:1017–1024. https://doi.org/10.1016/j.apgeochem.2007.02.011

Dong S, Xia J, Wang W, Liu H, Sheng L (2020) Review on impact factors and mechanisms of microplastic transport in soil and groundwater. Trans CSAE 36(14):1–8 (in Chinese with English abstract)

Dong Y, Jiang Z, Hu Y, Jiang Y, Tong L, Yu Y, Cheng J, He Y, Shi J, Wang Y (2024) Pathogen contamination of groundwater systems and health risks. Crit Rev Env Sci Technol 54(4):267–289. https://doi.org/10.1080/10643389.2023.2236486

Du Q, Wu J, Xu F, Yang Y, Li F (2024) Occurrence, species, and health effects of groundwater arsenic in typical rural areas along the Northern Foot of the Qinling Mountains, China. Expos Health 16(2):527–541. https://doi.org/10.1007/s12403-023-00576-7

Efstratiou A, Ongerth JE, Karanis P (2017) Waterborne transmission of protozoan parasites: review of worldwide outbreaks—An update 2011–2016. Water Res 114:14–22. https://doi.org/10.1016/j.watres.2017.01.036

Gao Y, Qian H, Zhou Y, Chen J, Wang H, Ren W, Qu W (2022) Cumulative health risk assessment of multiple chemicals in groundwater based on deterministic and Monte Carlo models in a large semiarid basin. J Clean Prod 352:131567. https://doi.org/10.1016/j.jclepro.2022.131567

Gu B, Ge Y, Chang SX, Luo W, Chang J (2013) Nitrate in groundwater of China: sources and driving forces. Glob Environ Chang 23:1112–1121. https://doi.org/10.1016/j.gloenvcha.2013.05.004

Guo H, Wen D, Liu Z, Jia Y, Guo Q (2014) A review of high arsenic groundwater in Mainland and Taiwan, China: distribution, characteristics and geochemical processes. Appl Geochem 41:196–217. https://doi.org/10.1016/j.apgeochem.2013.12.016

He X, Wu J, He S (2019) Hydrochemical characteristics and quality evaluation of groundwater in terms of health risks in Luohe aquifer in Wuqi County of the Chinese Loess Plateau, northwest China. Hum Ecol Risk Assess 25(1–2):32–51. https://doi.org/10.1080/10807039.2018.1531693

He X, Li P, Ji Y, Wang Y, Su Z, Elumalai V (2020) Groundwater arsenic and fluoride and associated arsenicosis and fluorosis in China: occurrence, distribution and management. Expos Health 12(3):355–368. https://doi.org/10.1007/s12403-020-00347-8

He X, Li P, Wu J, Wei M, Ren X, Wang D (2021) Poor groundwater quality and high potential health risks in the Datong Basin, northern China: research from published data. Environ Geochem Health 43:791–812. https://doi.org/10.1007/s10653-020-00520-7

Hu W, Liang J, Ju F, Wang Q, Liu R, Bai Y, Liu H, Qu J (2020) Metagenomics unravels differential microbiome composition and metabolic potential in rapid sand filters purifying surface water versus groundwater. Environ Sci Technol 54:5197–5206. https://doi.org/10.1021/acs.est.9b07143

Jia Y, Xi B, Jiang Y, Guo H, Yang Y, Lian X, Han S (2018) Distribution, formation and human-induced evolution of geogenic contaminated groundwater in China: a review. Sci Total Environ 643:967–993. https://doi.org/10.1016/j.scitotenv.2018.06.201

Karunanidhi D, Aravinthasamy P, Subramani T, Wu J, Srinivasamoorthy K (2019) Potential health risk assessment for fluoride and nitrate contamination in hard rock aquifers of Shanmuganadhi River basin, South India. Hum Ecol Risk Assess 25(1–2):250–270. https://doi.org/10.1080/10807039.2019.1568859

Koh D-H, Bhatti P, Coble JB et al (2014) Calibrating a population-based job-exposure matrix using inspection measurements to estimate historical occupational exposure to lead for a population-based cohort in Shanghai, China. J Expo Sci Environ Epidemiol 24:9–16. https://doi.org/10.1038/jes.2012.86
Lempart A, Kudlek E, Dudziak M (2020) The potential of the organic micropollutants emission from swimming accessories into pool water. Environ Int 136:105442. https://doi.org/10.1016/j.envint.2019.105442
Li J, Wang Y, Xie X, Zhang L, Guo W (2013) Hydrogeochemistry of high iodine groundwater: a case study at the Datong Basin, northern China. Environ Sci: Process Impacts 15:848–859. https://doi.org/10.1039/C3EM30841C
Li J, Wang Y, Guo W, Xie X, Zhang L, Liu Y, Kong S (2014) Iodine mobilization in groundwater system at Datong basin, China: evidence from hydrochemistry and fluorescence characteristics. Sci Total Environ 468–469:738–745. https://doi.org/10.1016/j.scitotenv.2013.08.092
Li P, He X, Li Y, Xiang G (2019) Occurrence and health implication of fluoride in groundwater of loess aquifer in the Chinese Loess Plateau: a case study of Tongchuan, northwest China. Expos Health 11(2):95–107. https://doi.org/10.1007/s12403-018-0278-x
Liu X, Wang Y, Li Y, Liu F, Shen J, Wang J, Xiao R, Wu J (2019) Multi-scaled response of groundwater nitrate contamination to integrated anthropogenic activities in a rapidly urbanizing agricultural catchment. Environ Sci Pollut Res 26:34931–34942. https://doi.org/10.1007/s11356-019-06595-9
Liu M, Zhuan Q, Zhang L, Meng L, Fu X, Hou Y (2022) Polystyrene microplastics induced female reproductive toxicity in mice. J Hazard Mater 424:127629. https://doi.org/10.1016/j.jhazmat.2021.127629
Liu Y, Fei Y, Li Y, Bao X, Zhang P (2024a) Pollution source identification methods and remediation technologies of groundwater: a review. China Geol 7:125−137. https://doi.org/10.31035/cg2022080
Liu Z, Fu J, Liu A, Zhang W (2024b) Influence of natural organic matter on nanoscale zero-valent iron for contaminants removal in water: a critical review. Chem Eng J 488:150836. https://doi.org/10.1016/j.cej.2024.150836
Ma L, Lu J, Zhao H (2018) Nitrate vulnerable zones and strategies of non-point pollution mitigation in China. J Agro-Environ Sci 37(11):2387–2391 (in Chinese with English abstract)
Mahmood Q, Wang J, Pervez A, Meryem SS, Waseem M, Ullah Z (2015) Health risk assessment and oxidative stress in workers exposed to welding fumes. Toxicol Environ Chem 97:634–639. https://doi.org/10.1080/02772248.2015.1059639
Mrema EJ, Rubino FM, Brambilla G, Moretto A, Tsatsakis AM, Colosio C (2013) Persistent organochlorinated pesticides and mechanisms of their toxicity. Toxicology 307:74–88. https://doi.org/10.1016/j.tox.2012.11.015
Muhammad HM, Sharifah SSS, Ummi MB, Zaleha MI (2022) Remediation technology inventions for soil and groundwater contamination. Med Health 17(1):13–30. https://doi.org/10.17576/MH.2022.1701.02
Mukherjee A, Coomar P, Sarkar S et al (2024) Arsenic and other geogenic contaminants in global groundwater. Nat Rev Earth Environ 5:312–328. https://doi.org/10.1038/s43017-024-00519-z
Murphy HM, Prioleau MD, Borchardt MA, Hynds PD (2017) Review: epidemiological evidence of groundwater contribution to global enteric disease, 1948–2015. Hydrogeol J 25:981–1001. https://doi.org/10.1007/s10040-017-1543-y
Nwadike BI, Falodun OI, Ogunjobi AA (2024) Bacterial and viral contaminants in drinking water: why do they really matter to us. In: Environmental pollution and public health. Elsevier, pp 3–28
Pandey PK, Kass PH, Soupir ML, Biswas S, Singh VP (2014) Contamination of water resources by pathogenic bacteria. AMB Expr 4:51. https://doi.org/10.1186/s13568-014-0051-x
Ran Z, Jia Y, Jiang Y, Shi Z, Shang C, Zang Y, Chen F, Lian X (2024) Research progress of groundwater pollution source analysis methods. Geol Bull China 43(1):153−162. https://doi.org/10.12097/gbc.2022.05.052 (in Chinese with English abstract)

Ravenscroft P (2007) Predicting the global extent of arsenic pollution of groundwater and its potential impact on human health. UNICEF Representative, pp 1–35

Rodriguez-Narvaez OM, Peralta-Hernandez JM, Goonetilleke A, Bandala ER (2017) Treatment technologies for emerging contaminants in water: a review. Chem Eng J 323:361–380. https://doi.org/10.1016/j.cej.2017.04.106

Sangkham S, Islam MA, Adhikari S, Kumar R, Sharma P, Sakunkoo P, Bhattacharya P, Tiwari A (2023) Evidence of microplastics in groundwater: a growing risk for human health. Groundw Sustain Dev 23:100981. https://doi.org/10.1016/j.gsd.2023.100981

Sarath KV, Shaji E, Nandakumar V (2023) Characterization of trace and heavy metal concentration in groundwater: a case study from a tropical river basin of southern India. Chemosphere 338:139498. https://doi.org/10.1016/j.chemosphere.2023.139498

Schreglmann K, Hoeche M, Steinbeiss S, Reinnicke S, Elsner M (2013) Carbon and nitrogen isotope analysis of atrazine and desethylatrazine at sub-microgram per liter concentrations in groundwater. Anal Bioanal Chem 405:2857–2867. https://doi.org/10.1007/s00216-012-6616-0

Selinus O, Alloway B, Centeno JA, Finkleman BR, Fuge R, Lindh U, Smedley P (2005) Essentials of medical geology: impacts of the natural environment on public health. Elsevier Academic Press, Amsterdam, pp 230–290

Sinclair CJ, Boxall ABA (2003) Assessing the ecotoxicity of pesticide transformation products. Environ Sci Technol 37:4617–4625. https://doi.org/10.1021/es030038m

Smedley PL, Kinniburgh DZ (2002) A review of the source, behavior and distribution of arsenic in natural waters. Appl Geochem 17:517–568. https://doi.org/10.1016/S0883-2927(02)00018-5

Subba Rao N (2017) Controlling factors of fuoride in groundwater in a part of South India. Arab J Geosci 10:524. https://doi.org/10.1007/s12517-017-3291-7

Sun G (2004) Arsenic contamination and arsenicosis in China. Toxicol Appl Pharm 198(3):268–271. https://doi.org/10.1016/j.taap.2003.10.017

Tang C, Chen J, Shindo S, Sakura Y, Zhang W, Shen Y (2004) Assessment of groundwater contamination by nitrates associated with wastewater irrigation: a case study in Shijiazhuang region, China. Hydrol Process 18:2303–2312. https://doi.org/10.1002/hyp.5531

Tang Q, Xu Q, Zhang F, Huang Y, Liu J, Wang X, Yang Y, Liu X (2013) Geochemistry of iodine-rich groundwater in the Taiyuan Basin of central Shanxi Province, North China. J Geochem Explor 135:117–123. https://doi.org/10.1016/j.gexplo.2012.08.019

Tsutsumi T, Ishihara A, Yamamoto A, Asaji H, Yamakawa S, Tokumura A (2014) The potential protective role of lysophospholipid mediators in nephrotoxicity induced by chronically exposed cadmium. Food Chem Toxicol 65:52–62. https://doi.org/10.1016/j.fct.2013.12.019

USEPA (2011) Exposure factors handbook, 2011 edn. U.S.

Vafeiadi M, Vrijheid M, Fthenou E, Chalkiadaki G, Rantakokko P, Kiviranta H, Kyrtopoulos SA, Chatzi L, Kogevinas M (2014) Persistent organic pollutants exposure during pregnancy, maternal gestational weight gain, and birth outcomes in the mother–child cohort in Crete, Greece (RHEA study). Environ Int 64:116–123. https://doi.org/10.1016/j.envint.2013.12.015

Van D-A, Ngo TH, Huynh TH, Nakada N, Ballesteros F, Tanaka H (2021) Distribution of pharmaceutical and personal care products (PPCPs) in aquatic environment in Hanoi and Metro Manila. Environ Monit Assess 193:847. https://doi.org/10.1007/s10661-021-09622-w

Wu X, Cobbina SJ, Mao G, Xu H, Zhang Z, Yang L (2016) A review of toxicity and mechanisms of individual and mixtures of heavy metals in the environment. Environ Sci Pollut Res 23:8244–8259. https://doi.org/10.1007/s11356-016-6333-x

Wang C, Gao Y, Wang W, Zhao J, Zhang W, Han H, Shi Y, Yu G, Sun D (2012) A national cross-sectional study on effects of fluoride-safe water supply on the prevalence of fluorosis in China. BMJ Open. https://doi.org/10.1136/bmjopen-2012-001564

Wang Y, Zheng C, Ma R (2018) Review: safe and sustainable groundwater supply in China. Hydrogeol J 26:1301–1324. https://doi.org/10.1007/s10040-018-1795-1

Wang Y, Wang Q, Deng Y, Chen Z, Cappellen PV, Yang Y, Goldscheider N (2020) Assessment of the impact of geogenic and climatic factors on global risk of urinary stone disease. Sci Total Environ 721:137769. https://doi.org/10.1016/j.scitotenv.2020.137769

Wang Y, Li J, Ma T, Xie X, Deng Y, Gan Y (2021) Genesis of geogenic contaminated groundwater: As, F and I. Crit Rev Env Sci Technol 51(24):2895–2933. https://doi.org/10.1080/10643389.2020.1807452

Wang Y, Yuan S, Shi J, Ma T, Xie X, Deng Y, Du Y, Gan Y, Guo Z, Dong Y, Zheng C, Jiang G (2023) Groundwater quality and health: making the invisible visible. Environ Sci Technol 57:5125–5136. https://doi.org/10.1021/acs.est.2c08061

Wang D, Li P, Mu D, Liu W, Chen Y, Fida M (2024) Unveiling the biogeochemical mechanism of nitrate in the vadose zone-groundwater system: Insights from integrated microbiology, isotope techniques, and hydrogeochemistry. Sci Total Environ 906:167481. https://doi.org/10.1016/j.scitotenv.2023.167481

Wen D, Zhang F, Zhang E, Wang C, Han S, Zheng Y (2013) Arsenic, fuoride and iodine in groundwater of China. J Geochem Explor 135:1–21. https://doi.org/10.1016/j.gexplo.2013.10.012

Xie X, Shi J, Pi K, Deng Y, Yan B, Tong L, Yao L, Dong Y, Li J, Ma L, Zheng C, Jiang G (2023) Groundwater quality and public health. Annu Rev Environ Resour 48:395–418. https://doi.org/10.1146/annurev-environ-112321-114701

Xu F, Li P (2024) Biogeochemical mechanisms of iron (Fe) and manganese (Mn) in groundwater and soil profiles in the Zhongning section of the Weining Plain (northwest China). Sci Total Environ 939:173506. https://doi.org/10.1016/j.scitotenv.2024.173506

Zamora-Ledezma C, Negrete-Bolagay D, Figueroa F, Zamora-Ledezma E, Ni M, Alexis F, Guerrero VH (2021) Heavy metal water pollution: a fresh look about hazards, novel and conventional remediation methods. Environ Technol Inno 22:101504. https://doi.org/10.1016/j.eti.2021.101504

Zhang E, Wang Y, Qian Y, Ma T, Zhang D, Zhan H, Zhang Z, Fei Y, Wang S (2013) Iodine in groundwater of the North China Plain: spatial patterns and hydrogeochemical processes of enrichment. J Geochem Explor 135:40–53. https://doi.org/10.1016/j.gexplo.2012.11.016

Zhang Y, Zheng X, Wei L, Sun R, Guo H, Liu X, Liu S, Li Y, Mai B (2018) The distribution and accumulation of phosphate flame retardants (PFRs) in water environment. Sci Total Environ 630:164–170. https://doi.org/10.1016/j.scitotenv.2018.02.215

Zhao H, Li P, He X, Ning J (2023) Microplastics pollution and risk assessment in selected surface waters of the Wei River Plain, China. Expos Health 15(4):745–755. https://doi.org/10.1007/s12403-022-00522-z

Zhao DL, Zhou W, Shen L, Li B, Sun H, Zeng Q, Tang CY, Lin H, Chung T-S (2024) New directions on membranes for removal and degradation of emerging pollutants in aqueous systems. Water Res 251:121111. https://doi.org/10.1016/j.watres.2024.121111

Effect of Groundwater-Lake Interaction on Water and Salinity Balance for Lakes in Arid Regions

Jie Chen, Ibrahim Ishag, Junzhu Yang, Zibo Xu, and Bin Zhao

Abstract Groundwater plays a crucial role in water and salinity balance for lakes in arid regions, essential for maintaining the water environment and ecosystem health of lakes. This chapter aims to provide a comprehensive review of the literature in the Web of Science core collection database concerning the research progress of groundwater on the water and salt evolution of lakes in arid regions. In arid areas, there has been a noticeable upward trajectory in the publication and citation of papers about the interaction between lakes and groundwater. The subject has a wide geographic range and offers research opportunities worldwide. In addition, we explored the research and various approaches regarding the effect of groundwater on the levels and salinity change of lake water. We stress the need for a heightened focus on elucidating the intricate interaction between groundwater and lakes and the related hydrogeochemical reactions. This offers valuable insights into assessing groundwater impact on water and salt equilibrium for lakes in arid regions. Furthermore, it seeks to broaden the scope of research within the lake water environment and scientific management.

Keywords Arid areas · Lake · Groundwater-lake interaction · Salinity · Hydrogeochemical reactions

J. Chen (✉) · I. Ishag · J. Yang · Z. Xu · B. Zhao
Key Laboratory of Subsurface Hydrology and Ecological Effect in Arid Area of Ministry of Education, Chang'an University, No. 126 Yanta Road, Xi'an 710054, Shaanxi, China
e-mail: jiechen17@chd.edu.cn

School of Water and Environment, Chang'an University, No. 126 Yanta Road, Xi'an 710054, Shaanxi, China

Key Laboratory of Eco-Hydrology and Water Security in Arid and Semi-Arid Regions of Ministry of Water Resources, Chang'an University, No. 126 Yanta Road, Xi'an 710054, Shaanxi, China

P. Li et al. (eds.), *Sustainable Groundwater and Environment: Challenges and Solutions*, Springer Hydrogeology, https://doi.org/10.1007/978-3-031-82194-3_5

1 Introduction

Lakes play a crucial role in the terrestrial hydrosphere and are intricately connected to the atmosphere, lithosphere, and biosphere (Li et al. 2017). The interaction between lake water and groundwater is a crucial component of the terrestrial hydrological cycle, widely observed in nature, and lake water and groundwater are hydraulically interconnected. The water exchange process between groundwater and lakes is a crucial component of the lake water cycle. Groundwater can also serve as a hidden pathway for transporting substances to lakes; therefore, the materials in groundwater (such as nutrients, heavy metals, carbon, and other components) are vital to lake ecosystems and material cycling (Burnett et al. 2013; Luo et al. 2018). Understanding the hydrodynamic behavior and material exchange between groundwater and lakes is a key factor in comprehending lake water chemistry and effectively protecting and managing lake water environments (Javadzadeh et al. 2020; Tirado-Conde et al. 2019).

Lakes are regarded as natural reservoirs, which are critical indicators of global climate change and regional environmental responses. China, with its diverse geography, is home to numerous lakes that are categorized into five major regions: the Qinghai-Tibet Plateau, the Mongolian-Xinjiang region, the Yunnan-Guizhou Plateau, the Northeast Plain and Mountains, and the Eastern Plain (Cao et al. 2022; Yu et al. 2021). Lakes are vital ecosystem components, particularly in arid regions like Mongolian Xinjiang (Chen et al. 2020; Han et al. 2024). In these areas, where water scarcity and ecological fragility are pronounced, maintaining the health and sustainability of lake ecosystems is essential for enhancing the overall quality and stability of regional environments (Wang et al. 2024). Recent research underscores these ecosystems' importance in mitigating environmental stressors' impacts, highlighting their significance in regional ecological resilience and sustainability.

Recent research underlines the critical role of groundwater in maintaining the water-salt balance in lakes, particularly in arid climate conditions where lake water evaporation is exceptionally high (Chen et al. 2020; Ding et al. 2024; Razavi-Termeh et al. 2024; Xiong et al. 2023). In such environments, evaporation often exceeds precipitation by 10 to 20, leading to significant salt concentration and enrichment in lake waters (Li et al. 2024b; Liang et al. 2024; Yan et al. 2024). Groundwater inflow is vital for replenishing lake water, facilitating salt excretion, and stabilizing the lake's hydrological and chemical balance (Jin et al. 2024; Meles et al. 2024; Yu et al. 2024a; Zhang et al. 2024). For instance, the intense evaporation rates combined with the interactions between lakes and groundwater systems significantly influence lake area and salinity levels on the Mengxin Plateau (Liu et al. 2023; Peng et al. 2020; Sun et al. 2022). However, recent observations indicate that many lakes in arid regions are experiencing a sharp decline in water levels, increased salinization, and degradation of ecological functions (Guo et al. 2024; Lv et al. 2024; Zhang et al. 2022). These changes are primarily attributed to the excessive extraction of environmental water and the rapid decline in groundwater levels.

The consensus in recent studies shows that the evolution of lake water volume and salinity in arid regions directly reflects the broader water-environmental balance within the basin (Chen et al. 2021). The core challenge in understanding this evolution lies in unraveling the complex interactions between groundwater and lake systems. Therefore, investigating the impact of groundwater on lake salinity dynamics is essential for revealing the mechanisms driving lake water environment changes and for developing informed strategies for scientific lake management (Dong et al. 2017).

This chapter presents a comprehensive review of recent domestic and international research to address the critical role of groundwater in the dynamics of lake water levels and salinity in arid regions. The review systematically examines the influence of groundwater on lake hydrology and salinity, focusing on quantitative methods used in current studies. Additionally, the chapter identifies future challenges and research directions in this field to contribute new insights and foundations for promoting the sustainable management and development of lake ecosystems in arid regions.

2 Water and Salt Balance for Lakes in Arid Regions

Lakes play a crucial role in the hydrological cycle, acting as significant reservoirs of water resources (Guo et al. 2023; Ji et al. 2024). The primary recharge sources for lakes include atmospheric precipitation, surface runoff, and subsurface inflows, while the main discharge mechanisms are evaporation, surface runoff outflow, and groundwater outflow (Jin et al. 2024; Zhou et al. 2024). In arid regions, lake-groundwater interactions are particularly critical for maintaining the water balance and the sustainability of these ecosystems (Chen et al. 2020; Mahamat Nour et al. 2022; Razavi-Termeh et al. 2024).

Lake-groundwater systems can generally be categorized into four interaction modes (Fig. 1): connected losing lakes, disconnected losing lakes, flow through lakes, and gaining lakes (Turner and Townley 2006). In arid regions, these lakes are characterized by minimal surface runoff, making stable groundwater recharge vital for sustaining their water levels (Befus et al. 2012). This stable recharge is often responsible for the relatively small seasonal water level fluctuations observed in arid region lakes, contributing to their long-term persistence.

Moreover, the interaction between lakes and groundwater in arid regions significantly impacts the salinity of lake waters (Gu et al. 2024; Lei et al. 2023; Qu et al. 2024). Subsurface runoff, which often carries away large quantities of salt, is crucial in maintaining the freshwater status of many arid regional lakes (Gong et al. 2016). Most lakes have no natural outlet. This dynamic is governed by the water balance equation, which in arid regions can be expressed as:

$$\Delta S = P + S_S + G_{in} - E - G_{out} \quad (1)$$

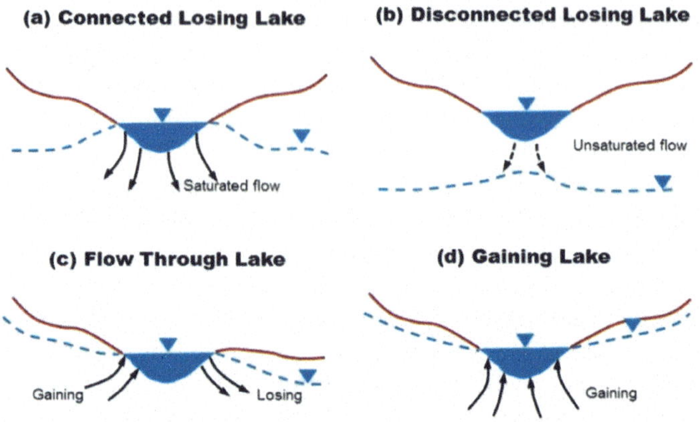

Fig. 1 Lake–groundwater interaction patterns

ΔS is change in storage (lake water volume), P is precipitation directly onto the lake, S_s is subsurface runoff, G_{in} is groundwater inflow, E is evaporation from the lake's surface, and G_{out} is groundwater outflow.

Therefore, the salinity balance in lakes, particularly in arid regions, can be described by the following equation:

$$\Delta C_L = C_{S_S} + C_P + C_{G_{in}} - C_{G_{out}} + R \tag{2}$$

where, ΔC_L is the change in ion concentration in lake water, C_{SS} is the ion concentration in surface runoff, C_P is the ion concentration in atmospheric precipitation, and C_{Gin} and C_{Gout} are the ionic concentrations of groundwater inflows and outflows, respectively. R represents the net effect of hydrogeochemical reactions (positive for dissolution and negative for precipitation).

Recent studies highlight the importance of interactions between lakes and groundwater in dry areas, especially regarding water and salinity equilibrium. These interactions are essential for maintaining the stability of lakes and the surrounding groundwater systems, especially in the context of changing climate conditions that could impact precipitation patterns, rates of evaporation, and groundwater replenishment (Wang and Zhou 2018). Understanding these dynamics is essential for developing effective water management strategies in arid regions, ensuring the long-term viability of these critical water resources.

3 Bibliometric Analysis

Bibliometric analysis is a powerful tool for quickly identifying research trends and hotspots by quantifying and analyzing the attributes of scientific publications. In recent years, this approach has become popular for assessing the development of research topics in various fields. To understand the current state of research on how groundwater impacts the evolution of water and salt dynamics in lakes located in arid regions, an advanced search was conducted using the terms "groundwater lake interaction" OR "groundwater-fed lake" AND "arid" as the search criteria.

The analysis of publications showed a noticeable rise in both the number of published articles and the frequency of references concerning interactions between lakes and groundwater in arid regions, particularly in the past ten years (Fig. 2). This pattern highlights the increasing global acknowledgment of the significance of ecological and environmental conservation in dry landscapes. The substantial increase in scholarly interest reflects the growing recognition of how groundwater significantly impacts the hydrological and chemical processes of lakes in these areas, which are often vital to local ecosystems and social development (Thelwall 2008). These researches emphasized the important patterns in research, including the investigation of salt buildup and alterations in water quality resulting from interactions between groundwater and lakes. These studies are becoming increasingly pivotal for comprehending the wider effects of climate change and human actions on dry areas. The increase in research attention suggests the growing significance placed on the sustainable management of water resources in arid and semi-arid regions, where water scarcity and environmental decline present notable challenges.

The analysis of search results indicates that researchers focusing on lake-groundwater interaction in arid regions represent over 100 countries and regions. Notably, the top 10 countries with the highest number of published papers collectively account for 96% of the global publication output, totaling 1008 papers. The United States leads with 295 articles, comprising 24.9% of the total, followed by China with 193 articles 24.1%, and Germany with 116 articles (11%). Subsequently, Canada, Australia, Italy, France, the United Kingdom, India, and Spain also made significant contributions, as depicted in Fig. 2. It is important to acknowledge the impact of economic factors on countries situated in Central Asia, North Africa, South Africa, and other arid and semi-arid regions in southern South America.

In dry regions, there are few investigations into lake-groundwater interactions due to political reasons and scientific and technological capacities. It's important to note that the Chinese Academy of Sciences and the China University of Geosciences are ranked fourth and sixth in the world in terms of the number of papers published, respectively. Moreover, the National Natural Science Foundation of China holds the top global position in the number of publications (Table 1). This demonstrates that China is leading research in this area. With a strategic direction for the development of ecological civilization, the government has allocated substantial resources to studying lake-groundwater in arid regions. The rise in the number of published papers in China

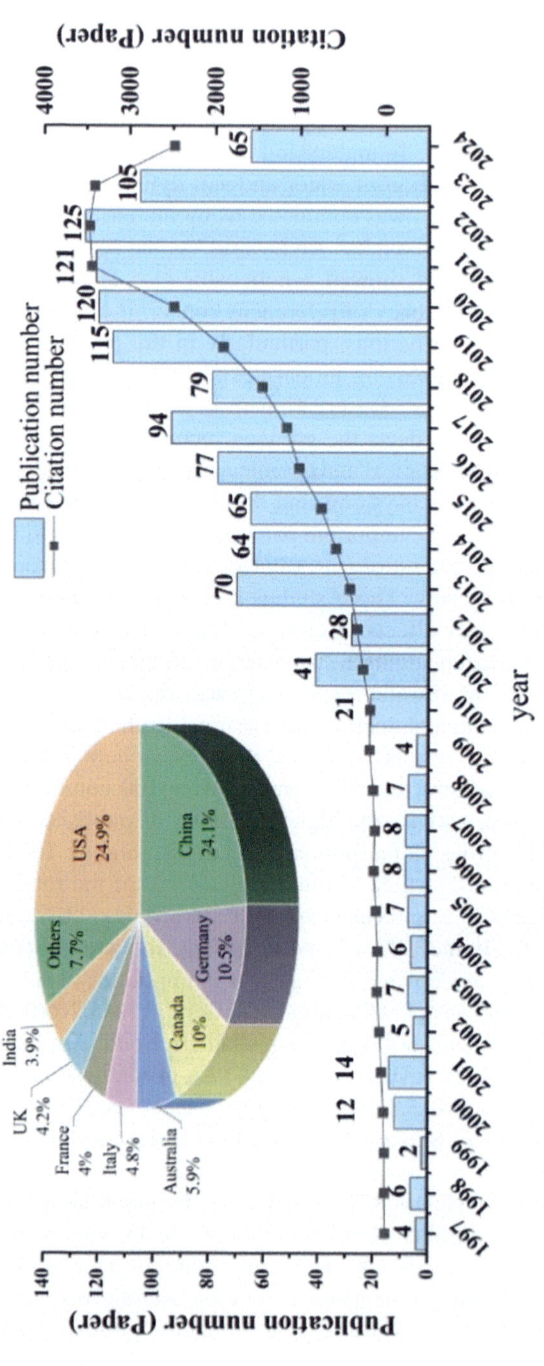

Fig. 2 The number of groundwater-lake interactions in arid regions publications published and their citations from 1997 to 2024

Table 1 Top 10 institutions publishing papers and funding agencies financially supporting groundwater-arid lake interaction in the world from 1997 to 2022

No	Mechanism	Number of articles published	Funding	Number of articles published
1	United States Department of the Interior	60	National Natural Science Foundation of China NSFC	133
2	United States Geological Survey	59	National Science Foundation NSF	68
3	Helmholtz Association	54	Natural Sciences and Engineering Research Council of Canada NSERC	44
4	Chinese Academy of Sciences	49	Fundamental Research Funds for The Central Universities	25
5	Utah System of Higher Education	39	United States Geological Survey	24
6	China University of Geosciences	38	European Commission	23
7	University of Utah	35	CGIAR	17
8	Helmholtz Centre for Environmental Research	28	United States Department of Energy Doe	16
9	Centre national de la recherche scientifique	27	Russian Foundation for Basic Research RFBR	15
10	Russian Academy of Sciences	26	UK Research Innovation UKRI	15

over the last ten years can be attributed to favorable natural geographical conditions, a robust scientific research environment, and ample funding.

4 Effect of Groundwater on Lake Water Balance in Arid Regions

Recent research has brought attention to the important role of underground water in maintaining water levels in lakes in dry areas. It is estimated that in certain instances, groundwater influx may make up over 80% of the overall volume of water in a lake (Gong et al. 2016; Stets et al. 2010; Zhu et al. 2023). Studies on lake-groundwater interactions use different methods, including direct, indirect, and advanced techniques that incorporate technological innovations and integrated modeling approaches (Rosenberry et al. 2015).

4.1 Direct Methods

Measuring and assessing changes in hydrological components at the physical intersection of lakes and groundwater involves the application of hydrodynamic methods, which are referred to as direct methods. The direct measurement of the water exchange rate between lakes and groundwater is often carried out using the seepage meter method (Lewandowski et al. 2015; Zhang et al. 2024). Utilizing seepage meters to gather real-time data on the interaction between lake water and groundwater provides valuable insights into their recharge and discharge dynamics (Kidmose et al. 2011; Ma et al. 2022). However, the accuracy of data collected through direct methods can be compromised by various factors, including the installation and operation of instruments, the spatial and temporal variability of hydrological elements, and the complex field conditions that can introduce significant errors.

4.2 Indirect Methods

In investigating lake-groundwater interaction, indirect methods such as analysis of water chemical composition, environmental isotopes, temperature characteristics, and tracers (including artificial and microbial tracers) play a vital role (Wu et al. 2017). These methods also involve the use of numerical simulations to model the interactions. The composition of water, including its chemical and isotopic makeup, provides essential insights into the processes of formation and migration of water bodies (Qu et al. 2024; Yang et al. 2024). Temperature signals that vary in response to the interplay between surface water and groundwater can be utilized to track and analyze the dynamic characteristics of this interaction (Rey et al. 2024; Saphores et al. 2024).

4.2.1 Geophysical Methods

Geophysical techniques have emerged as a potent means of studying the interactions between lakes and groundwater, especially in dry regions (Abu Salem et al. 2021; Dixit et al. 2024; Elewa et al. 2024; Geris et al. 2022). Cutting-edge methods like electrical resistivity tomography (ERT), ground-penetrating radar (GPR), and electromagnetic surveys provide researchers with the ability to visually represent subsurface formations and detect locations of groundwater influx and outflux (Bandini et al. 2023; Giampaolo et al. 2016). Electrical Resistivity Tomography (ERT) has proven to be effective in producing comprehensive spatial data on the distribution of groundwater around lakes. This complements the localized data obtained through direct measurement methods, offering a more holistic understanding of groundwater dynamics in these areas (Kazmierczak et al. 2016; Xu et al. 2022). GPR, or ground-penetrating radar, can identify variations in underground moisture levels, providing valuable information about the interconnection of lakes and nearby aquifers.

4.2.2 Remote Sensing and GIS Techniques

Remote sensing and Geographic Information System (GIS) techniques are invaluable for advancing our comprehension of lake-groundwater interactions. These methods provide a holistic, large-scale viewpoint, allowing for continuous monitoring of changes in lake surface area, volume, and water levels (Speldrich et al. 2021). These observations can be associated with groundwater dynamics, yielding essential prognostic insights for water resource management. Satellite imagery and aerial photography have proven crucial in evaluating the effects of groundwater extraction on lake contraction in dry regions. Furthermore, GIS-based hydrological modeling, which integrates diverse data sources such as climate data, topography, and land use patterns, can replicate the interplay between lakes and groundwater, thus enriching our insight into these intricate systems (Zerouali et al. 2024).

4.2.3 Isotopic and Hydrochemical Fingerprinting

Isotopic and hydrochemical fingerprinting utilizes the analysis of distinct isotopes and chemical tracers in water in order to trace the origins and movement routes of groundwater flowing into lakes (Feng et al. 2023; Gomez-Alday et al. 2022). Hydrochemical components such as chloride ions and environmental isotopes like $\delta^{18}O$, $\delta^{2}H$, and radon-222 (^{222}Rn) have played a crucial role in estimating the exchange flux between lakes and groundwater (Hajati et al. 2018; Zhao et al. 2018). These estimates are deemed vital, provided that the other equilibrium factors affecting the lake are well-determined and understood (Pan et al. 2018). The application of these methods has revealed significant information about the hydrological processes in arid environments, particularly concerning water balance and sources contributing to desert lakes. For instance, the stable isotopes of hydrogen and oxygen, denoted as δ^2H and $\delta^{18}O$, can be employed to distinguish between the input of groundwater and surface water to a lake (Lattaud et al. 2024; Salem et al. 2022; Yu et al. 2024b). The methods mentioned above are extremely important in dry regions where distinguishing between surface water and groundwater inflows is essential for understanding water balance and efficiently controlling limited water resources (Gong et al. 2016).

4.2.4 Lake-Groundwater Exchange Modeling

In recent years, remarkable progress has been made in using numerical models to simulate and predict the interactions between surface water and groundwater under varying climate change conditions and human activities (Zhan et al. 2016). For instance, Song et al. (2023) applied the Groundwater Modeling System (GMS) to study the dynamic changes in lake-groundwater exchange flux and groundwater levels in the Badain Jaran Desert, considering lakes as high-permeability units. Similarly, Wei (2018) utilized GMS software to simulate the interactions and exchanges between groundwater and surface water in lake regions, contributing to a better understanding of these processes. Additionally, Lu et al. (2022)utilized MODFLOW to simulate the water recharge requirements of Lake Hampen under different ecological protection scenarios, considering the dynamic exchange characteristics of lake and groundwater volumes (El-Zehairy et al. 2017). These studies emphasize the significance of hydrogeological conditions in governing the interaction between lakes and groundwater. A comprehensive understanding of the regional hydrogeological structure and water cycle processes is essential for successfully applying numerical simulations, which form the basis for accurately modeling the complex interactions within lake-groundwater systems.

However, the above-described methods have certain limitations. Many of the water chemistry, environmental isotope, and temperature tracer techniques are semi-quantitative, which can limit their precision in estimating exchange fluxes. Moreover, numerical models often require extensive data inputs and are highly sensitive to parameter variations, introducing uncertainties in the simulation results. These challenges underscore the need for a more integrated approach combining multiple methods and cross-validating the results to achieve a systematic and comprehensive understanding of lake-groundwater exchange mechanisms. To address these challenges, future research should focus on strengthening the joint application of various methods, including hydrological, hydrochemical, isotopic, and modeling approaches, which can enhance the reliability of the findings and provide a more robust framework for understanding and managing water resources in desert environments. Additionally, integrating advancements in remote sensing, geophysics, and machine learning with traditional methods could offer new opportunities for refining our understanding of these complex systems.

5 Effect of Groundwater on Lake Salinity in Arid Regions

The balance between natural runoff and high evaporation rates is critical for maintaining lakes' water levels and salinity in arid regions. When runoff is insufficient, lake water levels decline, increasing salinity (Wurtsbaugh et al. 2017). Over time, this process can transform freshwater lakes into saline lakes or even dry salt flats, severely degrading the ecological functions of these lakes. While the issue of declining lake water volumes has been extensively studied, the associated changes in lake salinity have received less attention. This gap is significant because numerous studies have shown that salinity imbalances can severely damage lake ecosystems, threatening biodiversity and sustainable use (Chen et al. 2019; Hassani et al. 2020; Kipriyanova et al. 2007). For example, Kipriyanova et al. (2007) documented how the salinity of

Lake Chany increased from 0.8 to 6.4 g/L, causing substantial damage to the lake's biodiversity and ecological environment. Similar findings by Jiang et al. (2022), Su et al. (2023), Zhang et al. (2023), and Gutierrez et al. (2018) in northwest China revealed that rising salinity levels have led to significant shifts in aquatic populations, reducing biodiversity and leading to the dominance of salt-tolerant species.

Recent research has expanded our understanding of the factors contributing to lake salinity in arid regions, focusing on the critical role of groundwater (Chen et al. 2020; Jodar et al. 2020; Tweed et al. 2011). Changes in groundwater flow patterns due to climate change and human activities can significantly influence lake salinity (Huo et al. 2007; Tang et al. 2022; Wang et al. 2021). Groundwater with high dissolved salt content can exacerbate the salinization process when discharged into lakes, especially in areas with high evaporation rates and where surface water inputs are limited (Kaushal et al. 2021; Salama and Fitzpatrick 1999). A study by Khogali et al. (2024) investigated the impact of groundwater on lake salinity in the Arabian Peninsula. It found that the inflow of saline groundwater and reduced surface water inflows due to decreased rainfall led to a sharp increase in lake salinity. This trend was particularly pronounced in smaller lakes that are more sensitive to changes in hydrological balance.

The evolution of lake salinity in arid areas is a complex interplay of factors, including artificial water replenishment, groundwater-lake interactions, and intense evaporation, alongside numerous hydrogeochemical reactions (Chen and Qian 2016; Gong et al. 2016; Zhang et al. 2005). Recent studies have employed advanced hydrogeochemical modeling tools like PHREEQC to simulate these processes (Mohamed et al. 2015). These models have been instrumental in identifying essential reactions, such as the dissolution of evaporite minerals, calcite precipitation, and ion exchange between groundwater and lake water, which are crucial in controlling salinity levels. For instance, Gu et al. (2023) used PHREEQC to model salinity changes in Qinghai Lake, China. They demonstrated that mixing fresh and saline groundwater and calcite precipitation played a significant role in the lake's salinization, particularly under scenarios of reduced surface water input due to climate change.

Human interventions, such as groundwater extraction and artificial recharge, also significantly influence lake salinity (Elewa et al. 2024; Lu et al. 2024). A recent study by Silva-Aguilera et al. (2024) highlighted the impact of groundwater pumping on the salinity of endorheic lakes. The study found that excessive groundwater extraction for agriculture led to declining lake levels and increased salinity due to the concentration of dissolved salts (Li et al. 2024a). The findings underscore the importance of sustainable groundwater management practices to mitigate these effects and preserve the ecological balance of lakes.

6 Future Directions and Recommendations

Currently, researchers and managers have overlooked aspects of these studies, such as the influence of climate variability and human activities on altering groundwater contributions to lakes. In arid regions, where water resources are limited, fluctuations in climate patterns, such as changes in precipitation and temperature, alongside increased water extraction for agricultural and industrial purposes, can significantly impact the equilibrium between groundwater and lake water levels. Therefore, future research efforts should prioritize integrating climate models and human impact assessments in the study of lake-groundwater interactions. This integrated approach will facilitate more accurate predictions and effective management of water resources in these fragile environments. Such understanding is paramount for researchers, policymakers, and water resource managers to conserve and manage water resources in arid regions.

The increasing recognition of groundwater's role in controlling lake salinity, particularly in climate change, underscores the need for integrated water resource management strategies. Future research should focus on developing coupled hydrological and hydrogeochemical models that incorporate the effects of climate variability, land-use changes, and human activities on lake salinity. These models could provide valuable insights for policymakers and water managers, helping design interventions that mitigate salinity and protect lake ecosystems in arid regions. Moreover, exploring innovative solutions such as managed aquifer recharge (MAR) and desalination technologies could offer sustainable ways to enhance freshwater availability and protect the ecological integrity of lakes in arid regions. In conclusion, while significant progress has been made in understanding the impact of groundwater on lake salinity, mainly through recent studies, ongoing research, and effective management strategies are essential to mitigate the adverse effects of salinization and ensure the sustainable use of these critical water resources.

With the in-depth study of lake-groundwater interaction in arid areas in recent years, researchers have found that groundwater plays a crucial role in the evolution of lake water environment in arid regions. It is vital to explore and reveal the lake-groundwater interaction relationship in arid areas more extensively and deeply in future related research, which is highly consistent with the strategic task direction of improving the quality and stability of ecosystems and has significant theoretical and practical value for the integrated protection and systematic management of mountains, rivers, forests, fields, lakes, grasslands and sands in arid areas.

Acknowledgements This research is supported by grants from the National Natural Science Foundation of China (42272289, 42341102), Youth Talents Promotion Plan of Shaanxi Association for Science and Technology (20220711), and the Fundamental Research Funds for the Central Universities in Chang'an University (300102293205, 300102294905, 300102294906, and 300102294902).

References

Abu Salem H, Gemail KS, Nosair AM (2021) A multidisciplinary approach for delineating wastewater flow paths in shallow groundwater aquifers: a case study in the southeastern part of the Nile Delta, Egypt. J Contam Hydrol 236:103701

Bandini F, Kooij L, Mortensen BK, Caspersen MB, Thomsen LG, Olesen D, Bauer-Gottwein P (2023) Mapping inland water bathymetry with ground penetrating radar (GPR) on board unmanned aerial systems (UASs). J Hydrol 616

Befus KM, Cardenas MB, Ong JB, Zlotnik VA (2012) Classification and delineation of groundwater–lake interactions in the Nebraska Sand Hills (USA) using electrical resistivity patterns. Hydrogeol J 20:1483–1495

Burnett WC, Peterson RN, Chanyotha S, Wattayakorn G, Ryan B (2013) Using high-resolution in situ radon measurements to determine groundwater discharge at a remote location: Tonle Sap lake, Cambodia. J Radioanal Nucl Chem 296:97–103

Cao C, Li N, Yue W, Wu L, Cao X, Zhai Y (2022) Analysis of the interaction between lake and groundwater based on water-salt balance method and stable Isotopic characteristics. Int J Environ Res Public Health 19

Chen J, Qian H (2016) Characterizing replenishment water, lake water and groundwater interactions by numerical modelling in arid regions: a case study of Shahu lake. Hydrol Sci J 1–10

Chen X, Liu X, Peng W, Dong F, Chen Q, Sun Y, Wang R (2019) Hydroclimatic influence on the salinity and water volume of a plateau lake in southwest China. Sci Total Environ 659:746–755

Chen J, Qian H, Gao Y, Wang H, Zhang M (2020) Insights into hydrological and hydrochemical processes in response to water replenishment for lakes in arid regions. J Hydrol 581

Chen Q, Wang S, Ni Z, Guo Y, Zhang H (2021) Dynamic and driving evolution of lake basin pressure in cold and arid regions based on a new method: a case study of three lakes in Inner Mongolia, China. J Environ Manag 298:113425

Ding M, Wang J, Song C, Sheng Y, Shawn Hutchinson JM, Langston AL, Marston L (2024) A framework of freshwater and saline lake typology classification through leveraging hydroclimate, spectral, and literature evidence. J Hydrol 632

Dixit M, Arora M, Jayaprakash HP, Achutha VR (2024) Achievements and significance of national hydrograph stations network and 55 years of continuous groundwater level monitoring in India. Groundw Sustain Dev

Dong S, Peng F, You Q, Guo J, Xue X (2017) Lake dynamics and its relationship to climate change on the Tibetan Plateau over the last four decades. Reg Environ Change 18:477–487

El-Zehairy AA, Lubczynski MW, Gurwin J (2017) Interactions of artificial lakes with groundwater applying an integrated MODFLOW solution. Hydrogeol J 26:109–132

Elewa HH, Nosair AM, Ibrahim A, Zelenakova M, Pietrucha-Urbanik K, Habib HM et al (2024) Use of remote sensing, spatial and geophysical modeling, and real recharging capabilities to identify suitable areas for groundwater exploitation in dry coastal areas. J Environ Manag 363:121243

Feng B, Zhong Y, He J, Sha X, Fang L, Xu Z, Qi Y (2023) Nitrogen sources and conversion processes in shallow groundwater around a plain lake (Northwest China): evidenced by multiple isotopes and water chemistry. Chemosphere 337:139322

Geris J, Comte J-C, Franchi F, Petros AK, Tirivarombo S, Selepeng AT, Villholth KG (2022) Surface water-groundwater interactions and local land use control water quality impacts of extreme rainfall and flooding in a vulnerable semi-arid region of Sub-Saharan Africa. J Hydrol 609

Giampaolo V, Capozzoli L, Grimaldi S, Rizzo E (2016) Sinkhole risk assessment by ERT: the case study of Sirino lake (Basilicata, Italy). Geomorphology 253:1–9

Gomez-Alday JJ, Hussein S, Arman H, Alshamsi D, Murad A, Elhaj K, Aldahan A (2022) A multi-isotopic evaluation of groundwater in a rapidly developing area and implications for water management in hyper-arid regions. Sci Total Environ 805:150245

Gong Y, Wang X, Hu BX, Zhou Y, Hao C, Wan L (2016) Groundwater contributions in water-salt balances of the lakes in the Badain Jaran Desert, China. J Arid Land 8:694–706

Gu X, Shi J, Duan R, Chang L, You X, Li X (2023) Hydrochemical characteristics and processes for groundwater in salt lake area: a case of Bieletan in the Qaidam Basin, Qinghai-Tibet Plateau. Research Square

Gu X, Xue M, Ye C, Jiao Y, Han Y, Wang W, et al. (2024) Evaluating heterogeneous lacustrine groundwater discharge of Hulun Lake using radium isotopes in Inner Mongolia, China. J Hydrol: Reg Stud 51

Guo W, Yang H, Ma Y, Hong F, Wang H (2023) Comprehensive evaluation of the hydrological health evolution and its driving forces in the river-lake system. Ecol Inform 75

Guo Q, Yang G, Wu X, Jin H, Tian Y (2024) Spatial-temporal evolution of plateau and plain lakes and climate driving factors based on Google Earth Engine platform. Front Environ Sci 12

Gutierrez MF, Tavşanoğlu ÜN, Vidal N, Yu J, Teixeira-de Mello F, Çakiroglu AI et al (2018) Salinity shapes zooplankton communities and functional diversity and has complex effects on size structure in lakes. Hydrobiologia 813:237–255

Hajati M-C, Frandsen M, Pedersen O, Nilsson B, Duque C, Engesgaard P (2018) Flow reversals in groundwater–lake interactions: a natural tracer study using $\delta^{18}O$. Limnologica 68:26–35

Han Y, Ma X, Yan W, Wang Y (2024) Windbreak and sand fixation service flow simulation in the terminal lake basin of inland rivers in arid regions: a case study of the Aral Sea basin. Sci Total Environ 945:174047

Hassani A, Azapagic A, D'Odorico P, Keshmiri A, Shokri N (2020) Desiccation crisis of saline lakes: a new decision-support framework for building resilience to climate change. Sci Total Environ 703:134718

Huo Z, Feng S, Kang S, Li W, Chen S (2007) Effect of climate changes and water-related human activities on annual stream flows of the Shiyang river basin in arid North-West China. Hydrol Process 22:3155–3167

Javadzadeh H, Ataie-Ashtiani B, Hosseini SM, Simmons CT (2020) Interaction of lake-groundwater levels using cross-correlation analysis: a case study of Lake Urmia Basin, Iran. Sci Total Environ 729

Ji K, Ouyang W, Lin C, He M, Liu X (2024) Eco-hydrological processes regulate lake riparian soil organic matter under dryness stress. Water Res 260:121938

Jiang X, Fan C, Liu K, Chen T, Cao Z, Song C (2022) Centenary covariations of water salinity and storage of the largest lake of Northwest China reconstructed by machine learning. J Hydrol 612

Jin Z, Tang S, Yuan L, Xu Z, Chen D, Liu Z et al (2024) Areal artificial recharge has changed the interactions between surface water and groundwater. J Hydrol 637

Jodar J, Rubio FM, Custodio E, Martos-Rosillo S, Pey J, Herrera C et al (2020) Hydrogeochemical, isotopic and geophysical characterization of saline lake systems in semiarid regions: the Salada de Chiprana lake, Northeastern Spain. Sci Total Environ 728:138848

Kaushal SS, Likens GE, Pace ML, Reimer JE, Maas CM, Galella JG et al (2021) Freshwater salinization syndrome: from emerging global problem to managing risks. Biogeochemistry 154:255–292

Kazmierczak J, Müller S, Nilsson B, Postma D, Czekaj J, Sebok E et al (2016) Groundwater flow and heterogeneous discharge into a seepage lake: combined use of physical methods and hydrochemical tracers. Water Resour Res 52:9109–9130

Khogali A, Chavanidis K, Stampolidis A, Kirmizakis P, Yassin M, Abu-Mahfouz IS et al (2024) Subsurface complexity controls the aquifer heterogeneity: a case study from the Al-Hassa oasis, Eastern Saudi Arabia. Groundw Sustain Dev 27

Kidmose J, Engesgaard P, Nilsson B, Laier T, Looms MC (2011) Spatial distribution of seepage at a flow-through lake: lake Hampen, Western Denmark. Vadose Zone J 10:110–124

Kipriyanova LM, Yermolaeva NI, Bezmaternykh DM, Dvurechenskaya SY, Mitrofanova EY (2007) Changes in the biota of Chany lake along a salinity gradient. Hydrobiologia 576:83–93

Lattaud J, Gondwe MJ, Saurer M, Helfter C, De Jonge C (2024) The preservation of photosynthetic and hydrological signals in the carbon and hydrogen isotope compositions of n-fatty acids in the seasonal wetland soils of the Okavango Delta (Botswana). Organic Geochem 195

Lei Y, Liu Y, Sun Z, Zou C, Ma R, Yin L, Pan H (2023) Influences of paleoclimatic environment and hydrogeochemical evolution on groundwater salinity in an arid inland plain in northwestern China. Appl Geochem 154

Lewandowski J, Meinikmann K, Nützmann G, Rosenberry DO (2015) Groundwater—The disregarded component in lake water and nutrient budgets. Part 2: effects of groundwater on nutrients. Hydrol Process 29:2922–2955

Li P, Feng W, Xue C, Tian R, Wang S (2017) Spatiotemporal variability of contaminants in lake water and their risks to human health: a case study of the Shahu lake tourist area, northwest China. Expo Health 9(3):213–225. https://doi.org/10.1007/s12403-016-0237-3

Li W, Kang S, Du T, Ding R, Zou M (2024a) Optimal groundwater depth and irrigation amount can mitigate secondary salinization in water-saving irrigated areas in arid regions. Agric Water Manag 302

Li Z, He M-Y, Li B, Wen X, Zhou J, Cheng Y et al (2024b) Multi-isotopic composition (Li and B isotopes) and hydrochemistry characterization of the Lakko Co Li-rich salt lake in Tibet, China: origin and hydrological processes. J Hydrol 630

Liang C, Yang B, Cao Y, Liu K, Wu J, Hao F, et al. (2024) Salinization mechanism of lakes and controls on organic matter enrichment: From present to deep-time records. Earth-Sci Rev 251

Liu B, Gao J, Xue M, Lu B, Ye C, Liu J et al (2023) High exogenous humus inhibits greenhouse gas emissions from steppe lakes. Environ Pollut 319:120946

Lu C, He X, Zhang B, Wang J, Kidmose J, Jarsjö J (2022) Comparison of numerical methods in simulating lake–groundwater interactions: lake Hampen, Western Denmark. Water 14

Lu C, Wu C, Sun Q, Wu X, Yan L, Qin T (2024) Seasonal river–lake-groundwater coupling simulation and groundwater overexploitation and ecological environment assessment in the Aiding lake Basin, NW China. J Hydrol 632

Luo X, Kuang XX, Jiao JJ, Liang SH, Mao R, Zhang XL, Li HL (2018) Evaluation of lacustrine groundwater discharge, hydrologic partitioning, and nutrient budgets in a proglacial lake in the Qinghai-Tibet Plateau: using ^{222}Rn and stable isotopes. Hydrol Earth Syst Sci 22:5579–5598

Lv Z, Li S, Xu X, Lei J, Peng Z (2024) Ecological risk assessment of landscape in arid area watersheds under ecological water conveyance: a case study of Taitema lake. Heliyon 10:e29575

Ma N, Tong L, Li Y, Yang C, Tan Q, He J (2022) Distribution of antibiotics in lake water-groundwater—Sediment system in Chenhu lake area. Environ Res 204:112343

Mahamat Nour A, Huneau F, Mahamat Ali A, Mahamat Saleh H, Ngo Boum-Nkot S, Nlend B et al (2022) Shallow Quaternary groundwater in the lake Chad basin is resilient to climate change but requires sustainable management strategy: results of isotopic investigation. Sci Total Environ 851:158152

Meles MB, Bradford S, Casillas-Trasvina A, Chen L, Osterman G, Hatch T et al (2024) Uncovering the gaps in managed aquifer recharge for sustainable groundwater management: a focus on hillslopes and mountains. J Hydrol 639

Mohamed EA, El-Kammar AM, Yehia MM, Abu Salem HS (2015) Hydrogeochemical evolution of inland lakes' water: A study of major element geochemistry in the Wadi El Raiyan depression, Egypt. J Adv Res 6:1031–1044

Pan G, Li X, Zhang J, Liu Y, Liang H (2018) Groundwater-flow-system characterization with hydrogeochemistry: a case in the lakes discharge area of the Ordos Plateau, China. Hydrogeol J 27:669–683

Peng J, Zheng B, Wang X, Wang M (2020) The driving force of basin water resource utilization to lake nutrients. Hydrol Sci J 65:855–866

Qu S, Zhao Y, Zhang K, Wang J, Li M, Yang X et al (2024) Multi-isotopes (deltaD, delta(18)O(water), (87)Sr/(86)Sr, delta(34)S and delta(18)O(sulfate)) as indicators for groundwater salinization genesis and evolution of a large agricultural drainage lake basin in Inner Mongolia, Northwest China. Sci Total Environ 946:174181

Razavi-Termeh SV, Sadeghi-Niaraki A, Farhangi F, Khiadani M, Pirasteh S, Choi SM (2024) Solving water scarcity challenges in arid regions: a novel approach employing human-based meta-heuristics and machine learning algorithm for groundwater potential mapping. Chemosphere 363:142859

Rey DM, Hare DK, Fair JH, Briggs MA (2024) Diel temperature signals track seasonal shifts in localized groundwater contributions to headwater streamflow generation at network scale. J Hydrol 639

Rosenberry DO, Lewandowski J, Meinikmann K, Nützmann G (2015) Groundwater—The disregarded component in lake water and nutrient budgets. Part 1: effects of groundwater on hydrology. Hydrol Process 29:2895–2921

Salama RB, Fitzpatrick ORW (1999) Contributions of groundwater conditions to soil and water salinization. Hydrogeol J 7:46–64

Salem ZE-S, Elnahrawy AM, Abdelrahman K, Fnais MS, Abu-Alam T (2022) Oxygen and hydrogen stable isotopes as recharge indicators, Central Nile Delta Quaternary aquifer, Egypt. J King Saud Univ Sci 34

Saphores E, Leray S, Su´arez F (2024) Groundwater–surface water exchange from temperature time series: a comparative study of heat tracer methods. J Hydrol 632

Silva-Aguilera RA, Escolero O, Alcocer J, Correa Metrio A, Vilaclara G, Lozano García S (2024) Long-term responses of maar lakes water level to climate and groundwater variability in central Mexico. J South Am Earth Sci 139

Song Y, Zhang Q, Melack JM, Li Y (2023) Groundwater dynamics of a lake-floodplain system: role of groundwater flux in lake water storage subject to seasonal inundation. Sci Total Environ 857:159414

Speldrich B, Gerla P, Tschann E (2021) Characterizing groundwater interaction with Lakes and Wetlands using GIS modeling and natural water quality measurements. Water 13

Stets EG, Winter TC, Rosenberry DO, Striegl RG (2010) Quantification of surface water and groundwater flows to open- and closed-basin lakes in a headwaters watershed using a descriptive oxygen stable isotope model. Water Resour Res 46

Su X, Lu C, Li M, Wang Y, Wang N (2023) Using (222)Rn temporal and spatial distributions to estimate the groundwater discharge rate and associated nutrient fluxes into high salinity lakes in Badain Jaran Desert, Northwest China. Sci Total Environ 857:159359

Sun Q, Sun J, Baidurela A, Li L, Hu X, Song T (2022) Ecological landscape pattern changes and security from 1990 to 2021 in Ebinur lake Wetland Reserve, China. Ecol Indic 145

Tang X, Xie G, Deng J, Shao K, Hu Y, He J et al (2022) Effects of climate change and anthropogenic activities on lake environmental dynamics: a case study in lake Bosten Catchment, NW China. J Environ Manag 319:115764

Thelwall M (2008) Bibliometrics to webometrics. J Inf Sci 34:605–621

Tirado-Conde J, Engesgaard P, Karan S, Müller S, Duque C (2019) Evaluation of temperature profiling and seepage meter methods for quantifying submarine groundwater discharge to coastal lagoons: impacts of saltwater intrusion and the associated thermal regime. Water 11

Turner JV, Townley LR (2006) Determination of groundwater flow-through regimes of shallow lakes and wetlands from numerical analysis of stable isotope and chloride tracer distribution patterns. J Hydrol 320:451–483

Tweed S, Leblanc M, Cartwright I, Favreau G, Leduc C (2011) Arid zone groundwater recharge and salinisation processes; an example from the lake Eyre Basin, Australia. J Hydrol 408:257–275

Wang T, Yang Z, Han F, Yu J, Ma X, Han J (2024) Assessment of tourism socio-ecological system resilience in arid areas: a case study of Xinjiang, China. Ecol Indic 159

Wang X-S, Zhou Y (2018) Investigating the mysteries of groundwater in the Badain Jaran Desert, China. Hydrogeol J 26:1639–1655

Wang Y, Gu X, Yang G, Yao J, Liao N (2021) Impacts of climate change and human activities on water resources in the Ebinur lake Basin, Northwest China. J Arid Land 13:581–598

Wei S (2018) Study on the interaction between groundwater and surface water in Kreuk lake Area, Qaidam Basin

Wu X, Wang X-S, Wang Y, Hu BX (2017) Origin of water in the Badain Jaran Desert, China: new insight from isotopes. Hydrol Earth Syst Sci 21:4419–4431

Wurtsbaugh W, Miller C, Null S et al (2017) Decline of the world's saline lakes. Nat Geosci 10:816–821

Xiong L, Jiang Y, Li X, Ren D, Huang G (2023) Long-term regional groundwater responses and their ecological impacts under agricultural water saving in an arid irrigation district, upper Yellow River basin. Agric Water Manag 288

Xu F, Li P, Du Q, Yang Y, Yue B (2022) Seasonal hydrochemical characteristics, geochemical evolution, and pollution sources of lake Sha in an Arid and Semiarid Region of Northwest China. Expo Health 15:231–244

Yan YN, Zhang JW, Wu N, Xia ZH, Liu L, Zhao ZQ (2024) Co-occurrence of elevated arsenic and fluoride concentrations in Wuliangsu lake: Implications for the genesis of poor-quality groundwater in the (paleo-)Huanghe (Yellow River) catchment, China. Water Res 258:121767

Yang S, Xie L, Yang K, Wu Y, Xiao P, Wei J, Zhong Y (2024) The lake and groundwater interaction based on water balance in Dongting lake, China. J Hydrol: Reg Stud 53

Yu C, Xu Y, Yan Y, Xiao W, Liu M, Cheng M et al (2021) Mercury and methylmercury in China's lake sediments and first estimation of mercury burial fluxes. Sci Total Environ 770:145338

Yu J, Wang Q, Gilfedder BS, Qian C, Xie Y, Kattel GR et al (2024a) Is the impact of groundwater on lake greenhouse gas dynamics underestimated? A comparative analysis of subsurface and ecological factors. J Hydrol 640

Yu X, Liu H, Wang Q, Kou X, Cao X, Xu Z et al (2024b) Hydrochemical and Stable Isotope characteristics of surface water and groundwater in Xiliugou and Wulagai River basin, North China. Ecohydrol Hydrobiol 24:62–72

Zerouali B, Bailek N, Islam ARMT, Katipoğlu OM, Ayek AAE, Santos CAG et al (2024) Enhancing groundwater potential zone mapping with a hybrid analytical method: the case of semiarid basin. Groundw Sustain Dev 26

Zhan L, Chen J, Zhang S, Li L, Huang D, Wang T. (2016) Isotopic signatures of precipitation, surface water, and groundwater interactions, Poyang Lake Basin, China. Environ Earth Sci 75

Zhang H, Yao B, Wang S, Huang Y (2022) Understanding the changes of optically active substances (OACs) in Hulun lake in the past 35 years and its indication to the degradation of aquatic ecology. J Clean Prod 377

Zhang K, Liu R, Bai E, Zhao Z, Peyrotty G, Fathy D et al (2023) Biome responses to a hydroclimatic crisis in an Early Cretaceous (Barremian–Aptian) subtropical inland lake ecosystem, Northwest China. Palaeogeogr Palaeoclimatol Palaeoecol 622

Zhang R, Guo C, Wu J, Wang S (2024) Analyzing the correlation between groundwater tables and the area of groundwater-dependent lakes in arid region. Ecol Indic 160

Zhang Y, Wu Y, Su J, Wen X, Liu F (2005) Groundwater replenishment analysis by using natural isotopes in Ejina Basin, Northwestern China. Environ Geol 48:6–14

Zhao D, Wang G, Liao F, Yang N, Jiang W, Guo L et al (2018) Groundwater-surface water interactions derived by hydrochemical and isotopic (^{222}Rn, deuterium, oxygen-18) tracers in the Nomhon area, Qaidam Basin, NW China. J Hydrol 565:650–661

Zhou Z, Zhou Z, Vanapalli SK (2024) A quantitative study of changing groundwater-surface water interactions of a large reservoir due to impoundment based on hydrochemistry and isotope characteristics. J Hydrol 634

Zhu Z, Fu C, Wu H, Wu H, Cao Y, Xia Y (2023) Decoding the hundred-year water level changes of the largest Saline lake in China: a joint lake-basin modeling study based on a revised SWAT+. J Hydrol: Reg Stud 49

Groundwater in Arid and Semi-arid Regions: Issues of Scarcity and Quality

Wenxu Hu, Zhen Zhao, Yuqing Zhang, Yong Xiao, Guangxiong Qin, Jie Wang, Huijuan Chen, Shaokang Yang, Liwei Wang, Hongjie Yang, and Dongyang Zhao

Abstract In arid regions where freshwater is a precious commodity, the assessment and management of groundwater levels and hydrochemical conditions are of utmost importance, particularly in the face of escalating climate change. The present study takes Delingha, a typical arid region of Qaidam Basin on Tibetan Plateau, as an example to get insight into the response of groundwater quantity and quality to climate change in arid and semi-arid regions based on the Standardized Precipitation Evapotranspiration Index (SPEI). The findings suggest the drought conditions in the Delingha Basin have generally been alleviated. The 3-month scale SPEI reveals that mild drought is the most common, particularly in winter, while severe and extreme droughts are more frequent in summer. On the 12-month scale, the years 1996, 1997, 2007, and 2023 were characterized by the most arid conditions. In terms of water quantity, the correlation between groundwater levels in inhabited areas and SPEI is weak, attributed to the significant impact of human activities. Water quality in the alluvial lacustrine plain is compromised due to the strong evaporative concentration effects resulting from sluggish flow rates. Considering water quantity and quality, the study identifies the northwestern sector of the area as the most suitable for sustainable groundwater development and utilization. This research offers valuable insights

W. Hu · Y. Zhang · Y. Xiao (✉) · J. Wang · L. Wang · H. Yang
Faculty of Geosciences and Engineering, Southwest Jiaotong University, Chengdu 611756, China
e-mail: xiaoyong@swjtu.edu.cn

Z. Zhao · G. Qin · H. Chen · S. Yang · D. Zhao
Key Lab of Geo-Environment of Qinghai Province, Xi'ning 810007, China

Bureau of Qinghai Environmental Geological Prospecting, Xi'ning 810007, China

Y. Xiao
MOE Key Laboratory of Groundwater Circulation and Environmental Evolution, China University of Geosciences (Beijing), Beijing 100083, P. R. China

Fujian Provincial Key Laboratory of Water Cycling and Eco-Geological Processes, Xiamen 361021, China

J. Wang · L. Wang · H. Yang
Sichuan Province Engineering Technology Research Center of Ecological Mitigation of Geohazards in Tibet Plateau Transportation Corridors, Chengdu 611756, China

P. Li et al. (eds.), *Sustainable Groundwater and Environment: Challenges and Solutions*, Springer Hydrogeology, https://doi.org/10.1007/978-3-031-82194-3_6

for the protection of water ecosystems and the sustainable management of water resources in arid and semi-arid regions.

Keywords Drought conditions · Climate change · SPEI · Groundwater quantity · Sustainable groundwater utilization

1 Introduction

Groundwater is one of the most precious natural resources on earth and plays a vital role in human health, socio-economic development and the balance of ecosystems (Aravinthasamy et al. 2020; Jiang et al. 2022). For arid and semi-arid regions where precipitation and surface water are scarce, groundwater resources are the main source of freshwater for domestic, industrial and agricultural activities (Nagaraj and Masilamani 2023; Xiao et al. 2022a). In many arid and semi-arid nations, including developing countries like India and Iran, the scarcity and degradation of groundwater resources pose significant existential challenges (Bahrami and Zarei 2023; Pandey et al. 2023; Wang et al. 2022). Therefore, monitoring the environmental quality of groundwater is crucial in arid and semi-arid regions (Alghamdi et al. 2023). However, groundwater responds to external changes with a lag, making it difficult to monitor these changes quickly and directly (Van Lanen and Peters 2000).

Climate change is a globally discussed and concerning issue among researchers and serves as a critical influence on the Earth's water cycle and water resources (Asif et al. 2023; Konapala et al. 2020; Qiu et al. 2023). Especially in arid and semi-arid regions with delicate ecosystems, climate change impacts the quantity and chemical composition of groundwater (Barbieri et al. 2023; Jiang et al. 2019). Climate change induces alterations in the magnitude, timing, and duration of groundwater recharge, thereby influencing the availability of groundwater in arid and semi-arid regions (Fallahi et al. 2023). As glaciers melt and permafrost degrades, mineral weathering and dissolution in mountainous and permafrost regions are intensifying (Han et al. 2023; Yi et al. 2023). This not only enhances hydraulic connectivity between water bodies but also complicates the water cycle (Kumar et al. 2022; Singh et al. 2024). Scientific research indicates that these changes profoundly affect the chemical composition and quality of groundwater, altering their hydrochemical characteristics and the spatial distribution of water quality (Dao et al. 2024). Therefore, it is critical to identify the responses of groundwater quantity and chemical quality to climate change in arid and semi-arid regions.

The main premise of drought research is to determine the intensity and frequency of drought in an area, which are important pieces of information for water resource management and planning (Balacco et al. 2022; Kamali and Asghari 2023; Vicente-Serrano et al. 2010). Numerous studies have focused on developing comprehensive universal drought indices. A significant milestone in this process is the development of the Palmer Drought Severity Index (PDSI), which quantifies wet and dry conditions using the water balance equation (Palmer 1965). However, it's fixed time scale limits its ability to capture short-term moisture variability (Alley 1984; Liu et al. 2017). Vicente-Serrano et al. (2010) proposed the Standardized Precipitation Evapotranspiration Index (SPEI), which is based on precipitation and potential evapotranspiration. It accounts for the accumulation of water deficits or surpluses across various timescales, providing a more nuanced approach to modeling aquifer behaviors and groundwater table fluctuations (Balacco et al. 2022). This index is a valuable tool for examining groundwater response to climate change in arid regions (Dong et al. 2023).

The Tibetan Plateau, known as the water tower of Asia, plays a crucial role in Asia's ecological security and water reserves, and its response to climate change is very sensitive (Wang et al. 2024; Yang et al. 2024). The present study takes Delingha, a typical arid region in the Qaidam Basin of the Tibetan Plateau, as an example to get insight into the response of groundwater quantity and quality to climate change in arid and semi-arid regions based on SPEI. Specifically, the aims are (1) to investigate the evolution patterns of the long sequence SPEI in arid regions and its coupling relationship with groundwater levels; (2) to clarify the spatial distribution of groundwater quality in the arid region; and (3) to propose strategies for the sustainable development and utilization of arid-area groundwater based on the quantity and quality of water in the context of climate change. This study could provide scientific guidance for the safety of water ecosystem and sustainable use of water resources in arid and semi-arid regions.

2 Material and Methods

2.1 Description of Study Area

The Delingha basin is located in the northeastern part of the Qaidam Basin on the Tibetan Plateau, with longitudes ranging between 96°49′33.6″E and 97°39′25.2″E, and latitudes ranging between 37°3′3.6″N and 37°26′56.4″N (Fig. 1). The basin area extends from the front of the Zongwulong Mountain in the north to the Denan hills in the south, from the Tuosu Lake and Keluke Lake in the west to the edge of the Buhete Mountain in the east, covering an area of about 1,500 km^2. The overall elevation exhibits a characteristic of being higher in the north and lower in the south, as well as higher in the east and lower in the west, with the maximum elevation difference being approximately 647 m. The study area is characterized by a typical plateau continental climate, with an average annual temperature of 5.3 °C and an average annual precipitation of 283.97 mm over the years from 1994 to 2024. The highest temperature and the maximum precipitation both occur in July, reaching 17.59 °C and 60.93 mm, respectively (Fig. 2). Precipitation throughout the year is predominantly concentrated in June to August, accounting for 61.1% of the total annual precipitation.

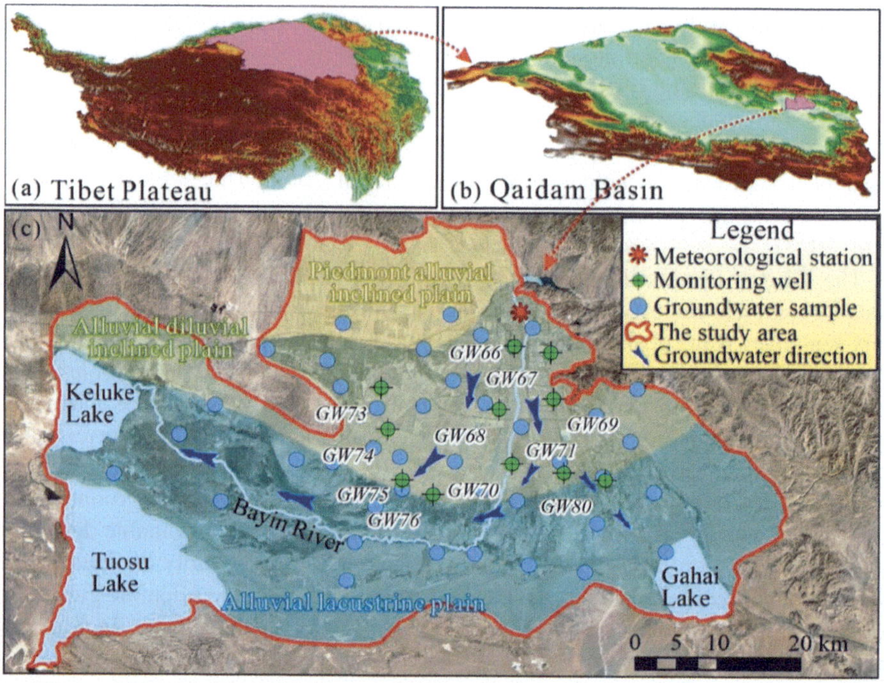

Fig. 1 Location of **a** the Qaidam Basin on Tibetan Plateau, **b** the Delingha Basin, and **c** the sampling and monitoring sites of groundwater

Fig. 2 Annual mean meteorological data of Delingha Basin

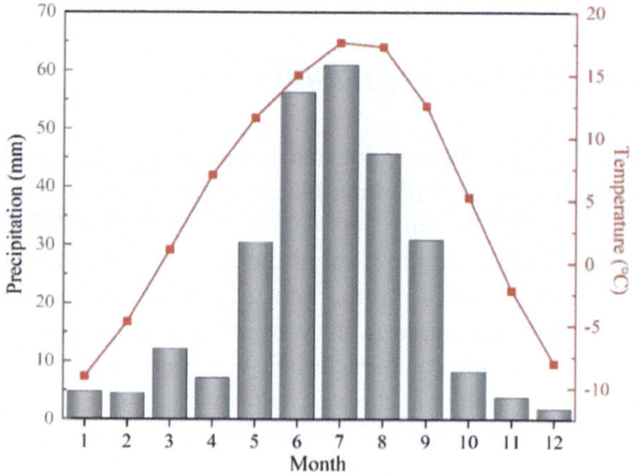

The study area features a primary river (Bayin River) and three inland lakes (Keluke Lake, Tuosu Lake and Gahai Lake). The Baying River originates from the southern foothill of the Wild Ox Ridge in the Qilian Mountain and flows sinuously from north to south, eventually entering the study area through the confines of Zongwulong Mountain and Buhete Mountain. Keluke Lake and Gahai Lake constitute the largest groundwater discharge areas within the study region. Keluke Lake is the sole freshwater lake in the study area, with a mineralization degree ranging from 0.64 to 0.79 g/L. Gahai Lake, situated in the southeastern part of the study area, is the terminal lake of the eastern end of the Delingha basin, with the average salinity of about 90 g/L. Tuosu Lake, the terminal lake of the western end of the Delingha basin, is connected to Keluke Lake via the Haoleng River, and its mineralization degree varies between 29.62 and 30.20 g/L.

The geomorphology of the study area, from north to south, is sequentially characterized by piedmont alluvial inclined plain, alluvial diluvial inclined plain and alluvial lacustrine plain. Additionally, the grain size of the lithology transitions from coarse to fine from north to south, and the terrain becomes less steep. When the Bayin River flows into the alluvial diluvial inclined plain, the coarser aquifer particles and the thicker storage spaces in the alluvial diluvial inclined plain allow for substantial infiltration of river water, forming two underground diverging flows to the east and west. A portion of the eastern diverging flow joins the Gahai Lake as groundwater, while another part, impeded by the Denan hills, re-enters the Bayin River as springs. The western diverging flow, after being obstructed and overflowing, gradually merges with the eastern Bayin River and flows westward into Keluke Lake. Due to the variations in lithology and topography, the aquifer structure of the Delingha basin also transitions from a single-layer to a multi-layer configuration from north to south.

2.2 *Methods*

2.2.1 Data Used

When conducting research on drought indices, ensuring a sufficient time scale is crucial for obtaining reliable and representative climatic information. Although a calculation time scale of no less than 30 years is generally recommended, a 20-year dataset is also acceptable in areas where data is scarce. This study selected meteorological data from the Delingha monitoring station for the period from 1994 to 2024, provided by the National Oceanic and Atmospheric Administration (NOAA) of the United States through the National Centers for Environmental Information (NCEI). The data covered is extensive and includes key meteorological parameters such as geographic location coordinates, altitude, average temperature, daily maximum and minimum temperatures, average wind speed, and precipitation. In addition, groundwater level data provided by the National Groundwater Monitoring Project Information Application Service System (from 2019 to 2024) was integrated to explore the response of groundwater volume to climatic drought.

2.2.2 Standardized Precipitation Evapotranspiration Index (SPEI)

The SPEI is a drought index that integrates both precipitation and potential evapotranspiration (PET), allowing for the assessment of the severity and duration of drought conditions (Vicente-Serrano et al. 2010). The main methods for calculating PET include the Penman-Monteith, Thornthwaite, and Hargreaves methods. Among these, the Penman-Monteith method is considered more accurate as it takes into account temperature, humidity, wind speed, and solar radiation. The Thornthwaite method is relatively simple to calculate, requiring only temperature, precipitation, and latitude data, making it suitable for situations where data acquisition is limited. The Hargreaves method is similar to the Thornthwaite method in its simplicity of data requirements but does not offer the same level of precision as the Penman-Monteith method. Based on the available meteorological data, this study employs the Thornthwaite method to calculate PET. The specific calculation formula is as follows.

$$i = \left(\frac{T}{5}\right)^{1.514} \tag{1}$$

where, i is the monthly heat index; T denotes the monthly average temperature in degrees Celsius.

$$PET = 16K\left(\frac{10T}{I}\right)^{m} \tag{2}$$

$$m = 6.75 \times 10^{-7} I^3 - 7.71 \times 10^{-5} I^2 + 1.79 \times 10^{-2} I + 0.492 \tag{3}$$

where, I represents the sum of the heat indices for 12 months; K stands for the correction factor determined by the latitude and the month; m is a coefficient that is solely related to I.

$$K = \left(\frac{N}{12}\right)\left(\frac{NDM}{30}\right) \tag{4}$$

where, N represents the maximum number of daylight hours; NDM denotes the number of days in a month.

$$N = \left(\frac{24}{\pi}\right)\omega_s \tag{5}$$

$$\omega_s = \arccos(-\tan\varphi \tan\delta) \tag{6}$$

where, ω_s stands for the angle at which the sun rises each hour; φ is the latitude of the meteorological station, measured in radians; δ denotes the solar declination angle, measured in radians.

$$\delta = 0.4093 sen\left(\frac{2\pi J}{365} - 1.405\right) \tag{7}$$

where, J is the average Julian day of the month.

2.2.3 Spearman Correlation Coefficient

The Spearman correlation coefficient is a non-parametric statistical method used to assess the monotonic relationship between two variables. Compared to the Pearson correlation coefficient, it is more suitable for handling data that is not normally distributed. Its characteristic is that it does not directly calculate the correlation between the original data values, but instead converts the data into rankings and then calculates the correlation of these rankings. The specific calculation formula is as follows.

$$\rho = 1 - \frac{6\sum d_i^2}{n\left(n^2 - 1\right)} \tag{8}$$

where, ρ is the Spearman correlation coefficient; d_i denotes as the difference in ranks between the two variables for the ith data point; n represents the number of data points.

The range of ρ is between -1 and 1. The closer ρ is to 1, the stronger the positive correlation between the two variables. Conversely, the closer ρ is to -1, the stronger the negative correlation between the two variables.

2.2.4 Entropy-Weighted Water Quality Index (EWQI)

The Entropy-Weighted Water Quality Index (EWQI) assesses comprehensive groundwater quality by integrating various physicochemical parameters (Zhang et al. 2024a). The EWQI is an enhanced evaluation method that builds upon the traditional Water Quality Index (WQI) by incorporating the concept of entropy weight (Unigwe and Egbueri 2023). This innovation significantly diminishes the potential for human-induced information loss and has gained widespread adoption worldwide (Alfaleh et al. 2023; Gao et al. 2024; Yang et al. 2023). Groundwater quality can be categorized into five classes based on the Entropy-Weighted Water Quality Index. The water is deemed excellent quality with EWQI values less than 50, good quality for values between 50 and 100, medium quality for values between 100 and 150, poor quality for values between 150 and 200, and extremely poor quality for values exceeding 200 (Xiao et al. 2022b). The specific steps for calculating the EWQI are shown in Fig. 3.

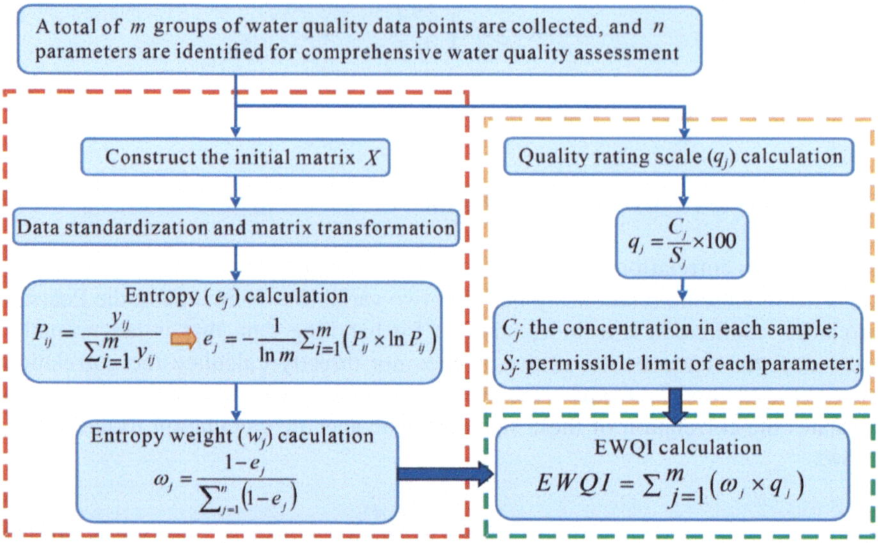

Fig. 3 Procedures of entropy-weighted water quality index calculation

3 Results and Discussion

3.1 *The Response of Water Quantity in the Study Area to the SPEI*

3.1.1 Temporal Evolution of SPEI at Different Representative Time Scales

The characteristics of the SPEI variations in the Delingha Basin from 1994 to 2024 across different time scales have been illustrated in a plot to reveal the long-term trends of precipitation and evaporation processes in this area (Fig. 4). In the graph, when the SPEI value is great than 0, it indicates that the Delingha Basin is in a state of water surplus, which is clearly marked in the form of blue bars. Conversely, when the SPEI value is less than 0, it indicates that the basin is experiencing drought, which is represented by red bars.

Due to the multi-timescale nature of the SPEI, its fluctuation patterns are highly sensitive to the time series. At shorter time scales (such as 1-month and 3-month scales), the frequency of flood and drought events is higher, but their duration is relatively short. In contrast, at longer time scales (such as 6-month and 12-month scales), the frequency of flood and drought events is lower, but the duration is longer. It is evident that across all time scales, from 1 to 3, 6, and 12 months, the SPEI generally shows an upward trend, indicating that over the past 30 years, the drought

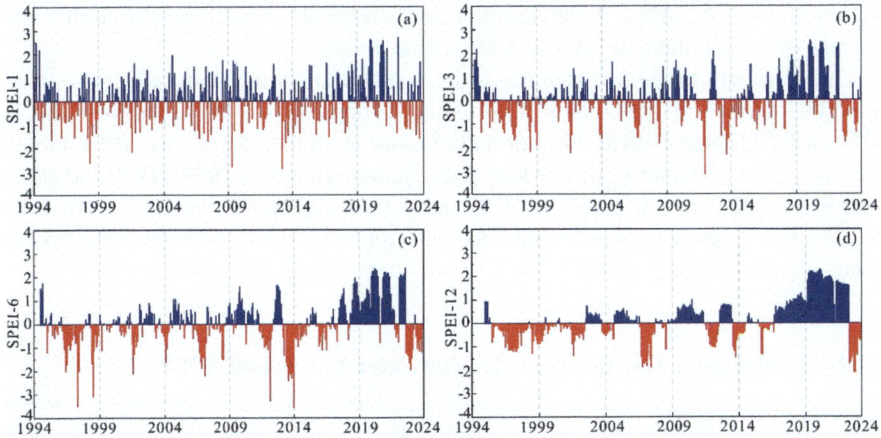

Fig. 4 Temporal evolution of SPEI at different time scales of **a** 1-month, **b** 3-months, **c** 6-months, and **d** 12-months at the Delingha meteorological station

conditions in the Delingha Basin have been alleviated. This climate change characteristic is consistent with the conclusion of humidification in western Qilian Mountains obtained by previous studies (Bai et al. 2023).

The uncertainty in monthly temperatures and precipitation leads to the most intense and frequent positive and negative fluctuations in the monthly SPEI. The SPEI-1 has experienced four extreme drought events, occurring in April 1998, July 2001, April 2009, and March 2013, with the SPEI of −2.64, −2.16, −2.78 and −2.86, respectively (Fig. 4a). For certain dates, the SPEI values are relatively high (for example, in January, February, October, November, and December of 2020, and January and December of 2021, etc.), all ranging between 2 to 3. Overall, from 1994 to 2024, mild droughts accounted for the highest proportion of total drought events, at 72.78%, followed by moderate droughts at 20%, while severe and extreme droughts accounted for 5% and 2.22%, respectively. Extreme droughts mainly occurred in the spring season.

The 3-month scale serves as a seasonal scale, and the variations in its SPEI can reflect the evolution patterns of seasonal droughts. From Fig. 4b, it can be seen that there are a total of 50 drought events for SPEI-3, with an average drought intensity of approximately 2.48, and an average drought duration of 3.54 months. Extreme drought events mainly occurred in the summer of 1998, summer of 2001, winter of 2011, and summer of 2013, with SPEI-3 values of −2.18, −2.24, −3.14, and −2.32, respectively. Winter has the highest number of drought occurrences (17 times), followed by summer, autumn, and spring (14, 11, and 8 times, respectively). Among them, summer is the frequent season for extreme drought events, accounting for 75% of the total number of drought events. In terms of drought severity, mild droughts are the primary type of drought event, accounting for 60% of the total drought events, and they mainly occur in winter. Moderate droughts account for 18% of the total, predominantly occurring in the summer and winter seasons. Severe droughts and extreme droughts make up 14% and 8%, respectively.

The 12-month time scale represents an annual scale, which can reflect the inter-annual variability characteristics of droughts. As depicted in Fig. 4d, the years 2020, 2021, and 2022 stand out as exceptionally humid, with their respective SPEI indices reaching 2.18, 1.83, and 1.55. The four driest years were 1996, 1997, 2007, and 2023, with SPEI values of −0.7, −0.7, −0.7, and −1.1, respectively. Overall, from 1994 to 2024, there were a total of 16 mild droughts and 1 moderate drought at the annual scale.

3.1.2 Correlation Test Between Groundwater Levels and SPEI

The impact of climate change on groundwater quantity is a significant research focus at present, but directly assessing the quantity of groundwater often entails complex technical challenges (Ascott et al. 2022). Leveraging the positive coupling relationship between groundwater levels and water quantity, this study ingeniously uses groundwater levels as a proxy variable to simplify the research process and enhance the feasibility of the study (Panday and Kumar 2024).

The long-term sequence of groundwater levels monitored by the national monitoring wells in the Delingha Basin is shown in Fig. 5. It is clearly observable that the groundwater levels at the national wells across the study area exhibit periodic fluctuations. These fluctuations are the result of the combined effects of natural factors and human activities. For instance, precipitation and evapotranspiration are key components in the natural cycle of groundwater resources, exhibiting significant seasonal variations. On the other hand, the extraction of groundwater is directly related to human demand, especially during peak water usage periods, where this demand is prominently manifested. GW67 is the national monitoring well with the most intense water level fluctuations, located around the urban area of Delingha. Intense periodic human activities, such as urban water extraction and agricultural irrigation, also justify the change in groundwater level at GW67 (Jiang et al. 2023). In general, the water levels at all monitoring wells show a slow downward trend over time, which is consistent with the pattern of intensified potential evapotranspiration of groundwater due to global warming. It is worth noting that the fluctuation of water levels in the monitoring wells has a cycle greater than 12 months, and there is a trend of continuous extension. This phenomenon may be due to the extension of the residence time in the process of surface water replenishing groundwater, caused by the continuous decline of groundwater levels.

The Kolmogorov–Smirnov test is used to determine whether the groundwater levels and SPEI values follow a normal distribution. The test results can be seen in Table 1, where the majority of the groundwater levels and SPEI do not conform to a normal distribution. Therefore, this study employs the Spearman correlation coefficient to analyze the correlation between groundwater levels and SPEI, as the Pearson correlation coefficient is suitable for data that conforms to a normal distribution (Balacco et al. 2022).

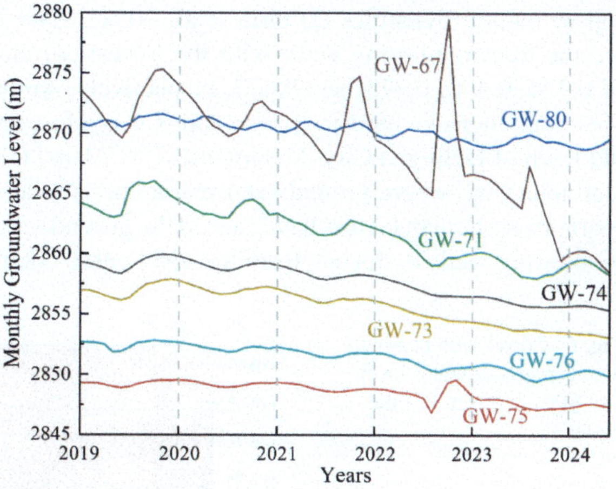

Fig. 5 Dynamic changes of groundwater level in Delingha Basin from 2019 to 2024

Table 1 The normality test of groundwater levels and SPEI at different time scales

Monitoring well	*p* value	SPEI time scale	*p* value
GW66	**<0.001**	SPEI-1	**0.046**
GW68	**0.026**	SPEI-3	0.198
GW70	**<0.001**	SPEI-6	**0.009**
GW69	**0.009**	SPEI-9	**0.007**
GW71	**0.024**	SPEI-12	0.054
GW73	**<0.001**	SPEI-18	0.2
GW74	**0.006**	SPEI-24	0.2
GW75	**0.002**	SPEI-48	**0.039**
GW76	**<0.001**		
GW67	**0.029**		
GW80	0.2		

The correlation coefficients between groundwater levels and SPEI are displayed in Fig. 6. It can be clearly seen that the correlation coefficients between the two variables show a trend of increasing first and then decreasing as the SPEI scale increases. The groundwater levels exhibit the most pronounced response to the SPEI series at the 6- and 9-month time scales, suggesting that the groundwater levels' response to climate variability takes approximately 6–9 months. Therefore, this study adopts the 6-month time scale SPEI index for spatial monitoring of groundwater levels to visually emphasize the relationship between groundwater levels and SPEI (Liu et al. 2013). The monitoring results, as depicted in Fig. 7, show that the spatial correlation coefficients exhibit a general pattern of being lower in the northeastern region and higher in the surrounding areas. This may be attributed to the sparser population in the front edge of the alluvial diluvial inclined plain and the alluvial lacustrine plain, which allows the groundwater levels in these areas to more closely follow natural dynamics (D'Oria et al. 2024). GW67, GW70, GW75, and GW69 are the four monitoring wells with the lowest correlation coefficients, with values of 0.324, 0.406, 0.473, and 0.477, respectively. Among them, GW67 is located in the comprehensive industrial zone and is very close to the commercial center of the old town of Delingha City. Meanwhile, GW70 and GW69 are the water sources for Delingha City, where groundwater extraction is relatively frequent. In other words, more severe human factors have caused the groundwater levels of the aforementioned monitoring wells to deviate from the ideal curve controlled by natural factors.

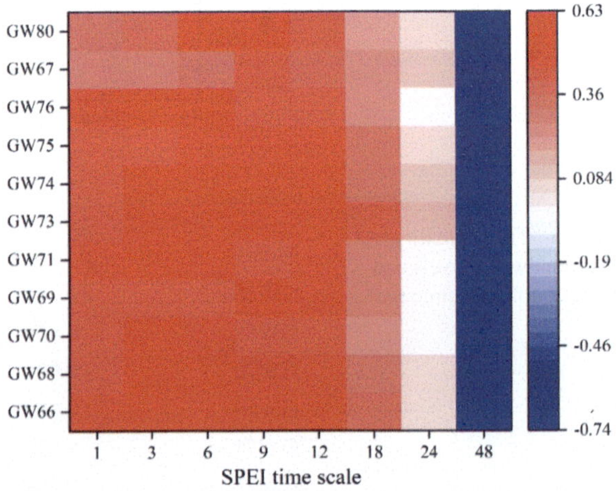

Fig. 6 Correlation heatmap between groundwater level and SPEI at different time scales

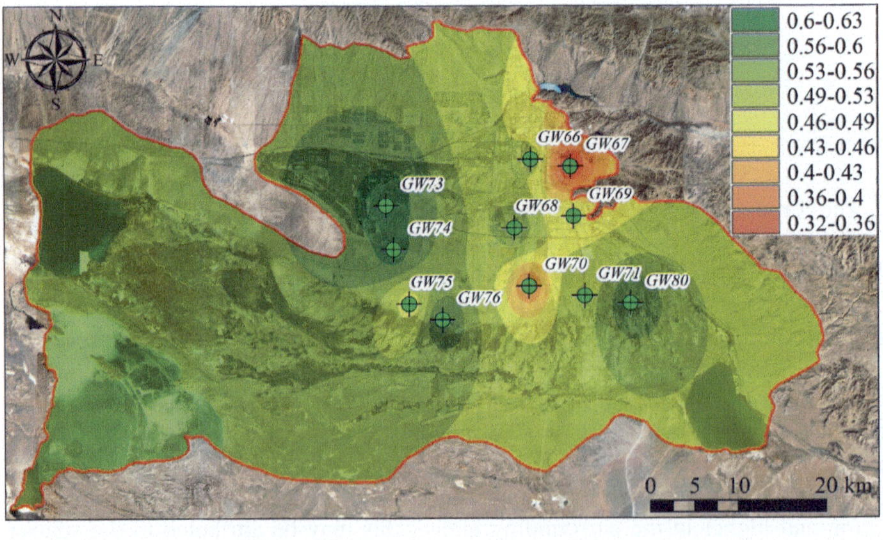

Fig. 7 Spatial interpolation plot of correlation coefficients between monitoring wells and 6-scale SPEI

3.2 The Overall Water Quality of Groundwater in the Study Area

3.2.1 Physicochemical Characteristics of Groundwater

To enhance the characterization of the hydrochemical quality of the phreatic groundwater in the Bayin River watershed, the results of the analytical tests for 13 parameters are statistically analysed as shown in Fig. 8. The drinking water quality limits of Chinese and WHO guidelines for these parameters are also illustrated (GAQS 2017; WHO 2017). The pH values of groundwater spanned from 7.18 to 8.54, averaging at 7.79, indicating an alkaline water quality. Only one sampled groundwater with the pH value exceeding the acceptable range of 6.5–8.5 as suggested by China. The Total Hardness (TH) of the phreatic groundwater ranged from 125.10 to 2,577.06 mg/L, with an average of 511.75 mg/L. The total dissolved solids (TDS) values varied between 170.48 and 9,318.09 mg/L, at a mean value of 1,282.14 mg/L. In comparison to the guidelines recommended by China for TH (450 mg/L) and TDS (1,000 mg/L), 34.15% and 31.71% of the sampled groundwaters exceeded the limit, respectively. This indicates that some of the sampled phreatic groundwaters in the study area exhibit brackish and hard properties, which may potentially lead to a decline in the overall groundwater quality (Hu et al. 2024; Zhang et al. 2024b).

The concentrations of the major anions and cations in the sampled phreatic groundwater vary widely and exceed the drinking water quality limit established by China and WHO to varying extents (Fig. 8). The mean concentrations of main cations in the groundwater are as follows: Na^+ (299.08 mg/L) > Ca^{2+} (104.91 mg/L) > Mg^{2+} (60.65 mg/L) > K^+ (2.5 mg/L). The concentration of Na^+ ranges from 13.0 mg/L to 3,304.92 mg/L. Ca^{2+} and Mg^{2+} concentrations vary from 26.05–551.10 mg/L to 2.92–358.43 mg/L, respectively. K^+ shows significantly lower concentrations compared to the other major cations, with value ranging from 0.00 to 23.50 mg/L. The mean concentration of major anions is Cl^- > $SO_4{}^{2-}$ > $HCO_3{}^-$. The concentration of Cl^- ranges from 3.68 to 3,545.00 mg/L, with a mean value of 357.92 mg/L. The concentrations of SO_4^{2-} and HCO_3^- varied from 3.26–1,998.05 mg/L to 6.00–1,671.95 mg/L, respectively, with mean values of 320.79 mg/L and 294.98 mg/L. Nitrate is a common groundwater pollutant originating from agricultural and industrial activities, serves as a valuable indicator for identify the impact of human activities on the hydrochemical quality of groundwater. The NO_3^- and NO_2^- concentrations in the phreatic groundwater of the study area ranges from 0–106.29 mg/L to 0.00–0.16 mg/L, with mean values of 16.23 and 0.02 mg/L. While the permissible limits for NO_3^- and NO_2^- for drinking purposes set by China and WHO are 20 and 0.02 mg/L. 19.51% of the phreatic groundwaters sampled in the study area exceed the prescribed limits for NO_3^- and NO_2^- respectively. The concentration of $NH_4{}^+$ in the phreatic groundwater is within the range of 0.0–0.2 mg/L, all these groundwater samples within the maximum acceptable limit of 0.2 mg/L suggested by WHO.

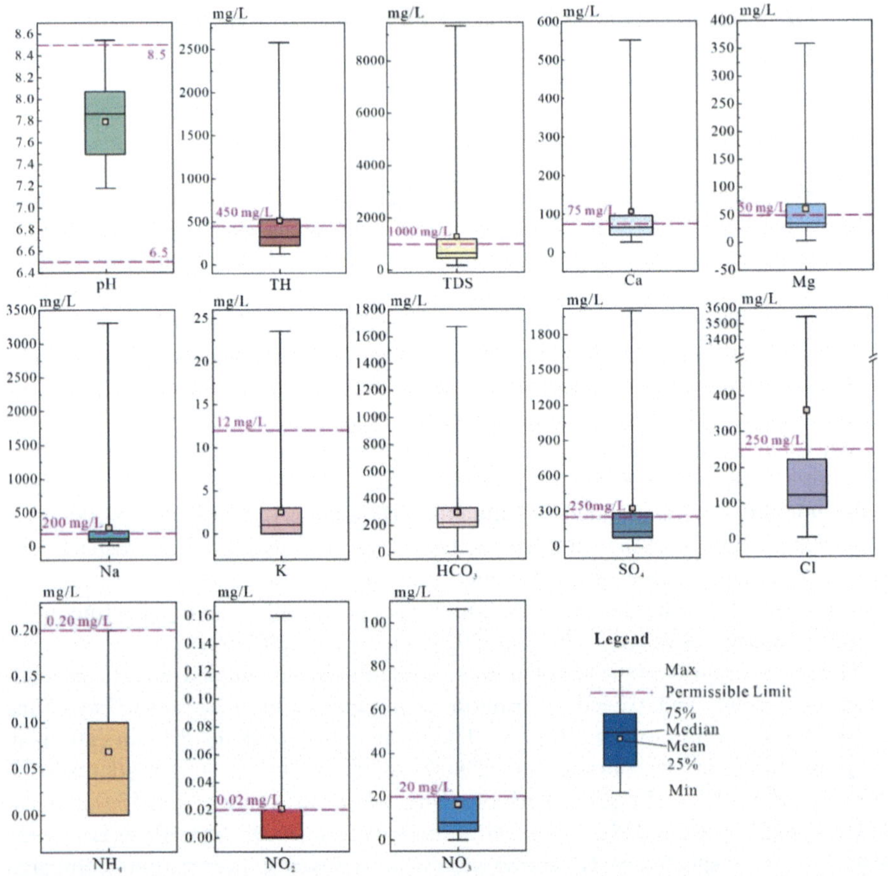

Fig. 8 Box plots of various indexes with respect to permissible limits

The variation of hydrochemical facies with TDS in the phreatic groundwater is illustrated in Fig. 9. The phreatic groundwater in the study area exhibits three predominant hydrochemical facies: the $SO_4 \cdot Cl$–$Ca \cdot Mg$ type, the $SO_4 \cdot Cl$–Na type, and the HCO_3–$Ca \cdot Mg$ type. Specifically, 46.34% and 36.59% of the sampled groundwaters are classified as $SO_4 \cdot Cl$–$Ca{\cdot}Mg$ type and $SO_4{\cdot}Cl$–Na type, respectively (Li et al. 2016; Qi et al. 2024). Only 17.07% of the groundwater samples are characterized as the HCO_3–$Ca \cdot Mg$ type. It can be seen that freshwater in the study area with TDS < 1,000 is predominantly characterized by the HCO_3–$Ca \cdot Mg$ and $SO_4 \cdot Cl$–$Ca{\cdot}Mg$ types. Brackish water with a TDS ranging from 1,000 to 3,000 mg/L and salt water with TDS > 3,000 mg/L are mostly of primarily associated with the $SO_4 \cdot Cl$–$Ca \cdot Mg$ and $SO_4 \cdot Cl$–Na types. The correlation between anion ratio and TDS content is more significant. As the TDS value of phreatic groundwater increased, the hydrochemical facies gradually evolved from HCO_3 dominated to Cl and SO_4 dominated. The spatial distribution map of TDS illustrates a progressive increase in groundwater

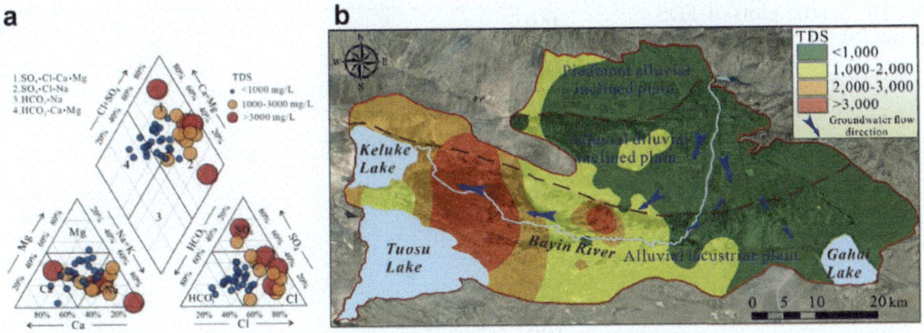

Fig. 9 **a** Piper diagram and **b** spatial distribution of TDS of groundwater in the study area

salinity from the mountain front area to the alluvial-lacustrine plain. In the middle and downstream of the Bayin River watershed, groundwater TDS contents have exceeded the prescribed drinking water limit, resulting in the groundwater becoming brackish or saline. Extensive research has demonstrated that the progressive trend of groundwater salinisation is predominantly attributed to the local arid climate and low precipitation levels (Zhang et al. 2020; Zheng et al. 2021). Particularly in the alluvial-lacustrine plain regions, intensified evaporation and water-rock interactions contribute to the salinisation of groundwater.

3.2.2 Spatial Distribution of Water Quality Based on EWQI

To further assess the groundwater quality in the watershed, 13 parameters are selected for a comprehensive water quality evaluation using the EWQI (Li et al. 2015). The results showed that the EWQI of phreatic groundwater in Bayin River watershed varying between 17.14 and 452.90, with an average value of 100.55. Based on the EWQI classification standards, the groundwater quality varied significantly from Rank 1 to Rank 5 (Fig. 10). Specifically, 36.59 and 29.27% of the phreatic groundwater quality is classified as excellent (Rank 1) and good quality (Rank 2), respectively. Almost all the groundwater in these two categories is fresh water with the TDS values less than 1,000 mg/L. 17.07% of the groundwater is designated as medium quality water (Rank 3) and their TDS distribution ranges from 530.04 to 2,056.98 mg/L. 4.88% and 9.76% of the groundwater is categorized as poor-quality water (Rank 4) and extremely poor-quality water (Rank 5), respectively. It is to be observed that all the groundwaters categorised as Rank 5 are salt water and their TDS values are greater than 3,000 mg/L. Overall, most groundwaters in the watershed is suitable for drinking or domestic purposes (Xiao et al. 2024). Groundwater quality is poor in some areas with high TDS.

Based on the EWQI calculations for groundwater, a spatial distribution map for basin synthesis and water quality assessment was obtained (Fig. 11). It can be seen that the groundwater in the Piedmont alluvial inclined plain, alluvial diluvial inclined plain, and the eastern alluvial lacustrine plain is categorised as excellent or good

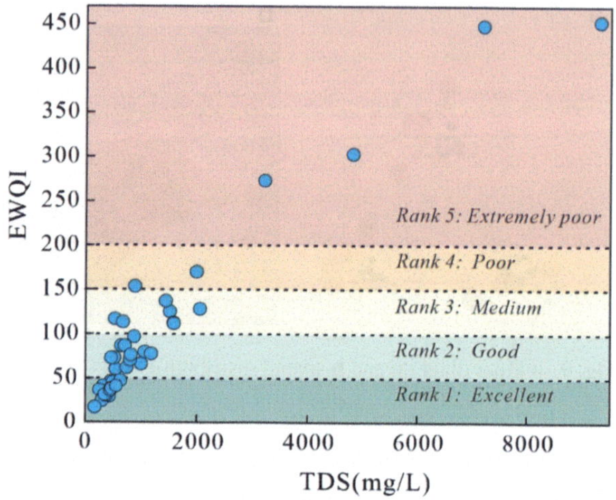

Fig. 10 Scatter plots of TDS versus EWQI of groundwater

quality water. Groundwater of medium quality is distributed in the central part of the alluvial lacustrine plain and also sporadically in the alluvial diluvial inclined plain. Poor and extremely poor-quality groundwater locations are predominantly distributed in the western part of the alluvial lacustrine plain in the tailrace lake area. Along the flow path of groundwater, water flowing southeast is more suitable for drinking purposes than that flowing southwest. Conversely, the overall water quality deteriorates as groundwater flows southwest. This pattern of change closely resembles the spatial distribution of TDS in the groundwater within the watershed, as well as the evolution pattern of hydrochemical facies. The population in the Delingha is sparse, and human activities have a very limited impact on the quality of groundwater. The poor water quality of the groundwater is mainly due to natural factors. Therefore, the salinization of groundwater caused by strong evaporation and prolonged water-rock interactions is the main reason for the decline in groundwater quality in the Bayin River watershed. In recent years, the degree of aridity in the Bayin River watershed has decreased, and the climate has gradually become more humid, which may lead to an improvement in the quality of groundwater within the watershed.

3.3 Sustainable Development and Utilization of Groundwater Based on Water Quantity and Quality Under Climate Change

Groundwater resources in arid regions are a vital water source for sustaining ecological systems and human activities, and their sustainable development and utilization are of significant importance to the region's economy, society, and environment. When selecting sites for the sustainable development and utilization of groundwater in arid areas, it is essential to comprehensively consider two key factors (water

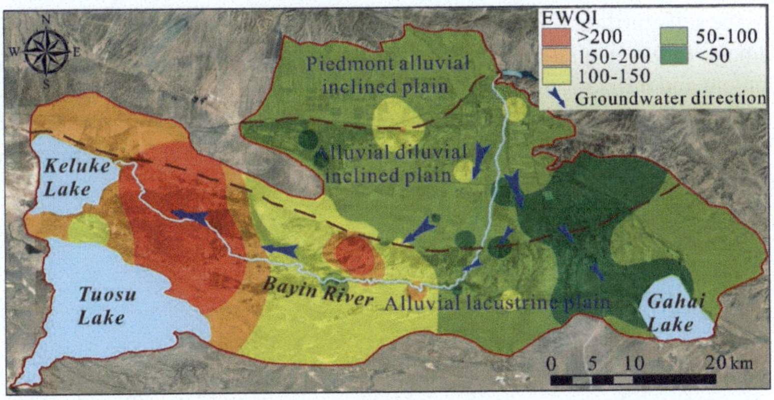

Fig. 11 Spatial distribution of water quality based on EWQI

quantity and water quality). In terms of water quantity, the stability of groundwater dynamics and the abundance of groundwater reserves are the primary factors in determining the selection of sites for sustainable groundwater development and utilization. Long-term stable groundwater resources ensure the long-term sustainable use of water resources, preventing the issue of resource depletion due to the continuous decline of groundwater levels. Regarding water quality, choosing groundwater with good quality as the object of utilization is a prerequisite for the sustainable development and utilization of groundwater. The quality of water is directly related to the health of water consumers. Moreover, good water quality can also reduce the cost of water safety treatment.

The spatial distribution of the correlation coefficient between groundwater levels and the SPEI indicates that groundwater tending towards natural dynamics is primarily concentrated on the west and southeast sides of the Delingha Basin. The frequent human activities in the urban area of Delingha have led to drastic changes in groundwater levels, making it unsuitable for long-term sustainable development and utilization of groundwater resources. Additionally, the spatial distribution of the EWQI illustrates that the groundwater in the study area is mainly influenced by strong water–rock interactions and evaporation, with poor water quality groundwater predominantly located on the west side of the basin. Therefore, the sustainable development and utilization sites for groundwater in the study area should be distributed on the west side of the alluvial diluvial inclined plain and the east side of the alluvial lacustrine plain. Compared to the former, the alluvial lacustrine plain in the east has finer lithology and more gentle terrain, which slows down the groundwater flow. In some areas, groundwater overflows, forming wetlands and marshes. The lower flow speed of groundwater also implies a slower replenishment rate. The replenishment of groundwater resources may not keep up with the pace of exploitation, leading to a continuous decline in groundwater levels and ultimately causing the depletion of groundwater resources. Moreover, the overflow of groundwater makes it more susceptible to pollution from external activities, and the slower flow rate also weakens the self-purification capacity of the water body, potentially leading to permanent damage to water quality. In summary, the optimal location for the sustainable development and utilization of groundwater in the study area is on the west side of the alluvial diluvial inclined plain, where the stable and abundant groundwater volume, better groundwater quality, and timely replenishment of groundwater resources all substantiate this fact.

4 Conclusions

With global climate change, the sustainable development and utilization of groundwater resources are becoming increasingly important for the livelihoods, economic development, and environmental protection of residents in arid regions. This study uses the Delingha Basin in the Qaidam Basin as a case study to deeply analyze the response of groundwater quantity to climate change. Finally, valuable recommendations are provided for the sustainable development and utilization of groundwater in this area, based on both water quantity and quality. The main findings of this study are as follows.

At shorter timescales (1-month and 3-month), the frequency of drought and flood events is higher, but their duration is shorter. In contrast, longer timescales (6-month and 12-month) show lower frequency but longer duration of these events. Overall, the drought conditions in the Delingha Basin have been alleviated. The 3-month SPEI, representing seasonal drought variations, highlights that mild droughts are most common, particularly in winter, while severe and extreme droughts are more frequent in summer. At the 12-month scale, representing annual drought variability, 2020–2022 were notably humid years, while 1996, 1997, 2007, and 2023 were the driest. Overall, mild droughts were most frequent across the years, with moderate droughts being relatively rare. The spatial distribution of the correlation coefficients between the 6-month scale SPEI index and groundwater levels indicates that the groundwater levels in the western and southeastern parts of the Delingha Basin are more inclined to follow natural dynamics and are less affected by human activities.

The pH of the groundwater in the study area ranges between 7.18 and 7.79. Due to the spatial variability of evaporation concentration, the TDS of the groundwater shows a characteristic of being higher downstream and lower upstream. The EWQI statistical results show that 36.59% and 29.27% of the groundwater fall into the excellent and good categories, respectively. 4.88% and 9.76% of the groundwater are categorized as poor and very poor water quality, respectively. There is a strong correlation between EWQI and TDS, and their spatial distribution patterns are similar, that is, the water quality on the east side of the alluvial lacustrine plain is poor.

This research, considering both the quantity and quality of water resources, posits that the western part of the alluvial diluvial inclined plain is the optimal site for the sustainable exploitation and utilization of groundwater in the region. This conclusion is drawn by weighing the benefits of the area's ample water supply coupled with its superior water quality, making it an ideal candidate for long-term groundwater management and development strategies.

Acknowledgements This research was funded by the National Natural Science Foundation of China (Grant No. 42477059; 42007183); the Key Lab of Geo-environment of Qinghai Province (Grant No. 2023-KJ-15; 2024-KJ-06); the Kunlun Talents—Science and Technology Leading Talents Program of Qinghai Province (Grant to Zhen Zhao); the Applied Basic Research Project of Science and Technology Program of Qinghai Province (Grant No. 2024-ZJ-771); and the Student Research Training Program of Southwest Jiaotong University (Grant No. 241925).

References

Alfaleh A, Ben KN, Alviz-Meza A (2023) Is the entropy-weighted water quality index a suitable index for evaluating the groundwater quality in Ha'il, Saudi Arabia? Water Sci Technol 88:778–797. https://doi.org/10.2166/wst.2023.230

Alghamdi AG, Aly AA, Majrashi MA, Ibrahim HM (2023) Impact of climate change on hydrochemical properties and quality of groundwater for domestic and irrigation purposes in arid environment: a case study of Al-Baha region, Saudi Arabia. Environ Earth Sci 82(1):39. https://doi.org/10.1007/s12665-022-10731-z

Alley WM (1984) The palmer drought severity index: limitations and assumptions. J Appl Meteorol Climatol 23(7):1100–1109. https://doi.org/10.1175/1520-0450(1984)023%3c1100:TPDSIL%3e2.0.CO;2

Aravinthasamy P, Karunanidhi D, Subba RN, Subramani T, Srinivasamoorthy K (2020) Irrigation risk assessment of groundwater in a non-perennial river basin of South India: implication from irrigation water quality index (IWQI) and geographical information system (GIS) approaches. Arab J Geosci 13(21):1125. https://doi.org/10.1007/s12517-020-06103-1

Ascott MJ, Macdonald DMJ, Sandwidi WJP, Black E, Verhoef A, Zongo G, Tirogo J, Cook P (2022) Time of emergence of impacts of climate change on groundwater levels in sub-Saharan Africa. J Hydrol 612:128107. https://doi.org/10.1016/j.jhydrol.2022.128107

Asif Z, Chen Z, Sadiq R, Zhu Y (2023) Climate change impacts on water resources and sustainable water management strategies in North America. Water Resour Manag 37(6):2771–2786. https://doi.org/10.1007/s11269-023-03474-4

Bahrami M, Zarei AR (2023) Assessment and modeling of groundwater quality for drinking, irrigation, and industrial purposes using water quality indices and GIS technique in fasarud aquifer (Iran). Model Earth Syst Environ 9(4):3907–3921. https://doi.org/10.1007/s40808-023-01725-2

Bai B, Yue P, Zhang Q, Yang J, Ma P, Han T, Jiang Y, Huang P, Ma Y (2023) Changing characteristics of ecosystem and water storage under the background of warming and humidification in the Qilian Mountains, China. Sci Total Environ 893:164959. https://doi.org/10.1016/j.scitotenv.2023.164959

Balacco G, Alfio MR, Fidelibus MD (2022) Groundwater drought analysis under data scarcity: the case of the Salento aquifer (Italy). Sustainability 14:707

Barbieri M, Barberio MD, Banzato F, Billi A, Boschetti T, Franchini S, Gori F, Petitta M (2023) Climate change and its effect on groundwater quality. Environ Geochem Health 45(4):1133–1144. https://doi.org/10.1007/s10653-021-01140-5

D'Oria M, Balacco G, Todaro V, Alfio MR, Tanda MG (2024) Assessing the impact of climate change on a coastal karst aquifer in a semi-arid area. Groundw Sustain Dev 25:101131. https://doi.org/10.1016/j.gsd.2024.101131

Dao PU, Heuzard AG, Le TXH, Zhao J, Yin R, Shang C, Fan C (2024) The impacts of climate change on groundwater quality: a review. Sci Total Environ 912:169241. https://doi.org/10.1016/j.scitotenv.2023.169241

Dong J, Xing L, Cui N, Zhao L, Guo L, Gong D (2023) Standardized precipitation evapotranspiration index (SPEI) estimated using variant long short-term memory network at four climatic zones of China. Comput Electron Agric 213:108253. https://doi.org/10.1016/j.compag.2023.108253

Fallahi MM, Shabanlou S, Rajabi A, Yosefvand F, IzadBakhsh MA (2023) Effects of climate change on groundwater level variations affected by uncertainty (case study: Razan aquifer). Appl Water Sci 13(6):143. https://doi.org/10.1007/s13201-023-01949-8

Gao M, Qian J, Li X, Wang Z, Hou X, Gui C, Bai Z, Li J, Zuo X, Zhao C (2024) Assessment of groundwater quality using entropy-weighted quality index (EWQI) and multivariate statistical approaches in Heilongdong Spring Basin, Northern China. Environ Earth Sci 83(7):196. https://doi.org/10.1007/s12665-024-11458-9

GAQS (2017) Standards for groundwater quality (GB/T 14848-2017). General Administration of Quality Supervision, Beijing

Han F, Liu T, Huang Y, Zan C, Pan X, Xu Z (2023) Response of water quality to climate warming and atmospheric deposition in an alpine lake of Tianshan Mountains, Central Asia. Ecol Indic 147:109949. https://doi.org/10.1016/j.ecolind.2023.109949

Hu W, Xiao Y, Wang L, Zhang Y, Feng M, Shi W, He C, Wen Y, Yang H, Han J, Wang J (2024) Spatial variability source identification and partitioning of groundwater constituents in a typical lakeside plain on Yungui Plateau. Process Saf Environ Protect 191:2402–2415. https://doi.org/10.1016/j.psep.2024.09.107

Jiang W, Meng L, Liu F, Sheng Y, Chen S, Yang J, Mao H, Zhang J, Zhang Z, Ning H (2023) Distribution, source investigation, and risk assessment of topsoil heavy metals in areas with intensive anthropogenic activities using the positive matrix factorization (PMF) model coupled with self-organizing map (SOM). Environ Geochem Health 45(8):6353–6370. https://doi.org/10.1007/s10653-023-01587-8

Jiang W, Sheng Y, Wang G, Shi Z, Liu F, Zhang J, Chen D (2022) Cl, Br, B, Li, and noble gases isotopes to study the origin and evolution of deep groundwater in sedimentary basins: a review. Environ Chem Lett 20(2):1497–1528. https://doi.org/10.1007/s10311-021-01371-z

Jiang W, Wang G, Sheng Y, Shi Z, Zhang H (2019) Isotopes in groundwater (2H, 18O, 14C) revealed the climate and groundwater recharge in the Northern China. Sci Total Environ 666:298–307. https://doi.org/10.1016/j.scitotenv.2019.02.245

Kamali S, Asghari K (2023) The effect of meteorological and hydrological drought on groundwater storage under climate change scenarios. Water Resour Manag 37(8):2925–2943. https://doi.org/10.1007/s11269-022-03268-0

Konapala G, Mishra AK, Wada Y, Mann ME (2020) Climate change will affect global water availability through compounding changes in seasonal precipitation and evaporation. Nat Commun 11(1):3044. https://doi.org/10.1038/s41467-020-16757-w

Kumar R, Manzoor S, Vishwakarma D, Al-Ansari N, Kushwaha NL, Elbeltagi A, Sushanth K, Prasad V, Kuriqi A (2022) Assessment of climate change impact on snowmelt runoff in Himalayan region. Sustainability 14:1150. https://doi.org/10.3390/su14031150

Li P, Wu J, Qian H (2015) Hydrochemical appraisal of groundwater quality for drinking and irrigation purposes and the major influencing factors: a case study in and around Hua County, China. Arab J Geosci 9(1):15. https://doi.org/10.1007/s12517-015-2059-1

Li P, Li X, Meng X, Li M, Zhang Y (2016) Appraising groundwater quality and health risks from contamination in a semiarid region of Northwest China. Expo Health 8(3):361–379. https://doi.org/10.1007/s12403-016-0205-y

Liu M, Qin P, Liu K, Xia Z, Wang M (2013) Response of lake water level of Honghu Lake to SPEI/SPI drought indices at different time scales. Meteorol Monthly 39(9):1163–1170. https://doi.org/10.7519/j.issn.1000-0526.2013.09.011

Liu Y, Zhu Y, Ren L, Singh V, Yang X, Yuan F (2017) A multiscalar Palmer drought severity index. Geophys Res Lett 44. https://doi.org/10.1002/2017GL073871

Nagaraj S, Masilamani US (2023) Hydrogeochemical and multivariate statistical approaches to investigate the characteristics of groundwater quality in fluoride-enriched hard rock region in Tirupathur district of Tamil Nadu, India. Environ Sci Pollut Res 30(44):99809–99829. https://doi.org/10.1007/s11356-023-29254-6

Palmer W (1965) Meteorological drought, pp 1–65

Pandey S, Mohapatra G, Arora R (2023) Groundwater quality, human health risks and major driving factors in arid and semi-arid regions of Rajasthan, India. J Clean Prod 427:139149. https://doi.org/10.1016/j.jclepro.2023.139149

Panday DP, Kumar M (2024) Decoding uranium variation over the Indian peninsula through leveraging standardized precipitation evapotranspiration indices and groundwater level fluctuations. Groundw Sustain Dev 26:101241. https://doi.org/10.1016/j.gsd.2024.101241

Qi Z, Wang S, Liu K (2024) Origin of mineral water in Heicigou, northern slope of eastern Kunlun Mountains. J Salt Lake Res 32(3):21–31. https://doi.org/10.3724/j.yhyj.2024003

Qiu J, Shen Z, Xie H (2023) Drought impacts on hydrology and water quality under climate change. Sci Total Environ 858:159854. https://doi.org/10.1016/j.scitotenv.2022.159854

Singh R, Kayastha SP, Shrestha SM, Sapkota RP (2024) Hydro-geochemical conditions under projected climate change scenarios of Marshyangdi River, Nepal. Theor Appl Climatol 155(6):5375–5387. https://doi.org/10.1007/s00704-024-04890-1

Unigwe CO, Egbueri JC (2023) Drinking water quality assessment based on statistical analysis and three water quality indices (MWQI, IWQI and EWQI): a case study. Environ Dev Sustain 25(1):686–707. https://doi.org/10.1007/s10668-021-02076-7

Van Lanen HAJ, Peters E (2000) Definition, effects and assessment of groundwater droughts. In: Vogt JV, Somma F (eds) Drought and drought mitigation in Europe. Springer, Netherlands, Dordrecht, pp 49–61

Vicente-Serrano SM, Beguería S, López-Moreno JI (2010) A multiscalar drought index sensitive to global warming: the standardized precipitation evapotranspiration index. J Clim 23(7):1696–1718. https://doi.org/10.1175/2009JCLI2909.1

Wang X, Zheng W, Tian W, Gao Y, Wang X, Tian Y, Li J, Zhang X (2022) Groundwater hydrogeochemical characterization and quality assessment based on integrated weight matter-element extension analysis in Ningxia, upper Yellow River, northwest China. Ecol Ind 135:108525. https://doi.org/10.1016/j.ecolind.2021.108525

Wang L, Xiao Y, Yang H, Zhang Y, Wang S, Qi Z, Han J, Hu W, Wang J, Hao Q, Senapthi V (2024) Formation mechanism of high-altitude glacial mineral water in the Kunlun Mountains of Tibetan Plateau: insights from isotopes and hydrochemistry. J Hydrol: Reg Stud 53:101789. https://doi.org/10.1016/j.ejrh.2024.101789

WHO (2017) Guidelines for drinking-water quality: fourth edition incorporating first addendum, 4th edn + 1st edn. World Health Organization, Geneva

Xiao Y, Hao Q, Zhang Y, Zhu Y, Yin S, Qin L, Li X (2022a) Investigating sources, driving forces and potential health risks of nitrate and fluoride in groundwater of a typical alluvial fan plain. Sci Total Environ 802:149909. https://doi.org/10.1016/j.scitotenv.2021.149909

Xiao Y, Liu K, Hao Q, Xiao D, Zhu Y, Yin S, Zhang Y (2022b) Hydrogeochemical insights into the signatures, genesis and sustainable perspective of nitrate enriched groundwater in the piedmont of Hutuo watershed, China. Catena 212:106020. https://doi.org/10.1016/j.catena.2022.106020

Xiao Y, Zhang Y, Yang H, Wang L, Han J, Hao Q, Wang J, Zhao Z, Hu W, Wang S, Fan Q, Qi Z (2024) Interaction regimes of surface water and groundwater in a hyper-arid endorheic watershed on Tibetan Plateau: insights from multi-proxy data. J Hydrol 644:132020. https://doi.org/10.1016/j.jhydrol.2024.132020

Yang Y, Li P, Elumalai V, Ning J, Xu F, Mu D (2023) Groundwater quality assessment using EWQI with updated water quality classification criteria: a case study in and around Zhouzhi County, Guanzhong Basin (China). Expo Health 15(4):825–840. https://doi.org/10.1007/s12403-022-00526-9

Yang H, Xiao Y, Zhang Y, Wang L, Wang J, Hu W, Liu G, Liu F, Hao Q, Wang C, Xu X (2024) Lithology controls on the mixing behavior and discharge regime of thermal groundwater in the

Bogexi geothermal field on Tibetan Plateau. J Hydrol 628:130523. https://doi.org/10.1016/j.jhydrol.2023.130523

Yi B, Liu J, He W, Lü X, Cao X, Chen X, Zeng X, Zhang Y (2023) Optical variations of dissolved organic matter due to surface water—Groundwater interaction in alpine and arid Datonghe watershed. Sci Total Environ 864:161036. https://doi.org/10.1016/j.scitotenv.2022.161036

Zhang B, Zhao D, Zhou P, Qu S, Liao F, Wang G (2020) Hydrochemical characteristics of groundwater and dominant water-rock interactions in the Delingha area, Qaidam Basin, Northwest China. Water 12:836

Zhang Y, Xiao Y, Yang H, Wang L, Wang J, Hu W, Wang N, Xu Z, Liu G, Chen F, Guo X (2024a) Hydrogeochemical features, genesis, and quality appraisal of confined groundwater in a typical large sedimentary plain. Water Environ Res 96:e11088. https://doi.org/10.1002/wer.11088

Zhang Y, Xiao Y, Yang H, Wang S, Wang L, Qi Z, Han J, Hao Q, Hu W, Wang J (2024b) Hydrogeochemical and isotopic insights into the genesis and mixing behaviors of geothermal water in a faults-controlled geothermal field on Tibetan Plateau. J Clean Prod 442:140980. https://doi.org/10.1016/j.jclepro.2024.140980

Zheng Q, Gu X, Yang B, Dang X, Jiang J, You X (2021) The quaternary groundwater quality evaluation of Delingha Basin in Qinghai Province. Northwest Geol 54(2):239–247. https://doi.org/10.19751/j.cnki.61-1149/p.2021.02.021

Machine Learning Techniques in Hydrogeological Research

Song He, Xiaoping Zhou, Yuan Liu, Xiaoguang Zhao, Zilong Guan, Yujie Ji, and Peiyue Li

Abstract In recent years, machine learning is more and more extensively applied to hydrogeological research for its reliability, efficiency and feasibility. This chapter presents a review on application of machine learning in hydrogeology, or groundwater science and summarizes the associated challenges and prospects. In this chapter, commonly used machine learning methods were firstly introduced, including naïve Bayes, artificial neural network, support vector machine, decision tree and random forest. Then, overview of applying machine learning to four main hydrogeological fields, including predicting groundwater variables, mapping hydrogeological variables, identifying groundwater contamination sources and assessing groundwater contamination risk, were summarized and presented. Finally, challenges and prospects of utilizing machine learning to hydrogeology were summarized. This chapter deepens understanding of application of machine learning to hydrogeology, and provides further insights for international scholars.

S. He · X. Zhou · Y. Liu · X. Zhao · Z. Guan
PowerChina Northwest Engineering Corporation Limited, No. 18 Zhangbadong Road, Xi'an 710065, Shaanxi, China

S. He
Key Laboratory of Water and Soil Resources Conservation and Restoration in the Middle and Lower Reaches of Yellow River Basin, MNR, Zhengzhou 450016, China

S. He · P. Li (✉)
School of Water and Environment, Chang'an University, No. 126 Yanta Road, Xi'an 710054, Shaanxi, China
e-mail: peiyueli@chd.edu.cn; lipy2@163.com

Y. Ji
Shaanxi Weihe Ecological Zone Protection Center (Shaanxi Weihe River Basin Control and Protection Center), No. 198 Xiqu Road, Xi'an 710004, Shaanxi, China

P. Li
Key Laboratory of Subsurface Hydrology and Ecological Effects in Arid Region of the Ministry of Education, Chang'an University, No. 126 Yanta Road, Xi'an 710054, Shaanxi, China

Key Laboratory of Eco-Hydrology and Water Security in Arid and Semi-arid Regions of the Ministry of Water Resources, Chang'an University, No. 126 Yanta Road, Xi'an 710054, Shaanxi, China

P. Li et al. (eds.), *Sustainable Groundwater and Environment: Challenges and Solutions*, Springer Hydrogeology, https://doi.org/10.1007/978-3-031-82194-3_7

Keywords Typical machine learning algorithms · Predicting groundwater variables · Groundwater mapping · Source identification · Contamination risk · Data sharing · Enhancing interdisciplinary cooperation

1 Introduction

Groundwater is a crucial resource for human survival, societal development, and ecological stability worldwide (He and Wu 2018; Jasechko and Perrone 2021; Wang and Li 2022), particularly in arid and semi-arid regions such as Northern China, Western United States, Middle East, India, South Africa, and Middle Australia (Adimalla and Wu 2019; Cerasoli and Porporato 2023; Elumalai et al. 2019; Li et al. 2019a; Rahimi-Feyzabad et al. 2022; Rutherford et al. 2021). Over recent decades, anthropogenic activities have led to significant groundwater pollution, posing serious threats to the sustainable development of natural ecosystems and human societies (Balogun et al. 2022; He et al. 2022a; Li et al. 2019b). To mitigate the impacts of groundwater pollution, various strategies have been implemented, such as China's "Regulations on Groundwater Management" announced in November 2021 (The State Council of the People's Republic of China 2021). Furthermore, a multitude of hydrogeological studies have been conducted to understand the evolution of groundwater environments and to provide a scientific basis for sustainable groundwater management (Jean-Baptiste et al. 2020; Li et al. 2018a; Pétré et al. 2019).

As we transition from the Information Age to the era of artificial intelligence and big data, machine learning, a significant technology for artificial intelligence, has been widely used in numerous fields, including environmental and ecological science (Jiang et al. 2022; Wellmann et al. 2020), geoscience (Silva et al. 2021), engineering and material science (Song et al. 2022; Wanigasekara et al. 2021), and clinical medicine (Myszczynska et al. 2020). Machine learning can generate accurate predictions based on hidden principles and relationships within known data (Vamathevan et al. 2019) and can address problems with multidimensional spaces and nonlinear properties without a complete understanding of the physical basis (Ghaffarian et al. 2022). Consequently, machine learning has been increasingly applied in hydrology and hydrogeology (Islam et al. 2021; Koch et al. 2019).

Given the increasing number of studies applying machine learning to hydrology and hydrogeology, it is essential to summarize current research to deepen our understanding of this field. Several reviews have summarized the application of machine learning methods, such as support vector machine (SVM), artificial neural network (ANN), fuzzy-logic base model, and random forest (RF), in hydrology or water science (Chen et al. 2020; Kambalimath and Deka 2020; Sujay Raghavendra and Deka 2014; Tyralis et al. 2019). For machine learning applications in hydrogeology, there are also several review papers (Adombi et al. 2021; Di Salvo 2022; Osman et al. 2022; Tao et al. 2022). However, a range of recent publications on various aspects of applying machine learning to hydrogeology have not been systematically summarized and reviewed, indicating a need for a comprehensive review.

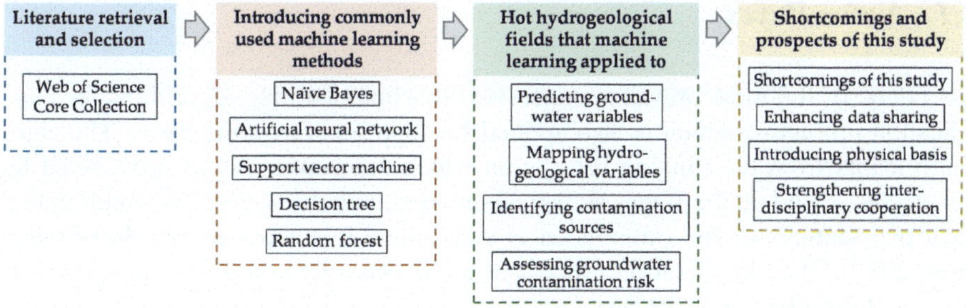

Fig. 1 Flowchart showing the steps of this chapter

This chapter provides a systematic review of the application of machine learning in various aspects of hydrogeology, based on representative studies indexed by the Web of Science Core Collection in recent years. We begin with an overview of machine learning, including supervised and unsupervised learning, dataset preparation, typical algorithms, and estimation of model performance. We then review the key hydrogeological topics where machine learning has been applied, such as predicting groundwater variables, mapping hydrogeological variables, identifying groundwater contaminant sources, and assessing groundwater contamination risk. We also discuss future challenges and prospects for applying machine learning to hydrogeology. This chapter will help deepen the understanding of the application of machine learning in hydrogeology, benefiting researchers worldwide. Figure 1 shows the flowchart of this chapter. This chapter is meaningful to help understand the application of machine learning in hydrogeology, which also benefits international scholars.

2 Introduction to Some Well-Known Machine Learning Techniques

Since being named by Arthur Samuel, a distinguished computer scientist at the time, in 1959, machine learning has come a long way and been extensively applied in numerous fields of science and engineering (Samuel 1959). Machine learning is a branch of artificial intelligence that semi-automatically executes variable prediction or pattern recognition on raw data after learning the underlying associations of given datasets, based on a series of mathematical theories and programed algorithms (Jordan and Mitchell 2015). It is particularly powerful when the dataset to be analyzed is too large or too complex, or when the internal mechanism of the research problem is unclear (Maxwell et al. 2018; Sun et al. 2022; Zhong et al. 2021). The overview of machine learning is briefly summarized as follows, and specialized publications are suggested for more details about machine learning (Hastie et al. 2009; Jebara 2004).

2.1 Naïve Bayes

NB is one of the most prevalent supervised machine learning algorithms for classification problems, which is also applied for regression in some cases. The algorithm learns from the training set, within which the sample points are labeled by the classes, based on the Bayes statistical theorem with the "naïve" assumption that each two features of the training set is conditional independent (Sánchez-Franco et al. 2019). The classification process of a NB classifier could be summarized as follows (Blanquero et al. 2021):

For feature vector $\boldsymbol{X} = (X_1, X_2, \ldots, X_m)$ containing m features, and class vector $\boldsymbol{Y} = (y_1, y_2, \ldots, y_n)$ with n classes, the probability of a sample $\boldsymbol{x} = (x_1, x_2, \ldots, x_m)$ belonging to the j-th class, y_i, could be calculated as:

$$P(y_j|\boldsymbol{x}) = \frac{P(y_j)P(\boldsymbol{x}|y_j)}{P(\boldsymbol{x})} \tag{1}$$

where, $P(y_j|\boldsymbol{x})$ denotes the probability of the sample $\boldsymbol{x}$ belonging to the j-th class, y_j. $P(y_j)$ and $P(\boldsymbol{x})$ denotes the probability of the j-th class, y_j, and the sample, $\boldsymbol{x}$. $P(\boldsymbol{x}|y_j)$ represents the conditional probability of sample $\boldsymbol{x}$ in the j-th class, y_j, which, according to the independent assumption introduced above, could be expressed as:

$$P(y_j|\boldsymbol{x}) = \frac{P(y_j)\prod_{i=1}^{m} P(x_i|y_j)}{P(\boldsymbol{x})} \tag{2}$$

where, $P(x_i|y_j)$ denotes the conditional probability of the i-th feature for the sample $\boldsymbol{x}$ in the j-th class, y_j. Because $P(\boldsymbol{x})$ is a constant value for a given input dataset, the following conclusion could be drawn:

$$P(y_j|\boldsymbol{x}) \propto P(y_j)\prod_{i=1}^{m} P(x_i|y_j) \tag{3}$$

Such that $P(y_j|\boldsymbol{x})$ could be computed as:

$$P(y_j|\boldsymbol{x}) = \arg\max_{y_j} P(y_j)\prod_{i=1}^{m} P(x_i|y_j) \tag{4}$$

where, argmax_{yj} denotes maximum posteriori estimation here used to determine $P(y_j)$ and $P(x_i|y_j)$. As $P(y_j)$ represents the frequency of class y_j in the training set, which is constant for a certain issue, the different NB classifier mainly differ from the assumptions of the distributions of $P(x_i|y_j)$. When the class vector becomes a numeric target variable, the NB could be applied to address regression problem using the same principle (Frank et al. 2000).

The advantages of NB are stable for classification problems, suitable for incremental training, and not sensitive to missing data. However, owing to the "naïve" assumption, dependencies between features are ignored by NB, and model performance depends on the hypothetic prior probability (Abdulrahman et al. 2020; Jain et al. 2023). NB is generally applied to map groundwater potential, and classify and map groundwater quality indices.

2.2 *Artificial Neural Network*

ANN is considered to originate with the computational model of mathematical logic coined by McCulloch and Pitts (1943), which was inspired by the connection and interaction of neurons in brain. Illustrated as Fig. 2, a typical ANN model comprises three layers, including input layer, hidden layer, and output layer, which are composed by a number of neurons (also known as nodes or units) (Zarra et al. 2019). The input layer contains the variables to be fed into the ANN model, like the receptor that receives external stimuli in neural reflexes. The hidden layer comprises "associative nodes" for feature detecting, which conduct computation and patterns recognition. The output layer export the results of the ANN model, the neuron number of which depends on the actual research (Hong 2017). The neurons in the adjacent layers are connected via synapse that determines connecting strength of neurons, through which the signals travel from one node to another, and transform into an output signal by adopting an activation function (Santos et al. 2021). The purpose to apply an activation function is to introduce nonlinearity. The most prevalent activation functions are logistic sigmoid and hyperbolic tangent function (Lee et al. 2020).

The advantages of ANN include: (1) strong generalization ability, (2) strong robustness and fault tolerance, (3) self-adjusting ability to a given dataset. Furthermore, artificial neural network is capable to execute parallel distribution process, which decreases the time consumption of the calculation. However, ANN requires data pretreatment, and cannot construct explicit models (Oyebode and Stretch 2019). Artificial neural network is commonly used to forecast groundwater level (or depth), assess groundwater quality, and predict groundwater quality.

2.3 *Support Vector Machine*

SVM is firstly proposed by Boser et al. (1992) to address binary classification based on statistical learning theory, with the principle of minimizing structure risk (Huang and Zhao 2018). Then, SVM is developed for multivariate classification and regression (Kok et al. 2021). For binary classification problem, the key point of SVM is to determine the optimal separating hyperplane with maximum margin (Mountrakis et al. 2011). Suppose there are numerous points belonging to two classes on a 2-demensional plane, which could be separated by many lines, shown as Fig. 3a.

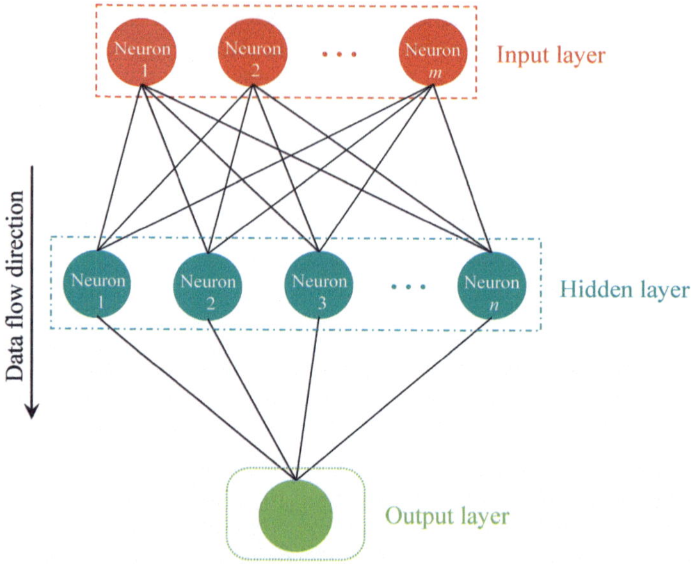

Fig. 2 Structure of a typical artificial neural network

These separating lines are called hyperplane in high-dimension space. The aim of SVM is to find out the hyperplane with the largest margin width among these hyperplane, illustrated as Fig. 3b. The points closest to the hyperplane is called support vector. The essence of SVM is to project the samples into a hyperspace using a kernel function, in which complicated non-liner patterns of the samples could be easily represented (Salcedo-Sanz et al. 2014). So that the data could be classified using a hyperplane with the maximum margin width, illustrated as Fig. 3c, or be fitted and predicted (Yao et al. 2008).

According to the existing publications, SVM is widely applied to forecast groundwater level (or depth), assess groundwater quality, and predict groundwater quality indices. It is capable to address high-dimensional and nonlinear problem with its kernel function. SVM also has strong ability of generalization and global optimization. However, it shows low efficiency to train large scale data. SVM is also sensitive to missing data. Traditional SVM has difficulties to address multi-classification problem (Cao et al. 2020; Westreich et al. 2010).

2.4 Decision Tree

Decision tree (DT) is a nonparametric machine learning algorithm extensively applied for classification and regression problems, which learns from a range of hierarchical "if/else" structures to acquire its predictions (López-Chau et al. 2013). Being a member of supervised learning algorithms, DT learns from the training set

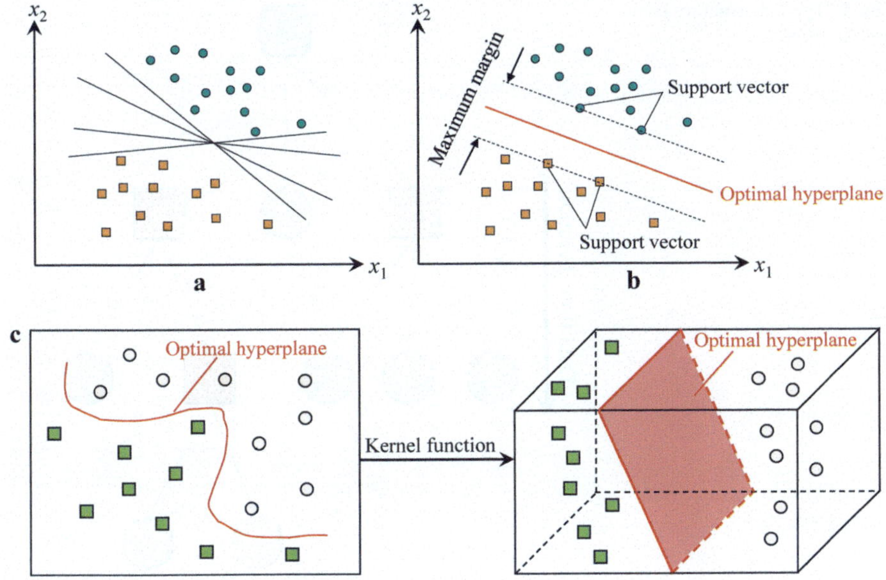

Fig. 3 Principle schematic of support vector machine

with labeled sample points, and makes prediction on unseen sample data. Provided by learning from high-quality training set, DT is able to make accurate prediction (Kingsford and Salzberg 2008). Implied by its name, the DT owns a tree-shaped structure illustrated by Fig. 4, which consists of a root node, branches (or called links), internal nodes, and leaves (also called terminal nodes). The root node is the topmost node of a tree where the data flow in the structure starts. The node in the tree structure could be divided into two types including parent node and child node. A parent node would be split into two or more child nodes according to certain rules, so that a branch is constructed with splitting a parent node. The splitting would be terminated on a leaf. That's to say, a child node would continue to be split into new child nodes as a new parent node if it is not a leaf. Typical DT algorithms include ID3, C4.5, C5.0, and CART (Kotsiantis 2011). DT is simple to be understood, interpreted and visualized, and capable to address the nonlinearity relationship between input and output variables of the dataset (Dabiri et al. 2022; Weber and Fajen 2015). It's drawbacks, however, which should be deliberated, are its overfitting tendency and inadequate capacity on classifying multiple output classes (Dabiri et al. 2022).

There are many advantages for DT. Firstly, the tree's structure is easy to be visualized based on the relationships among root node, internal node and terminal node (leaf). Secondly, various input data, including nominal, numeric and textual data, can be addressed using a DT. Thirdly, DT is capable to handle dataset with missing values or errors. In addition, the time cost to train a DT is relatively low. However, DT has relatively lower stability, and is easy to overfit the dataset (de Oña et al. 2017; Forner et al. 2021). Another disadvantage of DT is the replication problem (Rokach 2016).

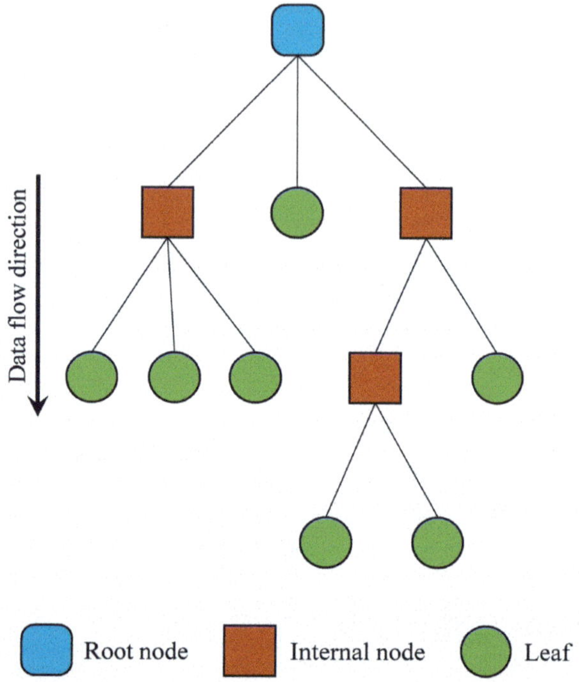

Fig. 4 Typical structure of a decision tree

DT is generally applied to classify groundwater quality, estimate spring potential, map groundwater potential and identify contamination site (source).

2.5 *Random Forest*

RF is one of the ensemble machine learning methods proposed by Breiman (2001) in 2001, which integrates a number of decision trees to diminish the drawback of overfitting and instability for a single tree (He et al. 2022b; Tyralis et al. 2019). The fundamental concept of RF is to independently construct a range of decision trees on the subsets randomly sampled from the original training set. The results of these decision trees are subsequently averaged to produce the results of the RF model (Antoniadis et al. 2021). As such, the robustness of RF model is improved compared to a single decision tree (Wu et al. 2020a). The training subset for each decision tree in a RF model is randomly extracted from the original training set of that RF model using a bootstrapping procedure, which divides the original training dataset in to two parts, "in-bag" subset and "out-of-bag (oob)" subset. The "in-bag" subset is used to training a decision tree in the forest, while the "oob" subset is not included in the training process for the tree (Wang et al. 2020b; Zounemat-Kermani et al. 2021). The randomness of RF reflects on two aspects. Not only the sample points but also the features in a subset to train a single tree are randomly extracted from the original training set. This sampling procedure is unique for each tree, which provides internal validation for the RF model. As the "oob" subset is not applied for training a tree, it can be used to evaluate the performance of the tree. The average of all "oob" predictions of a RF model provides a metric to assess the overall accuracy for the RF model (Nhu et al. 2020; Schonlau and Zou 2020). The ratio of samples in training and test sets is not fixed, and it ranges from 8:2 to 3:1 (He et al. 2022b; Nafouanti et al. 2021; Tesoriero et al. 2017).

As an ensemble algorithm of DT, RF possesses the advantages of DT, and also overcomes some shortcomings of DT. Furthermore, RF is able to address high-dimensional data, to prevent the overfitting, and to estimate the feature importance after training. Additionally, RF has high generalization capacity owing to unbiased estimation to generalization error. However, RF is more complex and has higher calculation cost than a single decision tree. RF is easy to overfit on dataset containing noise (Du et al. 2022; Galal et al. 2022; Tyralis et al. 2019). RF is usually applied to predict groundwater variables, assess groundwater quality, and map groundwater potential.

According to above presentation, the advantages, disadvantages and main applicative research fields of the aforesaid five typical machine learning approaches are summarized and listed as Table 1.

3 Application of Machine Learning to Hydrogeology

3.1 Overview

A bibliographic retrieval based on the publications from the dataset of Web of Science and consequent analysis have been carried out in this section to summarize the application of machine learning in hydrogeology. Five keywords, including "groundwater", "hydrogeology", "hydrogeologic", "hydrogeological" and "aquifer", were firstly searched in the Web of Science Core Collection with type of "Topic" through connective word of "Or" during the publication year from 2003 to 2023. The results were then refined using the keyword "machine learning", which screened 1710 publications. Finally, 1089 journal papers of article type and 28 of review type publish on the journals indexed by the Science Citation Index from 2010 to 2023 were selected to summarize application of machine learning in hydrogeology. The publication number of article and review in each year of 2010–2023 is illustrated as Fig. 5. During 2010–2017, papers of article type on application machine learning to hydrogeological research were less than 30 in each year, while number of review-type paper was zero in all this period. Since 2018, the number of article-type and review-type papers on application machine learning to hydrogeology has been increasing, and reached 321 and 10 in 2023, respectively. To get aware of the hot fields in which machine learning is applied to hydrogeology, the topics of aforesaid 1089 article papers were

Table 1 Advantages, disadvantages and main applied fields of typical machine learning methods

Method	Advantage	Disadvantage	Main applied field
Naïve Bayes (NB)	Stable efficiency for classification; suitable for incremental training; Simple form of algorithm; not sensitive to missing data	Dependencies between features are ignored; model performance depends on the hypothetic prior probability	Map groundwater potential; classify and map groundwater quality
Artificial neural network (ANN)	Strong capability to execute parallel distribution process; strong generalization ability; strong robustness and fault tolerance; self-adjusting ability to a given set of data	Requires numerous parameters; cannot produce explicit model; Susceptible to overparameterization and overfitting problems	Forecast groundwater level (or depth); assess groundwater quality; predict groundwater quality
Support vector machine (SVM)	Able to address high-dimensional problem; strong generalization ability; able to address nonlinear problem with kernel function; global optimization	Low efficiency to train large scale data; Sensitive to missing data; difficult to address multi-classification problem using traditional SVM	Forecast groundwater level (or depth); assess groundwater potential and quality; assess groundwater contamination risk
Decision tree (DT)	The tree's structure is easy to be visualized; able to handle various input data: nominal, numeric and textual; able to address datasets with errors or missing values; relatively low time consumption	Replication problem; easy to overfit the dataset; Relatively lower stability	Classify groundwater quality; estimate spring potential; map groundwater potential; contamination site identification (source identification);
Random forest (RF)	Be able to address high-dimensional data; able to handle various input data: nominal, numeric and textual; able to prevent overfitting; high generalization ability owing to unbiased estimation to generalization error; able to estimate feature importance after training	More complex and higher calculation cost than decision tree; Easy to overfit on datasets contain much noise	Predict groundwater variables; assess groundwater quality; map groundwater potential

summarized and their percentages were illustrated as the pie chart shown in Fig. 5. The main four hot topics of applying machine learning to hydrogeology are predicting groundwater variables, mapping hydrogeological variables, identifying contaminant sources and assessing groundwater contaminant risk, with percentages of 62.8%, 15.9%, 5.6% and 5.7%, respectively. In addition, there are another 109 publications focusing on other topics such as estimating hydrogeological parameters (Han et al. 2022; He et al. 2020; Moghaddam et al. 2021), designing groundwater monitoring

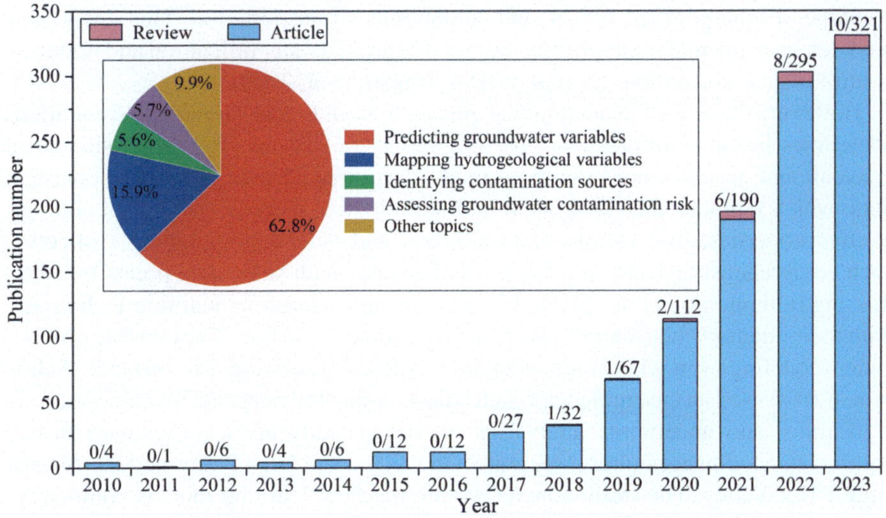

Fig. 5 Annual distribution of article-type and review-type publications on applying machine learning to hydrogeology during 2010–2023. The number on top of the columns represents quantity of article and review papers, respectively. The pie chart illustrates percentages of article-type publications on main hydrogeological fields to which machine learning is applied to

network (Chen et al. 2022; Khader and McKee 2014), solving partial differential equations (Wang et al. 2022b; Zhang et al. 2022b), seawater or saltwater intrusion (Etsias et al. 2020; Robinson et al. 2018) and etc. Following is the application of machine learning to the four main hydrogeological fields.

3.2 Prediciting Groundwater Variables

The application of machine learning methods in the field of hydrogeology has seen a significant rise over the recent decades. These methods have been increasingly utilized to predict groundwater variables, which include the depth of water levels, concentrations of physicochemical parameters such as arsenic, fluoride, pH, dissolved oxygen, total dissolved solids, and nitrate, along with comprehensive water quality indices like sodium adsorption ratio, permeability index, residual sodium carbonates, and water quality index (Elmeddahi and Ragab 2022; Erickson et al. 2018; Haggerty et al. 2023; Hikouei et al. 2023; Kouadri et al. 2022; Uddameri et al. 2020). The importance of modeling and predicting groundwater variables cannot be overstated, particularly in the context of sustainable groundwater resource management and protection. In the past, the primary tools for modeling and predicting groundwater levels and quality were mathematical physical models and corresponding numerical simulations. These methods have significantly contributed to our understanding of groundwater resources, shedding light on aspects like groundwater recharge, discharge, exploration, and contaminants transportation. This knowledge has been instrumental in supporting sustainable groundwater utilization and pollution control to a certain extent (Li et al. 2018b; Nagare et al. 2020).

However, the use of mathematical physical models and numerical simulations sometimes becomes impractical due to a variety of reasons. The establishment of conventional groundwater numerical models requires a range of highly consistent data, which includes meteorological and hydrological records, geological settings, aquifer properties, observations of water levels, and water quality indices. Collecting such comprehensive data can often be challenging, with some data points typically missing (Oikonomou et al. 2018). In such scenarios, machine learning techniques, with their inherent "black-box" characteristic, come to the fore. They enable groundwater modeling even when some data are unknown or missing. For instance, Sahoo et al. (2017) used an ensemble approach based on spectral analysis, machine learning algorithms, and uncertainty analysis to model groundwater level changes in two aquifers located in agricultural regions in the U.S.—the High Plains and the Mississippi River Valley. The study concluded that machine learning models could serve as an alternative method for groundwater simulation in regions where subsurface properties are unknown.

Machine learning's appeal isn't limited to its ability to handle missing data. It's also favored for its relatively lower time cost compared to conventional models. As data accumulation grows in hydrogeological studies, the data volume becomes considerably large. For instance, Koch et al. (2019) used a random forest approach to model the shallow groundwater level in Denmark, utilizing data observed from 14,916 wells. Moreover, many studies span national scales or cover multiple aquifer systems, which can be challenging to simulate using a single numerical model (Mohapatra et al. 2021). In such cases, machine learning is employed as a surrogate to traditional numerical models. For example, Ransom et al. (2022) predicted groundwater nitrate levels for drinking purposes across the conterminous U.S., using a three-dimensional extreme gradient boosting machine learning model based on observed data from 12,082 wells.

However, it's important to note that machine learning models lack the physical basis found in conventional models. To mitigate this drawback, some researchers have started to integrate machine learning algorithms with traditional models or develop physics-informed machine learning models (Xu et al. 2014; Yeung et al. 2022). These hybrid models represent an exciting frontier in hydrogeological studies that warrants further exploration.

3.3 *Mapping Hydrogeological Variables*

In addition to predicting groundwater levels and quality, machine learning has also been harnessed to map groundwater potential and quality indices, such as redox conditions, nitrate, arsenic, and fluoride (Knoll et al. 2020; Kumar et al. 2023; Podgorski et al. 2020, 2018). Traditionally, such mapping was conducted using geostatistical interpolation methods like inverse distance weighting, Kriging, and global polynomial interpolation (Cherif et al. 2021; Raco et al. 2021; Tsai et al. 2019). However, these methods heavily rely on the accuracy of field sampling data and consider limited spatial dependencies, often resulting in less than satisfactory performance (Shaji et al. 2021). Machine learning, with its ability to uncover hidden relationships between target variables and key influencing parameters, addresses this issue (Nath et al. 2022).

In many studies, a groundwater potential map illustrates the spatial distribution of the probability of groundwater occurrence (Díaz-Alcaide and Martínez-Santos 2019b). However, there is neither a universal definition for groundwater potential nor a standardized procedure for estimating it in existing research (Díaz-Alcaide and Martínez-Santos 2019b). When using machine learning to map groundwater potential, it is set as the target variable, generally categorized into several classes representing different probabilities of groundwater occurrence. Environmental factors influencing groundwater occurrence and recycling are selected as independent variables for the machine learning models. These typically include topography and geomorphology, geological settings, rainfall, soil properties, land use/land cover patterns, and hydrological conditions (Martinez-Santos and Renard 2020). For instance, Gómez-Escalonilla et al. (2022) mapped the groundwater potential in the Koulikoro and Bamako regions of Mali using twenty machine learning classifiers. They divided groundwater potential into five classes: very low, low, moderate, high, and very high. This classification of groundwater potential into five classes was also employed by Prasad et al. (2020), Wang et al. (2022a), and Bai et al. (2022). Kumar et al. (2021), on the other hand, classified groundwater potential into four grades based on the calculation results of the Groundwater Potential Index. Regardless of the classification method, the primary outcome of groundwater potential mapping is to determine the spatial distribution of the probability of groundwater occurrence, which is crucial for the utilization of groundwater resources.

Machine learning has also been extensively applied to map groundwater quality indices (DeSimone et al. 2020; Díaz-Alcaide and Martínez-Santos 2019a). Some groundwater quality parameters are mapped using machine learning to reveal fundamental chemical characteristics, such as pH, redox conditions, and salinity. For example, Rosecrans et al. (2017) mapped the redox condition of groundwater in Central Valley, California, using a boosted regression trees algorithm. The study produced maps illustrating the distributed probability of dissolved oxygen (DO) and manganese (Mn) below certain concentrations.

Besides mapping hydrogeochemical characteristics like redox condition and salinity, machine learning is also used to map harmful contaminants such as arsenic, fluoride, and nitrate, as well as associated health risks. Podgorski and Berg (2020, 2022) created global probability maps for arsenic and fluoride exceeding safe levels, respectively. These studies are significant in raising awareness about drinking water quality and identifying regions with safe groundwater worldwide.

Furthermore, some researchers have used machine learning to map human health risks caused by groundwater contaminants. For instance, Ruidas et al. (2022) conducted health hazard risk mapping in the Purulia district of West Bengal, India, using three machine learning methods. The study suggested that the central section of the study area was significantly susceptible to health hazards due to poor water quality, with parameters such as water table depth, pH, EC, arsenic, and nitrate contributing most to the health hazards.

In summary, the application of machine learning in mapping hydrogeological variables has proven to be an effective tool in understanding and managing groundwater resources. It offers a more nuanced and detailed perspective compared to traditional methods, providing critical insights for sustainable groundwater management.

3.4 Identifying Contamination Sources

With the rapid development of human society, groundwater environments worldwide are experiencing varying degrees of pollution, particularly in areas with intense anthropogenic activities such as large plains or basins where humans have settled (Li et al. 2022; Wang et al. 2021b; Xu et al. 2022). Polluted groundwater poses severe health risks when used for drinking and domestic purposes, leading to various endemic diseases (He and Wu 2019). Therefore, the effective control of groundwater pollution necessitates the identification of contaminant sources, which, theoretically, is an inverse problem.

Inverse problems involve determining groundwater pollution sources or deducing the transport processes of contaminants based on observational data (Mo et al. 2019). The identification of groundwater pollution sources, as an inverse problem, is primarily constrained by two difficulties: ill-posedness and uncertainty (Bonanno et al. 2022; Delsman et al. 2016). Ill-posedness means the problem cannot be mathematically solved, or its analytical solution is not unique. Uncertainty refers to the solution's sensitivity to subtle changes in the transportation process or minor errors in model parameters (Alzraiee et al. 2022; Todaro et al. 2021). Moreover, the traditional simulation-optimization method for groundwater pollution source identification is computationally expensive (Hou et al. 2021b).

In recent times, machine learning models combined with optimization algorithms, constructed as surrogate models, have been increasingly applied in identifying groundwater pollution sources due to their advantages in solving inverse problems and relatively lower time consumption (Hou et al. 2021a; Wang et al. 2020a). For instance, Mo et al. (2019) proposed a surrogate method based on a deep autoregressive neural network to replace the forward model for identifying groundwater contaminants in highly heterogeneous media. Li et al. (2020) proposed a hybrid model coupling particle swarm optimization (PSO) and extreme learning machine (ELM), termed PSO-ELM, to replace the numerical model embedded in the optimization algorithm for identifying groundwater contaminant sources. The results showed that the PSO-ELM not only alleviated the computational disaster but also accurately identified the contaminant sources of groundwater.

Although numerous studies have applied machine learning methods combined with optimizing algorithms to identify groundwater pollution sources, most of these methods were applied to hypothetical cases or ideal numerical models (An et al. 2022; Hou et al. 2021b; Wang et al. 2021a, 2020a). Only a few studies applied the proposed methods to practical field cases. For example, Zhao et al. (2020) developed a kernel-based extreme learning machine surrogate model together with four heuristic optimization algorithms to identify contaminant sources of groundwater. The method was validated on both hypothetical and actual cases to effectively reduce the computational cost and improve accuracy. However, the accuracy on the practical case was relatively lower than that on the hypothetical case. Pan et al. (2022) proposed an approach integrating a refined particle filter and a deep residual neural network surrogate model, named DRNN-RPF, for source estimation of groundwater contamination. The approach was tested on both ideal and practical cases with satisfactory performance.

Despite the numerous methods proposed for identifying groundwater contaminant sources using machine learning, their application or validation on hypothetical cases is more prevalent. The application to practical cases is still scarce. Moreover, the applicability of machine learning to identify groundwater contaminant sources under different spatial and temporal scales should be further discussed. The scale of field cases in existing studies ranges from several meters to several kilometers (Hou et al. 2021b; Kontos et al. 2022; Pan et al. 2022; Wang et al. 2020a; Zhao et al. 2020), which is only a small interval range for aquifer systems in the real world. This highlights the need for further research and development in this area.

3.5 Assessing Groundwater Contamination Risk

Besides the aforementioned applications, machine learning is also extensively used to assess groundwater contamination risk. The evaluation of groundwater contamination risk aims to determine the probability of groundwater being polluted under a specific set of circumstances (Morris and Foster 2000). This is seen as an extension of assessing groundwater vulnerability (Zhang et al. 2022a).

Groundwater vulnerability is traditionally evaluated using index methods, such as DRASTIC, OREADIC, and DRTSWI models (Aller et al. 1987; Li et al. 2022, 2017; Qian et al. 2012). However, the results derived from these index methods can be somewhat subjective, as the parameters are generally classified into several grades with different scores determined based on the professional knowledge of the researchers (Kazakis and Voudouris 2015; Wu et al. 2020b; Xiong et al. 2022). Furthermore, index methods only consider the natural properties of the aquifer, excluding external stresses on the aquifer, such as pollution loading and water consumption, from the assessment processes (Li et al. 2012).

To address these limitations, numerous studies have modified the index methods to extend the groundwater vulnerability assessment to groundwater contamination risk assessment using machine learning. These modified methods take into consideration the external stresses on the aquifer system. For instance, Barzegar et al.

(2018) assessed the groundwater contamination risk in the Marand plain of Northwest Iran by improving the DRASTIC model using five machine learning methods. In their study, the vulnerability index was modified using groundwater nitrate concentration, resulting in a new metric termed groundwater contamination risk. Similarly, Gharekhani et al. (2023) evaluated groundwater total contamination risk using machine learning in the Salmas plain of West Azerbaijan province in Northwest Iran. In their study, nitrate concentration was replaced by the contaminant concentration when calculating the conditioned risk indices.

Another approach to calculating groundwater contaminant risk is to integrate the groundwater vulnerability, contamination concentration, and pollution probability (Zhang et al. 2022a). Awais et al. (2021) assessed the groundwater contamination risk along the China Pakistan Economic Corridor in Northeast Pakistan using this approach. The final contamination risk was determined by the sum of groundwater vulnerability, nitrate concentration, and groundwater contamination probability. In contrast, Sajedi-Hosseini et al. (2018) evaluated groundwater contamination risk by integrating groundwater vulnerability, contamination concentration, and pollution probability. However, in their study, the final contamination risk was determined by the product of the three variables, rather than the sum used by Awais et al. (2021).

From the existing publications on the assessment of groundwater contamination risk, it is clear that both groundwater vulnerability reflecting the natural properties of the aquifer system and external stresses, such as pollution loading and water consumption, are considered. However, the calculation of groundwater contamination risk varies across studies. There is a need for a standard model with a certain procedure for the assessment of groundwater contamination risk, similar to the DRASTIC model for the assessment of groundwater vulnerability. This highlights the importance of further research and development in this area.

4 Limitations and Prospects

4.1 Limitation

This chapter summarizes the application of machine learning to hydrogeology based on 1089 journal articles and 28 review articles from the Web of Science dataset, spanning from 2003 to 2023. While previous reviews focused either on a single machine learning approach applied to hydrogeological research or on various machine learning methods applied to a single research field, this chapter introduces a variety of machine learning methods and summarizes the main research fields to which they have been applied.

However, this chapter has some limitations. Firstly, publications preprinted or online in 2024 were not included in the review, as parts of them are still not indexed in Web of Science. Secondly, this chapter does not cover the topic of predicting hydrogeological time series (such as time series of groundwater level), which is a significant application of machine learning in hydrogeology and has been reviewed in existing studies (Afrifa et al. 2022; Tao et al. 2022).

4.2 Prospects for Further Research

4.2.1 Enhancing Hydrological and Hydrogeological Data Sharing

The successful application of machine learning hinges on the availability of appropriate datasets for training and testing. Hydrogeological research in particular requires multidisciplinary data sharing. At present, various types of geoscientific data are successfully shared, but hydrological and hydrogeological data are often difficult to obtain for several reasons (Lin et al. 2023; Qiu et al. 2022).

In China, for example, hydrological and hydrogeological data are extensively collected by multiple official agencies, but they are not widely shared and are difficult to obtain (Lin et al. 2023). There are several reasons for this, including the practice of on-selling hydrological data, sensitivity and national security concerns, policy issues, and concerns about data quality (Lin et al. 2023). To meet the growing requirements for data sharing in hydrological and hydrogeological research, effective measures must be taken, such as digitizing non-sensitive data, implementing stronger regulations to prevent the selling of basic data, establishing platforms for automatic monitoring focused on data openness and transparency, and enhancing cooperation among different departments and teams to reduce duplication of work.

4.2.2 Introducing Physical Basis into Machine Learning models

The "black-box" nature of machine learning, while advantageous in modeling complex systems, can also be a limitation as most machine learning models ignore the physical mechanisms of the problem studied. To overcome this, physical knowledge is being introduced into machine learning. Some studies have combined conventional numerical models with machine learning, while others have used theory-guided machine learning, which integrates specific theoretical knowledge, laws, or principles of the field of study. However, more research is needed to assess the reliability and feasibility of applying theory-guided machine learning approaches to hydrogeology (Adombi et al. 2021).

4.2.3 Strengthening Interdisciplinary Cooperation

Applying machine learning to hydrogeological research requires multidisciplinary knowledge and techniques. It can be challenging for a single researcher to be proficient in all areas. For instance, a hydrogeologist with extensive knowledge of hydrogeological principles and mechanisms might not be adept at constructing machine learning models. Conversely, a researcher skilled in algorithm development may struggle with handling groundwater data to establish training and testing datasets. Therefore, promoting the application of machine learning to hydrogeological research necessitates increased interdisciplinary cooperation and communication between hydrogeologists and researchers in other fields.

5 Concluding Summary

Machine learning has been increasingly applied in hydrogeological research in recent years for its advantages of reliability, effectiveness and practicability. This chapter summarized the application of machine learning to hydrogeological research according to the existing publications. Common-used machine learning approaches and main hydrogeological fields that machine learning is applied to are introduced. In addition, challenges and prospects of machine learning for groundwater science were also presented. The main conclusions are as follows:

The current review selected 1089 article-type and 28 review-type journal papers indexed by the Web of Science Core Collection during 2010–2023 to summarized application of machine learning to hydrogeological researches. Quantity of research article on applying machine learning to groundwater science increased with high speed since 2018, in which the first review-type paper appeared. Four main hydrogeological fields using machine learning were recognized as predicting groundwater variables, mapping hydrogeological variables, identifying contamination sources and assessing groundwater contamination risk, with percentages of article-type papers of 62.8%, 15.9%, 5.6% and 5.7%, respectively.

To promote application of machine learning to hydrogeology in the future, hydrological and hydrogeological data sharing should be enhanced to provide solid data basis, as basic data sharing in groundwater science is not enough at present. Physical knowledge of hydrogeology is supposed to be introduced into machine learning models to overcome the drawback caused by the "black-box" characteristic of machine learning. Moreover, communication and cooperation among interdisciplinary researchers need to be further strengthened under the consideration of cross-disciplinary properties of applying machine learning to hydrogeology.

Acknowledgements This research was funded by the National Natural Science Foundation of China (42072286, 41761144059 and 42207084), the National Key Research and Development Program of China (2023YFC3706901), the Qinchuangyuan "Scientist + Engineer" Team Development Program of the Shaanxi Provincial Department of Science and Technology (2022KXJ-005), the Open Foundation of Key Laboratory of Water and Soil Resources Conservation and Restoration in the Middle and Lower Reaches of Yellow River Basin, MNR (WSRCR-2023-02), and the Natural Science Basic Research Program of Shaanxi (Program No.2023-JC-QN-0372). The authors are very grateful to the dedication of the editors and reviewers who helped us a lot to improve quality of the manuscript. The authors also thank Xuantong Lu for the help on literature retrieval.

References

Abdulrahman SA, Khalifa W, Roushdy M, Salem A-BM (2020) Comparative study for 8 computational intelligence algorithms for human identification. Comput Sci Rev 36:100237. https://doi.org/10.1016/j.cosrev.2020.100237

Adimalla N, Wu J (2019) Groundwater quality and associated health risks in a semi-arid region of south India: implication to sustainable groundwater management. Hum Ecol Risk Assess 25:191–216. https://doi.org/10.1080/10807039.2018.1546550

Adombi AVDP, Chesnaux R, Boucher M-A (2021) Review: theory-guided machine learning applied to hydrogeology—State of the art, opportunities and future challenges. Hydrogeol J 29:2671–2683. https://doi.org/10.1007/s10040-021-02403-2

Afrifa S, Zhang T, Appiahene P, Varadarajan V (2022) Mathematical and machine learning models for groundwater level changes: a systematic review and bibliographic analysis. Future Internet 14:259. https://doi.org/10.3390/fi14090259

Aller L, Bennett T, Lehr JH, Petty RJ, Hackett R (1987) DRASTIC: a standardized system for evaluating ground water pollution potential using hydrogeological settings. National Water Well Association, Dublin, Ohio

Alzraiee AH, White JT, Knowling MJ, Hunt RJ, Fienen MN (2022) A scalable model-independent iterative data assimilation tool for sequential and batch estimation of high dimensional model parameters and states. Environ Model Softw 150:105284. https://doi.org/10.1016/j.envsoft.2021.105284

An Y, Zhang Y, Yan X (2022) An Integrated Bayesian and machine learning approach application to identification of groundwater contamination source parameters. Water 14:2447. https://doi.org/10.3390/w14152447

Antoniadis A, Lambert-Lacroix S, Poggi J-M (2021) Random forests for global sensitivity analysis: a selective review. Reliab Eng Syst Saf 206:107312. https://doi.org/10.1016/j.ress.2020.107312

Awais M, Aslam B, Maqsoom A, Khalil U, Ullah F, Azam S, Imran M (2021) Assessing nitrate contamination risks in groundwater: a machine learning approach. Appl Sci 11:10034. https://doi.org/10.3390/app112110034

Bai Z, Liu Q, Liu Y (2022) Groundwater potential mapping in hubei region of china using machine learning, ensemble learning, deep learning and automl methods. Nat Resour Res 31:2549–2569. https://doi.org/10.1007/s11053-022-10100-4

Balogun MA, Anumah AO, Adegoke KA, Maxakato NW (2022) Environmental health impacts and controlling measures of anthropogenic activities on groundwater quality in Southwestern Nigeria. Environ Monit Assess 194:384. https://doi.org/10.1007/s10661-022-09805-z

Barzegar R, Moghaddam AA, Deo R, Fijani E, Tziritis E (2018) Mapping groundwater contamination risk of multiple aquifers using multi-model ensemble of machine learning algorithms. Sci Total Environ 621:697–712. https://doi.org/10.1016/j.scitotenv.2017.11.185

Blanquero R, Carrizosa E, Ramírez-Cobo P, Sillero-Denamiel MR (2021) Variable selection for Naïve Bayes classification. Comput Oper Res 135:105456. https://doi.org/10.1016/j.cor.2021.105456

Bonanno E, Blöschl G, Klaus J (2022) Exploring tracer information in a small stream to improve parameter identifiability and enhance the process interpretation in transient storage models. Hydrol Earth Syst Sci 26:6003–6028. https://doi.org/10.5194/hess-26-6003-2022

Boser BE, Guyon IM, Vapnik VN (1992) A training algorithm for optimal margin classifiers. In: Proceedings of the fifth annual workshop on computational learning theory, Pittsburgh, PA, USA, pp 144–152

Breiman L (2001) Decision forest. Mach Learn 45:5–32. https://doi.org/10.1023/a:1010933404324

Cao H, Zhang J, Yang F, An Q, Wang Y (2020) Extending capture range for piston error in segmented primary mirror telescopes based on wavelet support vector machine with improved particle swarm optimization. IEEE Access 8:111585–111597. https://doi.org/10.1109/access.2020.3002901

Cerasoli S, Porporato A (2023) California's groundwater overdraft: an environmental Ponzi scheme? J Hydrol 617:129081. https://doi.org/10.1016/j.jhydrol.2023.129081
Chen Y, Song L, Liu Y, Yang L, Li D (2020) A review of the artificial neural network models for water quality prediction. Appl Sci 10:5776. https://doi.org/10.3390/app10175776
Chen J, Dai Z, Dong S, Zhang X, Sun G, Wu J, Ershadnia R, Yin S, Soltanian MR (2022) Integration of deep learning and information theory for designing monitoring networks in heterogeneous aquifer systems. Water Resour Res 58:e2022WR032429. https://doi.org/10.1029/2022wr032429
Cherif K, Yamina B, Bilal B, Lahsen WK (2021) Delineation of groundwater potential zones in Wadi Saida Watershed of NW-Algeria using remote sensing, geographic information system-based AHP techniques and geostatistical analysis. J Groundw Sci Eng 9:45–64. https://doi.org/10.19637/j.cnki.2305-7068.2021.01.005
Dabiri H, Farhangi V, Moradi MJ, Zadehmohamad M, Karakouzian M (2022) Applications of Decision tree and random forest as tree-based machine learning techniques for analyzing the ultimate strain of spliced and non-spliced reinforcement bars. Appl Sci 12:4851. https://doi.org/10.3390/app12104851
de Oña J, de Oña R, Garrido C (2017) Extraction of attribute importance from satisfaction surveys with data mining techniques: a comparison between neural networks and decision trees. Transp Lett 9:39–48. https://doi.org/10.1080/19427867.2015.1136917
Delsman JR, Winters P, Vandenbohede A, Oude Essink GHP, Lebbe L (2016) Global sampling to assess the value of diverse observations in conditioning a real-world groundwater flow and transport model. Water Resour Res 52:1652–1672. https://doi.org/10.1002/2014wr016476
DeSimone LA, Pope JP, Ransom KM (2020) Machine-learning models to map pH and redox conditions in groundwater in a layered aquifer system, Northern Atlantic Coastal Plain, eastern USA. J Hydrol: Reg Stud 30:100697. https://doi.org/10.1016/j.ejrh.2020.100697
Di Salvo C (2022) Improving results of existing groundwater numerical models using machine learning techniques: a review. Water 14:2307. https://doi.org/10.3390/w14152307
Díaz-Alcaide S, Martínez-Santos P (2019a) Mapping fecal pollution in rural groundwater supplies by means of artificial intelligence classifiers. J Hydrol 577:124006. https://doi.org/10.1016/j.jhydrol.2019.124006
Díaz-Alcaide S, Martínez-Santos P (2019b) Review: advances in groundwater potential mapping. Hydrogeol J 27:2307–2324. https://doi.org/10.1007/s10040-019-02001-3
Du S, Gu C, Zhang W (2022) A review on the progresses in random forests theory and its applications in hydrogeology. China Environ Sci 42:4285–4295. https://doi.org/10.19674/j.cnki.issn1000-6923.20220419.006
Elmeddahi Y, Ragab R (2022) Prediction of the groundwater quality index through machine learning in Western Middle Cheliff plain in North Algeria. Acta Geophys 70:1797–1814. https://doi.org/10.1007/s11600-022-00827-2
Elumalai V, Nwabisa DP, Rajmohan N (2019) Evaluation of high fluoride contaminated fractured rock aquifer in South Africa—Geochemical and chemometric approaches. Chemosphere 235:1–11. https://doi.org/10.1016/j.chemosphere.2019.06.065
Erickson ML, Elliott SM, Christenson CA, Krall AL (2018) Predicting geogenic arsenic in drinking water wells in glacial aquifers, North-Central USA: accounting for depth-dependent features. Water Resour Res 54:10172–10187. https://doi.org/10.1029/2018wr023106
Etsias G, Hamill GA, Benner EM, Águila JF, McDonnell MC, Flynn R, Ahmed AA (2020) Optimizing laboratory investigations of saline intrusion by incorporating machine learning techniques. Water 12:2996. https://doi.org/10.3390/w12112996
Forner D, Hoit G, Noel CW, Eskander A, de Almeida JR, Rigby MH, Naimark D (2021) Decision modeling for economic evaluation in otolaryngology-head and neck surgery: review of techniques. Otolaryngol-Head Neck Surg 164:741–750. https://doi.org/10.1177/0194599820957288
Frank E, Trigg L, Holmes G, Witten IH (2000) Naive Bayes for regression. Mach Learn 41:5–25. https://doi.org/10.1023/a:1007670802811

Galal A, Talal M, Moustafa A (2022) Applications of machine learning in metabolomics: disease modeling and classification. Front Genet 13:1017340. https://doi.org/10.3389/fgene.2022.1017340

Ghaffarian S, van der Voort M, Valente J, Tekinerdogan B, de Mey Y (2022) Machine learning-based farm risk management: a systematic mapping review. Comput Electron Agric 192:106631. https://doi.org/10.1016/j.compag.2021.106631

Gharekhani M, Nadiri AA, Khatibi R, Nikoo MR, Barzegar R, Sadeghfam S, Moghaddam AA (2023) Quantifying the groundwater total contamination risk using an inclusive multi-level modelling strategy. J Environ Manag 332:117287. https://doi.org/10.1016/j.jenvman.2023.117287

Gómez-Escalonilla V, Martínez-Santos P, Martín-Loeches M (2022) Preprocessing approaches in machine-learning-based groundwater potential mapping: an application to the Koulikoro and Bamako regions, Mali. Hydrol Earth Syst Sci 26:221–243. https://doi.org/10.5194/hess-26-221-2022

Haggerty R, Sun J, Yu H, Li Y (2023) Application of machine learning in groundwater quality modeling—A comprehensive review. Water Res 233:119745. https://doi.org/10.1016/j.watres.2023.119745

Han Z, Kang X, Wu J, Shi X (2022) Characterization of the non-Gaussian hydraulic conductivity field via deep learning-based inversion of hydraulic-head and self-potential data. J Hydrol 610:127830. https://doi.org/10.1016/j.jhydrol.2022.127830

Hastie T, Tibshirani R, Friedman J (2009) The elements of statistical learning. Springer, New York, NY, USA

He S, Wu J (2018) Hydrogeochemical characteristics, groundwater quality, and health risks from hexavalent chromium and nitrate in groundwater of Huanhe Formation in Wuqi County, Northwest China. Expo Health 11:125–137. https://doi.org/10.1007/s12403-018-0289-7

He S, Wu J (2019) Relationships of groundwater quality and associated health risks with land use/land cover patterns: a case study in a loess area, Northwest China. Hum Ecol Risk Assess 25:354–373. https://doi.org/10.1080/10807039.2019.1570463

He Q, Barajas-Solano D, Tartakovsky G, Tartakovsky AM (2020) Physics-informed neural networks for multiphysics data assimilation with application to subsurface transport. Adv Water Resour 141:103610. https://doi.org/10.1016/j.advwatres.2020.103610

He S, Li P, Su F, Wang D, Ren X (2022a) Identification and apportionment of shallow groundwater nitrate pollution in Weining Plain, northwest China, using hydrochemical indices, nitrate stable isotopes, and the new Bayesian stable isotope mixing model (MixSIAR). Environ Pollut 298:118852. https://doi.org/10.1016/j.envpol.2022.118852

He S, Wu J, Wang D, He X (2022b) Predictive modeling of groundwater nitrate pollution and evaluating its main impact factors using random forest. Chemosphere 290:133388. https://doi.org/10.1016/j.chemosphere.2021.133388

Hikouei IS, Eshleman KN, Saharjo BH, Graham LLB, Applegate G, Cochrane MA (2023) Using machine learning algorithms to predict groundwater levels in Indonesian tropical peatlands. Sci Total Environ 857:159701. https://doi.org/10.1016/j.scitotenv.2022.159701

Hong Y (2017) Feasibility of using artificial neural networks to forecast groundwater levels in real time. Landslides 14:1815–1826. https://doi.org/10.1007/s10346-017-0844-5

Hou Z, Lao W, Wang Y, Lu W (2021a) Cyclic feedback updating approach and uncertainty analysis for the source identification of DNAPL-contaminated aquifers. J Water Resour Plan Manag 147:04020103. https://doi.org/10.1061/(asce)wr.1943-5452.0001322

Hou Z, Lao W, Wang Y, Lu W (2021b) Hybrid homotopy-PSO global searching approach with multi-kernel extreme learning machine for efficient source identification of DNAPL-polluted aquifer. Comput Geosci 155:104837. https://doi.org/10.1016/j.cageo.2021.104837

Huang Y, Zhao L (2018) Review on landslide susceptibility mapping using support vector machines. Catena 165:520–529. https://doi.org/10.1016/j.catena.2018.03.003

Islam ARMT, Talukdar S, Mahato S, Kundu S, Eibek KU, Pham QB, Kuriqi A, Linh NTT (2021) Flood susceptibility modelling using advanced ensemble machine learning models. Geosci Front 12:101075. https://doi.org/10.1016/j.gsf.2020.09.006

Jain S, Naicker D, Raj R, Patel V, Hu YC, Srinivasan K, Jen CP (2023) Computational intelligence in cancer diagnostics: a contemporary review of smart phone apps, current problems, and future research potentials. Diagnostics 13:1563. https://doi.org/10.3390/diagnostics13091563

Jasechko S, Perrone D (2021) Global groundwater wells at risk of running dry. Science 372:418–421. https://doi.org/10.1126/science.abc2755

Jean-Baptiste J, Le Gal La Salle C, Verdoux P (2020) Use of mixing models to explain groundwater quality time and space variation in a narrowed fluctuating alluvial aquifer. Appl Geochem 121:104700. https://doi.org/10.1016/j.apgeochem.2020.104700

Jebara T (2004) Machine learning. Springer, New York, NY, USA

Jiang J, Zhou H, Zhang T, Yao C, Du D, Zhao L, Cai W, Che L, Cao Z, Wu X (2022) Machine learning to predict dynamic changes of pathogenic Vibrio spp. abundance on microplastics in marine environment. Environ Pollut 305:119257. https://doi.org/10.1016/j.envpol.2022.119257

Jordan MI, Mitchell TM (2015) Machine learning: trends, perspectives, and prospects. Science 349:255–260. https://doi.org/10.1126/science.aaa8415

Kambalimath S, Deka PC (2020) A basic review of fuzzy logic applications in hydrology and water resources. Appl Water Sci 10:191. https://doi.org/10.1007/s13201-020-01276-2

Kazakis N, Voudouris KS (2015) Groundwater vulnerability and pollution risk assessment of porous aquifers to nitrate: modifying the DRASTIC method using quantitative parameters. J Hydrol 525:13–25. https://doi.org/10.1016/j.jhydrol.2015.03.035

Khader AI, McKee M (2014) Use of a relevance vector machine for groundwater quality monitoring network design under uncertainty. Environ Model Softw 57:115–126. https://doi.org/10.1016/j.envsoft.2014.02.015

Kingsford C, Salzberg SL (2008) What are decision trees? Nat Biotechnol 26:1011–1013. https://doi.org/10.1038/nbt0908-1011

Knoll L, Breuer L, Bach M (2020) Nation-wide estimation of groundwater redox conditions and nitrate concentrations through machine learning. Environ Res Lett 15:064004. https://doi.org/10.1088/1748-9326/ab7d5c

Koch J, Berger H, Henriksen HJ, Sonnenborg TO (2019) Modelling of the shallow water table at high spatial resolution using random forests. Hydrol Earth Syst Sci 23:4603–4619. https://doi.org/10.5194/hess-23-4603-2019

Kok ZH, Mohamed Shariff AR, Alfatni MSM, Khairunniza-Bejo S (2021) Support vector machine in precision agriculture: a review. Comput Electron Agric 191:106546. https://doi.org/10.1016/j.compag.2021.106546

Kontos YN, Kassandros T, Perifanos K, Karampasis M, Katsifarakis KL, Karatzas K (2022) Machine learning for groundwater pollution source identification and monitoring network optimization. Neural Comput Appl 34:19515–19545. https://doi.org/10.1007/s00521-022-07507-8

Kotsiantis SB (2011) Decision trees: a recent overview. Artif Intell Rev 39:261–283. https://doi.org/10.1007/s10462-011-9272-4

Kouadri S, Pande CB, Panneerselvam B, Moharir KN, Elbeltagi A (2022) Prediction of irrigation groundwater quality parameters using ANN, LSTM, and MLR models. Environ Sci Pollut Res 29:21067–21091. https://doi.org/10.1007/s11356-021-17084-3

Kumar R, Dwivedi SB, Gaur S (2021) A comparative study of machine learning and Fuzzy-AHP technique to groundwater potential mapping in the data-scarce region. Comput Geosci 155:104855. https://doi.org/10.1016/j.cageo.2021.104855

Kumar M, Singh P, Singh P (2023) Machine learning and GIS-RS-based algorithms for mapping the groundwater potentiality in the Bundelkhand region, India. Ecol Inform 74:101980. https://doi.org/10.1016/j.ecoinf.2023.101980

Lee D-H, Kim Y-T, Lee S-R (2020) Shallow landslide susceptibility models based on artificial neural networks considering the factor selection method and various non-linear activation functions. Remote Sensing 12:1194. https://doi.org/10.3390/rs12071194

Li Y, Li J, Chen S, Diao W (2012) Establishing indices for groundwater contamination risk assessment in the vicinity of hazardous waste landfills in China. Environ Pollut 165:77–90. https://doi.org/10.1016/j.envpol.2011.12.042

Li X, Gao Y, Qian H, Wu H (2017) Groundwater vulnerability and contamination risk assessment of the Weining Plain, using a modified DRASTIC model and quantized pollution loading method. Arab J Geosci 10:469. https://doi.org/10.1007/s12517-017-3255-y

Li P, He S, He X, Tian R (2018a) Seasonal hydrochemical characterization and groundwater quality delineation based on matter element extension analysis in a paper wastewater irrigation area, Northwest China. Expo Health 10:241–258. https://doi.org/10.1007/s12403-017-0258-6

Li L, Su C, Hao Q, Shao J (2018b) Numerical simulation of response of groundwater flow system in inland basin to density changes. J Groundw Sci Eng 6:7–17. https://doi.org/10.19637/j.cnki.2305-7068.2018.01.002

Li P, He X, Li Y, Xiang G (2019a) Occurrence and health implication of fluoride in groundwater of loess aquifer in the Chinese loess plateau: a case study of Tongchuan, Northwest China. Expo Health 11:95–107. https://doi.org/10.1007/s12403-018-0278-x

Li P, He X, Guo W (2019b) Spatial groundwater quality and potential health risks due to nitrate ingestion through drinking water: a case study in Yan'an City on the Loess Plateau of northwest China. Hum Ecol Risk Assess 25:11–31. https://doi.org/10.1080/10807039.2018.1553612

Li J, Lu W, Wang H, Fan Y, Chang Z (2020) Groundwater contamination source identification based on a hybrid particle swarm optimization-extreme learning machine. J Hydrol 584:124657. https://doi.org/10.1016/j.jhydrol.2020.124657

Li P, Wang D, Li W, Liu L (2022) Sustainable water resources development and management in large river basins: an introduction. Environ Earth Sci 81:179. https://doi.org/10.1007/s12665-022-10298-9

Lin J, Bryan BA, Zhou X, Lin P, Do HX, Gao L, Gu X, Liu Z, Wan L, Tong S, Huang J, Wang Q, Zhang Y, Gao H, Yin J, Chen Z, Duan W, Xie Z, Cui T, Liu J, Li M, Li X, Xu Z, Guo F, Shu L, Li B, Zhang J, Zhang P, Fan B, Wang Y, Zhang Y, Huang J, Li X, Cai Y, Yang Z (2023) Making China's water data accessible, usable and shareable. Nat Water 1:328–335. https://doi.org/10.1038/s44221-023-00039-y

López-Chau A, Cervantes J, López-García L, Lamont FG (2013) Fisher's decision tree. Expert Syst Appl 40:6283–6291. https://doi.org/10.1016/j.eswa.2013.05.044

Martinez-Santos P, Renard P (2020) Mapping groundwater potential through an ensemble of big data methods. Groundwater 58:583–597. https://doi.org/10.1111/gwat.12939

Maxwell AE, Warner TA, Fang F (2018) Implementation of machine-learning classification in remote sensing: an applied review. Int J Remote Sens 39:2784–2817. https://doi.org/10.1080/01431161.2018.1433343

McCulloch WS, Pitts W (1943) A logical calculus of the ideas immanent in nervous activity. Bull Math Biophys 5:115–133. https://doi.org/10.1007/bf02478259

Mo S, Zabaras N, Shi X, Wu J (2019) Deep autoregressive neural networks for high-dimensional inverse problems in groundwater contaminant source identification. Water Resour Res 55:3856–3881. https://doi.org/10.1029/2018wr024638

Moghaddam MA, Ferre PAT, Ehsani MR, Klakovich J, Gupta HV (2021) Can deep learning extract useful information about energy dissipation and effective hydraulic conductivity from gridded conductivity fields? Water 13:1668. https://doi.org/10.3390/w13121668

Mohapatra JB, Jha P, Jha MK, Biswal S (2021) Efficacy of machine learning techniques in predicting groundwater fluctuations in agro-ecological zones of India. Sci Total Environ 785:147319. https://doi.org/10.1016/j.scitotenv.2021.147319

Morris BL, Foster SD (2000) Cryptosporidium contamination hazard assessment and risk management for British groundwater sources. Water Sci Technol 41:67–77. https://doi.org/10.2166/wst.2000.0117

Mountrakis G, Im J, Ogole C (2011) Support vector machines in remote sensing: a review. ISPRS J Photogramm Remote Sens 66:247–259. https://doi.org/10.1016/j.isprsjprs.2010.11.001

Myszczynska MA, Ojamies PN, Lacoste AMB, Neil D, Saffari A, Mead R, Hautbergue GM, Holbrook JD, Ferraiuolo L (2020) Applications of machine learning to diagnosis and treatment of neurodegenerative diseases. Nat Rev Neurol 16:440–456. https://doi.org/10.1038/s41582-020-0377-8

Nafouanti MB, Li J, Mustapha NA, Uwamungu P, Al-Alimi D (2021) Prediction on the fluoride contamination in groundwater at the Datong Basin, Northern China: comparison of random forest, logistic regression and artificial neural network. Appl Geochem 132:105054. https://doi.org/10.1016/j.apgeochem.2021.105054

Nagare RM, Park YJ, Butterfield T, Belenky C, Scyrup S (2020) Analytical and numerical modeling of solute intrusion, recovery, and rebound in fractured bedrock. Groundwater 58:56–69. https://doi.org/10.1111/gwat.12892

Nath B, Chowdhury R, Ni-Meister W, Mahanta C (2022) Predicting the distribution of arsenic in groundwater by a geospatial machine learning technique in the two most affected Districts of Assam, India: the public health implications. Geohealth 6:e2021GH000585. https://doi.org/10.1029/2021GH000585

Nhu V-H, Shirzadi A, Shahabi H, Chen W, Clague JJ, Geertsema M, Jaafari A, Avand M, Miraki S, Talebpour Asl D, Pham BT, Ahmad BB, Lee S (2020) Shallow landslide susceptibility mapping by random forest base classifier and its ensembles in a Semi-Arid Region of Iran. Forests 11:421. https://doi.org/10.3390/f11040421

Oikonomou PD, Alzraiee AH, Karavitis CA, Waskom RM (2018) A novel framework for filling data gaps in groundwater level observations. Adv Water Resour 119:111–124. https://doi.org/10.1016/j.advwatres.2018.06.008

Osman AIA, Ahmed AN, Huang YF, Kumar P, Birima AH, Sherif M, Sefelnasr A, Ebraheemand AA, El-Shafie A (2022) Past, present and perspective methodology for groundwater modeling-based machine learning approaches. Arch Comput Methods Eng 29:3843–3859. https://doi.org/10.1007/s11831-022-09715-w

Oyebode O, Stretch D (2019) Neural network modeling of hydrological systems: a review of implementation techniques. Nat Resour Model 32:e12189. https://doi.org/10.1111/nrm.12189

Pan Z, Lu W, Bai Y (2022) Groundwater contamination source estimation based on a refined particle filter associated with a deep residual neural network surrogate. Hydrogeol J 30:881–897. https://doi.org/10.1007/s10040-022-02454-z

Pétré M-A, Rivera A, Lefebvre R (2019) Numerical modeling of a regional groundwater flow system to assess groundwater storage loss, capture and sustainable exploitation of the transboundary Milk River Aquifer (Canada, USA). J Hydrol 575:656–670. https://doi.org/10.1016/j.jhydrol.2019.05.057

Podgorski JE, Labhasetwar P, Saha D, Berg M (2018) Prediction modeling and mapping of groundwater fluoride contamination throughout India. Environ Sci Technol 52:9889–9898. https://doi.org/10.1021/acs.est.8b01679

Podgorski J, Berg M (2020) Global threat of arsenic in groundwater. Science 368:845–850. https://doi.org/10.1126/science.aba1510

Podgorski J, Wu R, Chakravorty B, Polya DA (2020) Groundwater arsenic distribution in india by machine learning geospatial modeling. Int J Environ Res Public Health 17:7119. https://doi.org/10.3390/ijerph17197119

Podgorski J, Berg M (2022) Global analysis and prediction of fluoride in groundwater. Nat Commun 13:4232. https://doi.org/10.1038/s41467-022-31940-x

Prasad P, Loveson VJ, Kotha M, Yadav R (2020) Application of machine learning techniques in groundwater potential mapping along the west coast of India. Gisci Remote Sens 57:735–752. https://doi.org/10.1080/15481603.2020.1794104

Qian H, Li P, Howard KWF, Yang C, Zhang X (2012) Assessment of groundwater vulnerability in the Yinchuan Plain, Northwest China using OREADIC. Environ Monit Assess 184:3613–3628. https://doi.org/10.1007/s10661-011-2211-7

Qiu W, Ma T, Wang Y, Cheng J, Su C, Li J (2022) Review on status of groundwater database and application prospect in deep-time digital earth plan. Geosci Front 13:101383. https://doi.org/10.1016/j.gsf.2022.101383

Raco B, Vivaldo G, Doveri M, Menichini M, Masetti G, Battaglini R, Irace A, Fioraso G, Marcelli I, Brussolo E (2021) Geochemical, geostatistical and time series analysis techniques as a tool to achieve the Water Framework Directive goals: an example from Piedmont region (NW Italy). J Geochem Explor 229:106832. https://doi.org/10.1016/j.gexplo.2021.106832

Rahimi-Feyzabad F, Yazdanpanah M, Gholamrezai S, Ahmadvand M (2022) Social network analysis of institutions involved in groundwater resources management: lessons learned from Iran. J Hydrol 613:128442. https://doi.org/10.1016/j.jhydrol.2022.128442

Ransom KM, Nolan BT, Stackelberg PE, Belitz K, Fram MS (2022) Machine learning predictions of nitrate in groundwater used for drinking supply in the conterminous United States. Sci Total Environ 807:151065. https://doi.org/10.1016/j.scitotenv.2021.151065

Robinson G, Moutari S, Ahmed AA, Hamill GA (2018) An advanced calibration method for image analysis in laboratory-scale seawater intrusion problems. Water Resour Manag 32:3087–3102. https://doi.org/10.1007/s11269-018-1977-6

Rokach L (2016) Decision forest: twenty years of research. Inf Fusion 27:111–125. https://doi.org/10.1016/j.inffus.2015.06.005

Rosecrans CZ, Nolan BT, Gronberg JM (2017) Prediction and visualization of redox conditions in the groundwater of Central Valley, California. J Hydrol 546:341–356. https://doi.org/10.1016/j.jhydrol.2017.01.014

Ruidas D, Pal SC, Towfiqul Islam ARM, Saha A (2022) Hydrogeochemical evaluation of groundwater aquifers and associated health hazard risk mapping using ensemble data driven model in a water scares Plateau Region of Eastern India. Expo Health 15:113–131. https://doi.org/10.1007/s12403-022-00480-6

Rutherford J, Ibrahimi T, Munday T, Markey A, Viezzoli A, Rapiti A, Paterson R (2021) An assessment of water sources for heritage listed organic mound springs in NW Australia using airborne geophysical (electromagnetics and magnetics) and satellite remote sensing methods. Remote Sens 13:1288. https://doi.org/10.3390/rs13071288

Sahoo S, Russo TA, Elliott J, Foster I (2017) Machine learning algorithms for modeling groundwater level changes in agricultural regions of the U.S. Water Resour Res 53:3878–3895. https://doi.org/10.1002/2016wr019933

Sajedi-Hosseini F, Malekian A, Choubin B, Rahmati O, Cipullo S, Coulon F, Pradhan B (2018) A novel machine learning-based approach for the risk assessment of nitrate groundwater contamination. Sci Total Environ 644:954–962. https://doi.org/10.1016/j.scitotenv.2018.07.054

Salcedo-Sanz S, Rojo-Álvarez JL, Martínez-Ramón M, Camps-Valls G (2014) Support vector machines in engineering: an overview. Wires Data Min Knowl Discovery 4:234–267. https://doi.org/10.1002/widm.1125

Samuel AL (1959) Some studies in machine learning using the game of checkers. IBM J Res Dev 3:210–229. https://doi.org/10.1147/rd.33.0210

Sánchez-Franco MJ, Navarro-García A, Rondán-Cataluña FJ (2019) A naive Bayes strategy for classifying customer satisfaction: a study based on online reviews of hospitality services. J Bus Res 101:499–506. https://doi.org/10.1016/j.jbusres.2018.12.051

Santos I, Castro L, Rodriguez-Fernandez N, Torrente-Patiño Á, Carballal A (2021) Artificial neural networks and deep learning in the visual arts: a review. Neural Comput Appl 33:121–157. https://doi.org/10.1007/s00521-020-05565-4

Schonlau M, Zou RY (2020) The random forest algorithm for statistical learning. Stata J: Promot Commun Stat Stata 20:3–29. https://doi.org/10.1177/1536867x20909688

Shaji E, Santosh M, Sarath KV, Prakash P, Deepchand V, Divya BV (2021) Arsenic contamination of groundwater: a global synopsis with focus on the Indian Peninsula. Geosci Front 12:101079. https://doi.org/10.1016/j.gsf.2020.08.015

Silva SJ, Ma P-L, Hardin JC, Rothenberg D (2021) Physically regularized machine learning emulators of aerosol activation. Geosci Model Dev 14:3067–3077. https://doi.org/10.5194/gmd-14-3067-2021

Song S, Wang Y, Chen F, Yan M, Zhang Q (2022) Machine learning-assisted high-throughput virtual screening for on-demand customization of advanced energetic materials. Engineering 10:99–109. https://doi.org/10.1016/j.eng.2022.01.008

Sujay Raghavendra N, Deka PC (2014) Support vector machine applications in the field of hydrology: a review. Appl Soft Comput 19:372–386. https://doi.org/10.1016/j.asoc.2014.02.002

Sun Z, Sandoval L, Crystal-Ornelas R, Mousavi SM, Wang J, Lin C, Cristea N, Tong D, Carande WH, Ma X, Rao Y, Bednar JA, Tan A, Wang J, Purushotham S, Gill TE, Chastang J, Howard D, Holt B, Gangodagamage C, Zhao P, Rivas P, Chester Z, Orduz J, John A (2022) A review of earth artificial intelligence. Comput Geosci 159:105034. https://doi.org/10.1016/j.cageo.2022.105034

Tao H, Hameed MM, Marhoon HA, Zounemat-Kermani M, Heddam S, Kim S, Sulaiman SO, Tan ML, Sa'adi Z, Mehr AD, Allawi MF, Abba SI, Zain JM, Falah MW, Jamei M, Bokde ND, Bayatvarkeshi M, Al-Mukhtar M, Bhagat SK, Tiyasha T, Khedher KM, Al-Ansari N, Shahid S, Yaseen ZM (2022) Groundwater level prediction using machine learning models: a comprehensive review. Neurocomputing 489:271–308. https://doi.org/10.1016/j.neucom.2022.03.014

Tesoriero AJ, Gronberg JA, Juckem PF, Miller MP, Austin BP (2017) Predicting redox-sensitive contaminant concentrations in groundwater using random forest classification. Water Resour Res 53:7316–7331. https://doi.org/10.1002/2016wr020197

The State Council of the People's Republic of China (2021) Regulations on groundwater management. http://www.gov.cn/zhengce/content/2021-11/09/content_5649924.htm. Accessed 30 Apr 2024

Todaro V, D'Oria M, Tanda MG, Gómez-Hernández JJ (2021) Ensemble smoother with multiple data assimilation to simultaneously estimate the source location and the release history of a contaminant spill in an aquifer. J Hydrol 598:126215. https://doi.org/10.1016/j.jhydrol.2021.126215

Tsai J-P, Chang P-Y, Yeh T-CJ, Chang L-C, Hsiao C-T (2019) Constructing the apparent geological model by fusing surface resistivity survey and borehole records. Groundwater 57:590–601. https://doi.org/10.1111/gwat.12830

Tyralis H, Papacharalampous G, Langousis A (2019) A brief review of random forests for water scientists and practitioners and their recent history in water resources. Water 11:910. https://doi.org/10.3390/w11050910

Uddameri V, Silva A, Singaraju S, Mohammadi G, Hernandez E (2020) Tree-based modeling methods to predict nitrate exceedances in the Ogallala aquifer in texas. Water 12:1023. https://doi.org/10.3390/w12041023

Vamathevan J, Clark D, Czodrowski P, Dunham I, Ferran E, Lee G, Li B, Madabhushi A, Shah P, Spitzer M, Zhao S (2019) Applications of machine learning in drug discovery and development. Nat Rev Drug Discov 18:463–477. https://doi.org/10.1038/s41573-019-0024-5

Wang H, Lu W, Li J (2020a) Groundwater contaminant source characterization with simulation model parameter estimation utilizing a heuristic search strategy based on the stochastic-simulation statistic method. J Contam Hydrol 234:103681. https://doi.org/10.1016/j.jconhyd.2020.103681

Wang Q, Zhou Y, Ding W, Zhang Z, Muhammad K, Cao Z (2020b) Random forest with self-paced bootstrap learning in lung cancer prognosis. ACM Trans Multimed Comput Commun Appl 16:1–12. https://doi.org/10.1145/3345314

Wang H, Lu W, Chang Z (2021a) Simultaneous identification of groundwater contamination source and aquifer parameters with a new weighted-average wavelet variable-threshold denoising method. Environ Sci Pollut Res 28:38292–38307. https://doi.org/10.1007/s11356-021-12959-x

Wang L, Li P, Duan R, He X (2021b) Occurrence, controlling factors and health risks of Cr^{6+} in groundwater in the Guanzhong Basin of China. Expo Health 14:239–251. https://doi.org/10.1007/s12403-021-00410-y

Wang Y, Li P (2022) Appraisal of shallow groundwater quality with human health risk assessment in different seasons in rural areas of the Guanzhong Plain (China). Environ Res 207:112210. https://doi.org/10.1016/j.envres.2021.112210

Wang D, Qian J, Ma L, Zhao W, Gao D, Hou X, Ma H (2022a) Characterizing groundwater distribution potential using GIS-based machine learning model in Chihe River basin, China. Environ Earth Sci 81:324. https://doi.org/10.1007/s12665-022-10444-3

Wang J, Pang X, Yin F, Yao J (2022b) A deep neural network method for solving partial differential equations with complex boundary in groundwater seepage. J Petrol Sci Eng 209:109880. https://doi.org/10.1016/j.petrol.2021.109880

Wanigasekara C, Oromiehie E, Swain A, Prusty BG, Nguang SK (2021) Machine learning-based inverse predictive model for AFP based thermoplastic composites. J Ind Inf Integr 22:100197. https://doi.org/10.1016/j.jii.2020.100197

Weber RM, Fajen BR (2015) Decision-tree analysis of control strategies. Psychon Bull Rev 22:653–672. https://doi.org/10.3758/s13423-014-0732-0

Wellmann T, Lausch A, Scheuer S, Haase D (2020) Earth observation based indication for avian species distribution models using the spectral trait concept and machine learning in an urban setting. Ecol Ind 111:106029. https://doi.org/10.1016/j.ecolind.2019.106029

Westreich D, Lessler J, Funk MJ (2010) Propensity score estimation: neural networks, support vector machines, decision trees (CART), and meta-classifiers as alternatives to logistic regression. J Clin Epidemiol 63:826–833. https://doi.org/10.1016/j.jclinepi.2009.11.020

Wu C, Fang C, Wu X, Zhu G (2020a) Health-risk assessment of arsenic and groundwater quality classification using random forest in the Yanchi Region of Northwest China. Expo Health 12:761–774. https://doi.org/10.1007/s12403-019-00335-7

Wu W, Liao R, Hu Y, Wang H, Liu H, Yin S (2020b) Quantitative assessment of groundwater pollution risk in reclaimed water irrigation areas of northern China. Environ Pollut 261:114173. https://doi.org/10.1016/j.envpol.2020.114173

Xiong Y, Liu J, Yuan W, Liu W, Ma S, Wang Z, Li T, Wang Y, Wu J (2022) Groundwater contamination risk assessment based on groundwater vulnerability and pollution loading: a case study of typical Karst Areas in China. Sustainability 14:9898. https://doi.org/10.3390/su14169898

Xu T, Valocchi AJ, Choi J, Amir E (2014) Use of machine learning methods to reduce predictive error of groundwater models. Ground Water 52:448–460. https://doi.org/10.1111/gwat.12061

Xu F, Li P, Chen W, He S, Li F, Mu D, Elumalai V (2022) Impacts of land use/land cover patterns on groundwater quality in the Guanzhong Basin of northwest China. Geocarto Int 37:16769–16785. https://doi.org/10.1080/10106049.2022.2115153

Yao X, Tham LG, Dai FC (2008) Landslide susceptibility mapping based on support vector machine: a case study on natural slopes of Hong Kong, China. Geomorphology 101:572–582. https://doi.org/10.1016/j.geomorph.2008.02.011

Yeung YH, Barajas-Solano DA, Tartakovsky AM (2022) Physics-informed machine learning method for large-scale data assimilation problems. Water Resour Res 58:e2021WR031023. https://doi.org/10.1029/2021wr031023

Zarra T, Galang MG, Ballesteros F Jr, Belgiorno V, Naddeo V (2019) Environmental odour management by artificial neural network—A review. Environ Int 133:105189. https://doi.org/10.1016/j.envint.2019.105189

Zhang Q, Li P, Lyu Q, Ren X, He S (2022a) Groundwater contamination risk assessment using a modified DRATICL model and pollution loading: a case study in the Guanzhong Basin of China. Chemosphere 291:132695. https://doi.org/10.1016/j.chemosphere.2021.132695

Zhang X, Zhu Y, Wang J, Ju L, Qian Y, Ye M, Yang J (2022b) GW-PINN: a deep learning algorithm for solving groundwater flow equations. Adv Water Resour 165:104243. https://doi.org/10.1016/j.advwatres.2022.104243

Zhao Y, Qu R, Xing Z, Lu W (2020) Identifying groundwater contaminant sources based on a KELM surrogate model together with four heuristic optimization algorithms. Adv Water Resour 138:103540. https://doi.org/10.1016/j.advwatres.2020.103540

Zhong S, Zhang K, Bagheri M, Burken JG, Gu A, Li B, Ma X, Marrone BL, Ren ZJ, Schrier J, Shi W, Tan H, Wang T, Wang X, Wong BM, Xiao X, Yu X, Zhu JJ, Zhang H (2021) Machine learning: new ideas and tools in environmental science and engineering. Environ Sci Technol 55:12741–12754. https://doi.org/10.1021/acs.est.1c01339

Zounemat-Kermani M, Batelaan O, Fadaee M, Hinkelmann R (2021) Ensemble machine learning paradigms in hydrology: a review. J Hydrol 598:126266. https://doi.org/10.1016/j.jhydrol.2021.126266

Groundwater Salinization in Coastal Regions and the Control Mechanisms: Insights for Sustainable Groundwater Development and Management

Johnson C. Agbasi, Mahamuda Abu, Chaitanya B. Pande, Henry C. Uwajingba, Sani I. Abba, and Johnbosco C. Egbueri

Abstract Groundwater salinization in coastal regions poses a significant threat to water security, ecosystem health, and socio-economic development worldwide. This chapter provides a comprehensive analysis of the concepts, mechanisms, impacts, and management strategies related to coastal groundwater salinization. It examines case studies from diverse geographical contexts, including the Mediterranean coastal areas, South Asian coastal aquifers, and the Gulf Coast of the United States, to illustrate the global scope and regional specificities of the issue. The chapter analyzes the interaction between natural hydrogeological processes and anthropogenic influences that drive salinization, emphasizing the need for integrated management approaches. It assesses the environmental, socio-economic, and health impacts of groundwater salinization, highlighting the far-reaching consequences for coastal communities and ecosystems. The various control mechanisms were evaluated, including monitoring systems, engineering solutions, and policy interventions, and sustainable groundwater development strategies within the framework of integrated water resources management were explored. Emerging technologies and innovative approaches for

J. C. Agbasi · J. C. Egbueri (✉)
Department of Geology, Chukwuemeka Odumegwu Ojukwu University, Uli 431124, Nigeria
e-mail: johnboscoegbueri@gmail.com; jc.egbueri@coou.edu.ng

M. Abu
Department of Geological Engineering, University for Development Studies, Tamale, Ghana

C. B. Pande
Indian Institute of Tropical Meteorology, Pune, Maharashtra 411008, India

H. C. Uwajingba
Department of Surveying and Geoinformatics, Federal Polytechnic, Ugep, Cross River State, Nigeria

S. I. Abba
Department of Civil Engineering, Prince Mohammad Bin Fahd University, Al Khobar 31952, Saudi Arabia

J. C. Egbueri
Research Management Office (RMO), Chukwuemeka Odumegwu Ojukwu University, Uli, Anambra State, Nigeria

P. Li et al. (eds.), *Sustainable Groundwater and Environment: Challenges and Solutions*, Springer Hydrogeology, https://doi.org/10.1007/978-3-031-82194-3_8

enhancing coastal aquifer management are discussed, along with recommendations for future research and practice. By synthesizing current knowledge and identifying key challenges and opportunities, this chapter contributes to the broader understanding of coastal groundwater management and promotes sustainable water resource practices in the face of growing global challenges.

Keywords Coastal Aquifers of South Asia · Desalination · Gulf Coast of the United States · Mediterranean coastal areas · Salinity hazard · Water quality assessment

1 Introduction

Coastal regions worldwide face unprecedented challenges in managing their groundwater resources. This crisis is exacerbated by the convergence of population growth, rapid urbanization, and climate change (Dodd and Rishworth 2023; Meng et al. 2023). These areas, where land and sea intersect, harbor complex hydrogeological systems that are uniquely vulnerable to salinization. The intricate interplay between natural processes and anthropogenic activities has rendered groundwater salinization in coastal aquifers a multifaceted issue of global concern (Dodd and Rishworth 2023). The significance of groundwater in coastal areas cannot be overstated. In many regions, surface water resources are limited or unreliable, making groundwater an indispensable lifeline for sustaining human activities and ecosystems. Coastal aquifers play a crucial role in supporting diverse ecosystems, including wetlands and estuaries, by maintaining baseflow to streams and rivers. These groundwater-dependent ecosystems are not only essential for biodiversity but also provide invaluable ecosystem services that underpin local and regional economies (Boulton et al. 2023). The agricultural sector in coastal regions heavily relies on groundwater irrigation, enabling crop cultivation in areas with limited rainfall and contributing significantly to food security. Industries along coastlines depend on groundwater for various operations, from manufacturing processes to cooling systems. Moreover, tourism, a major economic driver in many coastal areas, hinges on reliable freshwater supplies for hotels, resorts, and recreational facilities. The availability of clean groundwater is paramount for maintaining the attractiveness and sustainability of these destinations.

However, the very attributes that make coastal aquifers invaluable also render them vulnerable to salinization. The Mediterranean coastal areas, stretching across southern Europe, northern Africa, and the Middle East, exemplify the challenges of managing coastal aquifers in semi-arid to arid climates (Bouhout et al. 2024). These regions often rely heavily on groundwater for domestic, agricultural, and industrial uses, making them susceptible to overexploitation and subsequent salinization (Ezzaouy et al. 2023). Similarly, in South Asia, rapid industrialization and urbanization along coastlines have led to unprecedented groundwater demand. Countries like Bangladesh, India, and Vietnam are grappling with the consequences of saltwater intrusion in their coastal aquifers, affecting millions of people's access to freshwater

(Rakib et al. 2020). The Gulf Coast of the United States presents a unique case where natural geological processes, such as the presence of salt domes, combine with anthropogenic factors like hydrocarbon exploration to influence groundwater salinity (Chaudhuri and Ale 2014). This region highlights the complexity of managing coastal aquifers in areas with intensive industrial activities and diverse geological settings. Small island developing states (SIDS) in the Pacific and Caribbean face perhaps the most acute challenges. Their limited land area and freshwater resources make them particularly vulnerable to salinization, with sea-level rise posing an existential threat to these nations' water security (Thomas et al. 2020). The plight of SIDS serves as a stark reminder of the global implications of climate change on coastal groundwater resources.

Recent research has shed light on the intricate relationship between groundwater salinization and climate change. Rising sea levels and increased frequency of extreme weather events, such as storm surges and prolonged droughts, are exacerbating saltwater intrusion in coastal aquifers (Etukudoh et al. 2024). This climate-induced salinization is compounded by human activities, creating a complex feedback loop that threatens water security in coastal regions worldwide. Emerging technologies are offering new avenues for understanding and managing coastal aquifers. Remote sensing and machine learning algorithms are revolutionizing the monitoring and prediction of groundwater dynamics in coastal areas (Springer et al. 2023). These advanced tools allow for more accurate mapping of saltwater intrusion fronts and enable proactive management strategies to mitigate salinization risks.

The socio-economic impacts of groundwater salinization extend far beyond water scarcity. Recent studies have highlighted the cascading effects on public health, agriculture, and local economies. In Bangladesh, for instance, increased salinity in drinking water has been linked to higher incidences of hypertension and other health issues among coastal populations (Rahman et al. 2023). The salinization of agricultural lands not only affects crop yields but also alters soil microbial communities, potentially disrupting entire ecosystems (Khondoker et al. 2023). As coastal communities worldwide grapple with these multifaceted challenges, it becomes increasingly clear that conventional water management approaches are inadequate. The complexity of groundwater salinization demands solutions that are not only technologically advanced but also environmentally sustainable and socially equitable. From improved monitoring systems and innovative extraction techniques to policy reforms and community engagement (Barbieri 2023), a wide array of strategies must be considered and tailored to specific regional contexts.

This chapter aims to provide a comprehensive overview of groundwater salinization in coastal regions and explore sustainable management strategies. The specific objectives are: (1) to crystalize the concepts and mechanisms underlying groundwater salinization in coastal aquifers, (2) to analyze case studies from diverse geographical contexts, identifying common challenges and region-specific issues, (3) to assess the environmental, socio-economic, and health impacts of groundwater salinization in coastal communities, (4) to evaluate various control mechanisms and management strategies for mitigating and preventing groundwater salinization, and (5) to explore sustainable groundwater development and management approaches that balance human needs with ecosystem preservation. By addressing these objectives, this chapter contributes to the critical discourse on global water resource management, offering insights that are both timely and essential for policymakers, researchers, and practitioners working towards sustainable coastal aquifer management in an era of unprecedented environmental change.

2 Groundwater Salinization: Concepts and Mechanisms

2.1 Definition and Types of Groundwater Salinization

Groundwater salinization is a complex phenomenon characterized by an increase in dissolved salt content within aquifers. This process is typically quantified through measurements of total dissolved solids (TDS) or electrical conductivity (EC). According to the World Health Organization, freshwater is defined as having TDS less than 1000 mg/L, while water exceeding this threshold is classified as brackish or saline (WHO 2017).

The origins of groundwater salinization are diverse and can be categorized into several types. Primary salinization occurs naturally due to geological formations, ancient marine deposits, or long-term water-rock interactions (Greggio et al. 2012). This process is particularly prevalent in arid regions where evaporative concentration plays a significant role. Secondary salinization, on the other hand, results from anthropogenic activities that alter the hydrological cycle or introduce salts into the environment (Greggio et al. 2012). This includes agricultural practices, industrial effluents, and urban runoff. Seawater intrusion represents a specific form of salinization in coastal aquifers, where saltwater displaces freshwater due to density differences and hydraulic gradients (Cantelon et al. 2022). This process is exacerbated by over-extraction of groundwater and sea-level rise associated with climate change (Dieu et al. 2022). Upconing, another salinization mechanism, involves the vertical movement of saline water from deeper aquifers into freshwater zones, often triggered by excessive pumping (Cantelon et al. 2022). Recent research has highlighted the role of upconing in the deterioration of water quality in many coastal regions worldwide (García-Menéndez et al. 2016).

2.2 Mechanisms of Salinization

2.2.1 Natural Causes of Salinization

Natural processes contributing to groundwater salinization are diverse and interconnected (Fig. 1). Seawater intrusion in coastal aquifers is influenced by sea-level changes, tidal fluctuations, and long-term climatic variations (Ketabchi et al. 2016).

Recent studies have emphasized the role of coastal aquifer heterogeneity in controlling the dynamics of seawater intrusion, highlighting the need for high-resolution characterization of aquifer properties (Ahmed et al. 2022). Saltwater upwelling, where deep saline groundwater naturally migrates upwards through faults or permeable layers, has been identified as a significant source of salinization in various geological settings (Chabab et al. 2022). This process can be exacerbated by tectonic activity and changes in hydraulic gradients. Evaporative concentration plays a crucial role in arid and semi-arid regions, where high evaporation rates concentrate salts in surface water bodies and shallow aquifers. Climate change is expected to intensify this process in many regions, potentially leading to increased salinization of water resources (Egbueri et al. 2023). Geochemical processes, including water-rock interactions such as mineral dissolution and ion exchange, contribute to the natural increase in groundwater salinity over time (Abugu et al. 2024). Recent advances in isotope geochemistry have provided new insights into these processes, allowing for better discrimination between natural and anthropogenic sources of salinity (Ammar et al. 2020). Atmospheric deposition, particularly in coastal areas, contributes to the accumulation of salts in soils and shallow groundwater (Ortiz and Jin 2021). Sea spray and wind-blown salt particles can travel considerable distances inland, affecting water quality in coastal aquifers.

2.2.2 Anthropogenic Causes of Salinization

Human activities have emerged as dominant drivers of groundwater salinization in many regions. Over-extraction of groundwater, often driven by increasing water demand for agriculture and urban development, can lower water tables and alter hydraulic gradients. This process induces seawater intrusion in coastal areas and can lead to upcoming of deeper saline water (Dieu et al. 2022). Agricultural activities contribute significantly to groundwater salinization through various mechanisms. Irrigation with saline water, excessive fertilizer use, and poor drainage practices can lead to soil salinization and subsequent groundwater contamination (Mohanavelu et al. 2021). The adoption of precision agriculture techniques and improved irrigation management has shown promise in reducing these impacts (Abioye et al. 2020). Industrial activities, including the disposal of industrial effluents, mining operations, and oil and gas extraction, introduce various contaminants, including salts, into aquifers (Li et al. 2020).

Urbanization and infrastructure development have profound impacts on groundwater systems. Coastal land reclamation, construction of underground structures, and changes in land use patterns can alter natural groundwater flow systems and enhance salinization processes (Seibert et al. 2024). Green infrastructure and low-impact development approaches are being explored as potential solutions to mitigate these effects. Climate change, while not directly anthropogenic, exacerbates salinization through sea-level rise, changes in precipitation patterns, and increased

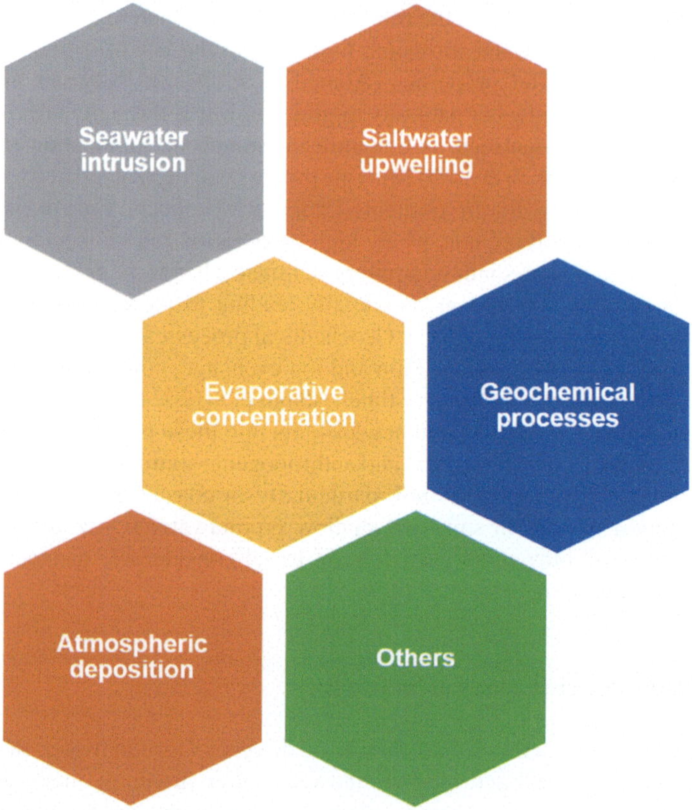

Fig. 1 Schematic diagram illustrating natural mechanisms of groundwater salinization in coastal aquifers

frequency of extreme events (Dieu et al. 2022). Recent modeling studies have highlighted the complex interactions between climate change and groundwater salinization, emphasizing the need for integrated management approaches (Dieu et al. 2022).

2.3 *Factors Influencing Salinization*

Salinization is coastal regions is influenced by three key factors (Fig. 2).

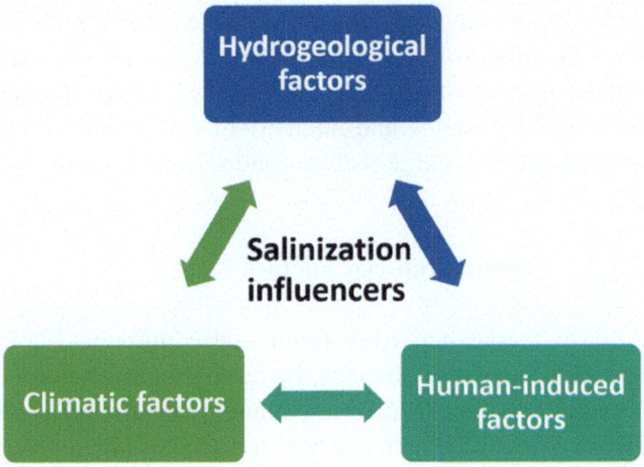

Fig. 2 Key influential factors of coastal groundwater salinization

2.3.1 Hydrogeological Factors

The susceptibility of coastal aquifers to salinization is largely determined by their hydrogeological characteristics. Aquifer geometry and stratigraphy play a crucial role in influencing groundwater flow patterns and the potential for saltwater intrusion by creating preferential flow paths, barriers, and varying mixing zones, which affect the extent and severity of intrusion (Yu and Michael 2019; O'Connor et al. 2019; Geng and Michael 2020). Recent advancements in 3D geological modeling and hydrogeophysical techniques have improved the understanding of these complex systems (Hermans et al. 2023). Hydraulic properties, including aquifer transmissivity, storativity, and hydraulic conductivity, affect the rate and extent of saline water movement (Rezaei et al. 2021). High-resolution characterization of these properties is essential for accurate prediction of salinization processes and effective management strategies. Coastal topography significantly influences the vulnerability of aquifers to seawater intrusion and flooding. Low-lying coastal areas are particularly susceptible to these processes, with studies highlighting the importance of considering local-scale topographic variations in risk assessments (Michael et al. 2013). Groundwater-surface water interactions, including the presence of coastal lagoons, estuaries, and tidal creeks, can influence salinity patterns in adjacent aquifers (Mondal et al. 2024). These interactions are increasingly recognized as important factors in coastal aquifer management, requiring integrated approaches that consider both surface and subsurface processes.

2.3.2 Climatic Factors

Climate plays a crucial role in groundwater salinization processes (Seibert et al. 2024). Precipitation patterns affect groundwater recharge rates and the flushing of salts from the unsaturated zone. Evapotranspiration, particularly high rates in arid and semi-arid regions, can concentrate salts in soils and shallow groundwater. Sea-level fluctuations, driven by long-term climate change, increase the risk of seawater intrusion in coastal aquifers (Dieu et al. 2022). Extreme weather events, including increased frequency and intensity of storms and droughts, can significantly impact groundwater recharge patterns and salinity dynamics.

2.3.3 Human-Induced Factors

Anthropogenic activities significantly influence salinization processes in coastal aquifers. Land-use changes, including urbanization, deforestation, and agricultural expansion, alter surface runoff and groundwater recharge patterns (Michael et al. 2013). Water management practices, including groundwater extraction policies, irrigation methods, and drainage systems, have profound effects on the water balance and salt distribution in coastal aquifers (Greggio et al. 2012). Coastal engineering projects, such as the construction of dams, levees, and sea walls, can modify natural hydrological processes and impact groundwater–seawater interactions (Chen et al. 2024). Waste disposal practices, including improper management of municipal and industrial waste, can bring in contaminants, including salts, into groundwater systems (Li et al. 2020). Global environmental change, driven by human activities, exacerbates natural salinization processes through its impacts on sea level and weather patterns (Michael et al. 2013).

3 Case Studies of Groundwater Salinization in Coastal Regions

3.1 Case Study 1: Mediterranean Coastal Areas

Groundwater salinization in Mediterranean coastal regions, particularly in Morocco, has been extensively studied using various methods, including hydrogeochemical, isotopic, and multivariate statistical techniques. Research in the Essaouira Coastal region of Morocco has provided significant insights into the processes and controls of groundwater salinization (Ouarani et al. 2020, 2021, 2023; Bahir et al. 2021; Ouzerbane et al. 2022). Bahir et al. (2020) employed the hydrochemical facies evolution diagram to evaluate seawater intrusion effects on groundwater aquifers. Their findings revealed that 77% of the collected samples indicated seawater intrusion modifying and contributing to aquifer salinization, even in aquifers located up to 18 km inland. A multi-method approach by Ouarani et al. (2023) concluded that groundwater salinization in Moroccan Mediterranean coastal aquifers results from both seawater intrusion and salt dissolution from a salt diapir within the coastal catchment. In Tunisia, Kammoun et al. (2021) assessed the controls of coastal groundwater salinization in the Grombalia aquifers using chemical and isotopic analyses (δ^2H,

$\delta^{18}O$, ^{3}H, ^{14}C, and $\delta^{13}C$). Their study concluded that groundwater salinization, especially in shallow aquifers, is largely due to returned irrigated waters. Bouzourra et al. (2015) reported that salty marine water intrusion and returned irrigated waters have been the main causes of groundwater salinization in northeastern Tunisia since 1972. Similar conclusions were reached by Ferchichi et al. (2018) in the Medjerda Lower Valley Aquifer and by Trabelsi et al. (2012) in the Djeffara plain of southeastern Tunisia. In Libya, groundwater salinization due to seawater intrusion has been documented since 1930 (Alfarrah and Walraevens 2018). The Jifarah Plain in northwestern Libya, particularly around Tripoli, has been well-studied, indicating continuous seawater intrusion as the primary cause of groundwater salinization, exacerbated by over-extraction due to increased demand.

3.2 *Case Study 2: Coastal Aquifers of South Asia*

Due to geographical privileges given out to inhabitants of these catchments and industrialization in and around coastal areas within South Asia (Xiong et al. 2020), there has been a continuous population increase, leading to increased pressure on groundwater resources. Groundwater in these regions has, due to over-extraction, low recharge, seawater intrusion, and other latent intra-aquifer natural processes as well as anthropogenic processes rendered this highly demanded resource unsafe for usage in these regions (Sun et al. 2011; Werner et al. 2013). In Thailand and Bangladesh, seawater intrusion is the primary cause of saline groundwater, with consequent impacts on groundwater fluorine content (Xiong et al. 2020). Rakib et al. (2020) reported significant seawater intrusion in the southwestern coastal aquifers of Bangladesh, rendering groundwater unsuitable for drinking. Also, in Vietnam, the Binh Thuan province in the South-central part, a multi-methodological approach in the assessment of the freshening and salinization of the coastal aquifers using cation exchange code and chloride conservative element, the ionic delta, the HFE-diagram, and geochemical modeling by Dieu et al. (2022), concluded that salinization is intense in deeper aquifers and the dry season. The salinization also extends far into the inland aquifers with an accompanying sharp variation in the groundwater salt content and has been attributed to the over-extraction of groundwater and low/no recharge (Dieu et al. 2022). In the last 5 decades, according to Han and Currell (2018), groundwater salinization in the coastal aquifers of northern China has been on the rise. To understand the groundwater salinization processes in the region, hydrochemistry and stable isotopic ($\delta^{18}O$ and $\delta^{2}H$) analysis of different water bodies e.g. groundwater, surface water, geothermal water, and seawater were analyzed (Han and Currell 2018). Seawater intrusion was found to be the main cause of groundwater salinization in these parts of China and has been catalyzed by over-extraction of groundwater and low aquifer recharge.

3.3 Case Study 3: Gulf Coast of the United States

Groundwater salinization in the US Gulf Coast aquifers can be traced back to 1985 (Blanco et al. 2013). The Ogallala aquifer underlies 49 counties and about 2 groundwater water management areas in Texas, US. Groundwater salinization in the aquifer from data compiled between 1960 to 2010, is linked to a number of natural and anthropogenic within aquifer processes such as saline water intrusion from saline plumes, upwelling of highly mineralized groundwater from underlying geological formations, as well as anthropogenic activities of return of irrigated waters into groundwater, aquifers, over-pumping of groundwater and hydrocarbon exploration activities (Chaudhuri and Ale 2014; Han et al. 2019). In the coastal areas of Oso Bay of south Texas (Bighash and Murgulet 2015), multivariate statistical analysis applications on hydrochemical data, and multitemporal electrical resistivity tomography surveys, suggest that the changing salinity regimes and the saline groundwater in the area can be attributed to a number of factors including the recirculation of hypersaline groundwaters, trapped oil field brines, and surface outflows (Bighash and Murgulet 2015).

3.4 Comparative Analysis of Case Studies

A comparative analysis of the case studies reveals both commonalities and unique features in groundwater salinization across different coastal regions. While seawater intrusion emerges as the primary natural process responsible for coastal groundwater salinization globally, the specific mechanisms and contributing factors vary by region. In the Mediterranean coastal areas, seawater intrusion is compounded by salt dissolution from geological formations, particularly in Morocco. The South Asian coastal aquifers face similar challenges, with seawater intrusion exacerbated by over-extraction and low recharge rates. The Gulf Coast of the United States presents a more complex scenario, where hydrocarbon exploration and exploitation activities contribute significantly to groundwater salinization, in addition to the common factors observed in other regions.

The case studies highlight the effectiveness of multi-methodological approaches in characterizing coastal groundwater salinization. However, there is a notable gap in the application of advanced computational techniques, particularly machine learning methods, which have proven reliable in other aspects of groundwater characterization (Egbueri 2023; Abu et al. 2024; Agbasi et al. 2024). The integration of machine learning techniques such as artificial neural networks and multilinear regression modeling could provide deeper insights into coastal groundwater salinization processes and enhance predictive capabilities.

To address the worsening groundwater situation in these coastal regions, artificial recharge of aquifers emerges as a potential remedial measure to inhibit seawater intrusion. Future research should focus on identifying suitable zones for artificial recharge and determining the specific recharge quantities needed to counteract seawater intrusion effectively. Additionally, long-term monitoring programs and the development of region-specific management strategies are essential for sustainable groundwater management in coastal areas.

4 Impacts of Groundwater Salinization

Groundwater salinization has several impacts. The section focuses on three main areas (Fig. 3).

4.1 Environmental Impacts

Groundwater salinization in coastal regions has several environmental consequences with far-reaching implications. The degradation of wetlands and estuaries emerges as a primary concern, as elevated salinity levels disrupt the ecological equilibrium of these critical habitats (Herbert et al. 2015). Concurrently, terrestrial ecosystems face significant challenges, notably soil salinization resulting from irrigation with saline groundwater. This process markedly diminishes soil fertility and accelerates land degradation (Khondoker et al. 2023), culminating in substantial biodiversity loss across both terrestrial and aquatic domains.

Alterations in groundwater salinity significantly impact biogeochemical cycles, perturbing nutrient cycling and carbon sequestration in coastal ecosystems (Röthig et al. 2023). These modifications reverberate through the coastal system, altering fundamental ecological processes. Furthermore, the salinization of coastal aquifers contributes to land subsidence (Adem et al. 2024), a phenomenon that exacerbates coastal erosion and heightens vulnerability to sea-level rise. This creates a positive feedback loop of environmental degradation, wherein each impact amplifies the others, leading to progressively severe ecological consequences.

Fig. 3 Impacts of groundwater salinization

4.2 Socio-economic Impacts

The socio-economic implications of groundwater salinization are both extensive and multifaceted, permeating various aspects of coastal communities. Agricultural productivity in saline-affected areas experiences significant decline, posing substantial threats to food security and rural livelihoods (Sultan et al. 2023). The repercussions extend beyond agricultural sectors, cascading through local and regional economies. Water supply costs escalate as additional treatment becomes imperative to mitigate increased salinity, imposing a considerable burden on both public resources and individual households (Venkatesan et al. 2011).

Infrastructure suffers from the corrosive nature of saline water, necessitating frequent maintenance and replacements that deplete economic resources (Venkatesan et al. 2011). In coastal regions where tourism constitutes a primary economic driver, the degradation of freshwater resources can precipitate substantial economic losses, undermining a crucial source of income (Stavros et al. 2016). Property values in saline-affected areas often experience significant depreciation, eroding individual wealth and weakening local economic foundations (Venkatesan et al. 2011). The fishing industry, vital to many coastal and estuarine communities, faces challenges as changing groundwater salinity disrupts aquatic ecosystems, directly impacting local livelihoods and food security (Muruganandam et al. 2023).

4.3 Health Impacts

Groundwater salinization presents a complex array of health risks to coastal populations, with consequences ranging from chronic diseases to acute conditions. Long-term consumption of high-salinity water demonstrates a strong correlation with increased incidences of hypertension and cardiovascular diseases, adding a significant burden to public health systems (Rahman et al. 2023). The impacts manifest in daily life, with numerous individuals experiencing a range of gastrointestinal problems and dermatological conditions linked to the use of saline water for drinking and bathing (Rahman et al. 2023).

Of particular concern are the potential impacts on maternal and child health. Evidence suggests a link between high-salinity drinking water and an increased risk of preeclampsia in pregnant women, accentuating the far-reaching consequences of this environmental issue (Rahman et al. 2023). The unpalatable nature of saline water may lead to reduced water consumption, potentially causing dehydration, especially in hot climates. This affects all human populations, especially the most vulnerable (elderly and young children).

Beyond direct consumption, soil salinization leads to reduced crop yields and compromised nutritional quality, potentially contributing to malnutrition in affected communities (Mukhopadhyay et al. 2021). This nutritional deficit can have long-lasting impacts on community health, particularly for children whose development may be impaired by inadequate nutrition. Moreover, the psychological toll associated with water insecurity and loss of livelihoods due to salinization in coastal communities warrants consideration, adding another dimension to the complex web of health impacts engendered by this environmental crisis (Rahman and Ahmad 2018).

5 Control Mechanisms for Groundwater Salinization

5.1 Monitoring and Early Warning Systems

Groundwater monitoring and early warning systems is commonly done using the methods such as groundwater quality monitoring, geophysical techniques, remote sensing and machine learning techniques (Fig. 4).

5.1.1 Groundwater Quality Monitoring

Effective groundwater quality monitoring forms the cornerstone of successful salinization management in coastal aquifers. The systematic collection and analysis of key water quality parameters provide crucial insights into the spatial and temporal dynamics of salinization processes. Recent technological advancements have significantly enhanced the capacity to detect and predict salinization trends, revolutionizing the approach to coastal aquifer management. While electrical conductivity (EC) and chloride concentration remain primary indicators of groundwater salinity, the application of stable isotope analysis has gained prominence in recent years. Multi-isotope approaches, combining elements such as strontium, boron, and water isotopes, offer powerful tools for distinguishing between various salinity sources, including seawater intrusion, agricultural return flows, and geogenic sources (Kammoun et al. 2021; Ammar et al. 2020; Han and Currell 2018). This enhanced source identification

Fig. 4 Monitoring and early warning systems for groundwater salinization

capability allows for more targeted and effective management strategies. The advent of high-frequency monitoring systems, utilizing in-situ sensors and telemetry (Reljić et al. 2023), has transformed data collection practices. These systems enable near real-time monitoring of water quality parameters, facilitating rapid detection of salinization events and short-term fluctuations related to tidal cycles or storm surges. Novel technologies, such as fiber-optic distributed temperature sensing (FO-DTS), provide high-resolution data on the spatial distribution of the freshwater-saltwater interface, offering unprecedented insights into the dynamic behavior of coastal aquifers (Folch et al. 2020).

5.1.2 Geophysical Techniques

Non-invasive geophysical methods have proven highly effective in delineating the freshwater-saltwater interface in coastal aquifers. These techniques allow for high-resolution mapping of saltwater intrusion and monitoring of salinization fronts over time (Dey and Prakash 2020). Key methods include electrical resistivity tomography (ERT), which provides detailed subsurface resistivity images (Akingboye and Osazuwa 2021), and electromagnetic surveys (EM) for rapid, large-scale assessments of saltwater distribution (Nenna et al. 2013). Ground penetrating radar (GPR) offers high-resolution imaging of shallow subsurface structures (Tao et al. 2022), while seismic methods help characterize aquifer geometry (Tinivella et al. 2022). These complementary techniques, often used in combination, provide comprehensive insights into coastal aquifer dynamics and saltwater intrusion processes.

5.1.3 Remote Sensing Applications

Satellite-based technologies enable large-scale monitoring of groundwater resources and related phenomena in coastal areas. These methods facilitate the assessment of regional-scale changes in groundwater storage and identification of areas at risk of salinization (Springer et al. 2023). The Gravity Recovery and Climate Experiment (GRACE) measures changes in Earth's gravity field, allowing for estimation of regional-scale groundwater storage variations (Chang et al. 2020), while Interferometric Synthetic Aperture Radar (InSAR) provides high-resolution measurements of land surface deformation (González 2024). These techniques have been successfully applied to assess aquifer depletion and identify subsidence hotspots in various coastal regions (Springer et al. 2023). Additionally, multispectral and hyperspectral imaging offer insights into surface indicators of groundwater conditions, aiding in the mapping of groundwater-dependent ecosystems (Murray et al. 2022). These remote sensing approaches complement ground-based observations by providing continuous, wide-area coverage, thus enhancing the understanding of coastal aquifer dynamics and vulnerability to salinization.

5.1.4 Machine Learning Applications

Machine learning techniques have emerged as powerful tools for enhancing groundwater monitoring and prediction in coastal aquifers (BinMakhashen and Benaafi 2024). These advanced computational methods can process and analyze complex datasets from various sources, including monitoring wells, geophysical surveys, and remote sensing observations (Hussain et al. 2024; Singh et al. 2024; Nwazelibe et al. 2023). Machine learning algorithms, such as artificial neural networks and support vector machines, excel at predictive modeling, pattern recognition, and data integration (Singh et al. 2024; Kurani et al. 2023). They can forecast groundwater levels, salinity concentrations, and saltwater interface movements (Panthi et al. 2022), while also detecting early warning signs of water quality issues. Moreover, these techniques can optimize monitoring network designs (Musumeci et al. 2018) and enable real-time analysis of incoming data streams (Mehmood and Anees 2020). While machine learning offers significant potential, it should complement rather than replace traditional hydrogeological expertise and field observations. Effective application requires careful model selection, rigorous validation, and ongoing refinement (Agbasi and Egbueri 2024; Ahmed et al. 2023a) to ensure reliable and actionable results for coastal aquifer management.

5.2 Management Strategies

5.2.1 Saltwater Intrusion Barriers

Engineered solutions for creating hydraulic barriers against seawater intrusion have evolved significantly, with a focus on enhancing efficiency and sustainability. While traditional techniques such as injection wells and subsurface dams remain effective (Hussain et al. 2019), innovative approaches are being developed to address site-specific challenges and improve long-term performance. Recent advancements in injection well systems have focused on optimizing their design and operation through multi-objective approaches that consider both cost-effectiveness and long-term sustainability (Zhuang et al. 2024). These optimized configurations have demonstrated significant reductions in saltwater intrusion compared to conventional designs, while minimizing energy consumption and environmental impacts (Saad et al. 2023). The success of these systems accentuates the importance of site-specific optimization in maximizing the effectiveness of hydraulic barriers. Subsurface dams have gained renewed interest due to their potential for minimal environmental impact and long-term effectiveness (Abd-Elaty et al. 2024). Recent studies have highlighted the importance of comprehensive site-specific hydrogeological investigations and the use of advanced numerical modeling techniques in the design and implementation of these structures (Abd-Elaty et al. 2019, 2024). When properly implemented, subsurface dams can effectively reduce seawater intrusion while maintaining groundwater flow, offering a sustainable solution for coastal aquifer management.

5.2.2 Sustainable Groundwater Extraction and Pumping Optimization

Optimizing well placement and pumping rates is crucial for maintaining the balance between freshwater and saltwater in coastal aquifers. Decision support systems (DSS) can aid in developing sustainable extraction strategies that minimize the risk of salinization. Examples of such systems include GIS-based tools that integrate hydrogeological models with multi-criteria decision analysis for optimizing pumping schedules (Gebre et al. 2021), and simulation-optimization frameworks combining groundwater flow models (e.g., MODFLOW) with evolutionary algorithms to determine optimal pumping rates (Norouzi Khatiri et al. 2023). These DSS tools typically incorporate factors such as aquifer heterogeneity, seasonal variations, and socio-economic constraints to provide robust management solutions for coastal groundwater resources (Torresan et al. 2016). By leveraging these advanced computational methods, water managers can make more informed decisions to balance water demand with long-term aquifer sustainability.

5.2.3 Agricultural Best Practices

Agricultural practices play a crucial role in managing salinization risks in coastal areas. Recent research has focused on developing integrated approaches that combine improved irrigation techniques, crop selection, and soil management strategies to mitigate salt accumulation in agricultural lands (Abioye et al. 2020). Advanced irrigation technologies, such as subsurface drip irrigation and precision sprinkler systems, have shown significant potential in reducing soil salinization while maintaining crop yields (Wang et al. 2023). These systems, when coupled with irrigation scheduling based on real-time soil moisture and salinity sensors, optimize water use efficiency and salt leaching (Ahmed et al. 2023b), contributing to more sustainable agricultural practices in coastal regions. The selection and development of salt-tolerant crops have gained increasing attention as a strategy to maintain agricultural productivity in saline environments (Egea et al. 2023). Genetic engineering and selective breeding techniques have led to the development of crop varieties with enhanced salt tolerance, capable of maintaining yields under moderate saline conditions where conventional varieties fail (Padmavathi et al. 2024; Atta et al. 2023). The combination of these genetic improvements with appropriate management practices offers a promising approach to sustainably cultivating crops in salt-affected coastal areas. Soil management practices, including the use of organic amendments and crop rotation, have been shown to play a critical role in mitigating salinization effects (Tedeschi et al. 2023). Long-term studies have demonstrated that integrated soil management approaches, such as the application of biochar combined with crop rotation, can significantly improve soil structure and reduce salt accumulation over time (Yuan et al. 2023).

5.2.4 Artificial Recharge Techniques

Managed aquifer recharge (MAR) offers promising solutions for augmenting freshwater resources and mitigating salinization in coastal aquifers. Various MAR techniques have shown potential for enhancing water security in areas at risk of salinization (Kazakis 2018). These include infiltration basins, which allow surface water to percolate into the aquifer; injection wells, used to directly introduce freshwater into specific aquifer zones; and subsurface dams, which create underground barriers to saltwater intrusion (Sherif et al. 2023). In coastal areas, MAR can utilize diverse water sources such as treated wastewater, desalinated water, or harvested stormwater (Olabode and Comte 2024). The choice of MAR technique depends on local hydrogeological conditions, available water sources, and management objectives (Meles et al. 2024), with successful implementations demonstrating improved groundwater quality and increased freshwater availability in coastal regions worldwide (Olabode and Comte 2024; Sherif et al. 2023).

5.2.5 Desalination Technologies

Desalination technologies play a crucial role in addressing groundwater salinization in coastal regions by converting saline or brackish water into freshwater. These methods help alleviate pressure on existing freshwater aquifers and mitigate seawater intrusion impacts. Reverse osmosis (RO) is widely adopted due to its scalability and efficiency (Harby et al. 2024). This process forces saline water through a semi-permeable membrane under high pressure, separating dissolved salts from water molecules (Harby et al. 2024). Thermal desalination techniques, such as multi-stage flash distillation (MSF) and multiple-effect distillation (MED), remain significant in energy-rich regions (Narayan et al. 2010). These methods involve heating saline water to produce vapor, which is then condensed to yield freshwater. Emerging technologies like electrodialysis (ED) and capacitive deionization (CDI) offer promising alternatives for brackish water desalination (Tauk et al. 2024). ED uses an electric field to remove charged particles through ion-exchange membranes, while CDI employs electrically charged electrodes to adsorb and desorb ions. Forward osmosis (FO) represents a novel approach utilizing natural osmotic pressure gradients, though challenges in draw solute development persist (Reddy et al. 2024). Hybrid systems combining multiple desalination technologies optimize performance and reduce costs. Integration of renewable energy sources with desalination plants is becoming increasingly common (Zakariazadeh et al. 2024), addressing energy consumption concerns. Advancements in desalinization focuses on improving energy efficiency, reducing environmental impacts, and enhancing brine management strategies (Hu and Chen 2024).

5.2.6 Use of Alternative Water Sources

The incorporation of alternative water sources has emerged as a crucial strategy for reducing pressure on coastal aquifers and mitigating salinization risks. Recent advancements in water treatment technologies and resource management approaches have expanded the potential for utilizing non-conventional water sources in coastal regions. Treated wastewater reuse has gained significant traction as a sustainable water source for irrigation and aquifer recharge (Olabode and Comte 2024). Comprehensive reviews of managed aquifer recharge using treated wastewater in coastal areas have highlighted successful case studies demonstrating its potential to combat seawater intrusion while providing a reliable water source for agriculture (Olabode and Comte 2024; Sherif et al. 2023). These studies emphasize the importance of robust treatment processes and public acceptance in the successful implementation of wastewater reuse projects. Rainwater harvesting systems have been increasingly integrated into urban water management strategies in coastal cities. Large-scale implementation of these systems has shown potential to significantly reduce dependence on groundwater while simultaneously mitigating urban flooding risks (Xu et al. 2023). This multi-functional approach to water management exemplifies the synergies that can be achieved through integrated urban water planning. Case studies in coastal cities have demonstrated how the integration of rainwater harvesting, stormwater management, and wastewater reuse can significantly reduce pressure on groundwater resources (Olabode and Comte 2024; Sherif et al. 2023), which reduces the likelihood of saltwater intrusion and enhances urban livability.

6 Sustainable Groundwater Development and Management

6.1 Integrated Water Resources Management (IWRM) Approach

The Integrated Water Resources Management (IWRM) approach provides a comprehensive framework for holistic water management. This approach is particularly relevant to coastal aquifers facing salinization challenges, as it emphasizes the coordinated development and management of water, land, and related resources (GWP 2000).

6.2 Stakeholder Engagement, Public Awareness and Community Participation

Involving local communities, water users, and other stakeholders in decision-making processes is crucial for ensuring the effectiveness and legitimacy of management strategies (Almulhim and Abubakar 2023). Public awareness campaigns and educational programs play a vital role in fostering a culture of water conservation in coastal communities.

6.3 Policy and Regulatory Frameworks

Comprehensive groundwater legislation addressing issues such as water rights, extraction limits, and water quality standards is essential for effective management (Pino-Vargas et al. 2023). Adaptive policy frameworks that can incorporate new scientific knowledge and changing environmental conditions are particularly important in the context of coastal aquifer management.

6.4 Recommendations for Future Research and Practice

Future research should focus on developing adaptation strategies that anticipate and address the impacts of climate change on coastal aquifers. The application of artificial intelligence and machine learning techniques shows promise for enhancing monitoring and early warning systems in groundwater management. Integrated, adaptive approaches that can address the complex and dynamic nature of salinization processes in coastal aquifers should be a priority for future research and practice.

7 Conclusions

This chapter has demonstrated that groundwater salinization in coastal regions represents a complex and pressing challenge that threatens water security, ecosystem health, and socio-economic development globally. Through detailed analysis, a comprehensive overview of the concepts, mechanisms, impacts, and management strategies related to coastal groundwater salinization were provided. The investigation revealed that salinization processes, involving intricate interactions between natural hydrogeological factors and anthropogenic influences, necessitates an integrated approach to management. Examined case studies from the Mediterranean coastal areas, South Asian coastal aquifers, and the Gulf Coast of the United States has illustrated the global scope of the issue and the diversity of challenges faced by different regions. Significantly, the impacts of groundwater salinization extend far beyond the immediate hydrological realm, affecting environmental systems, economic activities, and human health. These wide-ranging consequences accentuate the urgency of implementing effective control mechanisms and sustainable management practices. Recent advances in monitoring technologies, including geophysical techniques, remote sensing and machine learning applications, offer new opportunities for early detection and mapping of salinization processes. These tools, combined with innovative management strategies such as managed aquifer recharge and optimized extraction practices, provide a foundation for addressing salinization challenges. The path towards sustainable groundwater development in coastal regions requires a commitment to integrated water resources management, stakeholder engagement, and adaptive policy frameworks. Looking ahead, decision support systems and emerging technologies like artificial intelligence show promise in enhancing the capacity to manage complex coastal aquifer systems in the face of uncertainty and change. As coastal regions worldwide grapple with the impacts of population growth, urbanization, and climate change, the importance of sustainable groundwater management becomes increasingly critical. Future research and practice should focus on developing holistic, adaptive approaches that can address the dynamic nature of coastal groundwater systems while ensuring their long-term sustainability.

References

Abd-Elaty I, Abd-Elhamid HF, Nezhad MM (2019) Numerical analysis of physical barriers systems efficiency in controlling saltwater intrusion in coastal aquifers. Environ Sci Pollut Res 26(35):35882–35899. https://doi.org/10.1007/s11356-019-06725-3

Abd-Elaty I, Kuriqi A, Ahmed A (2024) Effectiveness of different mixed physical barriers in controlling seawater intrusion in homogeneous and layered coastal aquifers. Groundw Sustain Dev 27:101318. https://doi.org/10.1016/j.gsd.2024.101318

Abioye EA, Abidin MS, Mahmud MS, Buyamin S, Ishak MH, Abd Rahman MK, Otuoze AO, Onotu P, Ramli MS (2020) A review on monitoring and advanced control strategies for precision irrigation. Comput Electron Agric 173:105441. https://doi.org/10.1016/j.compag.2020.105441

Abu M, Musah R, Zango MS (2024) A combination of multivariate statistics and machine learning techniques in groundwater characterization and quality forecasting. Geosyst Geoenviron 3:100261. https://doi.org/10.1016/j.geogeo.2024.100261

Abugu HO, Egbueri JC, Agbasi JC, Ezugwu AL, Omeka ME, Ucheana IA, Aralu CC (2024) Hydrochemical characterization of ground and surface water for irrigation application in Nigeria: a review of progress. Chem Afr 7:3011–3036. https://doi.org/10.1007/s42250-024-00917-2

Adem E, Shults R, Ukasha M, Elfeki A, Alqahtani F, Elhag M (2024) Land subsidence and groundwater storage change assessment using InSAR and GRACE in the arid environment of Saudi Arabia. Nat Hazards. https://doi.org/10.1007/s11069-024-06733-8

Agbasi JC, Abu M, Egbueri JC (2024) Towards sustainable industrial development: modelling the quality, scaling potential and corrosivity of groundwater using GIS, spatial statistics, soft computing and index-based methods. Environ Dev Sustain. https://doi.org/10.1007/s10668-024-05105-3

Agbasi JC, Egbueri JC (2024) Prediction of potentially toxic elements in water resources using MLP-NN, RBF-NN, and ANFIS: a comprehensive review. Environ Sci Pollut Res 31:30370–30398. https://doi.org/10.1007/s11356-024-33350-6

Ahmed A, Robinson G, Hamill G, Etsias G (2022) Seawater intrusion in extremely heterogeneous laboratory-scale aquifer: steady-state results. Water 14(7):1069. https://doi.org/10.3390/w14071069

Ahmed SF, Alam MS, Hassan M, Rozbu MR, Ishtiak T, Rafa N, Mofijur M, Shawkat Ali AB, Gandomi AH (2023a) Deep learning modelling techniques: current progress, applications, advantages, and challenges. Artif Intell Rev 56(11):13521–13617. https://doi.org/10.1007/s10462-023-10466-8

Ahmed Z, Gui D, Murtaza G, Yunfei L, Ali S (2023b) An overview of smart irrigation management for improving water productivity under climate change in drylands. Agronomy 13(8):2113. https://doi.org/10.3390/agronomy13082113

Akingboye AS, Osazuwa IB (2021) Subsurface geological, hydrogeophysical and engineering characterisation of Etioro-Akoko, southwestern Nigeria, using electrical resistivity tomography. NRIAG J Astron Geophys 10(1):43–57. https://doi.org/10.1080/20909977.2020.1868659

Alfarrah N, Walraevens K (2018) Groundwater overexploitation and seawater intrusion in coastal areas of arid and semi-arid regions. Water 10:143. https://doi.org/10.3390/w10020143

Almulhim AI, Abubakar IR (2023) Developing a sustainable water conservation strategy for Saudi Arabian cities. Groundw Sustain Dev 23:101040. https://doi.org/10.1016/j.gsd.2023.101040

Ammar SB, Taupin JD, Alaya MB, Zouari K, Patris N, Khouatmia M (2020) Using geochemical and isotopic tracers to characterize groundwater dynamics and salinity sources in the Wadi Guenniche coastal plain in northern Tunisia. J Arid Environ 178:104150. https://doi.org/10.1016/j.jaridenv.2020.104150

Atta K, Mondal S, Gorai S, Singh AP, Kumari A, Ghosh T, Roy A, Hembram S, Gaikwad DJ, Mondal S, Bhattacharya S (2023) Impacts of salinity stress on crop plants: improving salt tolerance through genetic and molecular dissection. Front Plant Sci 14:1241736. https://doi.org/10.3389/fpls.2023.1241736

Bahir M, EL Mountassir O, Ouazar D, Chehbouni A, Carreira PM (2021) Stable isotope and quality of groundwater around Ksob sub-basin, Essaouira, Morocco. Sustain Water Resour Manag 7(5):73. https://doi.org/10.1007/s40899-021-00553-5

Bahir M, Ouazar D, Ouhamdouch S, Zouari K (2020) Assessment of groundwater mineralization of alluvial coastal aquifer of essaouira basin (Morocco) using the hydrochemical facies evolution diagram (HFE-Diagram). Groundw Sustain Dev 11:100487. https://doi.org/10.1016/j.gsd.2020.100487

Barbieri M (2023) Groundwater salinity: origin, impact, and potential remedial measures and management solutions. Front Water 5:1202576. https://doi.org/10.3389/frwa.2023.1202576

Bighash P, Murgulet D (2015) Application of factor analysis and electrical resistivity to understand groundwater contributions to coastal embayments in semi-arid and hypersaline coastal settings. Sci Total Environ 532:688–701

BinMakhashen GM, Benaafi M (2024) Predicting seawater intrusion in coastal areas using machine learning: a case study of arid coastal aquifers, Saudi Arabia. Groundw Sustain Dev 26:101300. https://doi.org/10.1016/j.gsd.2024.101300

Blanco RI, Naja GM, Rivero RG, Price RM (2013) Spatial and temporal changes in groundwater salinity in South Florida. Appl Geochem 38:48–58

Bouhout S, Haboubi K, El Hammoudani Y, Haboubi C, El Abdouni A, Elyoubi MS, Dimane F (2024) Comparative assessment of groundwater quality challenges confronting Mediterranean coastal aquifers: a synoptic review. In: BIO web of conferences, vol 109. EDP Sciences, p 01002

Boulton AJ, Bichuette ME, Korbel K, Stoch F, Niemiller ML, Hose GC, Linke S (2023) Recent concepts and approaches for conserving groundwater biodiversity. In: Malard F, Griebler C, Rétaux S (eds) Groundwater ecology and evolution. Elsevier, pp 525–50. https://doi.org/10.1016/B978-0-12-819119-4.00001-9

Bouzourra H, Bouhlila R, Elango L, Slama F, Ouslati N (2015) Characterization of mechanisms and processes of groundwater salinization in irrigated coastal area using statistics, GIS, and hydrogeochemical investigations. Environ Sci Pollut Res 22:2643–2660. https://doi.org/10.1007/s11356-014-3428-0

Cantelon JA, Guimond JA, Robinson CE, Michael HA, Kurylyk BL (2022) Vertical saltwater intrusion in coastal aquifers driven by episodic flooding: a review. Water Resour Res 58(11):e2022WR032614. https://doi.org/10.1029/2022WR032614

Chabab E, Kühn M, Kempka T (2022) Upwelling mechanisms of deep saline waters via Quaternary erosion windows considering varying hydrogeological boundary conditions. Adv Geosci 58:47–54. https://doi.org/10.5194/adgeo-58-47-2022

Chang LL, Yuan R, Gupta HV, Winter CL, Niu GY (2020) Why is the terrestrial water storage in dryland regions declining? A perspective based on gravity recovery and climate experiment satellite observations and Noah land surface model with multiparameterization schemes model simulations. Water Resour Res 56(11):e2020WR027102. https://doi.org/10.1029/2020WR027102

Chaudhuri S, Ale S (2014) Long term (1960–2010) trends in groundwater contamination and salinization in the Ogallala aquifer in Texas. J Hydrol 513:376–390

Chen Y, Su Z, Zhu X (2024) The impact of climate change on water resources: mechanism analysis, challenge assessment, and response strategies. Adv Resour Res 4(4):806–818. https://doi.org/10.50908/arr.4.4_806

Dey S, Prakash O (2020) Management of saltwater intrusion in coastal aquifers: an overview of recent advances. In: Singh et al (eds) Environmental processes and management. Water science and technology library, vol 91. Springer, Cham. https://doi.org/10.1007/978-3-030-38152-3_17

Dieu LP, Cong-Thi D, Segers T, Ho HH, Nguyen F, Hermans T (2022) Groundwater salinization and freshening processes in the Luy River coastal aquifer, Vietnam. Water 14:2358. https://doi.org/10.3390/w14152358

Dodd C, Rishworth GM (2023) Coastal urban reliance on groundwater during drought cycles: opportunities, threats and state of knowledge. Cambridge Prisms: Coast Futures 1:e11. https://doi.org/10.1017/cft.2022.11

Egbueri JC (2023) A multi-model study for understanding the contamination mechanisms, toxicity and health risks of hardness, sulfate, and nitrate in natural water resources. Environ Sci Pollut Res 30(22):61626–61658. https://doi.org/10.1007/s11356-023-26396-5

Egbueri JC, Ighalo JO, Pande CB (2023) Climate change impacts on Nigeria. Springer International Publishing AG. https://doi.org/10.1007/978-3-031-21007-5

Egea I, Estrada Y, Faura C, Egea-Fernández JM, Bolarin MC, Flores FB (2023) Salt-tolerant alternative crops as sources of quality food to mitigate the negative impact of salinity on agricultural production. Front Plant Sci 14:1092885. https://doi.org/10.3389/fpls.2023.1092885

Etukudoh EA, Ilojianya VI, Ayorinde OB, Daudu CD, Adefemi A, Hamdan A (2024) Review of climate change impact on water availability in the USA and Africa. Int J Sci Res Arch 11(1):942–951

Ez-zaouy Y, Bouchaou L, Schreiber H, Montcoudiol N, Kalberkamp U, Danni SO, Touab A, Abourrig F, Hssaisoune M (2023) Combined geophysical methods to investigate seawater intrusion in the Souss-Massa coastal area, Morocco. Groundw Sustain Dev 21:100915. https://doi.org/10.1016/j.gsd.2023.100915

Ferchichi H, Ben Hamouda MF, Farhat B, Ben Mammou A (2018) Assessment of groundwater salinity using GIS and multivariate statistics in a coastal Mediterranean aquifer. Int J Environ Sci Technol. https://doi.org/10.1007/s13762-018-1767-y

Folch A, del Val L, Luquot L, Martínez-Pérez L, Bellmunt F, Le Lay H, Rodellas V, Ferrer N, Palacios A, Fernández S, Marazuela MA (2020) Combining fiber optic DTS, cross-hole ERT and time-lapse induction logging to characterize and monitor a coastal aquifer. J Hydrol 588:125050. https://doi.org/10.1016/j.jhydrol.2020.125050

García-Menéndez O, Morell I, Ballesteros BJ, Renau-Prunonosa A, Renau-Llorens A, Esteller MV (2016) Spatial characterization of the seawater upconing process in a coastal Mediterranean aquifer (Plana de Castellón, Spain): evolution and controls. Environ Earth Sci 75:728. https://doi.org/10.1007/s12665-016-5531-7

Gebre SL, Cattrysse D, Van Orshoven J (2021) Multi-criteria decision-making methods to address water allocation problems: a systematic review. Water 13(2):125. https://doi.org/10.3390/w13020125

Geng X, Michael HA (2020) Preferential flow enhances pumping-induced saltwater intrusion in volcanic aquifers. Water Resour Res 56(5):e2019WR026390. https://doi.org/10.1029/2019WR026390

Global Water Partnership (GWP) (2000) Integrated water resources management. Technical Advisory Committee (TAC) Background Paper no.4. Stockholm, Sweden

González PJ (2024) Interferometric synthetic aperture radar (InSAR). In: Chaussard E, Jones C, Chen JA, Donnellan A (eds) Remote sensing for characterization of geohazards and natural resources. Springer Remote Sensing/Photogrammetry. Springer, Cham, pp 53–73. https://doi.org/10.1007/978-3-031-59306-2_3

Greggio N, Mollema P, Antonellini M, Gabbianelli G (2012) Irrigation management in coastal zones to prevent soil and groundwater salinization. IntechOpen, London, UK. https://doi.org/10.5772/50534

Han D, Currell MJ (2018) Delineating multiple salinization processes in a coastal plain aquifer, northern China: hydrochemical and isotopic evidence. Hydrol Earth Syst Sci 22:3473–3491. https://doi.org/10.5194/hess-22-3473-2018

Han X, Wang D, Medeiros SC, Bilskie MV, Hagen SC, Hall CR (2019) Exploration of the effects of storm surge on the extent of saltwater intrusion into the surficial aquifer in coastal east-central Florida (USA). Sci Total Environ 648:1002–1017

Harby K, Emad M, Benghanem M, Abolibda TZ, Almohammadi K, Aljabri A, Alsaiari A, Elgendi M (2024) Reverse osmosis hybridization with other desalination techniques: an overview and opportunities. Desalination 581:117600. https://doi.org/10.1016/j.desal.2024.117600

Herbert ER, Boon P, Burgin AJ, Neubauer SC, Franklin RB, Ardón M, Hopfensperger KN, Lamers LP, Gell P (2015) A global perspective on wetland salinization: ecological consequences of a growing threat to freshwater wetlands. Ecosphere 6(10):1–43. https://doi.org/10.1890/ES14-00534.1

Hermans T, Goderniaux P, Jougnot D, Fleckenstein JH, Brunner P, Nguyen F, Linde N, Huisman JA, Bour O, Lopez Alvis J, Hoffmann R (2023) Advancing measurements and representations of subsurface heterogeneity and dynamic processes: towards 4D hydrogeology. Hydrol Earth Syst Sci 27(1):255–287. https://doi.org/10.5194/hess-27-255-2023

Hu Z, Chen Y (2024) Advancements in sustainable desalination with ocean thermal energy: a review. Desalination 586:117770. https://doi.org/10.1016/j.desal.2024.117770

Hussain MS, Abd-Elhamid HF, Javadi AA, Sherif MM (2019) Management of seawater intrusion in coastal aquifers: a review. Water 11(12):2467. https://doi.org/10.3390/w11122467

Hussain W, Ali M, Sadaf R, Al-Khafaji HF, Sadiq I, Hu C, Daud H, Ahmed SA (2024) Advanced AI approach for enhanced predictive modeling in reservoir characterization within complex geological environments. Model Earth Syst Environ 10:5043–5061. https://doi.org/10.1007/s40808-024-02049-5

Kammoun S, Trabelsi R, Re V, Zouari K (2021) Coastal aquifer salinization in semi-arid regions: the case of Grombalia (Tunisia). Water 13:129. https://doi.org/10.3390/w13020129

Kazakis N (2018) Delineation of suitable zones for the application of managed aquifer recharge (MAR) in coastal aquifers using quantitative parameters and the analytical hierarchy process. Water 10(6):804. https://doi.org/10.3390/w10060804

Ketabchi H, Mahmoodzadeh D, Ataie-Ashtiani B, Simmons CT (2016) Sea-level rise impacts on seawater intrusion in coastal aquifers: review and integration. J Hydrol 535:235–255. https://doi.org/10.1016/j.jhydrol.2016.01.083

Khondoker M, Mandal S, Gurav R, Hwang S (2023) Freshwater shortage, salinity increase, and global food production: a need for sustainable irrigation water desalination—A scoping review. Earth 4(2):223–240. https://doi.org/10.3390/earth4020012

Kurani A, Doshi P, Vakharia A, Shah M (2023) A comprehensive comparative study of artificial neural network (ANN) and support vector machines (SVM) on stock forecasting. Ann Data Sci 10(1):183–208. https://doi.org/10.1007/s40745-021-00344-x

Li C, Gao X, Li S, Bundschuh J (2020) A review of the distribution, sources, genesis, and environmental concerns of salinity in groundwater. Environ Sci Pollut Res 27:41157–41174. https://doi.org/10.1007/s11356-020-10354-6

Mehmood E, Anees T (2020) Challenges and solutions for processing real-time big data stream: a systematic literature review. IEEE Access 8:119123–119143. https://doi.org/10.1109/ACCESS.2020.3005268

Meles MB, Bradford S, Casillas-Trasvina A, Chen L, Osterman G, Hatch T, Ajami H, Crompton O, Levers L, Kisekka I (2024) Uncovering the gaps in managed aquifer recharge for sustainable groundwater management: a focus on hillslopes and mountains. J Hydrol 639:131615. https://doi.org/10.1016/j.jhydrol.2024.131615

Meng Y, Wu J, Li P, Wang Y (2023) Distribution characteristics, source identification and health risk assessment of trace metals in the coastal groundwater of Taizhou City, China. Environ Res 238:117085. https://doi.org/10.1016/j.envres.2023.117085

Michael HA, Russoniello CJ, Byron LA (2013) Global assessment of vulnerability to sea-level rise in topography-limited and recharge-limited coastal groundwater systems. Water Resour Res 49(4):2228–2240. https://doi.org/10.1002/wrcr.20213

Mohanavelu A, Naganna SR, Al-Ansari N (2021) Irrigation induced salinity and sodicity hazards on soil and groundwater: an overview of its causes, impacts and mitigation strategies. Agriculture 11(10):983. https://doi.org/10.3390/agriculture11100983

Mondal M, Mukherjee A, Kumar P, Ismaeel NM, Das K (2024) A process-based impact of tropical cyclone and hurricane on surface water-groundwater interaction process and contaminant mobilization of coastal aquifers: a semi analytical review. Progr Disaster Sci 22:100318. https://doi.org/10.1016/j.pdisas.2024.100318

Mukhopadhyay R, Sarkar B, Jat HS, Sharma PC, Bolan NS (2021) Soil salinity under climate change: challenges for sustainable agriculture and food security. J Environ Manag 280:111736. https://doi.org/10.1016/j.jenvman.2020.111736

Murray C, Larson A, Goodwill J, Wang Y, Cardace D, Akanda AS (2022) Water quality observations from space: a review of critical issues and challenges. Environments 9(10):125. https://doi.org/10.3390/environments9100125

Muruganandam M, Rajamanickam S, Sivarethinamohan S, Reddy MK, Velusamy P, Gomathi R, Ravindiran G, Gurugubelli TR, Munisamy SK (2023) Impact of climate change and anthropogenic activities on aquatic ecosystem—A review. Environ Res 238:117233. https://doi.org/10.1016/j.envres.2023.117233

Musumeci F, Rottondi C, Nag A, Macaluso I, Zibar D, Ruffini M, Tornatore M (2018) An overview on application of machine learning techniques in optical networks. IEEE Commun Surv Tutor 21(2):1383–1408. https://doi.org/10.1109/COMST.2018.2880039

Narayan GP, Sharqawy MH, Summers EK, Lienhard JH, Zubair SM, Antar MA (2010) The potential of solar-driven humidification–dehumidification desalination for small-scale decentralized water production. Renew Sustain Energy Rev 14(4):1187–1201. https://doi.org/10.1016/j.rser.2009.11.014

Nenna V, Herckenrath D, Knight R, Odlum N, McPhee D (2013) Application and evaluation of electromagnetic methods for imaging saltwater intrusion in coastal aquifers: seaside Groundwater Basin, California. Geophysics 78(2):B77-88. https://doi.org/10.1190/geo2012-0004.1

Norouzi Khatiri K, Nematollahi B, Hafeziyeh S, Niksokhan MH, Nikoo MR, Al-Rawas G (2023) Groundwater management and allocation models: a review. Water 15(2):253. https://doi.org/10.3390/w15020253

Nwazelibe VE, Egbueri JC, Unigwe CO, Agbasi JC, Ayejoto DA, Abba SI (2023) GIS-based landslide susceptibility mapping of Western Rwanda: an integrated artificial neural network, frequency ratio, and Shannon entropy approach. Environ Earth Sci 82(19):439. https://doi.org/10.1007/s12665-023-11134-4

O'Connor MT, Cardenas MB, Neilson BT, Nicholaides KD, Kling GW (2019) Active layer groundwater flow: the interrelated effects of stratigraphy, thaw, and topography. Water Resour Res 55(8):6555–6576. https://doi.org/10.1029/2018WR024636

Olabode OF, Comte JC (2024) Water scarcity in the fast-growing megacity of Lagos, Nigeria and opportunities for managed aquifer recharge. Wiley Interdiscip Rev Water 11(5):e1733. https://doi.org/10.1002/wat2.1733

Ortiz AC, Jin L (2021) Chemical and hydrological controls on salt accumulation in irrigated soils of southwestern US. Geoderma 391:114976. https://doi.org/10.1016/j.geoderma.2021.114976

Ouarani M, Bahir M, Mulla DJ, Ouazar D, Chehbouni A, Dhiba D, Ouhamdouch S, El Mountassir O (2020) Groundwater quality characterization in an overallocated semi-arid coastal area using an integrated approach: case of the Essaouira Basin, Morocco. Water 12(11):3202. https://doi.org/10.3390/w12113202

Ouarani M, Bahir M, Mulla DJ, Ouazar D, Chehbouni A, Dhiba D (2021) Groundwater quality deterioration and recharge modes in a semi-arid coastal aquifer: Essaouira Aquifer case study (Morocco). Arab J Geosci 14(19):1989. https://doi.org/10.1007/s12517-021-08267-w

Ouarani M, Brahim YA, Mulla D, Rafik A, Azennoud K, Bouchaou L, Chehbouni A (2023) A comprehensive overview of groundwater salinization and recharge processes in a semi-arid coastal aquifer (Essaouira, Morocco). J Hydrol: Reg Stud 49:101501

Ouzerbane Z, Loulida S, El Hmaidi A, Essahlaoui A, Boughalem M, Ousmana H, Berrada M (2022) Study of the salinity of groundwater in the HAHA syncline by the Kohonen self-organized classification (Essaouira, Morocco). Environ Sci Pollut Res 29:1–20. https://doi.org/10.1007/s11356-021-16598-0

Padmavathi G, Bangale U, Rao KN, Balakrishnan D, Arun MN, Singh RK, Sundaram RM (2024) Progress and prospects in harnessing wild relatives for genetic enhancement of salt tolerance in rice. Front Plant Sci 14:1253726. https://doi.org/10.3389/fpls.2023.1253726

Panthi J, Pradhanang SM, Nolte A, Boving TB (2022) Saltwater intrusion into coastal aquifers in the contiguous United States—A systematic review of investigation approaches and monitoring networks. Sci Total Environ 836:155641. https://doi.org/10.1016/j.scitotenv.2022.155641

Pino-Vargas E, Espinoza-Molina J, Chávarri-Velarde E, Quille-Mamani J, Ingol-Blanco E (2023) Impacts of groundwater management policies in the Caplina Aquifer, Atacama Desert. Water 15(14):2610. https://doi.org/10.3390/w15142610

Rahman MM, Ahmad S (2018) Health, livelihood and well-being in the coastal delta of Bangladesh. In: Nicholls et al (eds) Ecosystem services for well-being in deltas. Palgrave Macmillan, Cham. https://doi.org/10.1007/978-3-319-71093-8_7

Rahman MS, Hossain KM, van Loenhout J, Wallemacq P, Guha-Sapir D (2023) Effects of salinity on health due to environmental exposure: experiences from Bangladesh. In: Nasreen M et al (eds) Coastal disaster risk management in Bangladesh. Taylor and Francis, Routledge, London, pp 15–43. https://doi.org/10.4324/9781003253495

Rakib MA, Sasaki J, Matsuda H, Quraishi SB, Mahmud MJ, Bodrud-Doza M, Atique Ullah AKM, Fatema KJ, Newaz MA, Bhuiyan MAH (2020) Groundwater salinization and associated co-contamination risk increase severe drinking water vulnerabilities in the southwestern coast of Bangladesh. Chemosphere 246:125646

Reddy S, Dey K, Rufuss DD, Arulvel S, Akinaga T (2024) Forward osmosis desalination: a critical review focussing on recent advancements in draw solution recovery techniques for enhanced efficiency and regeneration. J Environ Chem Eng 12(5):113968. https://doi.org/10.1016/j.jece.2024.113968

Reljić M, Romić M, Romić D, Gilja G, Mornar V, Ondrasek G, Bubalo Kovačić M, Zovko M (2023) Advanced continuous monitoring system—Tools for water resource management and decision support system in salt affected delta. Agriculture 13(2):369. https://doi.org/10.3390/agriculture13020369

Rezaei M, Shahbazi K, Shahidi R, Davatgar N, Bazargan K, Rezaei H, Saadat S, Seuntjens P, Cornelis W (2021) How to relevantly characterize hydraulic properties of saline and sodic soils for water and solute transport simulations. J Hydrol 598:125777

Röthig T, Trevathan-Tackett SM, Voolstra CR, Ross C, Chaffron S, Durack PJ, Warmuth LM, Sweet M (2023) Human-induced salinity changes impact marine organisms and ecosystems. Glob Change Biol 29(17):4731–49. https://doi.org/10.1111/gcb.16859

Saad S, Javadi AA, Abd-Elhamid HF, Farmani R (2023) Mitigating seawater intrusion in coastal aquifers: novel approach with treated wastewater injection and groundwater circulation. J Hydrol 626:130139. https://doi.org/10.1016/j.jhydrol.2023.130139

Seibert SL, Greskowiak J, Oude Essink GH, Massmann G (2024) Understanding climate change and anthropogenic impacts on the salinization of low-lying coastal groundwater systems. Earth's Future 12(8):e2024EF004737. https://doi.org/10.1029/2024EF004737

Sherif M, Sefelnasr A, Al Rashed M, Alshamsi D, Zaidi FK, Alghafli K, Baig F, Al-Turbak A, Alfaifi H, Loni OA, Ahamed MB (2023) A review of managed aquifer recharge potential in the Middle East and North Africa Region with examples from the Kingdom of Saudi Arabia and the United Arab Emirates. Water 15(4):742. https://doi.org/10.3390/w15040742

Singh G, Singh J, Wani OA, Egbueri JC, Agbasi JC (2024) Assessment of groundwater suitability for sustainable irrigation: a comprehensive study using indexical, statistical, and machine learning approaches. Groundw Sustain Dev 24:101059. https://doi.org/10.1016/j.gsd.2023.101059

Springer A, Lopez T, Owor M, Frappart F, Stieglitz T (2023) The role of space-based observations for groundwater resource monitoring over Africa. Surv Geophys 44(1):123–172. https://doi.org/10.1007/s10712-022-09759-4

Stavros S, Fani S, Stergios T, Ioannis S, Olga C (2016) The environmental pressures and perspectives of tourism on coastal and insular zone. The case of Greece. Nat Environ Pollut Technol 15(3):1009–1020

Sultan MT, Mahmud U, Khan MZ (2023) Addressing soil salinity for sustainable agriculture and food security: innovations and challenges in coastal regions of Bangladesh. Future Foods 8:100260. https://doi.org/10.1016/j.fufo.2023.100260

Sun WC, Clement TP, Simpson MJ, Lee KK (2011) Does sea-level rise have an impact on saltwater intrusion? Adv Water Resour 34:1283–1291

Tao M, Chen X, Cheng Q, Binley A (2022) Evaluating the joint use of GPR and ERT on mapping shallow subsurface features of karst critical zone in southwest China. Vadose Zone J 21(1):e20172. https://doi.org/10.1002/vzj2.20172

Tauk M, Cretin M, Bechelany M, Sistat P, Zaviska F (2024) Capacitive deionization: a promising water treatment and desalination technology. In: Jlassi K, Oturan MA, Ismail AF, Chehimi MM (eds) Clean water: next generation technologies. Advances in science, technology & innovation. Springer, Cham, pp 25–40. https://doi.org/10.1007/978-3-031-48228-1_2

Tedeschi A, Schillaci M, Balestrini R (2023) Mitigating the impact of soil salinity: recent developments and future strategies. Ital J Agron 18(2):2173. https://doi.org/10.4081/ija.2023.2173

Thomas A, Baptiste A, Martyr-Koller R, Pringle P, Rhiney K (2020) Climate change and small island developing states. Annu Rev Environ Resour 45(1):1–27. https://doi.org/10.1146/annurev-environ-012320-083355

Tinivella U, Giustiniani M, Picotti S, Accaino F (2022) Seismic methods for aquifer analysis: the Basso Livenza case study. In: Di Mauro A, Scozzari A, Soldovieri F (eds) Instrumentation and measurement technologies for water cycle management. Springer Water. Springer, Cham, pp 437–459. https://doi.org/10.1007/978-3-031-08262-7_18

Torresan S, Critto A, Rizzi J, Zabeo A, Furlan E, Marcomini A (2016) DESYCO: a decision support system for the regional risk assessment of climate change impacts in coastal zones. Ocean Coast Manag 120:49–63. https://doi.org/10.1016/j.ocecoaman.2015.11.003

Trabelsi R, Abid K, Zouari K, Yahyaoui H (2012) Groundwater salinization processes in shallow coastal aquifer of Djeffara plain of Medenine, Southeastern Tunisia. Environ Earth Sci. https://doi.org/10.1007/s12665-011-1273-8

Venkatesan AK, Ahmad S, Johnson W, Batista JR (2011) Salinity reduction and energy conservation in direct and indirect potable water reuse. Desalination 272(1–3):120–127. https://doi.org/10.1016/j.desal.2011.01.007

Werner AD, Bakker M, Post VEA, Vandenbohede A, Lu C, Ataie-Ashtiani B, Simmons CT, Barry DA (2013) Seawater intrusion processes, investigation and management: recent advances and future challenges. Adv Water Resour 51:3–26

Wang Y, Gao M, Chen H, Fu X, Wang L, Wang R (2023) Soil moisture and salinity dynamics of drip irrigation in saline-alkali soil of Yellow River basin. Front Environ Sci 11:1130455. https://doi.org/10.3389/fenvs.2023.1130455

WHO (2017) Guidelines for drinking-water quality. Fourth Edition Incorporating the First Addendum. World Health Organization, Geneva

Xiong G, Chen G, Xu X, Liu W, Fu T, Khokiattiwong S, Kornkanitnan N, Seddique AA, Shi X, Liu S, Su Q, Xu X (2020) A comparative study on hydrochemical evolution and quality of groundwater in coastal areas of Thailand and Bangladesh. J Asian Earth Sci 195:104336

Xu J, Dai J, Wu X, Wu S, Zhang Y, Wang F, Gao A, Tan Y (2023) Urban rainwater utilization: a review of management modes and harvesting systems. Front Environ Sci 11:1025665. https://doi.org/10.3389/fenvs.2023.1025665

Yu X, Michael HA (2019) Mechanisms, configuration typology, and vulnerability of pumping-induced seawater intrusion in heterogeneous aquifers. Adv Water Resour 128:117–128. https://doi.org/10.1016/j.advwatres.2019.04.013

Yuan Y, Liu Q, Zheng H, Li M, Liu Y, Wang X, Peng Y, Luo X, Li F, Li X, Xing B (2023) Biochar as a sustainable tool for improving the health of salt-affected soil. Soil Environ Health 1(3):100033. https://doi.org/10.1016/j.seh.2023.100033

Zakariazadeh A, Ahshan R, Al Abri R, Al-Abri M (2024) Renewable energy integration in sustainable water systems: a review. Clean Eng Technol 18:100722. https://doi.org/10.1016/j.clet.2024.100722

Zhuang X, Wang W, Su Y, Yan B, Li Y, Li L, Hao Y (2024) Multi-objective optimization of reservoir development strategy with hybrid artificial intelligence method. Expert Syst Appl 241:122707. https://doi.org/10.1016/j.eswa.2023.122707

Characteristics of Groundwater Level Dynamics in Xi'an (2010–2020) and Groundwater Level Prediction for 2030 Using Regression Analysis and BP Neural Network: Implication to Groundwater Resources Management

Junyan Yang, Jianhua Wu, Yan Tian, Xiaomei Kou, and Song He

Abstract The chapter focuses on the dynamics of groundwater levels in Xi'an, aiming to evaluate groundwater resources and enhance their utilization. Using groundwater level monitoring data from 2010 to 2020, this research systematically analyzed the interannual and annual changes in groundwater levels through statistical analysis methods. The study investigated two primary factors influencing groundwater level dynamics: precipitation and artificial exploitation, employing mathematical and statistical approaches. Furthermore, prediction models based on regression analysis and the BP neural network method were developed to forecast groundwater level trends and provide a rational framework for groundwater development and utilization. Measures and suggestions for groundwater resources management were also proposed in this chapter. The findings indicate an overall decreasing trend in groundwater levels from 2010 to 2016, followed by a rebounding trend from 2016 to 2020 due to reduced groundwater exploitation and improved water supply infrastructure. Both precipitation and anthropogenic exploitation significantly impact

J. Yang · J. Wu (✉)
School of Water and Environment, Chang'an University, No. 126 Yanta Road, Xi'an 710054, Shaanxi, China
e-mail: wujianhua@chd.edu.cn; wjh2005xy@126.com

Key Laboratory of Subsurface Hydrology and Ecological Effects in Arid Region of the Ministry of Education, Chang'an University, No. 126 Yanta Road, Xi'an 710054, Shaanxi, China

Key Laboratory of Eco-Hydrology and Water Security in Arid and Semi-Arid Regions of the Ministry of Water Resources, Chang'an University, No. 126 Yanta Road, Xi'an 710054, Shaanxi, China

Y. Tian
PowerChina Sinohydro Bureau 3 Co.,LTD, No. 4069 Expo Avenue, Chanba Ecological District, Xi'an 710024, Shaanxi, China

X. Kou · S. He
PowerChina Northwest Engineering Corporation Limited, No. 18 Zhangbadong Road, Xi'an 710065, China

P. Li et al. (eds.), *Sustainable Groundwater and Environment: Challenges and Solutions*, Springer Hydrogeology, https://doi.org/10.1007/978-3-031-82194-3_9

groundwater level changes in Xi'an, with different landforms exhibiting varying responses to precipitation. Correlation analysis reveals correlation coefficients of 0.673 and 0.843 between precipitation, exploitation, and groundwater levels, respectively, indicating a stronger correlation between groundwater exploitation and water table depth compared to precipitation. The BP neural network model demonstrates higher prediction accuracy than the regression model. Based on these models, the projected average depths of groundwater level in Xi'an for 2030 are estimated to be 14.778 m and 14.823 m, respectively, indicating a rebound to levels observed a decade ago. Measures on sustainable groundwater resources management in Xi'an are further proposed to ensure a stabilized groundwater level. These results provide valuable insights for groundwater resource management and sustainable development in Xi'an.

Keywords Groundwater level dynamics · Precipitation · Exploitation · Regression analysis · BP neural network prediction

1 Introduction

Groundwater, a vital component of global water resources, profoundly affects human survival and production (Li et al. 2024a). Groundwater dynamics encompass the changes in groundwater level, quantity, quality, and temperature over time and space, influenced by meteorological, hydrological, geological, and anthropogenic factors (Liu et al. 2018). Among these dynamics, groundwater level serves as a crucial indicator, reflecting the water balance affected by natural conditions such as precipitation, evaporation, water flow, geography, geomorphology, lithology, and anthropogenic exploitation (Zagana et al. 2007). Consequently, groundwater level dynamics are often referred to as the "pulse of the earth" (Yang et al. 2017).

In terms of research methods, predicting groundwater level dynamics can be approached using deterministic methods like numerical and simulation techniques, as well as non-deterministic methods such as regression analysis and time series analysis (He et al. 2021). Zagana et al. (2007) employed geostatistical analysis to establish a correlation between groundwater level and precipitation, demonstrating a strong relationship between precipitation changes and groundwater level fluctuations. Similarly, Wang et al. (2018) integrated multivariate techniques, wavelet analysis, and base flow index to assess the combined impact of climate change and human activities on the groundwater system in the Guanzhong Basin of northwest China. Li et al. (2020) analyzed groundwater level changes and their main contributing factors in the Yao Ba Oasis from 1982 to 2018, utilizing Morlet wavelet analysis, Principal Component Analysis (PCA), and the Mann-Kendall (MK) trend test. Zhang et al. (2023) combined hybrid data-driven models and remote sensing data to propose a new strategy for real-time groundwater level prediction with improved accuracy and interpretability, which significantly broadened the scope of application of the prediction strategy. Furthermore, Artificial Neural Networks (ANN) were initially applied to water science analysis in the 1990s and have since found widespread use in various fields, including water level and water quality analysis (Nourani et al. 2015), with the Backpropagation (BP) neural network models being the most commonly employed (Hosseini et al. 2016). For instance, Guzman et al. (2017) simulated daily groundwater level fluctuations in an alluvial aquifer in the southeastern United States using an ANN-based model. Li et al. (2019) predicted groundwater levels in an arid oasis in Northwest China using the artificial bee colony algorithm and a double hidden layer backpropagation neural network. Similarly, Shi et al. (2023) innovatively combined correlation analysis and machine learning to study groundwater level fluctuations and develop a groundwater resource analysis and prediction model. Zhang et al. (2021) developed the WA-ARMA-BP model to accurately predict annual mean groundwater depth, demonstrating the model's high feasibility in predicting non-stationary time series data. Li et al. (2024b) adopted the GM (1, 1) and BP Neural Network models to predict the groundwater levels in Xi'an for the year from 2025 to 2035, and suggested that BP Neural Network had higher prediction accuracy and reliability. This research demonstrated the wide application of artificial intelligence technology in hydrogeological research.

Groundwater serves as an invaluable water resource for the residents of Xi'an, situated in the Weihe Plain, and plays a critical role in industrial and agricultural production (Li and Hou 2020; Li et al. 2018; Gao et al. 2022). However, with the rapid urbanization and development since the 1970s, the exploitation and utilization of groundwater resources have steadily increased, leading to a continuous decline in groundwater levels (Taylor et al. 2013; Wada et al. 2010). In recent years, the dynamic evolution of groundwater in the Xi'an area has been influenced by meteorological, hydrological, and human activities (Wang et al. 2019; Li and Hou 2020). Despite numerous studies applying statistical and neural network methods to groundwater level prediction, limited research has been conducted on groundwater level prediction specifically in Xi'an. Therefore, it is of great significance to study the dynamics and predict the trend of groundwater in this region (He et al. 2020; Li et al. 2018). In light of the correlation between precipitation, anthropogenic exploitation, and groundwater level dynamics, this study aims to achieve the following objectives: (1) unveil the patterns of groundwater level changes in Xi'an from 2010 to 2020, (2) analyze the dominant factors influencing groundwater level dynamics in Xi'an during the period of 2010–2020, and (3) employ regression analysis and BP neural network models to predict groundwater levels in 2030, aligning with the principles of establishing a resource-saving society. The findings of this research will provide valuable insights for the rational exploitation and utilization of water resources in Xi'an.

2 Study Area

2.1 Physical Geography

Xi'an, located at coordinates 107°40′–109°49′E and 33°39′–34°45′N, is the largest central city in northwestern China, situated in the middle of the Guanzhong Plain in Shaanxi Province (Fig. 1) (Gao et al. 2022; Li et al. 2018). It encompasses 13 administrative regions and has a total area of 1.01×10^4 km^2. The city is bordered by the Weihe River in the north, and its northern geomorphological units include river floodplains, river terraces, and loess tableland, arranged from northwest to southeast (Li et al. 2024b; Mu et al. 2020). The Qinling Mountains to the south are excluded from this study due to their complex hydrogeological conditions and relatively low exposure to human activity.

From 2010 to 2020, the study area experienced an average annual precipitation of 786 mm. Xi'an has a temperate continental climate (Shi et al. 2020), characterized by significant variations in annual precipitation and distinct seasons of cold, warm, dry, and wet. Precipitation distribution within the year is uneven (Wang et al. 2024). Moreover, the spatial distribution of precipitation shows an increasing trend from the northern plains to the southern mountains, with noticeable differences (Wang et al. 2023).

The study area includes water-bearing formations primarily composed of alluvial pore water-bearing formations and loess water-bearing formations. These formations consist mainly of Quaternary sediments, including silt, loess, and sand. The thickness of the water-bearing layer typically ranges from 20 to 70 m (Shi et al. 2020; Wang et al. 2019). The lithology in the area mainly consists of coarse and fine sand, gravel pebbles, and loess. The water abundance varies greatly across different lithology, with unit surges ranging from 1000 to 5000 m^3/d for coarse and fine sand and gravel pebbles and only 5–100 m^3/d for loess (Wang et al. 2019).

2.2 Groundwater Development and Utilization

The centralized exploitation of groundwater in Xi'an began in 1956, and until the implementation of the Hei River Drinking Water Project in 2000, groundwater served as the sole source of centralized water supply in the city (Li et al. 2024b). However, due to the imbalance between groundwater revenues and expenditures and the uneven distribution of precipitation over time and space, groundwater levels in Xi'an have experienced a significant decline (Mu et al. 2020; Wang et al. 2019).

Groundwater exploration in Xi'an primarily focuses on centralized water supply wells and self-provided wells. The Bahe River and Fenghe River water sources have aquifers predominantly composed of sand and pebbles, exhibiting wide and thick distribution. These are important water supply sources for Xi'an City. In recent years, with the continuous improvement of urban facilities and water supply projects,

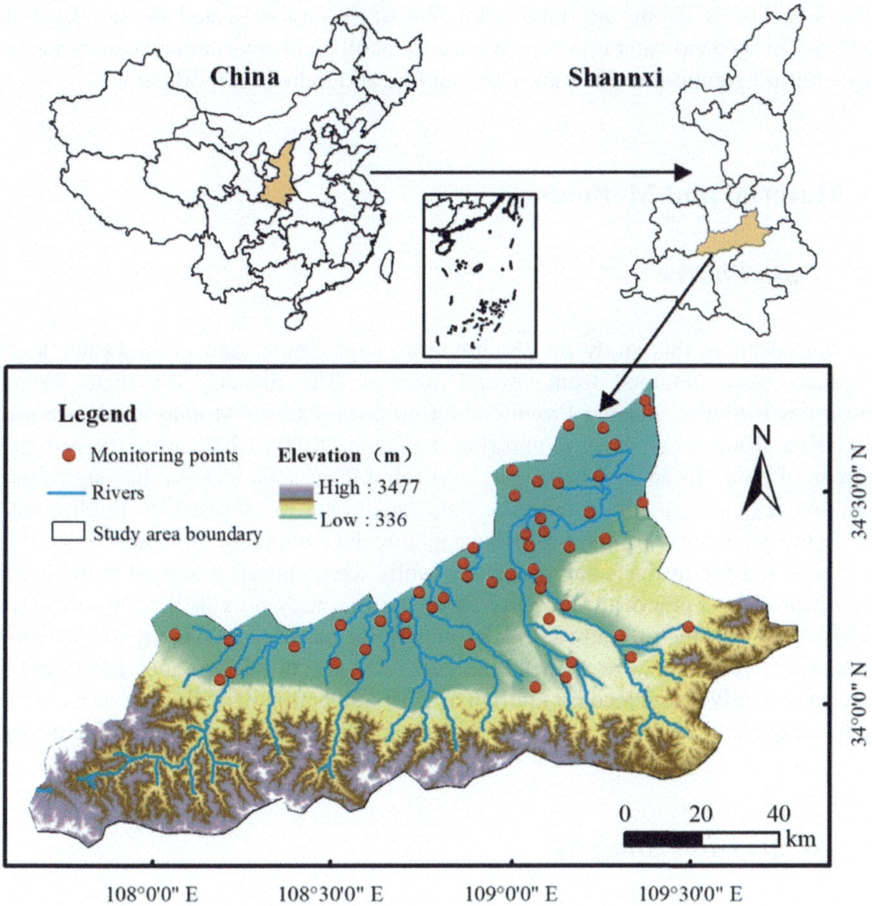

Fig. 1 Location of the study area and distribution of monitoring sites

the surface water supply in Xi'an has gradually become the dominant source (Zhou et al. 2021), leading to a reduction in annual groundwater extraction. This shift has improved the previous reliance on groundwater for water supply (Li and Hou 2020; Yao et al. 2021). However, groundwater remains highly valuable for exploitation, geological information, medical treatment, and its crucial role in maintaining the environmental ecosystem (Iglesias et al. 2015; Yuan et al. 2019). The unique resource and ecological attributes of groundwater make it an important water resource for the city of Xi'an.

Nevertheless, groundwater development faces various constraints and may become unsustainable under reduced precipitation, decreased runoff, and multiple precipitation events (Bhowmick et al. 2013). Therefore, it is necessary to analyze and study the temporal and spatial variations of groundwater levels and predict future trends. From 2010 to 2020, the average amount of shallow groundwater extracted in

Xi'an was 8.62×10^8 m^3, accounting for 36% of the total extracted amount (Li et al. 2024b). This groundwater is primarily used for ecological environment maintenance, agricultural irrigation, residential water supply, and industrial production.

3 Material and Methods

3.1 Data Source

The data used in this study on precipitation, exploration, and groundwater level dynamics were obtained from several sources. The Shaanxi Provincial Water Resources Bulletin, Shaanxi Provincial Groundwater Level Monitoring Yearbook, and Xi'an groundwater level monitoring data from 2010 to 2020 were the primary sources of data. To ensure the quality and reliability of the data, a thorough analysis, and review was conducted. The data underwent pre-processing to eliminate any inconsistencies or errors. From the available data, a total of 50 monitoring wells were selected for further analysis. These wells were chosen based on their representativeness and practical reliability, ensuring that they provide a comprehensive understanding of the groundwater dynamics in Xi'an (Fig. 1). Among the selected monitoring wells, 15 were at the state-level, 30 were at the province-level, and 5 were specifically designated for civilian use. This selection of wells aims to meet the study's requirements while adhering to the principles of authenticity and objectivity.

3.2 Regression Analysis

Regression analysis is a statistical method used to establish a relationship between a dependent variable and one or more independent variables based on available data (Tiryaki and Aydin 2014; Yang et al. 2017). Prior to conducting regression analysis, it is common to perform a correlation analysis between the variables (Li et al. 2021; Wu et al. 2020). Bivariate correlation analysis (Aloe 2015) typically employs either the Pearson correlation coefficient or the Spearman correlation coefficient to quantify the degree of association between the independent and dependent variables (Song and Park 2020). The Spearman correlation coefficient is a statistical measure of the strength and direction of the relationship between two sets of variables, calculated based on their ranks. The Pearson correlation coefficient is suitable for assessing the correlation between two variables that follow a bivariate normal distribution, while the Spearman correlation coefficient is applicable when dealing with rank data, non-bivariate normally distributed data, or data with uncertain distributions (Xiao et al. 2016; Zhang et al. 2016). In this study, the Spearman correlation coefficient was used to evaluate the correlation of precipitation and exploration with groundwater level depth, considering the characteristics of the available data. It can be calculated as:

$$\rho = 1 - \frac{6\sum d^2}{n^3 - n} \quad (1)$$

where, ρ is the Spearman correlation coefficient, d represents the difference in the rank of per pair of samples, and n denotes the number of parameters.

A multiple linear regression model for forecasting was developed with the mathematical expression:

$$Y = \beta_0 + \beta_1 X_1 + \beta_2 X_2 + \cdots + \beta_p X_p \quad (2)$$

where, Y denotes the dependent variable, β_0 is the constant term, $X_1, X_2, \ldots X_n$ are independent variables, and $\beta_1, \beta_2, \ldots \beta_p$ represent the regression parameters.

SPSS software, recognized as a comprehensive statistical analysis tool, is lauded for its user-friendly interface, extensive functionalities, and simplicity of use (Duricki et al. 2016; Pan et al. 2021; Zou et al. 2019). This software was employed in this study to scrutinize the correlation between shifts in groundwater levels and their associated factors, leading to the development of a robust multiple linear regression model. Based on the regression analysis, the quantitative relationship between groundwater level and precipitation and exploration was established, and the dominant factors affecting groundwater level were analyzed and explained, providing a basis for the selection of experimental data for the BP neural network.

3.3 BP Neural Network Prediction Model

The BP neural network is a multilayer forward network that uses an error backpropagation algorithm (Ding et al. 2011; Zhang et al. 2018). It is a representative example of artificial neural networks (Yan and Zeng 2013). One of its most notable features is its ability to process problems by simulating the structure of the human brain and the human thought process. This involves three components: the input layer, the hidden layer, and the output layer (Fig. 2). MATLAB provides numerous functions for implementing the BP network, including functions for designing, training, and simulating the Neural Network (Yan and Zeng 2013). In this study, the MATLAB neural network toolbox was used to design the BP neural network and derive a predictive model for groundwater levels.

Figure 2 illustrates the basic unit of a neural network (Li et al. 2024b). In this figure, $X_1, X_2, \ldots, X_n$ represent the input values, Y denotes the output value, W_{k1} signifies the weight between the input layer and the hidden layer, W_{k2} represents the weight between various hidden layers, and W_{k3} indicates the weight between the hidden layer and the output layer. The process of constructing the BP neural network model is depicted in Fig. 3. Importantly, a neural network does not necessitate the creation of an exact mathematical model. Instead, it learns from the input and output training sample data, generalizing the relationships inherent in the system's input and output. When using the BP network to develop a groundwater level prediction

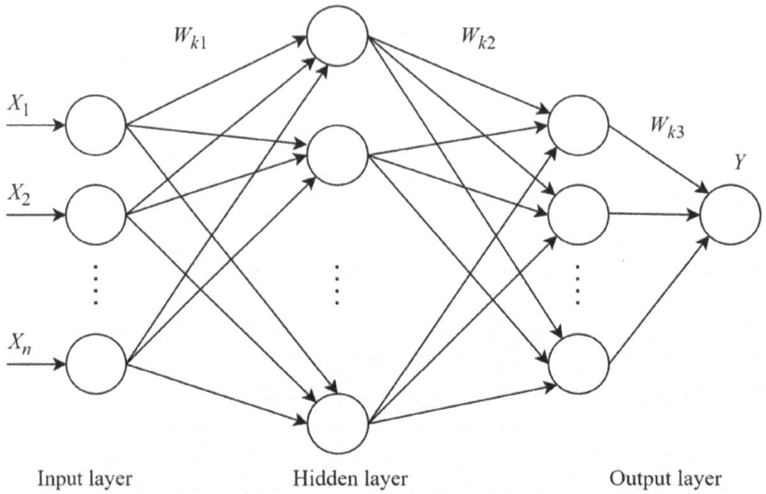

Fig. 2 Structure of the BP neural network

model, the groundwater level and its influencing factors are considered as a system. The influencing factors constitute the system's input, and the groundwater level is the system's output. The learning efficiency and training frequency are set, and after multiple experiments and comparative analyses, the optimal number of units in the hidden layer is selected based on the training and testing errors derived from the training results. This process ultimately determines the final structure of the neural network.

4 Results and Discussion

4.1 Dynamic Characteristics of Interannual Groundwater Levels and Dominant Factors

Based on the groundwater monitoring data, we obtained precipitation, exploitation, and groundwater level depth in Xi'an from 2010 to 2020, as presented in Table 1. The change curves for these variables are depicted in Fig. 4. The groundwater table's fluctuations can be divided into two stages: In the first stage, from early 2010 to 2016, the groundwater level showed a general downward trend, decreasing at an average annual rate of 0.23 m. The only exception was in 2011, when the groundwater level rose due to increased precipitation. The maximum rate of decrease, 4.4%, was observed in 2013. In the second stage, from late 2016 to 2020, the groundwater level increased at a rate of 0.19 m per year due to a relative rise in precipitation and the

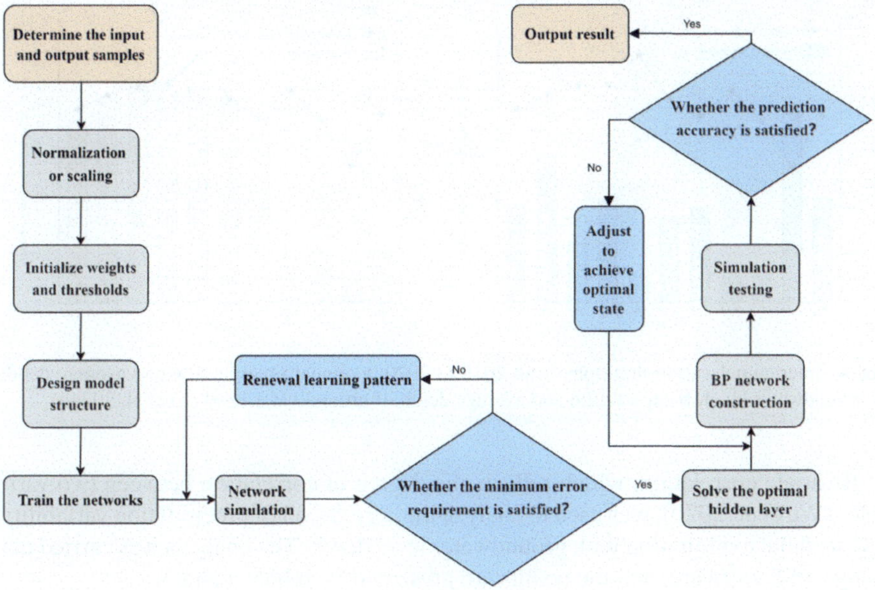

Fig. 3 BP neural network model calculation process

implementation of agricultural water conservation and diversion measures (Li et al. 2024b). The maximum rate of increase, 2.6%, was recorded in 2017.

Table 1 Precipitation, groundwater exploitation and groundwater level depths in Xi'an from 2010 to 2020

Year	Precipitation (mm)	Difference between precipitation and multi-year average precipitation (mm)	Groundwater exploitation (billion m^3)	Groundwater level depth (m)
2010	819	33	6.24	15.15
2011	1002	216	6.01	14.42
2012	659	−127	6.23	15.05
2013	656	−130	6.37	15.48
2014	802	16	9.57	15.11
2015	779	−7	10.01	15.36
2016	656	−130	10.26	15.99
2017	823	37	10.26	15.57
2018	728	−58	10.37	15.56
2019	910	124	10.05	15.44
2020	809	23	9.49	15.23

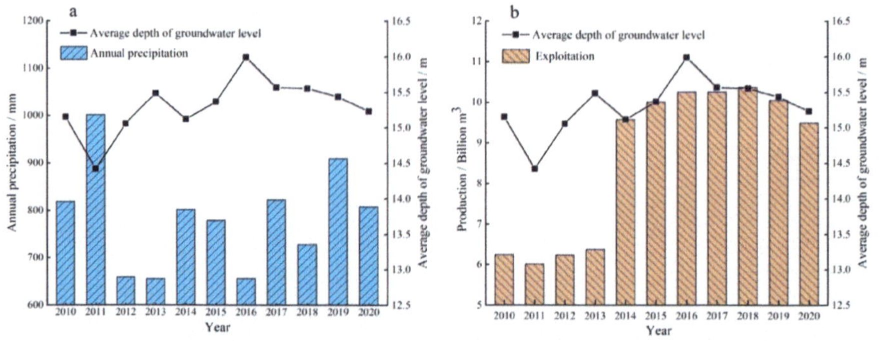

Fig. 4 Interannual variation graphs from 2010 to 2020, **a** annual precipitation and average depth of groundwater level, **b** exploration and average depth of groundwater level (Li et al. 2024b)

Bivariate correlation, which reflects the degree of correlation between two variables (Wu et al. 2020), was used to analyze the correlation of precipitation variability and artificial exploitation with groundwater level depth. The analysis was carried out using SPSS software, and the results are presented in Tables 2 and 3.

According to the literature, when the p-value from a two-tailed test is less than 0.05, the two variables are significantly correlated. The closer the absolute value of the Spearman correlation coefficient is to 1, the stronger the correlation between the independent and dependent variables. As shown in Table 2, the correlation between precipitation variability and groundwater level depth is 0.673 ($p = 0.023 < 0.05$), and the correlation between exploitation and groundwater level depth is 0.843 ($p = 0.001 < 0.01$). These results are consistent with those reported by Li et al. (2024b), and they demonstrate that the correlations of precipitation variability and exploitation with groundwater level depth are significant, with the correlation between exploitation and groundwater level depth being the strongest. This indicates that both precipitation and groundwater extraction significantly influence groundwater

Table 2 Correlation of precipitation with groundwater level depth

Variables	Coefficients	Difference between precipitation and multi-year average precipitation	Groundwater level depth
Difference between precipitation and multi-year average precipitation	Correlation coefficient	1	−0.673*
	Sig. (Double tails)	–	0.023
	N	11	11
Groundwater level depth	Correlation coefficient	−0.673*	1
	Sig. (Double tails)	0.023	–
	N	11	11

Note * Significant correlation at 0.05 level (two-tailed)

Table 3 Correlation between groundwater exploitation and groundwater level

Variables	Coefficients	Groundwater exploitation	Groundwater level depth
Groundwater exploitation	Correlation coefficient	1	0.843**
	Sig. (Double tails)	–	0.001
	N	11	11
Groundwater level depth	Correlation coefficient	0.843**	1
	Sig. (Double tails)	0.001	–
	N	11	11

Note ** Significant correlation at 0.01 level (two-tailed)

level dynamics in Xi'an. However, among the primary factors affecting groundwater levels, anthropogenic exploitation has the most substantial impact.

4.2 *Dynamic Characteristics of Annual Groundwater Level in Different Landforms*

Changes in groundwater levels are the product of both natural factors and human activities (Holman et al. 2012; Wada et al. 2012), particularly atmospheric precipitation and anthropogenic exploitation (Wang et al. 2018). These changes in groundwater levels exhibit considerable variability within a single year. Considering the topography and hydrogeological conditions of Xi'an, the dynamic changes in groundwater levels across five types of landforms in the study area were analyzed: river floodplain, primary terrace, secondary terrace, tertiary terrace, and loess tableland. The annual dynamic change of shallow groundwater level is positively correlated with precipitation, which generally increases in rainy season and decreases in dry season, and produces the corresponding seasonal change of "low–high–low" with the change of precipitation. Therefore, based on precipitation and monitoring data from 2010 to 2020, representative monitoring wells were selected for each landform, and then the monthly variation curves of precipitation and groundwater depth for a wet year (2011), an average year (2015) and a drought year (2016) were obtained. These curves are presented in Fig. 5.

According to Fig. 5, the variation in groundwater levels in the wet year (2011) can be divided into three stages:

Stage I: From the beginning of the year to mid-April, the groundwater level depth increased continuously. During this period, precipitation was less than 100 mm. The groundwater level was influenced by increasing evapotranspiration and the impact of agricultural irrigation, leading to a continual rise in groundwater level depth. This trend was most apparent in the primary terraces and river floodplains.

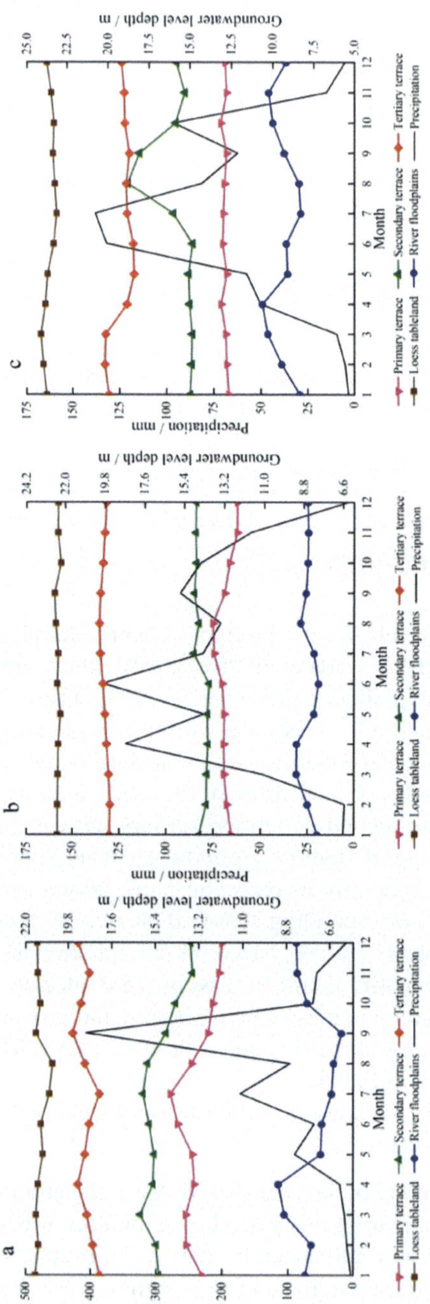

Fig. 5 Monthly variation curve of precipitation and groundwater level, **a** in 2011 (wet year), **b** in 2015 (average year), **c** in 2016 (drought year)

Stage II: From early June to the end of September, the groundwater level depth decreased. The groundwater discharged through evaporation and exploitation was significantly less than the amount recharged by precipitation. As shown in the Fig. 5, maximum precipitation occurred in September, coinciding with the minimum groundwater level depth in the river floodplain.
Stage III: From the beginning of October to December, the groundwater level depth in the terrace region decreased due to the lagging effect of precipitation recharge by infiltration. Conversely, the groundwater level depth in the river floodplain increased because the amount of precipitation recharge was less than the amount discharged by evaporation and exploitation.

Similarly, the variation in groundwater levels in the average year (2015) can be divided into three stages:

Stage I: From early January to the end of April, the overall regional groundwater level fluctuated slowly and decreased.
Stage II: From early May to mid-September, the fluctuation of groundwater level was more obvious, and the groundwater level depth in the river floodplain became significantly lower with the increase of precipitation; the fluctuation of groundwater level in the primary and secondary terraces was more obvious, and the lowest water level of the year occurred in July and August, respectively.
Stage III: From October to the end of December, regional groundwater levels overall rebounded, except in the river floodplain, due to the lagging effect of precipitation infiltration and the reduced impact of evapotranspiration anthropogenic exploitation.

On the contrary, the variation in groundwater levels in the drought year (2016) can be divided into four stages:

Stage I: From the beginning of the year to April, the groundwater level depth tended to increase. During this period, precipitation was less than 20 mm. The lack of effective precipitation and the impact of agricultural irrigation led to the maximum annual groundwater level depth in both the river floodplains and the terrace areas.
Stage II: From the end of April to the middle of July, the groundwater level depth recovered due to the infiltration of precipitation recharging the groundwater.
Stage III: From the end of July to September, the groundwater level depth in the secondary terrace decreased significantly, and the groundwater level depth in other landscape types also showed a tendency to increase. This was caused by increasing evaporation and the influence of anthropogenic exploitation, with discharge exceeding recharge.
Stage IV: From early October to December, the groundwater level depth remained relatively unchanged. Even though there was some recharge by precipitation in October, it could not effectively recharge the groundwater due to the influence of previous exploitation.

It is interesting to observe that in both wet and dry years, the loess tableland area, due to its large groundwater level depth, was not significantly affected by precipitation recharge and anthropogenic activities. The groundwater level remained stable and fluctuated throughout the year. In contrast, the river floodplains were notably affected by precipitation, with the groundwater level increasing as precipitation increased and decreasing as precipitation declined.

5 Groundwater Level Prediction

In accordance with socio-economic development planning and groundwater conservation and exploitation policy in Xi'an (Wang et al. 2019), it is expected that after 2020, industrial water and ecological environment water will no longer be exploited from groundwater, and they will mainly rely on large and medium-sized reservoirs, and diversion projects, and the agricultural water will mostly rely on shallow groundwater. In consequence, the groundwater exploitation is expected to be reduced to 480 million/m^3 in 2030. The predicted water level depths were obtained using the linear regression equation and BP neural network, respectively.

5.1 Regression Prediction Model

The linear regression model is available to predict the dependent variable on the basis of satisfying the model coefficient of determination and correlation tests (Fang and Lahdelma 2016). Therefore, according to the basic principle of stepwise regression analysis, with the difference between precipitation and multi-year average precipitation and groundwater extraction as the independent variables (respectively X_1 and X_2) and groundwater level depth as dependent variables (Y), based on SPSS software calculation, the model significance test and regression coefficient tables were obtained (Tables 4 and 5).

The correlation coefficient, *R*, is a quantity that reflects the degree of correlation between the independent variable and the dependent variable, and the absolute value is closer to 1, which indicates a greater linear correlation between the two quantities; R^2 is the coefficient of determination, which maps the explained variance as

Table 4 Variance and model significance test results

R	R^2	Adjusted R^2	Sum	df	Mean square	F	Significance
0.876[a]	0.767	0.709	1.202	2	0.601	13.182	0.003[b]

[a] Dependent variable: groundwater level, [b] Predicted variables: constants, difference between precipitation and multi-year average precipitation, and groundwater exploitation

Table 5 Regression coefficients and significance

Model	Unstandardized coefficient		Standardization coefficient		
	B	Standard error	Beta	T	Significance
Constants	14.223	0.308	–	46.112	0.000
Difference between precipitation and multi-year average precipitation	−0.002	0.001	−0.599	−3.507	0.008
Groundwater exploitation	0.125	0.035	0.612	3.584	0.007

a percentage of the variance of the dependent variable; whereas the adjusted R^2 is the corrected coefficient of determination, which is a correction to the coefficient of determination R^2 after taking into account the interaction between the independent variables. As can be seen from Table 4, the correlation coefficient R reaches 0.876, which suggests a significant linear correlation between precipitation and the difference in average multi-year precipitation, extraction and groundwater level; R^2 is 0.767, suggesting that the independent variables explain 76.7% of the dependent variable; the adjusted R^2 is 0.709 > 0.5, which indicates that the linear regression equation is well fitted and the overall regression effect is remarkable. As shown in Table 5, β_0 is 14.223, β_1 is −0.002 and β_2 is 0.125, so that the linear regression prediction equation: $Y = 14.223 - 0.002X_1 + 0.125X_2$. From this equation, the predicted value of groundwater level in 2030 will be 14.823 m.

5.2 BP Neural Network Prediction Model

On the basis of the preliminary analysis of hydrogeological conditions and selection of all possible groundwater level influencing factors, the model was effectively screened using regression analysis, which concluded that precipitation and exploration had a significant effect on groundwater levels, thus identifying the optimal input factors for the model. Following this, the data from 2010 to 2018 were selected as the training model and the data from 2019 to 2020 were used as the validation model, with a total of 11 samples of observed values year by year, a training number of 1000 times, a target minimum error of 0.000001 and a learning rate of 0.01, in which the network model was trained. The training results show that the Mean Square Error (MSE) is $1.1151e^{-15}$ when the number of units in the middle implicit layer is 8, which is the optimal number of nodes in the implicit layer and the network model achieves the desired results. This is consistent with the results reported by Li et al. (2024b). Finally, a three-layer network model consisting of two input units corresponding to precipitation variability and extraction, and one output unit corresponding to the groundwater level depth was developed. Results of the BP neural network predictions and regression predictions were analyzed and compared in Table 6.

By comparing and analyzing the measured and predicted values of the BP neural network model, the groundwater level process is well fitted and able to synchronize the dynamics of the rise and decline changes, which basically reflects the dynamic change law of the groundwater level in Xi'an. On the basis of clarification of precipitation and exploitation volumes, the predicted value of groundwater level in Xi'an in 2030 is obtained as 14.778 m.

From Table 6, it can be seen that the minimum absolute deviation of the prediction model of BP neural network is 0.013 and the minimum relative deviation is 0.086%, while the minimum absolute deviation of the prediction model of linear regression is 0.075 and the minimum relative deviation is 0.484%, indicating that the predicted value of BP neural network is much closer to the actual value. Meanwhile, the MSEs of the BP neural network and linear regression prediction models are 0.235 and 0.004, respectively. Although both values are less than 1, the MSE value of the linear regression model is much less than 0.1, which indicates that the model is over fitted.

When the difference between the precipitation and the multi-year average is known to be a fixed value, the groundwater level is stabilized by controlling the amount of exploitation. Therefore, based on both regression prediction equation and BP neural network prediction model, the predicted value of groundwater level is available. However, since the error accuracy and fitting degree of BP neural network model are better than that of linear regression model, the BP neural network prediction results are more reliable.

Table 6 Comparison of prediction results between neural network and regression models

Year	Measured groundwater level depths (m)	BP network model		Regression equation	
		Predicted values (m)	Relative errors (%)	Predicted values (m)	Relative errors (%)
2010	15.15	14.766	−1.903	14.937	−1.406
2011	14.42	15.205	−1.807	14.542	0.848
2012	15.05	15.127	0.086	15.256	1.367
2013	15.48	15.143	−1.443	15.279	−1.297
2014	15.11	15.205	−4.906	15.387	1.835
2015	15.36	14.758	−5.196	15.488	0.835
2016	15.99	15.201	−2.273	15.766	−1.404
2017	15.57	14.732	−4.558	15.432	−0.890
2018	15.56	15.106	−0.805	15.635	0.484
2019	15.44	15.046	−2.553	15.231	−1.352
2020	15.23	14.833	−2.607	15.363	0.875
2030		14.78		14.82	

In this study, assuming that the precipitation is stable within a reasonable range, the groundwater exploitation is predicted, and then the prediction results of the groundwater level are 14.778 m and 14.823 m by the BP neural network and regression equations, respectively. It can be found that the prediction value is similar to the groundwater level depth in 2010, meaning that the groundwater level rebounds to the level of over ten years ago, which indicates that the control of groundwater exploitation will help to alleviate the perennial downward trend of groundwater level and the contradiction between the supply and demand of water resources and other problems.

6 Groundwater Resources Management

According to the prediction results of the average depth of groundwater in 2030 obtained by BP neural network and linear regression model, it is projected that the overall groundwater level will rebound to the level observed over ten years ago. This rebound is expected to continue if groundwater extraction is controlled within a reasonable range.

While measures to limit groundwater extraction can help control the decline in regional groundwater levels, it's important to remember that groundwater is a dynamic system, with supply and use constantly changing. To ensure water supply stability, ecological security, and societal development, scientific management of groundwater is crucial.

However, the rebound in groundwater levels may induce new environmental problems, such as damage to metro trains and soil salinization. Therefore, in addition to adhering to the principles of ecological priority, hierarchical monitoring, intelligent monitoring, and collaborative governance, it is essential to conduct research to understand the mechanisms behind these new problems and identify the factors influencing them. This will allow us to develop effective mitigation strategies.

- **Collaborative Operation Management**: Establish a groundwater information monitoring list and clarify the control indicators of groundwater level and quantity. Implement graded management in areas with high irrigation and exploitation demand to ensure groundwater levels are generally stabilized and reasonably controlled in the region through synergistic cooperation between various departments.
- **Intelligent Management**: Utilize big data and information technology to establish an intelligent management platform for groundwater resources control. Establish an information-sharing platform to monitor groundwater changes and set up an early warning mechanism to better cope with the impact of ecological changes and anthropogenic water use demands on groundwater resources.
- **Decision-making System Platform**: Strengthen the monitoring and measurement of groundwater withdrawals and ensure the completeness and comprehensiveness of the density of monitoring wells. Implement control measures such as compensated use and feasibility demonstration of permits and groundwater

use plan management to achieve a balance between groundwater exploitation and replenishment.

- **Research on Emerging Environmental Problems**: Conduct research to understand the mechanisms behind the new environmental problems caused by the rebound in groundwater levels, such as damage to metro trains and soil salinization. Identify the factors influencing these problems and develop effective mitigation strategies. This could include structural reinforcement of metro tunnels, implementation of drainage systems, or the use of soil amendments to prevent salinization.

By adopting these measures, researchers and policy makers can not only manage the rebound of groundwater levels but also effectively address the potential environmental problems that may arise from it.

7 Conclusions

(1) The changes in groundwater levels in Xi'an from 2010 to 2020 are diverse. From 2010 to 2016, the overall trend for groundwater levels was a decline, with an average annual decrease of 0.23 m. However, from 2016 to 2020, the groundwater levels showed a consistent rebound, rising on average by 0.19 m annually. Based on the correlation test, the values of the correlation coefficients of precipitation, exploitation and groundwater level are 0.673 and 0.843, respectively, which indicates that groundwater exploitation is significantly and strongly correlated with the changes in groundwater level, and to a certain extent, it shows that the dominant factor that causes the dynamic changes in groundwater level in Xi'an from 2010 to 2020 is anthropogenic activities, and the influence of precipitation, climate and other factors on the dynamic changes in groundwater level is secondary to anthropogenic activities.

(2) Precipitation under the indicator of natural factor characteristics and groundwater exploitation under the domination of anthropogenic factors are closely related to the dynamic changes of groundwater level, and the response degree of groundwater level dynamics under different landforms in the study area is different. The analysis of the selected characteristic years shows that the river floodplain is most significantly affected by the infiltration and recharge of precipitation; there is an obvious lag in the recharge of precipitation in the year of abundant water in the terrace area, which is mainly used for agricultural irrigation and industrial exploitation, and the groundwater level shows a continuous decreasing trend in the drought year, which is most significant in the secondary terrace; the loess tableland is not significantly affected due to the greater depth of the groundwater level.

(3) Based on the data from 2010 to 2020, the BP neural network and linear regression model validate that the average depth of groundwater level in 2030 will be 14.778 m and 14.823 m, respectively, and the groundwater level will be

rebounded to the level of over ten years ago. This indicates that the control of groundwater exploitation is beneficial to control the downward trend of regional groundwater level and alleviate the contradiction between supply and demand of water resources. On this basis, it is pointed out that a single limitation of exploitation cannot maintain the dynamic balance of groundwater, and long-term recharge over discharge may cause a series of new ecological problems and threaten the safety of life and property. In this regard, a proposal was made to establish a scientifically sound groundwater monitoring system.

Acknowledgements This study received financial support from the National Natural Science Foundation of China (42272302 and 42072286), the National Key Research and Development Program of China (2023YFC3706901) and the Qinchuangyuan "Scientist + Engineer" Team Development Program of the Shaanxi Provincial Department of Science and Technology (2022KXJ-005). The authors would like to express their gratitude towards the help rendered by Dr. Xiaodong He and Prof. Peiyue Li during field investigation and data analysis.

References

Aloe AM (2015) Inaccuracy of regression results in replacing bivariate correlations. Res Synth Methods 6:21–27. https://doi.org/10.1002/jrsm.1126

Bhowmick S, Nath B, Halder D, Biswas A, Majumder S, Mondal P, Chakraborty S, Nriagu J, Bhattacharya P, Iglesias M, Roman-Ross G, Mazumder DG, Bundschuh J, Chatterjee D (2013) Arsenic mobilization in the aquifers of three physiographic settings of West Bengal, India: understanding geogenic and anthropogenic influences. J Hazard Mater 262:915–923. https://doi.org/10.1016/j.jhazmat.2012.07.014

Ding SF, Su CY, Yu JZ (2011) An optimizing BP neural network algorithm based on genetic algorithm. Artif Intell Rev 36:153–162. https://doi.org/10.1007/s10462-011-9208-z

Duricki DA, Soleman S, Moon LDF (2016) Analysis of longitudinal data from animals with missing values using SPSS. Nat Protoc 11:1112–1129. https://doi.org/10.1038/nprot.2016.048

Fang TT, Lahdelma R (2016) Evaluation of a multiple linear regression model and SARIMA model in forecasting heat demand for district heating system. Appl Energy 179:544–552. https://doi.org/10.1016/j.apenergy.2016.06.133

Gao J, Zhou WB, Li SW, Li CH, Chen HY (2022) The evaluation of groundwater carrying capacity in Xi'an. Water 14(22):3756. https://doi.org/10.3390/w14223756

Guzman SM, Paz JO, Tagert MLM (2017) The use of NARX neural networks to forecast daily groundwater levels. Water Resour Manag 31:1591–1603. https://doi.org/10.1007/s11269-017-1598-5

He L, Hou M, Chen S, Zhang J, Chen J, Qi H (2021) Construction of a spatio-temporal coupling model for groundwater level prediction: a case study of Changwu area, Yangtze River Delta region of China. Water Supply 21(7):3790–3809. https://doi.org/10.2166/ws.2021.140

He S, Li P, Wu J, Elumalai V, Adimalla N (2020) Groundwater quality under land use/land cover changes: a temporal study from 2005 to 2015 in Xi'an, Northwest China. Hum Ecol Risk Assess 26:2771–2797. https://doi.org/10.1080/10807039.2019.1684186

Holman IP, Allen DM, Cuthbert MO, Goderniaux P (2012) Towards best practice for assessing the impacts of climate change on groundwater. Hydrogeol J 20:1–4. https://doi.org/10.1007/s10040-011-0805-3

Hosseini Z, Gharechelou S, Nakhaei M, Gharechelou S (2016) Optimal design of BP algorithm by ACOR model for groundwater-level forecasting: a case study on Shabestar plain, Iran. Arab J Geosci 9(6):436. https://doi.org/10.1007/s12517-016-2454-2

Iglesias O, de Dios MAF, Tavares T, Sanroman MA, Pazos M (2015) Heterogeneous electro-Fenton treatment: preparation, characterization and performance in groundwater pesticide removal. J Ind Eng Chem 27:276–282. https://doi.org/10.1016/j.jiec.2014.12.044

Li H, Hou EK (2020) Groundwater dynamic response mechanism and quantity vulnerability assessment under the influence of human activities. Environ Sci Pollut Res 27:22046–22064. https://doi.org/10.1007/s11356-020-08645-z

Li H, Ma C, Zhou W, Yan Q, Song Y (2018) Characterizing the evolution of groundwater flow field and its driving forces in Xi'an, China. J Hydrol Eng 23:05018017. https://doi.org/10.1061/(ASCE)HE.1943-5584.0001668

Li HH, Lu YD, Zheng C, Yang M, Li SL (2019) Groundwater level prediction for the arid oasis of Northwest China based on the artificial bee colony algorithm and a back-propagation neural network with double hidden layers. Water 11:860. https://doi.org/10.3390/w11040860

Li H, Lu Y, Zheng C, Zhang X, Zhou B, Wu J (2020) Seasonal and inter-annual variability of groundwater and their responses to climate change and human activities in arid and desert areas: a case study in Yaoba Oasis, Northwest China. Water 12:303. https://doi.org/10.3390/w12010303

Li W, Wu J, Zhou C, Nsabimana A (2021) Groundwater pollution source identification and apportionment using PMF and PCA-APCS-MLR receptor models in Tongchuan City, China. Arch Environ Contam Toxicol 81(3):397–413. https://doi.org/10.1007/s00244-021-00877-5

Li P, Kou X, Wang Y, Niu L (2024a) Building a more sustainable Chinese Loess Plateau. J Earth Sci. https://doi.org/10.1007/s12583-024-1970-3

Li P, Liang H, Yang J, Tian Y, Kou X (2024b) Dynamic characteristics and trend prediction of groundwater level in Xi'an City, China based on GM (1,1) and BP neural network models. Northwestern Geology. https://doi.org/10.12401/j.nwg.2024118 (in Chinese)

Liu J, Gao Z, Wang M, Li Y, Ma Y, Shi M, Zhang H (2018) Study on the dynamic characteristics of groundwater in the valley plain of Lhasa City. Environ Earth Sci 77:646. https://doi.org/10.1007/s12665-018-7833-4

Mu EL, Yan L, Ding AZ, Deng W, Wang Y (2020) Determination of controlled limit value of groundwater level depth and management practice in Xi'an, China. Sci Rep 10:15505. https://doi.org/10.1038/s41598-020-72523-4

Nourani V, Alami MT, Vousoughi FD (2015) Self-organizing map clustering technique for ANN-based spatiotemporal modeling of groundwater quality parameters. J Hydroinf 18(2):288–309. https://doi.org/10.2166/hydro.2015.143

Pan C, Zhang ZS, Sun WK, Shi J, Wang HB (2021) Development of ship weather routing system with higher accuracy using SPSS and an improved genetic algorithm. J Mar Sci Technol 26:1324–1339. https://doi.org/10.1007/s00773-021-00800-6

Shi BW, Wan C, Hu WW, Xu CC, Liu D, Li MX, Li R, Feng CP, Xi BD (2023) Analysis and prediction of the changes in groundwater resources under heavy precipitation and ecological water replenishment. J Water Clim Change 14(6):1762–1778. https://doi.org/10.2166/wcc.2023.348

Shi W, Chen G, Meng XM, Jiang WY, Chong Y, Zhang Y, Dong Y, Zhang MS (2020) Spatial-temporal evolution of land subsidence and rebound over Xi'an in Western China revealed by SBAS-InSAR analysis. Remote Sens 12(22):3756. https://doi.org/10.3390/rs12223756

Song HY, Park S (2020) An analysis of correlation between personality and visiting place using Spearman's rank correlation coefficient. KSII Trans Internet Inf Syst 14:1951–1966. https://doi.org/10.3837/tiis.2020.05.005

Taylor RG, Scanlon B, Döll P, Rodell M, van Beek R, Wada Y, Longuevergne L, Leblanc M, Famiglietti JS, Edmunds M, Konikow L, Green TR, Chen J, Taniguchi M, Bierkens MFP, MacDonald A, Fan Y, Maxwell RM, Yechieli Y, Gurdak JJ, Allen DM, Shamsudduha M, Hiscock

K, Yeh PJF, Holman I, Treidel H (2013) Ground water and climate change. Nat Clim Chang 3:322–329. https://doi.org/10.1038/nclimate1744

Tiryaki S, Aydin A (2014) An artificial neural network model for predicting compression strength of heat treated woods and comparison with a multiple linear regression model. Constr Build Mater 62:102–108. https://doi.org/10.1016/j.conbuildmat.2014.03.041

Wada Y, van Beek LPH, van Kempen CM, Reckman JWTM, Vasak S, Bierkens MFP (2010) Global depletion of groundwater resources. Geophys Res Lett 37(20):L20402. https://doi.org/10.1029/2010GL044571

Wada Y, van Beek LPH, Weiland FCS, Chao BF, Wu YH, Bierkens MFP (2012) Past and future contribution of global groundwater depletion to sea-level rise. Geophys Res Lett 39(9):L09402. https://doi.org/10.1029/2012gl051230

Wang WK, Zhang ZY, Duan L, Wang ZF, Zhao YQ, Zhang Q, Dai ML, Liu HZ, Zheng XY, Sun YB (2018) Response of the groundwater system in the Guanzhong Basin (Central China) to climate change and human activities. Hydrogeol J 26:1429–1441. https://doi.org/10.1007/s10040-018-1757-7

Wang YQ, Wang ZF, Cheng WC (2019) A review on land subsidence caused by groundwater withdrawal in Xi'an, China. Bull Eng Geol Env 78:2851–2863. https://doi.org/10.1007/s10064-018-1278-6

Wang D, Li P, He X, He S (2023) Exploring the response of shallow groundwater to precipitation in the northern piedmont of the Qinling Mountains, China. Urban Clim 47:101379. https://doi.org/10.1016/j.uclim.2022.101379

Wang D, Li P, Mu D, Liu W, Chen Y, Fida M (2024) Unveiling the biogeochemical mechanism of nitrate in the vadose zone-groundwater system: insights from integrated microbiology, isotope techniques, and hydrogeochemistry. Sci Total Environ 906:167481. https://doi.org/10.1016/j.scitotenv.2023.167481

Wu J, Li P, Wang D, Ren X, Wei M (2020) Statistical and multivariate statistical techniques to trace the sources and affecting factors of groundwater pollution in a rapidly growing city on the Chinese Loess Plateau. Hum Ecol Risk Assess 26(6):1603–1621. https://doi.org/10.1080/10807039.2019.1594156

Xiao CW, Ye JQ, Esteves RM, Rong CM (2016) Using Spearman's correlation coefficients for exploratory data analysis on big dataset. Concurr Comput-Pract Exp 28:3866–3878. https://doi.org/10.1002/cpe.3745

Yan ZH, Zeng L (2013) The BP neural network with MATLAB. In: International conference on electrical, control and automation engineering (ECAE), Hong Kong, China, pp 565–569

Yang H, Feng Q, Guo X, Li Y, Huang S, Li H (2017) Analysis on the variation of groundwater level and its influence factors in Minqin Oasis based on the regression model. J Arid Land Resour Environ 31:98–103. https://doi.org/10.13448/j.cnki.jalre.2017.051

Yao YY, Zhang MS, Deng YK, Dong Y, Wu XS, Kuang XX (2021) Evaluation of environmental engineering geology issues caused by rising groundwater levels in Xi'an, China. Eng Geol 294. https://doi.org/10.1016/j.enggeo.2021.106350

Yuan L, Sun HG, Zhang Y, Zhang YK, Gu XF, Dawley S (2019) Temporal scaling analytical method to identify multi-fractionality in groundwater head fluctuations. Groundwater 57:485–491. https://doi.org/10.1111/gwat.12831

Zagana E, Kuells C, Udluft P, Constantinou C (2007) Methods of groundwater recharge estimation in eastern Mediterranean—A water balance model application in Greece, Cyprus and Jordan. Hydrol Process 21:2405–2414. https://doi.org/10.1002/hyp.6392

Zhang D, Lin JQ, Peng QD, Wang DS, Yang TT, Sorooshian S, Liu XF, Zhuang JB (2018) Modeling and simulating of reservoir operation using the artificial neural network, support vector regression, deep learning algorithm. J Hydrol 565:720–736. https://doi.org/10.1016/j.jhydrol.2018.08.050

Zhang MF, Wang YM, Wang X, Zhou WB (2021) Groundwater depth forecasting using a coupled model. Discret Dyn Nat Soc 2021:6614195. https://doi.org/10.1155/2021/6614195

Zhang Q, Li P, Ren X, Ning J, Li J, Liu C, Wang Y, Wang G (2023) A new real-time groundwater level forecasting strategy: coupling hybrid data-driven models with remote sensing data. J Hydrol 625:129962. https://doi.org/10.1016/j.jhydrol.2023.129962

Zhang WY, Wei ZW, Wang BH, Han XP (2016) Measuring mixing patterns in complex networks by Spearman rank correlation coefficient. Physica A 451:440–450. https://doi.org/10.1016/j.physa.2016.01.056

Zhou W, Chai J, Xu Z, Wang Y, Wei K, Dang Y (2021) Historical evolution of urban water conservancy projects in Xi'an, China in the past 3,000 years and its revelations. Water Supply 21(5):2173–2190. https://doi.org/10.2166/ws.2021.043

Zou DJ, Lloyd JEV, Baumbusch JL (2019) Using SPSS to analyze complex survey data: a primer. J Modern Appl Stat Methods 18(1):eP3253. https://doi.org/10.22237/jmasm/1556670300

Evaluating Uranium Toxicity in Groundwater and Associated Health Risks in a Hard Rock Aquifer, South India

B. Preethi, T. Subramani, P. Gopinathan, and R. Saravannan

Abstract Groundwater contamination by uranium has emerged as a significant public health concern, primarily due to its dual radiological and chemical toxicities. This study assesses uranium concentration levels in groundwater samples from the Raichur district in Karnataka state, India, and examines their implications for human health. A total of 54 groundwater samples were collected from various sources, including shallow wells, domestic wells, and irrigation wells, to evaluate uranium concentrations and associated health risks. Uranium levels ranged from 0.05 to 158.32 parts per billion (ppb), with an average concentration of 23.07 ppb, indicating that approximately 20.3% of the samples exceed the World Health Organization's (WHO) permissible limit of 30 ppb. The study adopted geospatial analysis to map uranium distribution and conducted a comprehensive health risk assessment, calculating Excess Cancer Risk (ECR), Lifetime Average Daily Dose (LADD), and Hazard Quotient (HQ) to evaluate both carcinogenic and non-carcinogenic risks. The finding reveals that children, particularly those aged 4–8 years, and infants are at heightened risk due to elevated ECR and LADD values, highlighting the increased vulnerability of younger age groups to uranium exposure. Adult males, as well as pregnant and lactating women, also show significant ECR and HQ values, indicating their susceptibility to uranium's adverse health effects. The high uranium concentrations are primarily due to local geological formations and anthropogenic activities, such as the use of phosphate fertilizers and agricultural practices prevalent in the region. These factors contribute to the leaching of uranium into groundwater, posing serious health risks. The results emphasize the need for stringent regulatory controls, effective groundwater monitoring, and the implementation of remediation strategies to mitigate uranium contamination. This study provides serious concern into the health implications of uranium contamination in groundwater and emphasizes the

B. Preethi · T. Subramani (✉) · R. Saravannan
Department of Geology, College of Engineering Guindy (CEG), Anna University, Chennai 600025, India
e-mail: geosubramani@gmail.com

P. Gopinathan
Resource Quality Assessment Division, CSIR-Central Institute of Mining and Fuel Research, Dhanbad, Jharkhand 828108, India

P. Li et al. (eds.), *Sustainable Groundwater and Environment: Challenges and Solutions*, Springer Hydrogeology, https://doi.org/10.1007/978-3-031-82194-3_10

importance of coordinated efforts in public health, environmental monitoring, and policy-making to address the challenges posed by uranium exposure.

Keywords Uranium · Groundwater · Health risk · Geological · Anthropogenic · South India

1 Introduction

The issue of human health is intricately linked to environmental and groundwater pollution, with significant implications worldwide (Anand et al. 2020; Preethi et al. 2023). The deterioration of water quality has emerged as a pressing concern, exacerbated by environmental changes and escalating human activities (Hema et al. 2010; Subramani et al. 2013; Karunanidhi et al. 2021; Li et al. 2019). Various factors, including natural processes, anthropogenic activities, and climate change, affect water quality, leading to increased levels of pollutants and contaminants in water bodies (Karunanidhi et al. 2020; Vennila et al. 2007, 2008; Muthamilselvan 2020). The presence of these pollutants poses serious risks to human health and the environment, emphasize the need for comprehensive water quality monitoring and the implementation of measures to prevent further degradation (WHO 2008). Globally, extensive research has been conducted to examine groundwater quality and its associated health hazards, recognizing that consuming contaminated water can have severe health implications (Zamora et al. 1998; Singh et al. 2012; Gopinathan et al. 2019). For example, a study by Karunanidhi et al. (2021) assessed the suitability of groundwater for food production and sustainable drinking water supply in a semi-urban region of South India (Bartlett et al. 1998). The study found that the groundwater was saline and slightly alkaline, with significant concentrations of Total Dissolved Solids (TDS), chloride (Cl), and sodium (Na), which were linked to anthropogenic activities contaminating the groundwater (Srinivasan et al. 2015). The Total Hazard Index (THI), used to evaluate the health risks associated with fluoride and nitrate in groundwater, indicated that a substantial number of samples exceeded the threshold limit, posing risks to infants, adolescents, teens, and adults (Brown et al. 2019). The study highlighted the importance of effective groundwater management to achieve Sustainable Development Goals (SDGs) 3, 6, and 11, focusing on good health and well-being, clean water and sanitation, and sustainable cities and communities (Bhaskar Rao et al. 1996).

Furthermore, the health risks posed by radionuclides in groundwater have been documented in regions with specific geological formations, such as limestone mining zones in South India (Kumar et al 2022). These studies have shown that the estimated average absorbed dose rate from soil samples is slightly higher than the global average, with increased values of excess lifetime cancer risk (ELCR) observed at sample sites (Kim et al. 2004). These findings suggest that post-mining land recovery and careful consideration in planting should be prioritized to mitigate prolonged exposure and associated health risks (Kurttio et al. 2002). Uranium contamination in drinking water has become a significant public health concern due to its dual nature as a radiological and chemical hazard. The World Health Organization (WHO) has established a permissible limit of 30 parts per billion (ppb) for uranium concentrations in drinking water to minimize health risks (WHO 2011a). Uranium primarily enters the human body through ingestion and inhalation, with its concentration in water sources largely influenced by regional geological contexts, such as soil and bedrock composition (Health Canada 1999; Zamora et al. 1998). Groundwater contamination tends to be more severe in regions where deeper wells are used, as these wells are more likely to access uranium-rich geological strata, leading to higher uranium concentrations compared to surface water or shallower wells (Muthamilselvan 2020). Several global studies have reported the prevalence of uranium in drinking water and its associated health risks. Singh et al. (2012) found uranium concentrations ranging from 0.48 to 571.7 $\mu g\ L^{-1}$ in the Bathinda district of Punjab, India, with radiological risks varying significantly based on local geological conditions and water usage patterns. Similarly, research by Kim et al. (2004) in Korea estimated excess cancer risks in groundwater, highlighting low yet non-negligible health risks due to uranium exposure (Domingo et al. 1989; Gladys Mathews et al. 2015).

The health impacts of uranium exposure are categorized into carcinogenic (radiological) and non-carcinogenic (chemical) effects (Pandit et al. 1988; Kumar et al. 2011). Uranium isotopes emit alpha particles that can cause DNA damage and increase cancer risk when ingested or inhaled (WHO 2008). Non-carcinogenic effects are primarily due to uranium's chemical toxicity, particularly its nephrotoxic effects, as demonstrated by research showing changes in kidney function biomarkers at varying exposure levels (Kurttio et al. 2002, 2006). The ability of uranium to accumulate in the renal cortex underscores the potential for long-term health impacts, necessitating stringent regulation and monitoring of uranium levels in drinking water. Animal studies have further elucidated the biological impact of uranium, showing that uranium can deposit in bones, serving as an indicator of chronic exposure (Larivière et al. 2007). These findings support the need for continued research and monitoring to mitigate the health risks associated with uranium exposure, emphasizing the importance of understanding the mechanisms of uranium mobilization in natural environments (Muthamilselvan 2020).

2 Study Area

Raichur district, located in the north-eastern part of Karnataka state within the Northern Maidan region, spans between 15°33′ and 16°34′ North latitudes and 76°14′ to 77°36′ East longitudes (Fig. 1). It lies between the Krishna and Tungabhadra rivers, which form its natural boundaries to the north and south, respectively, and are perennial, contributing to the region's hydrological richness. Geomorphologically, the district features a diverse landscape, including a northern rugged plateau, southern lower plains with inselbergs and isolated hillocks, and valley fills, with the terrain sloping towards the Krishna River in the north and the Tungabhadra in the south.

Raichur experiences a semi-arid climate characterized by mild winters and extremely hot summers, with December being the coldest month (average minimum 17.7 °C) and May the hottest (average maximum 39.8 °C, peaking at 45.0 °C). The region sees high relative humidity during the monsoon, exceeding 75%, and wind speeds surpassing 15 km/h in June and July. Annual rainfall averages 721 mm, predominantly from the southwest monsoon (67%) and northeast monsoon (24%), with an annual potential evaporation of about 1950 mm, indicating the significant influence of monsoonal patterns on the district's climate and agricultural practices.

2.1 Geology of the Study Area

Raichur district's geology is predominantly composed of granites, gneisses, and Dharwarschists, which are classified as 'hard rock' formations due to their lack of primary porosity, inherently limiting their ability to store and transmit groundwater (Jayananda et al. 1995; Krishnamurthy, 1977). However, secondary porosity is present in these rocks due to geological processes such as faulting, fracturing, and weathering, which enhance their permeability and water-yielding capacity (Smith and Jones 2017; Gopinathan et al. 2020). Groundwater occurs under water table conditions in the weathered and jointed portions of these hard rocks and under confined to semi-confined conditions within the fractured rocks, making them critical sources for groundwater extraction (Kumar and Rao 2016; Gopinathan et al. 2022). The district's surface is largely covered by black cotton soils, characterized by low permeability, thus impeding water percolation and circulation. This results in limited groundwater recharge and the formation of pockets of poor-quality groundwater in certain areas (Patel 2018). The hydrological behavior of the district is heavily influenced by its geological structure, which not only dictates the availability and movement of groundwater but also poses challenges for agricultural and water management practices (Anand et al. 2017; Gupta and Sharma 2020).

3 Methodology

A comprehensive and systematic approach was employed to collect and analyze groundwater samples to assess uranium concentrations and associated health risks in the Raichur region (Fig. 2).

3.1 Sample Collection

A total of 54 groundwater samples were collected from various sources, including shallow-dug wells located in mining areas, domestic wells, and irrigation wells within

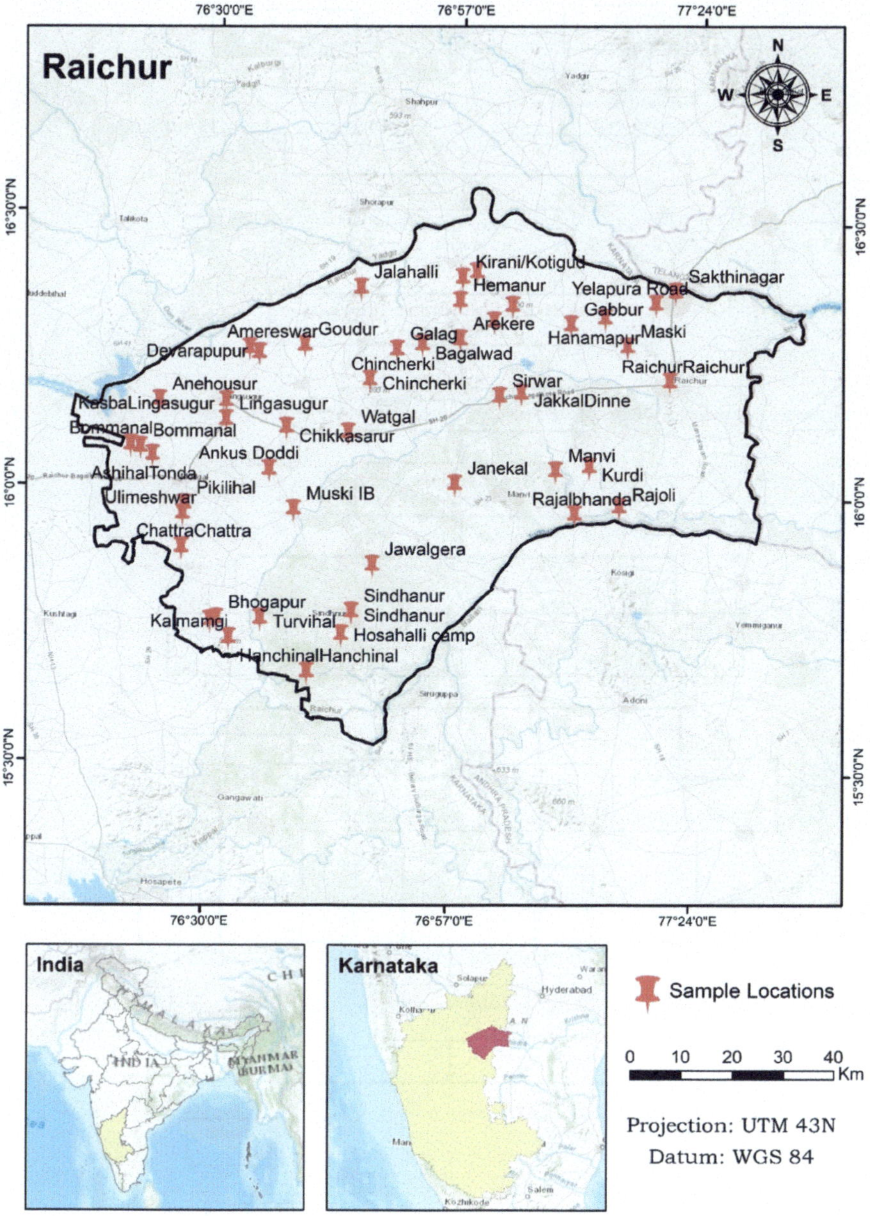

Fig. 1 Base map of the study area

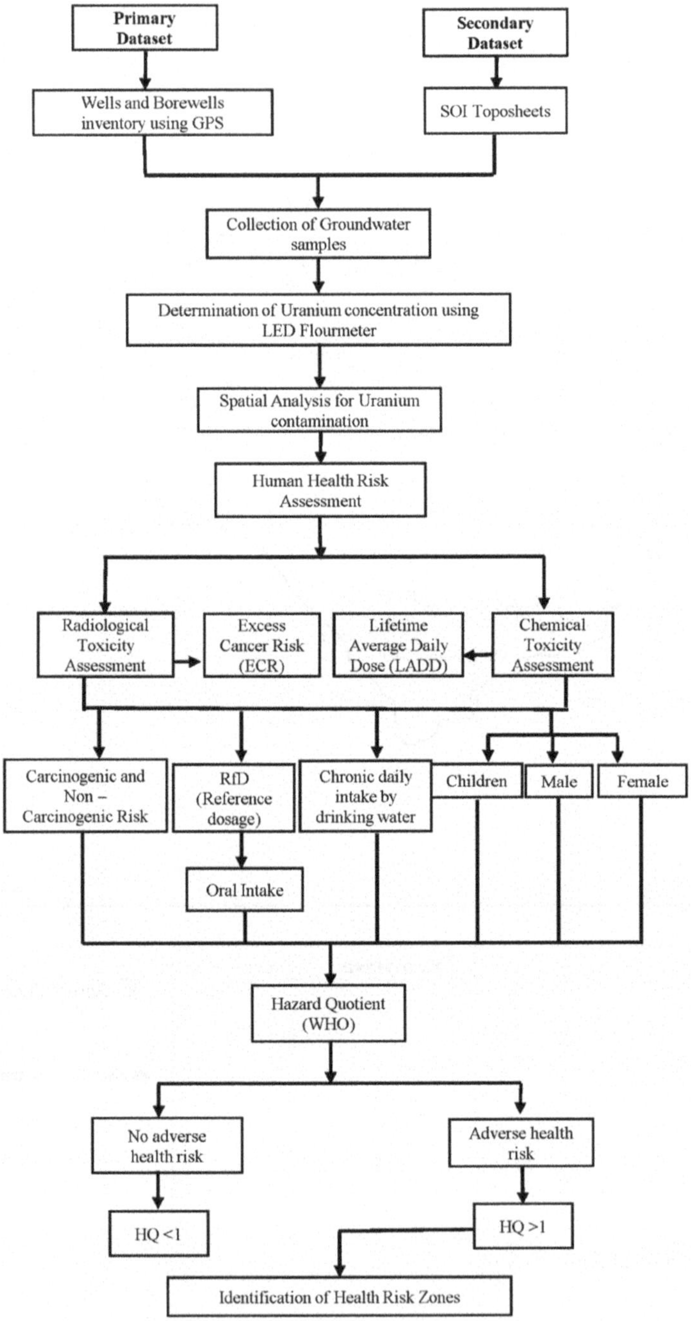

Fig. 2 Flowchart showing methodology adopted in this study

the vicinity of the Raichur region (as illustrated in Fig. 2). The sampling strategy was designed to cover diverse locations to capture variability in uranium concentrations across different land uses and geological settings. The spatial coordinates of each sampling point were meticulously recorded using handheld Global Positioning System (GPS) devices to ensure precise geographic mapping of sampling locations (Divya and Gopinathan 2019; Divya et al. 2021).

3.2 Sample Preservation and Preparation

To maintain the integrity of the samples and prevent contamination, all collected water samples were stored in polythene bottles that had been meticulously cleaned and acid-washed. Each bottle was securely sealed with double caps to ensure that no external contaminants could enter. Upon collection, the water samples were filtered using Whatman 42 filter paper to remove any particulate matter that could interfere with subsequent analyses.

3.3 Uranium Analysis

The concentration of uranium in the groundwater samples was determined using a laser-induced fluorimeter, a highly sensitive method suitable for detecting low concentrations of uranium. To ensure the accuracy and reliability of the fluorimeter, a rigorous calibration procedure was conducted using uranium standards of known concentrations (2 parts per billion (ppb), 4 ppb, and 6 ppb). The use of these standards facilitated the generation of a calibration curve, enabling precise quantification of uranium concentrations in the groundwater samples. The calibration process was repeated periodically to maintain the accuracy of the measurements (Fig. 3).

3.4 Health Risk Assessment

To evaluate the potential health risks associated with uranium exposure through groundwater consumption, the annual effective dose and the associated risks of radiological and chemical toxicity were calculated. These assessments were performed for different demographic groups, including various age categories and genders, to provide a comprehensive understanding of the population at risk. The calculations considered uranium ingestion rates and the specific health implications related to both radiological and chemical toxicity.

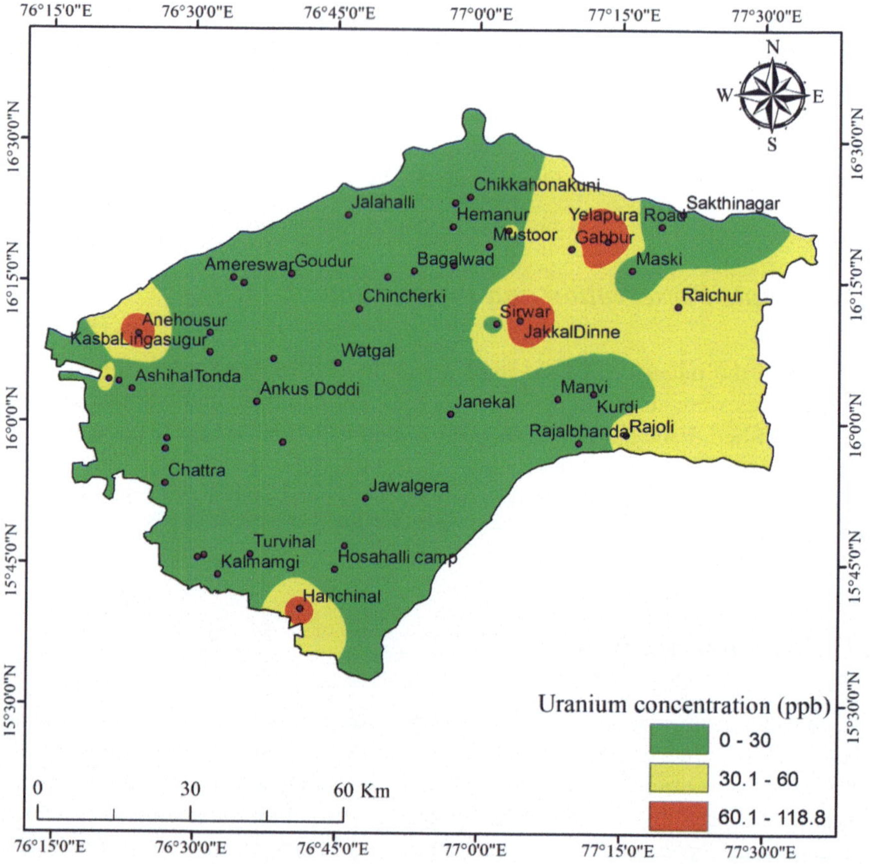

Fig. 3 Uranium concentration and distribution map

3.5 Geospatial Analysis

Uranium distribution maps and health risk assessments were developed using the ArcGIS 10.8 software environment. Spatial analysis tools were employed to visualize the distribution of uranium concentrations across the study area, allowing for the identification of potential hotspots and areas of concern. These geospatial analyses provided critical insights into the spatial variability of uranium contamination and facilitated a targeted approach to risk mitigation.

3.6 Data Analysis and Interpretation

The data obtained from uranium concentration measurements and health risk calculations were analyzed to draw meaningful conclusions regarding the extent of uranium contamination and its potential impact on public health. Statistical analyses were conducted to identify significant trends and correlations between uranium concentrations and environmental factors, such as land use patterns, proximity to mining activities, and agricultural practices.

4 Assessment of Uranium Toxicity

Uranium presents dual hazards as both a chemical and radiological toxicant. Concentrations of uranium exceeding 15 μg/L in drinking water have been associated with adverse health outcomes (WHO 2008). The chemical toxicity of uranium, particularly its nephrotoxic effects, is of significant concern as it primarily targets the kidneys (Keith et al. 1999; Leggett 1989). Extensive studies have demonstrated that uranium exposure can impair kidney function, even at relatively low concentrations, due to its ability to interfere with renal processes (Guglielmotti et al. 1989; Kurttio et al. 2002; Larivière et al. 2007; Zamora et al. 1998; Kurttio et al. 2006). Safe concentrations of uranium in drinking water are generally considered to range between 2 and 30 μg/L, according to guidelines by the World Health Organization (WHO) and studies focusing on uranium toxicity (Kurttio et al. 2002; WHO 2008). Beyond chemical toxicity, uranium's radiological effects, mainly due to alpha particle emissions, pose significant concerns for DNA damage, which may lead to cancer and developmental abnormalities (Amakom and Jibiri 2010). As a result, evaluating both radiological and chemical toxicity is essential for comprehensive risk assessments. Since regulatory agencies like the World Health Organization (WHO) and the International Commission on Radiological Protection (ICRP) do not provide specific slope factors and reference doses for radionuclides, the U.S. Environmental Protection Agency (US-EPA) values are commonly employed in these calculations. Health effects from uranium ingestion through drinking water are categorized into excess lifetime cancer risk (radiological) and non-carcinogenic (chemical) risks.

4.1 Radiological Risk Assessment

Radiological risk assessment focuses on estimating the probability of radiogenic cancer mortality or morbidity per unit intake or dose of radionuclides. This assessment utilizes risk coefficients that account for age- and gender-specific radiation risks, as outlined in Federal Guidance Report No. 13 (US-EPA 1999). These coefficients provide the probability of cancer per unit activity inhaled or ingested for internal exposure, or per unit time-integrated activity concentration in air or soil for external exposure. They are applicable to both chronic and acute exposure scenarios, under the assumption of constant radionuclide concentration over a lifetime. The excess lifetime carcinogenic risk (ECR) is calculated by multiplying the average daily dose by the slope factor and the lifespan duration, with a slope factor for uranium set at 6.4 $\times 10^{-11}$ (US-EPA 1989). The concept of a 1-in-1-million (10^{-6}) incremental cancer risk level was introduced by the FDA in the 1970s to manage environmental contaminants and is widely endorsed by the US-EPA as a benchmark for acceptable risk (Kelly 1991). Acceptable cancer risk levels typically range from 1 in 10,000 (10^{-4}) to 1 in 1,000,000 (10^{-6}), depending on exposure scenarios, although higher risk levels up to 10^{-3} have been deemed acceptable in certain contexts (Graham 1993; Kelly 1991; Lohner 1997; Travis and Kelly 1987; US-EPA 1991; Ye-shin et al. 2004).

4.1.1 Risk Co-efficient Mortality

Radiological risk which is also expressed as Excess cancer risk (ECR) is evaluated using the Eq. 1:

$$\mathrm{ECR} = \mathrm{U} \times \mathrm{R(m)} \tag{1}$$

where, U = U concentration in groundwater (Bq/L), R(m) = Risk Factor (Mortality) (Per Bq/L). U = M(U) × C, where, M(U) = Measured Value of U (μg/L), C = Conversion Factor (0.025 Bq/L), R(m) = R(Co) × IR × TED, *where*, R(Co) = Risk Coefficient, (Bq^{-1}), IR = Ingestion Rate (L/Day), TED = Total Exposure Duration (Days).

4.1.2 Risk Co-efficient Morbidity

Morbidity refers to the state of being diseased or unhealthy within a population. Radiological risk which is also expressed as Excess cancer risk is evaluated using the Eq. 2.

$$\mathrm{ECR} = \mathrm{U} \times \mathrm{R(M)} \tag{2}$$

where, ECR = Excess Cancer Risk, U = U concentration in groundwater (Bq/L), R(M) = Risk Factor (Morbidity) (Per Bq/L). U = M(U) × C, where, M(U) = Measured Value of U (μg/L), C = Conversion Factor (0.025 Bq/L), R(M) = R(Co) × IR × TED, where, R(Co) = Risk Coefficient (Bq^{-1}), IR = Ingestion Rate (L/Day), TED = Total exposure duration (Days).

4.2 Chemical Toxicity Risk Assessment

Chemical toxicity risk refers to the potential for adverse health effects, such as cancer, organ damage, or other disorders, resulting from exposure to chemical substances. The assessment of chemical toxicity often involves parameters such as the reference dose (RfD), which represents the estimated daily exposure level considered unlikely to cause harmful effects over a lifetime (US-EPA 2002). Factors such as inherent chemical toxicity, exposure pathways (ingestion, inhalation, dermal contact), exposure duration, and population vulnerability (age, gender, health conditions) are critical in evaluating chemical toxicity risks (National Research Council 1983). Chemical toxicity assessments rely on toxicological data from animal studies, human epidemiological studies, and predictive modeling to determine safe exposure levels, which guide regulatory limits and public health guidelines (WHO 2009).

4.2.1 Lifetime Average Daily Dose (LADD)

The chemical toxicity risk is quantified in terms of the Lifetime Average Daily Dose (LADD) using Eq. (3) (US-EPA 2000). The LADD values are compared against the reference dose (RfD) calculated based on the maximum permissible uranium concentrations of 60 μg/L (AERB, India) and 30 μg/L (WHO guidelines) in drinking water (AERB 2004; WHO 2011b). The equation for LADD is as follows:

$$\text{LADD} = \frac{\text{EPC} \times \text{IR} \times \text{EF} \times \text{LE}}{\text{AT} \times \text{BW}} \tag{3}$$

where, EPC = Exposure point concentration of uranium (mg/L), IR = Water ingestion rate (L/day), EF = Exposure frequency (350 days/year), LE = Life expectancy (years), AT = Average time (years), BW = Body weight for Indian standard (Dang et al. 1994)

4.3 Hazard Quotient (HQ)

The Hazard Quotient (HQ) is a risk metric used to evaluate the potential for non-carcinogenic effects from chemical exposure. It is calculated by dividing the estimated exposure level, usually expressed as the average daily dose, by the reference dose (RfD), as described in Eq. 4. The HQ provides a quantitative measure of risk, with values less than one indicating that exposure is unlikely to result in adverse health effects, while values greater than one suggest potential health risks and may necessitate further investigation or regulatory intervention (US-EPA 1989; Hallenbeck 1993). The HQ is commonly employed in environmental risk assessments to prioritize risks associated with contaminants in air, water, soil, or food, informing public health policies and actions (Lentz et al. 2001).

$$HQ = \frac{LADD}{RfD} \tag{4}$$

4.4 Permissible Limits

The assessment of uranium toxicity in groundwater is guided by permissible limits established by the World Health Organization (WHO) and the Atomic Energy Regulatory Board (AERB). These guidelines address both radiological and chemical risks associated with uranium exposure. For radiological risk toxicity assessment, the recommended permissible limits include a risk factor of less than 10^{-4} BqL^{-1} which helps in minimizing the exposure to radioactivity from uranium isotopes present in water sources. Furthermore, the Excess Cancer Risk, which quantifies the potential risk of developing cancer over a lifetime due to exposure, should be kept below10^{-3} $mSvY^{-1}$. This limit is crucial as it provides a safety threshold to reduce the probability of radiation-induced carcinogenesis (WHO 2017; AERB 2018). In addition to radiological parameters, chemical risk toxicity assessment is also critical. The Lifetime Average Daily Dose (LADD), which measures the average amount of uranium ingested daily over a lifetime, should not exceed 0.8 $\mu g\ kg^{-1}\ day^{-1}$. This metric helps in assessing the chronic exposure levels to uranium, ensuring they remain within safe limits to avoid chemical toxicity (WHO 2017). Another essential parameter is the Hazard Quotient (HQ), which evaluates the non-carcinogenic risk by comparing the potential exposure to a reference dose. Both WHO and AERB recommend maintaining the Hazard Quotient below 1 $mg^{-1}\ kg^{-1}\ day^{-1}$ An HQ value below one indicates that the exposure is unlikely to pose significant health risks (AERB 2018). These guidelines collectively provide a comprehensive framework for evaluating the safety of groundwater with respect to uranium contamination, balancing the need to control both radiological and chemical hazards to safeguard public health (WHO 2017; AERB 2018).

5 Results and Discussion

5.1 Spatial Concentration of Uranium

Uranium concentrations in groundwater were assessed across a specific region using 54 water samples collected from open wells, hand pumps, and bore wells. The concentrations ranged from 0.05 to 158.32 parts per billion (ppb), with an average concentration of 23.07 ppb, which is below the World Health Organization (WHO) recommended safe drinking water limit of 30 ppb. However, 11 samples, accounting for 20.3% of the total, exceeded the WHO standard. The highest uranium concentration was found in sample number 1 from Hanchinal, followed by samples from Hanamapur, Jakkal Dinne, Sindhanur, Anehousur, Gabbur, Bommanal, Raichur, Rajoli, and Kakargal, all of which were above the WHO-prescribed limit.

The spatial distribution of uranium concentrations demonstrates variability throughout the region in Fig. 3. Most areas have concentrations ranging from 0 to 30 ppb, indicating compliance with international safety standards for drinking water. However, certain locations exhibit concentrations between 30.1 and 60 ppb, which covers nearly 30% of the study area, which exceed the WHO recommended limit, suggesting potential health risks associated with long-term consumption of untreated groundwater. In addition, several areas, particularly around Sigwar, Gabbur, Lingasugur, and Hanchinal, display significantly elevated uranium concentrations between 60.1 and 118.8 ppb. These values surpass the safety thresholds set by the Atomic Energy Regulatory Board (AERB) of India, indicating a substantial health risk due to high uranium levels.

5.2 Excess Cancer Risk (ECR)

The assessment of Excess Cancer Risk (ECR) indicates varying levels of carcinogenic exposure across different demographic groups in the Raichur district (Fig. 4). For children, significant peaks in ECR are observed in samples 1, 31, and 46, especially within the 4–8 years age group, highlighting a heightened cancer risk due to increased interaction with the environment. These risks are associated with high uranium concentrations in groundwater, influenced by local geological conditions and anthropogenic activities, such as the use of phosphate fertilizers and agricultural practices. In males, the highest ECR values are seen in adults, with notable peaks at samples 1, 2, 5, 19, 21, 32, and 46, and 33 water samples exceeding the ECR limit for those above 18 years. This suggests that occupational exposure and the consumption of contaminated water contribute to elevated cancer risks in this group. For females, significant ECR values are recorded at samples 1, 5, 19, 31, and 46, with 30 water samples exceeding the threshold for those above 18 years and 23 for pregnant and lactating women. These findings imply serious cancer risks linked to uranium exposure, with potential adverse effects on reproductive health and fetal development. The consistent pattern of high ECR across these groups underscores the need for stringent monitoring of uranium levels in groundwater, regulation of agricultural practices, and targeted interventions to protect vulnerable populations from carcinogenic exposure.

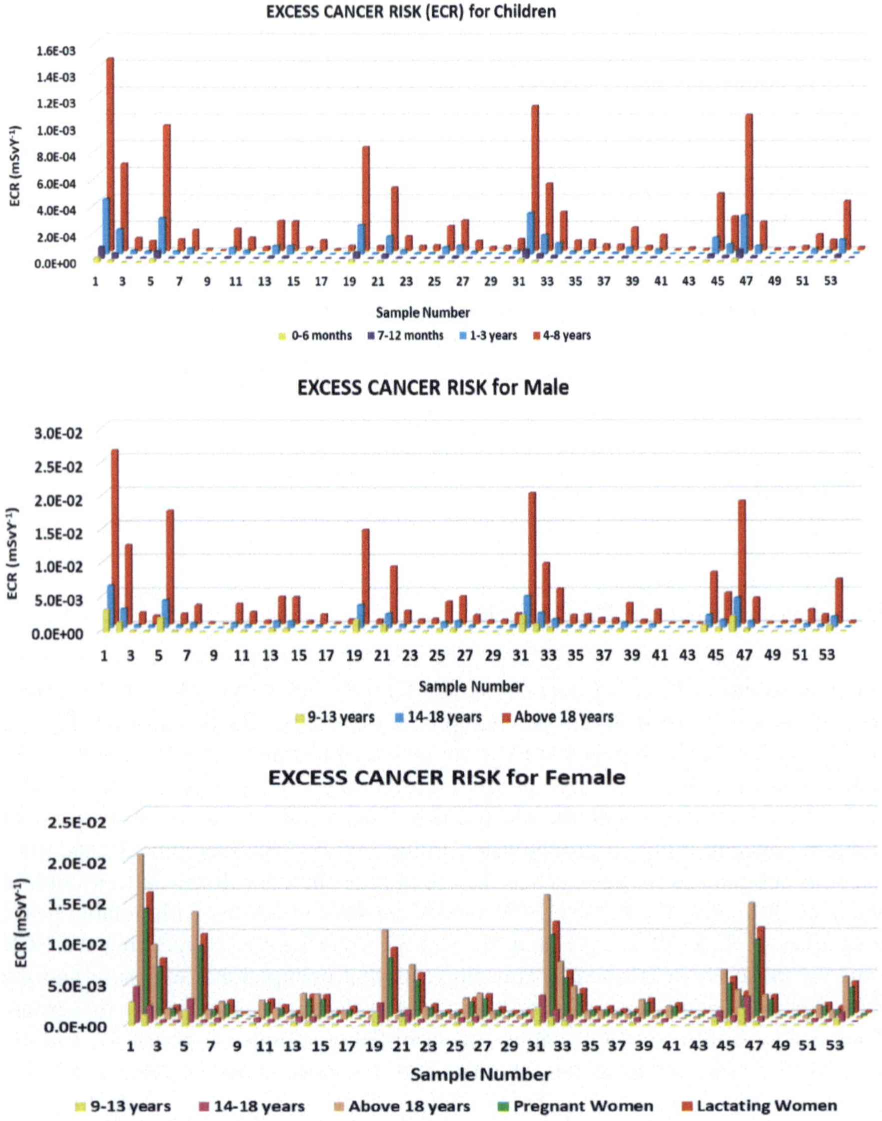

Fig. 4 ECR for children, male and female

5.3 *Lifetime Average Daily Dose*

The Lifetime Average Daily Dose (LADD) analysis reveals chronic exposure to uranium and other potentially toxic substances among children, males, and females across various age groups in the Raichur district (Fig. 5). For children, particularly vulnerable infants aged 0–6 and 7–12 months, elevated LADD values are evident, with 33 water samples exceeding threshold values, showing significant peaks at sample numbers 1, 2, 5, 19, 31, and 46. This heightened exposure, largely due to high uranium concentrations in groundwater from intensive agricultural areas, underscores the risk of ingestion through breastfeeding or formula preparation. Males across all age groups also demonstrate consistent exposure, with significant LADD peaks corresponding to the same sample numbers, indicating chronic low-level exposure that can accumulate and pose long-term health risks, especially in the 9–13 years and 14–18 years categories, where 23 and 21 samples, respectively, exceed threshold values. For females, pregnant and lactating women exhibit the highest LADD values, with 25 and 27 water samples surpassing safe limits, reflecting the risk of uranium transference to the fetus or through breast milk, which poses significant health concerns. While females above 18 years show relatively lower LADD risk, those in the 9–13- and 14–18-years age groups have 20 and 19 water samples, respectively, exceeding threshold values. These findings emphasize the critical need for stringent monitoring of uranium levels in groundwater, particularly in regions of intensive agriculture, as well as the implementation of regulatory measures to control fertilizer use and public health initiatives aimed at providing safe drinking water (Aravinthasamy et al. 2020). Ensuring that vulnerable populations, such as infants, children, and pregnant or lactating women, have access to uncontaminated water is essential for mitigating potential health risks and protecting public health.

5.4 Hazard Quotient

The Hazard Quotient (HQ) analysis for non-carcinogenic health risks due to uranium exposure in the Raichur district reveals critical vulnerabilities among children, males, and females (Fig. 6). For children, the highest HQ values are found in infants, particularly those aged 0–6 months, with 25 water samples exceeding the unity threshold, and in infants aged 7–12 months, with 22 samples exceeding this value. Among children aged 1–3 years, 20 samples exceed the unity value, while 16 samples exceed it for those aged 4–8 years. Significant peaks in HQ are observed at sample numbers 1, 2, 5, 19, 21, 31, 32, 46, and 53, correlating with regions of intensive agricultural practices and high uranium concentrations in groundwater. These elevated HQ values suggest considerable risks of neurological, renal, or developmental health issues, particularly in infants, underscoring the urgent need for regulation of agricultural inputs and rigorous water quality testing. For males, elevated HQ values across all age groups indicate non-carcinogenic risks, with significant peaks in the same sample numbers. In males aged 9–13 years, 17 water samples exceed the unity threshold; for those aged 14–18 years, 16 samples exceed it, and for those above 18 years, 12 samples do. These findings highlight the potential health impacts from chronic exposure to uranium through contaminated water and agricultural products, necessitating comprehensive environmental policies to manage agricultural practices and mitigate contamination. For females, pregnant and lactating women are at significant non-carcinogenic risk, with 20 water samples each exceeding the unity threshold,

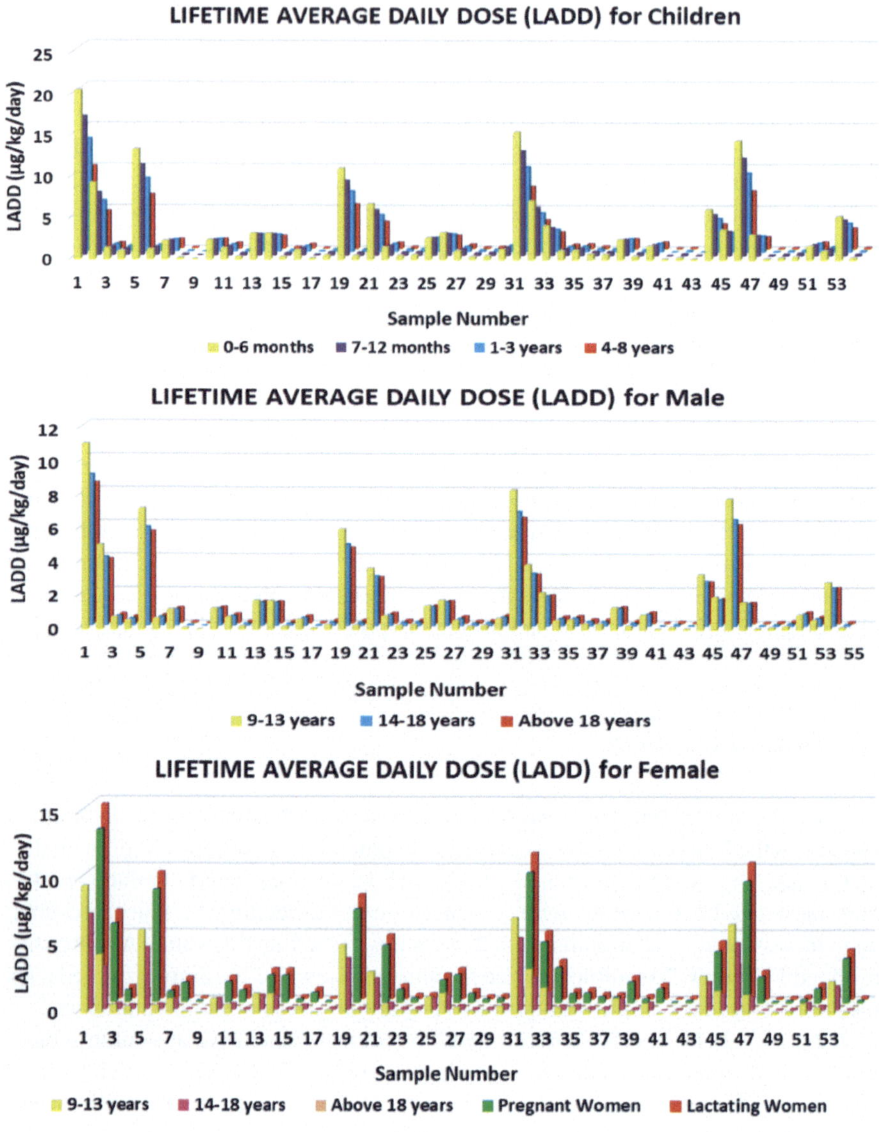

Fig. 5 LADD for children, male and female

suggesting potential developmental toxicity and endocrine disruption risks. Adolescent females (9–13 years) have 16 water samples, and those aged 14–18 years have 11 samples exceeding the unity threshold. These data emphasize the need for stringent environmental monitoring, regulation, and targeted public health interventions to safeguard vulnerable populations from uranium exposure and its associated health risks.

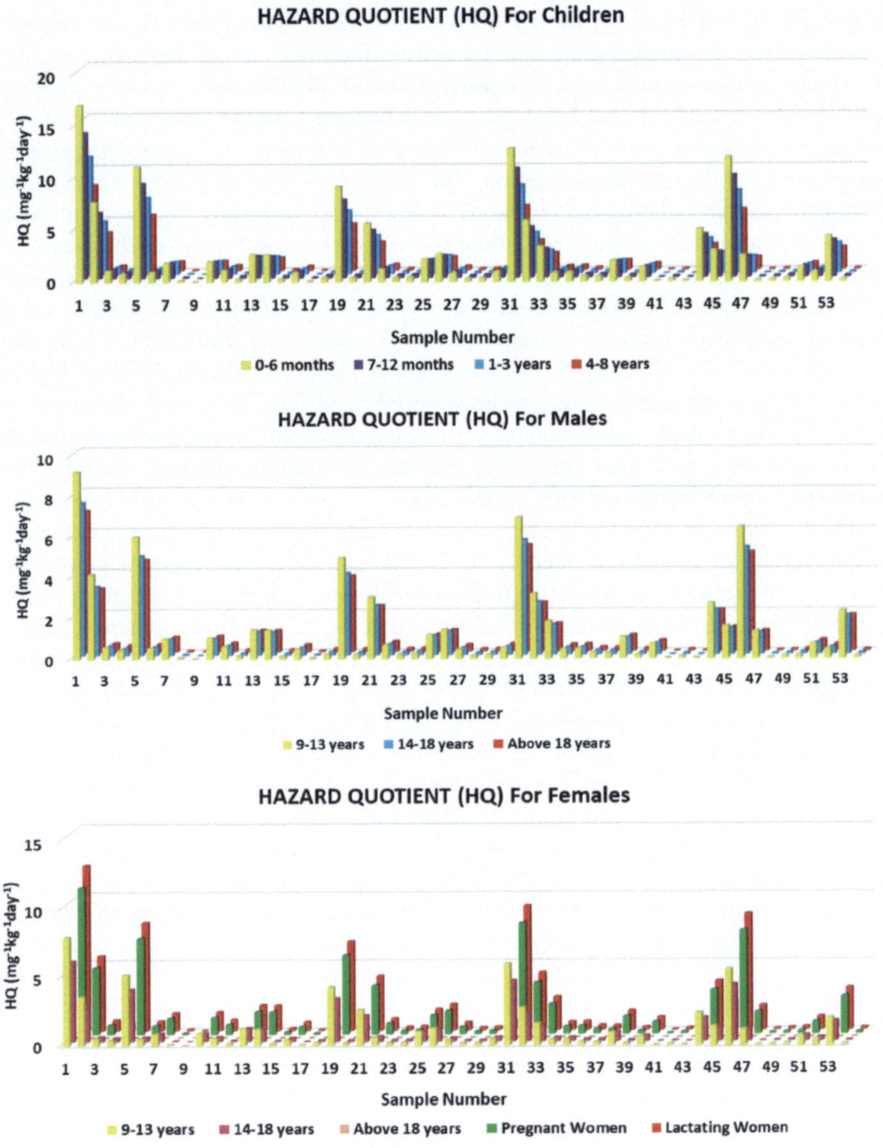

Fig. 6 HQ for children, male and female

6 Conclusions

This inclusive study highlights the substantial public health risks posed by uranium contamination in groundwater, with specific emphasis on the Raichur region. The research has identified uranium concentrations in groundwater samples that frequently exceed the safety thresholds recommended by the World Health Organization (WHO), highlighting the dual hazards of radiological and chemical toxicity. These elevated uranium levels, primarily observed in areas with intensive agricultural activities, suggest that local geological factors, coupled with anthropogenic influences such as the widespread use of phosphate fertilizers, are contributing significantly to groundwater contamination. The analysis of Excess Cancer Risk (ECR) demonstrates that radiological exposure to uranium significantly heightens the risk of cancer, particularly among older children, adult males, and women who are pregnant or lactating. This finding indicates that these populations are more vulnerable to the carcinogenic effects of uranium due to physiological and lifestyle factors that increase their exposure to contaminated water sources. The study's data show that children, especially those aged 4–8 years, are at increased risk due to their higher water consumption rates relative to body weight and their increased likelihood of ingesting soil or dust. Adult males, particularly those involved in agricultural or mining activities, may have prolonged exposure to uranium through occupational pathways, further increasing their cancer risk.

The Lifetime Average Daily Dose (LADD) calculations reveal chronic exposure levels that are of concern, especially for infants and pregnant or lactating women. Infants, being more susceptible due to their developing organs and lower body mass, are at heightened risk of uranium toxicity, which can lead to developmental delays and other health issues. For pregnant and lactating women, the transfer of uranium through the placenta or breast milk poses significant risks to fetal and neonatal development, including potential teratogenic effects and impacts on growth and cognitive function. This chronic exposure underscores the need for regular monitoring of water quality, especially in regions known for high uranium levels, and the implementation of interventions to reduce exposure, such as providing access to alternative water sources and advocating for dietary adjustments that limit uranium intake. The assessment of the Hazard Quotient (HQ) further emphasizes the potential non-carcinogenic health risks associated with uranium exposure. High HQ values, particularly in infants and pregnant women, indicate significant risks of adverse health effects such as nephrotoxicity, which can lead to renal dysfunction, and neurotoxicity, potentially causing developmental and cognitive impairments. The kidneys, being the primary site for uranium accumulation and processing, are particularly vulnerable, with evidence showing that uranium exposure can lead to tubular damage and altered renal function. This risk is heightened in populations with pre-existing health conditions or compromised renal function, making them more susceptible to the toxic effects of uranium.

To mitigate these substantial health risks, it is imperative to implement comprehensive strategies for managing uranium contamination. This includes establishing stringent regulatory controls over the use of uranium-bearing materials and fertilizers, enhancing groundwater monitoring programs to detect and manage uranium levels, and developing community education initiatives to inform the public about the risks associated with uranium exposure. Remediation efforts should focus on decontaminating affected water sources and rehabilitating contaminated sites, particularly in areas of intensive agriculture and mining. Additionally, promoting the use of advanced water treatment technologies, such as reverse osmosis and ion exchange, can effectively reduce uranium levels in drinking water, providing safer alternatives for affected communities.

References

AERB, DAE, India (2004) Atomic Energy Regulatory Board, Department of atomic energy, India, Drinking water specification

Amakom CA, Jibiri NN (2010) A review of some chemical radiological toxicological properties of uranium. J Radiat Res Appl Sci 3(1–2):189–199

Anand B, Karunanidhi D, Subramani T (2020) Promoting artificial recharge to enhance groundwater potential in the lower Bhavani River basin of South India using geospatial techniques. Enviorn Sci Pollut Res. https://doi.org/10.1007/s11356-020-09019-1

Anand B, Karunanidhi D, Subramani T, Srinivasamoorthy K, Raneesh KY (2017) Prioritization of sub-watersheds based on quantitative morphometric analysis in lower Bhavani Basin, Tamil Nadu, India using DEM and GIS techniques. Arab J Geo sci 24(10):1–18. https://doi.org/10.1007/s12517-017-3312-6.

Aravinthasamy P, Karunanidhi D, Subba Rao N, Subramani T, Srinivasamoorthy K (2020) Irrigation risk assessment of groundwater in a non-perennial river basin of South India: implication from irrigation water quality index (IWQI) and geographical information system (GIS) approaches. Arab J Geosci https://doi.org/10.1007/s12517-020-06103-1

Atomic Energy Regulatory Board (AERB) (2004) Safety Standards for Disposal of Radioactive Waste. Mumbai, India: AERB

Atomic Energy Regulatory Board (AERB) (2018) Safety code on radiation protection in mining and processing of Uranium. Atomic Energy Regulatory Board, India

Bartlett JM, Dougherty-Page JS, Harris NBW, Hawkesworth CJ, Santosh M (1998) The application of single Harris zircon evaporation and model Nd ages to the interpretation of polymetamorphic terrains: an example from the Proterozoic mobile belt of south India. Contrib Mineral Petrol 131:181–195

Bhaskar Rao YJ, Chetty TRK, Janardhan AS, Gopalan K (1996) Sm-Nd and Rb-Srages and P-T history of the Archaean Sittampundi and Bhavani layered meta anorthosite complexes in Cauvery Shear zone, South India: Evidence for neoproterozoic reworking of Archaean crust. Contrib Mineral Petrol 125:237–250

Brown L, Smith J, Jones M (2019) Hydrological and geological characteristics of granitic regions. J Environ Geol 45(2):345–358

Dang HS, Tran VD, Le HH (1994) Studies on water consumption and water use rates in some regions of Vietnam. J Prev Med Public Health 29(3):32–36

Divya Y, Gopinathan P (2019) Soil water content measurement using hyper-spectral remote sensing techniques—A case study from north-western part of Tamil Nadu, India. Remote Sens Appl: Soc Environ 14:1–7. https://doi.org/10.1016/j.rsase.2019.01.005

Divya Y, Gopinathan P, Jayachandran K, Ayad M (2021) Fadhil Al-Quraishi Color slices analysis of land use changes due to urbanization in a city environment of Miami Area, South Florida, USA. Model Earth Syst Environ 7:537–546. https://doi.org/10.1007/s40808-020-00883-x

Domingo JL et al (1989) The development toxicity of uranium in mice. Toxicology 55:143–152

Hema S, Subramani T, Elango L (2010) GIS study on vulnerability assessment of water quality in a part of Cauvery River. Int J Environ Sci 1(1):1–17

Gladys Mathews N, Nagaiah MB, Kumar K, Ambika MR (2015) Radiological and chemical toxicity due to ingestion of uranium through drinking water in the environment of Bangalore, India. J Radiol Protect 35(2015):447–455

Gopinathan P, Nandini CV, Parthiban S, Sathish S, Singh AK, Singh PK (2019) A geo-spatial approach to perceive the groundwater regime of hard rock terrain- a case study from Morappur area, Dharmapuri district, South India. Groundw Sustain Dev 100316. https://doi.org/10.1016/j.gsd.2019.100316

Gopinathan P, Parthiban S, Magendran T, Al-Quraishi AMF, Singh AK, Singh PK (2020) Mapping of ferric and ferrous iron oxides distribution using band ratio techniques with ASTER data and geochemistry of Kanjamalai and Godumalai, Tamil Nadu, south India. Remote Sens Appl: Soc Environ 100306. https://doi.org/10.1016/j.rsase.2020.100306

Gopinathan P, Roy P, Subramani T, Karunanidhi D (2022) Detection of iron-bearing mineral assemblages in Nainarmalai granulite region, south India, based on satellite image processing and geochemical anomalies. Environ Monit Assess 194:866. https://doi.org/10.1007/s10661-022-10570-2

Graham JD (1993) EPA cancer risk assessment guidelines: do they effectively incorporate recent science? Risk Anal 13(5):511–522

Guglielmotti MB, Cabrini RL, Anzano MA, Gimenez MS (1989) Pathology of uranium nephrotoxicity in rats after acute intoxication with uranyl acetate. Toxicol Pathol 17(2):375–383

Gupta R, Sharma V (2020) Impacts of soil permeability on groundwater quality. Agric Sci Technol 32(4):789–802

Hallenbeck WH (1993) Quantitative risk assessment for environmental and occupational health. Lewis Publishers

Health Canada (1999) Uranium in drinking water, document for public comment prepared by federal provincial subcommittee on drinking water

International Commission on Radiation Protection 2007 The 2007 Recommendations of the International Commission on Radiological Protection (ICRP Publication103, Annals of the ICRP) vol 37. Elsevier, Amsterdam

Jayananda M, Janardhan AS, Sivasubramanian P, Peucat JJ (1995) Geochronologic and Isotopic constraints on granulite formation in the Kodaikanal area, Southern India. Mem Geol Soc India 34:373–390

Karunanidhi D, Aravinthasamy P, Subramani T, Setia R (2021) Groundwater suitability estimation for sustainable drinking water supply and food production in a semi-urban area of south India: a special focus on risk evaluation for making healthy society. Sustain Cities Soc 73. https://doi.org/10.1016/j.scs.2021.103077

Karunanidhi D, Aravinthasamy P, Deepali M, Subramani T, Sunkari ED (2020) Appraisal of subsurface hydrogeochemical processes in a geologically heterogeneous semi-arid region of south India based on mass transfer and fuzzy comprehensive modeling. Environ Geochem Health 43(2):1009–1028. https://doi.org/10.1007/s10653-020-00676-2

Keith S, Faroon O, Fowler B, Roland G (1999) Toxicological profile for uranium. Agency for Toxic Substances and Disease Registry (US), Atlanta, GA

Kelly B (1991) Incremental cancer risk: a concept adopted by regulatory agencies. Environ Health Perspect 96:211–219

Kim Y-S, Park H-S, Kim J-Y, Park S-K, Cho B-W, Sung I-H, Shin D-C (2004) Health risk assessment for uranium in Korean groundwater. J Environ Radioact 77:77–85

Krishnamurthy P (1977) On some geochemical aspects of Sevathur carbonatite complex, North Arcot District, Tamil Nadu. J Geol Soc India 18:265–274

Kumar A, Usha N, Sawant PD, Tripathi RM, Raj SS, Mishra M, Rout S, Supreeta P, Singh J, Kumar S, Kushwaha HS (2011) Risk assessment for natural uranium in subsurface water of Punjab State, India. Human Ecol Risk Assess 17:381–393

Kumar A, Rao S (2016) Groundwater dynamics in fractured rock aquifers. J Hydrol 57(3):234–246

Kumar E, Subramani T, Li P, Karunanidhi D (2022) Quantification of health and environmental risks due to radionuclides in limestone mining regions of Ariyalur, South India. Hum Ecol Risk Assess Int J 28:243–264

Kurttio P, Harmoinen A, Saha H, Salonen L, Karpas Z, Komulainen H (2002) Renal effects of uranium in drinking water. Environ Health Perspect 110(4):337–342

Kurttio P, Harmoinen A, Saha H, Salonen L, Karpas Z, Komulainen H, Auvinen A (2006) Kidney toxicity of ingested uranium from drinking water. Am J Kidney Dis 47:976–982

Larivière D, Harvey PJ, Mann KK, Pelletier É (2007) Uranyl nitrate-induced developmental abnormalities in zebrafish: a model to study the impact of uranium mining discharges. Aquat Toxicol 85(3):230–240

Leggett RW (1989) Uranium: chemical toxicity profile. Office of Health and Environmental Assessment, Office of Research and Development, US Environmental Protection Agency, Washington, DC

Lentz TJ et al (2001) Application of the hazard quotient approach in risk assessment. J Occup Environ Hyg 1(1):46–55

Li P, He X, Guo W (2019) Spatial groundwater quality and potential health risks due to nitrate ingestion through drinking water: a case study in Yan'an City on the Loess Plateau of northwest China. Hum Ecol Risk Assess 25(1–2):11–31. https://doi.org/10.1080/10807039.2018.1553612

Lohner TW (1997) Derivation of risk-based cleanup levels for soil and groundwater in California. Risk Anal 17(1):23–35

Muthamilselvan A (2020) Anomalous concentration of uranium in groundwater in parts of Salem District, Central Tamil Nadu, India, Adv Earth Environ Sci 1(1):1–9

National Research Council (1983) Risk assessment in the Federal Government: managing the process. National Academy Press

Pandit MK, Sialt AN, Sukumaran GB, Ramanathan S, Ferreira VP (1988). Carbon and oxygen isotopic variation in Tamil Nadu carbonatites, south India; Subramanian V (1983) Geology and geochemistry of the carbonatites of Tamil Nadu, India. Ph.D. thesis, Indian Institute of Science, Bangalore 83:175–182

Patel N (2018) Soil and water interactions in arid regions. Int J Soil Sci 21(1):123–137

Preethi B, Subramani T, Karunanidhi D, Gopinathan P (2023) Human health risks due to exposure to water pollution: a review. Water 15:2532. https://doi.org/10.3390/w15142532

Singh L, Kumar R, Kumar S, Bajwa BS, Singh S (2012) Health risk assessments due to uranium contamination of drinking water in Bathinda region, Punjab State, India. Radioprotection 48:191–202

Smith J, Jones M (2017) Geological features of the Raichur district. Indian J Geol 12(4):401–415

Srinivasan R, Verma K, Singh P (2015) Influence of geological formations on groundwater flow. Geol Surv J 34(2):201–215

Subramani T, Anandakumar S, Kannan R, Elango L (2013) Identification of major hydrogeochemical processes in a hard rock terrain by NETPATH modeling. Earth Resour Environ 365–370

Travis CC, Kelly B (1987) Cancer risk assessment: time for a change? Science 238(4833):1628–1630

US Environmental Protection Agency (US-EPA) (1989) Risk assessment guidance for superfund: volume I—Human health evaluation manual (Part A). US-EPA, Washington, DC

US Environmental Protection Agency (USEPA) (1993) Integrated risk information system (IRIS). USEPA, Washington, DC

US Environmental Protection Agency (USEPA) (1994) Health effects assessment summary tables. USEPA, Washington, DC

US-EPA (US Environmental Protection Agency) (1999) Federal Guidance Report No. 13: risk assessment of radon in drinking water. US-EPA, Washington, DC

United States Environmental Protection Agency (UP-ESA) (1999) Preliminary health risk reduction and cost analysis: revised national primary drinking water standards for radionuclides review draft (United States Environmental Protection Agency, Washington, DC)

United States Environmental Protection Agency (2000) Preliminary Health risk reduction and cost analysis: revised national primary drinking water standards for radionuclides review draft. United States Environmental Protection Agency, Washington, DC

US-EPA (2002) A review of the reference dose and reference concentration processes. U.S. Environmental Protection Agency

Vennila G, Subramani T, Elango L (2007) Rainfall variation analysis of Vattamalaikarai sub-basin, Tamil Nadu, India. Appl Hydrol 20(3):50–59

Vennila G, Subramani T, Elango L (2008) GIS based groundwater quality assessment of Vattamalaikarai basin, Tamil Nadu, India. Nat Environ Pollut Technol 7(4):585–592

WHO (World Health Organization) (2008) Uranium in drinking-water: background document for development of WHO guidelines for drinking-water quality. World Health Organization, Geneva, Switzerland

World Health Organization (WHO) (2009) Principles for evaluating health risks in children associated with exposure to chemicals. WHO Press

World Health Organization (WHO) (2011a) Guidelines for drinking-water quality. WHO, Geneva, Switzerland

World Health Organization (2011b) Uranium in drinking water who guidelines for drinking water quality, 4th edn. World Health Organization, Geneva

World Health Organization (WHO) (2017) Guidelines for drinking-water quality: fourth edition incorporating the first addendum. World Health Organization

Ye-shin K, Kim S, Yoo S, Kim S (2004) Setting acceptable risk level of exposure to environmental radionuclides. Environ Health Toxicol 19(4):331–338; US Environmental Protection Agency (USEPA) (2000) Framework for ecological risk assessment. USEPA, Washington, DC

Zamora ML et al (1998) Chronic ingestion of uranium in drinking water: a study of kidney bioeffects in humans. Toxicol Sci 43:31–45

Groundwater Resources Availability in the Shangzhou Basin Area, China: A Modeling Research

Pengbin Zhang, Jianhua Wu, Fei Xu, and Weichao Liu

Abstract The Shangzhou Basin, a principal groundwater development region in the Shangzhou District of China, is also a vital water source protection zone for the South-to-North Water Diversion Project. In recent years, escalating groundwater exploitation has posed significant and irreversible risks to the local ecology. This study introduces a numerical model, based on MODFLOW, constructed specifically for this region. This model serves to assess the permissible groundwater extraction volume, thereby offering robust support for the sustainable utilization of these resources. Under current extraction practices, the total annual groundwater recharge and discharge in the study area are 25.81 million m^3 and 27.80 million m^3, respectively. The balance discrepancy between recharge and discharge stands at -1.31 million m^3/a, suggesting that the region is in a state of generally balanced state. Water exchange between groundwater and other water bodies predominantly takes place in the valley terrace area. The simulation results align closely with the existing water resource utilization status in the study area. Building on this, various extraction scenarios were established to forecast groundwater changes over the next decade (2020–2030). The maximum allowable extraction volume for the Shangzhou Basin area is determined to be 16 million m^3/a. This research fundamentally supports the sustainable development and management of groundwater resources in the Shangzhou Basin.

Keywords Maximum allowable extraction · Groundwater level prediction · MODFLOW model · Shangzhou Basin

P. Zhang · J. Wu (✉) · F. Xu · W. Liu
School of Water and Environment, Chang'an University, No. 126 Yanta Road, Xi'an 710054, Shaanxi, China
e-mail: wujianhua@chd.edu.cn; wjh2005xy@126.com

Key Laboratory of Subsurface Hydrology and Ecological Effects in Arid Region of the Ministry of Education, Chang'an University, No. 126 Yanta Road, Xi'an 710054, Shaanxi, China

Key Laboratory of Eco-Hydrology and Water Security in Arid and Semi-Arid Regions of the Ministry of Water Resources, Chang'an University, No. 126 Yanta Road, Xi'an 710054, Shaanxi, China

P. Li et al. (eds.), *Sustainable Groundwater and Environment: Challenges and Solutions*, Springer Hydrogeology, https://doi.org/10.1007/978-3-031-82194-3_11

1 Introduction

Groundwater is an indispensable component of our global water supply, crucial for sustaining ecosystems, bolstering agricultural and industrial activities, and providing potable water (Bierkens and Wada 2019). Traditionally, it has been viewed as a renewable resource (Wang et al. 2019). However, with societal advancements, groundwater extraction has escalated, and its renewable nature has gradually diminished. This has led to emerging issues such as declining groundwater levels, deteriorating water quality, and damaging ecosystems (Qu et al. 2021; Shen and Xu 2011). These challenges underscore the importance of sustainable groundwater management, necessitating a comprehensive understanding of groundwater dynamics.

Traditional research methods often fall short in comprehensively and promptly understanding groundwater dynamics and storage status. Thus, numerical simulation models have become invaluable for analyzing groundwater's dynamic changes within aquifers (Cedrick et al. 2021). Scholars have utilized various software tools, such as GMS, MODFLOW, MIKE SHE, and FEFLOW, to simulate groundwater movement, yielding significant advancements in the field (Colins et al. 2016; Jesudhas et al. 2017). Among these, MODFLOW has emerged as one of the most widely used software in recent groundwater research (Akter and Ahmed 2021). Through numerical models, researchers like Rezaei and Sargazi (2009), Lavigne et al. (2010), Biswas and Melesse (2016), and Narula and Gosain (2013) have studied various aspects of groundwater, from artificial recharge influences to groundwater flow in different formations. Numerous methods have also been proposed to estimate recoverable groundwater amounts, including remote sensing technology, water balance method, and isotope technology (Zareian and Eslamian 2021). However, these methods often require considerable human and material resources. In contrast, numerical simulations based on monitoring data can not only capture the laws of groundwater movement but also yield more reasonable and accurate estimates of extractable groundwater.

The Shangzhou District in Shaanxi, an important water resource protection area, primarily relies on surface water for its supply, with groundwater being relatively underutilized. The Shangzhou Basin, located within this district, is a densely populated urban planning area experiencing gradual growth. Given the district's topography and the basin's role as a primary groundwater source, understanding the basin's hydrogeological conditions and groundwater distribution and dynamics is vital. The basin's per capita available water resources are slightly below the internationally recognized absolute water shortage line (Li and Qian 2018) and the rapid industrial development in the area is exacerbating water resource shortages (De Graaf et al. 2014; Jia et al. 2019). Thus, there is an urgent need to expand groundwater exploitation knowledgeably and sustainably (Hanasaki et al. 2018; Wada et al. 2016).

Reasonable groundwater exploitation aligns with national strategic policies, protects local ecology, and effectively prevents environmental geological hazards (Döll et al. 2014; Van Asselen et al. 2011; Richey et al. 2015; Wada et al. 2012, 2010). Previous studies on the Shangzhou Basin have primarily focused on surface water, leaving a gap in groundwater research. This chapter aims to fill this gap by analyzing the water balance under current extraction conditions, predicting groundwater dynamic changes under different extraction conditions, and determining the exploitable amount of groundwater in the Shangzhou Basin. The findings will guide rational groundwater development and utilization in the Shangzhou Basin and provide a scientific basis for the basin's future contribution to the South-to-North Water Diversion Project.

2 Study Area

2.1 *Location and Geography*

The Shangzhou Basin area is located in the southeast of the Shangzhou District of Shangluo City, Shaanxi Province, China (109°53′43″–110°02′15″E, 33°47′45″–33°54′21″N). It is the central urban planning area of the Shangzhou District. The study area is the transition zone from mountains to river valley terraces in the Shangzhou District. As a whole, it is high in the northwest and low in the southeast, and the Danjiang River runs through the entire basin area (Fig. 1). The study area belongs to the monsoon and semi-humid mountain climate in the transition zone of the southern margin of the warm temperate zone. The annual average water surface evaporation is above 1000 mm, and it is mainly concentrated from April to September, accounting for 70–80% of the annual evaporation. The average annual precipitation is 699.44 mm, and the inter-annual precipitation is unevenly distributed and varies greatly. It belongs to the Danjiang River system and is a secondary tributary of the Yangtze River. The river network density is 0.69–1.28 km/km^2. The average annual runoff depth is 230 mm. From the perspective of geomorphology, the study can be divided into mountains, hills and valley terraces. The surface elevation increases from the valley plain to the mountain (from 641 to 1005 m).

2.2 *Hydrogeological Settings*

The groundwater in the study area is mainly Quaternary loose rock pore water. It is mainly distributed in the floodplain and first-order terraces on both sides of Danjiang River and its main tributaries. It is rich in water and is the only groundwater source available for exploitation in the territory (Figs. 2 and 3).

The aquifer is mainly composed of sand and gravel, with a thickness of 10–15 m. The water level depth in this area is about 1–10 m, and the floodplain is the best for water abundance. The water specific yield is generally greater than 1000 m^3/d in the floodplain area, but the water yield of the first terrace is quite diverse, and its water specific yield of a single well is 10–400 m^3/d. The position of the second

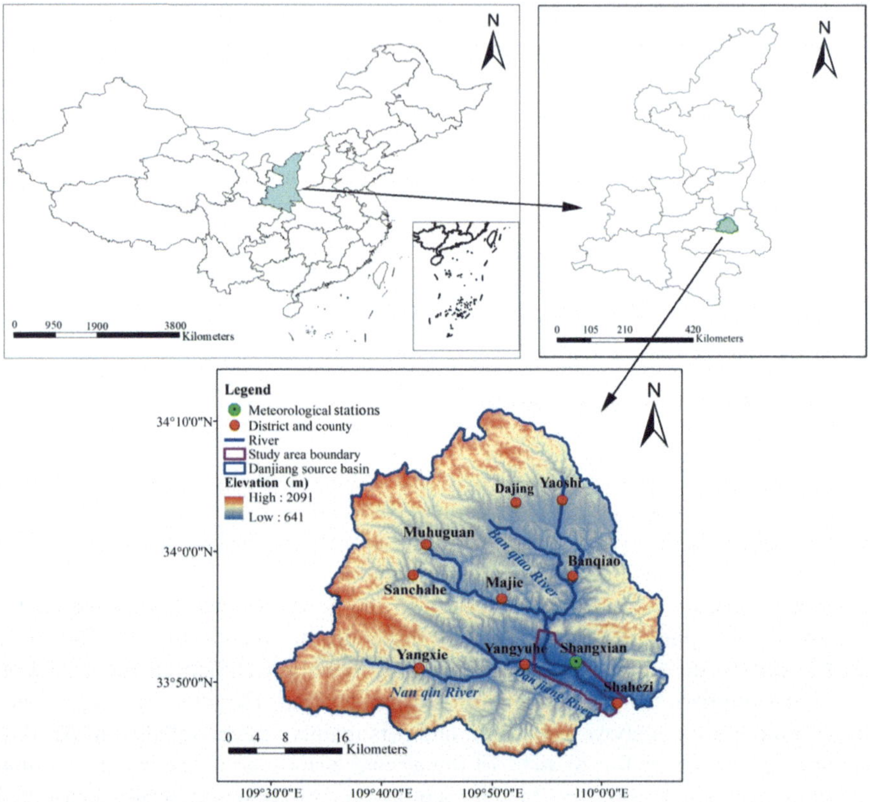

Fig. 1 Location and topography of the study area

terrace is high, and only the low portion of it contains water. The aquifer in the remaining areas is thin, which has no actual water supply significance. In the vertical direction, the aquifer is a loose rock pore aquifer, with sandy pebble gravel as the main lithology. The lower part of the aquifer is the Tertiary strata, which is sandstone and sandy mudstone interbedded, with dense lithology and weak water permeability. The aquifer has no obvious aquifuge in the vertical direction, the hydraulic connection is close and the groundwater level is unified.

The natural conditions in the area determine the formation and movement of groundwater. The rock strata around the study area are inclined from the periphery to the hinterland, so the lateral recharge and discharge of groundwater are formed between the middle and low mountain areas. Danjiang River is the only surface river in the area. Because of its large vertical drop, there is a complementary relationship between the river sections and groundwater, and discharge to the river is one of the main channels for groundwater discharge. However, in some sections where small rivers exist, river water may percolate into the phreatic aquifer. Vertically, it is mainly

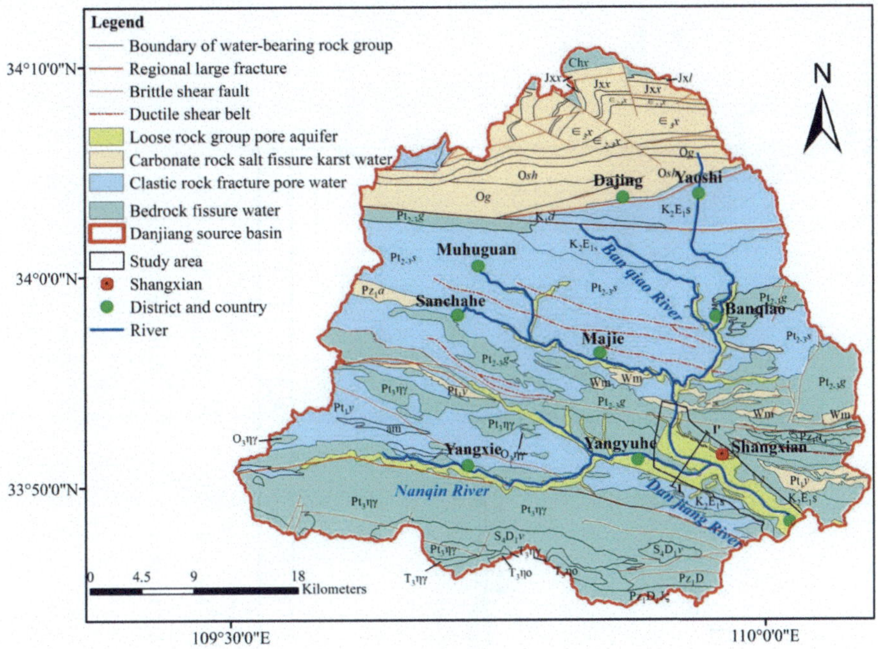

Fig. 2 Hydrogeological map of the study area

manifested as atmospheric precipitation, river and irrigation water infiltration. The regional groundwater flow is generally from northwest to southeast.

3 Material and Methods

3.1 Data Sources

Data collection is an essential part of the research process. In order to ensure the accuracy and reliability of the research, this study adopts a strict data collection method. The research mainly relies on the data from the 'Shangluo Water Resources Bulletin' from 2018 to 2020 and the data provided by Shangluo Water Conservancy Bureau of Shaanxi Province. The collected and organized data include hydrological, extraction and observation well information. In addition, the evaporation discharge is obtained through previous investigation and monitoring. Rainfall data were obtained by Shangxian meteorological station. In order to obtain more accurate data for a comprehensive study, a field survey was also conducted to monitor the surface water and groundwater levels in the study area to determine the conversion relationship between surface water and groundwater. The conversion amount of river water and

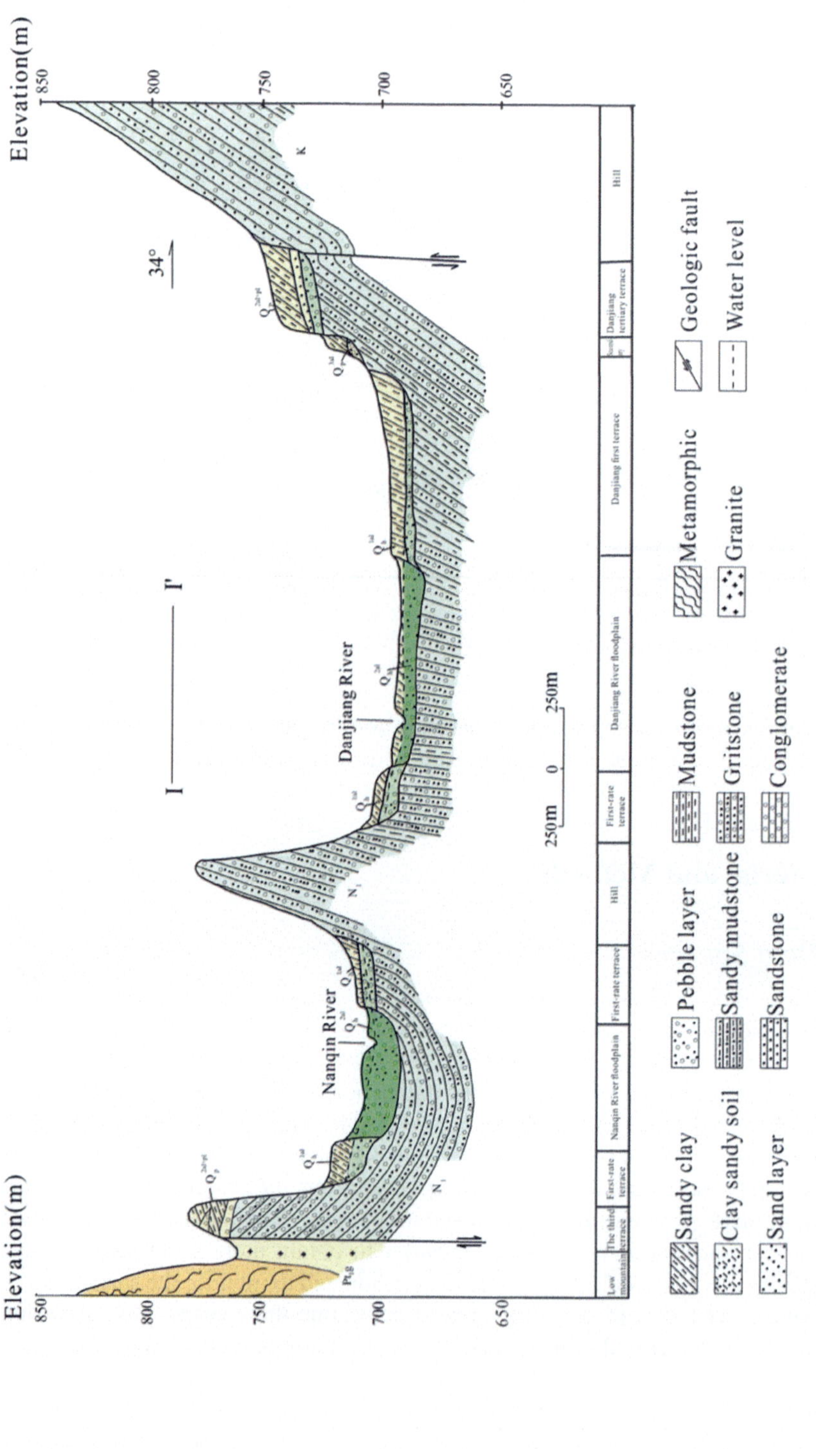

Fig. 3 Hydrogeological profile of the study area

groundwater was estimated according to Darcy's law. The hydraulic conductivity of aquifer is determined according to the previous drilling data and the empirical value.

3.2 *Conceptual Model*

3.2.1 Aquifer Generalization

Based on the hydrogeological analysis of the study area, it is identified that the aquifer in this region is the Quaternary phreatic aquifer, serving as the primary groundwater source. Consequently, this study develops a numerical groundwater model for the area (Gebere et al. 2021; Kumar 2013; Lachaal and Gana 2016). Given the absence of distinct aquifuges and aquitards between the Quaternary aquifers, and the bottom of this aquifer is comprised of the Tertiary strata characterized by sandstone and sandy mudstone with notably low permeability, the entire Quaternary phreatic aquifer is treated as a unified phreatic aquifer group (Li et al. 2022; Maity and Mandal 2019). The analysis of hydrogeological surveys and drilling data reveals a consistent variation in stratum thickness across the simulation area, with the aquifer typically around 15 m thick. Following Darcy's law, the groundwater system is conceptualized as a two-dimensional, heterogeneous, anisotropic, and transient flow system, as discussed in studies by Ahmed and Umar (2009), Gropius et al. (2022), and Mohamed et al. (2016).

3.2.2 Generalization of Boundary Conditions

According to the analysis of the flow field characteristics and stratum structure in the study area, the lateral boundaries of the study area are identified as watershed boundaries with distinct properties (Singh and Shukla 2016). In the whole study area, groundwater is discharged to the outside of the area in the southeast corner of the study area, and there is lateral outflow through this boundary. The other surrounding boundaries accept the lateral inflow from the mountainous. Therefore, the surrounding boundaries of the whole study area are generalized into a constant flow boundary. The flow rate of the lateral inflow boundary is set positive, and the flow rate of the lateral outflow boundary is set negative. The upper boundary of the system is defined by the water table of the phreatic aquifer, and it is a free surface boundary. The primary segment of the Danjiang River within the study area is designated as a river boundary, and it can have river water percolation into the aquifer. The bottom of the phreatic aquifer is composed of sandy mudstone with limited water permeability, and thus is conceptualized as a water-resisting boundary for the model.

3.2.3 Hydrogeological Parameters and Source-Sink Treatment

The hydrogeological parameters required to construct the model primarily encompass the rainfall infiltration coefficient, hydraulic conductivity, and specific yield (Chakraborty et al. 2019; Gobinath et al. 2022). After extensive data collection and field investigation, the study area has been partitioned into four units: valley terrace, first-order terrace, second-order terrace, and hills. The hydrogeological parameters from hydrogeological survey were used as the initial input for each unit. However, the hydrological parameters for each unit will undergo adjustment during the calibration period to yield observed groundwater levels matching the observed groundwater levels.

The sources and sinks within the study area are principally constituted by two elements, namely recharge and discharge (Sathe and Mahanta 2019; Walraevens et al. 2015). The aquifer primarily undergoes recharge through precipitation infiltration, agricultural irrigation infiltration, lateral runoff from the boundary, and leakage from certain sections of the river. Discharge primarily occurs through evaporation from groundwater, leakage to rivers, lateral discharge at the boundary, and artificial extraction.

3.3 *Mathematical Model*

Based on the hydrogeological conceptualization, the mathematical expression of the heterogeneous, anisotropic two-dimensional unsteady flow groundwater model is constructed as follows (Mohanty et al. 2013; Sathish et al. 2018; Zhang et al. 2023):

$$\begin{cases} \frac{\partial}{\partial x}\left(K_x(H-B)\frac{\partial H}{\partial x}\right)+\left(K_y(H-B)\frac{\partial H}{\partial y}\right)+W=\mu\frac{\partial H}{\partial t} & \\ H(x,y,0)=H_0(x,y) & (x,y)\in D \\ H(x,y,t)|_{\Gamma_1}=H_1(x,y) & t>0 \\ -KH\frac{\partial H}{\partial n}\Big|_{\Gamma_2}=q & t>0 \\ \lim\limits_{r\to r_w}\left(r\int_0^{2\pi}KH\frac{\partial H}{\partial r}d\theta\right)=Q_i & t>0 \\ K'\frac{H_d-H}{M'}A=Q_d & t>0\ (\text{canals}) \end{cases} \tag{1}$$

where, H represents the elevation of the groundwater table, in meters (m); B stands for the elevation of the aquifer base, in meters (m); K_x and K_y denote the hydraulic conductivities of the aquifer in the horizontal x and y directions, respectively, in meters per day (m/d); W designates the strength of vertical water exchange intensity expressed in cubic meters per day per square meter (m^3/d m^2); $h_0(x, y)$ represents the initial water level at the location (x, y) within the simulation range, in meters (m); μ is the specific yield of the phreatic aquifer, dimensionless; K' is the vertical hydraulic conductivity at the bottom of the water diversion canal, in meters per day (m/d); M'

represents the vertical thickness at the bottom of the water diversion canal, in meters (m); A is the area of the water diversion canal, in square meters (m^2); q denotes the flow rate per unit width for the second type of boundary, in square meters per day (m^2/d); Q_i is the average daily extraction volume for the ith well, in cubic meters per day (m^3/d); r is the diameter of the extraction well, in meters (m); Γ_1 indicates the river boundary; Γ_2 signifies the second type of boundary; Q_d represent the volume of water diversion expressed in cubic meters per day (m^3/d), and D represents the range of the simulation area.

3.4 Discretization of Model

The aquifer in the study area has been spatially discretized based on the existing geological, hydrological, and hydrogeological conditions (Berehanu et al. 2017; Devarajan and Sindhu 2015; Mirlas et al. 2022; Niswonger et al. 2011). Vertically, the study area consists of a single unconfined aquifer, so it is modeled as a single-layer structure. Horizontally, the domain was discretized by employing a rectangular grid with dimensions of 100 m × 100 m, resulting in a total of 5647 active grids. Moreover, based on the geological and hydrogeological investigation in the study area, the area is divided into four parameter zones, namely I, II, III, and IV, in accordance with its topography, geomorphology, and lithological conditions (as depicted in Fig. 4).

The calibration period is 365 days from January 1, 2018 to December 31, 2018, which is divided into 12 stress periods. In order to improve the calculation accuracy, each stress period is further divided into 10 sub-stages.

3.5 Recharge and Discharge

Recharge and discharge are important elements in hydrogeological studies, especially in groundwater resources availability calculation. According to the principle of water budget calculation, the following relationship in any area at any time exists:

$$Q_r - Q_p = \mu \cdot A \cdot \frac{h_t - h_0}{\Delta t} \tag{2}$$

where, Q_r is the groundwater recharge in the water budget calculation area (m^3/d); Q_p is the groundwater discharge in the water budget calculation area (m^3/d); Δt is the period considered for the water budget calculation (d); h_0 is the average groundwater level (m) at the initial time of the water budget calculation period; h_t is the average groundwater level (m) at the end of the water budget calculation period; A is the area of water budget calculation area (m^2); μ is the specific yield of the phreatic aquifer.

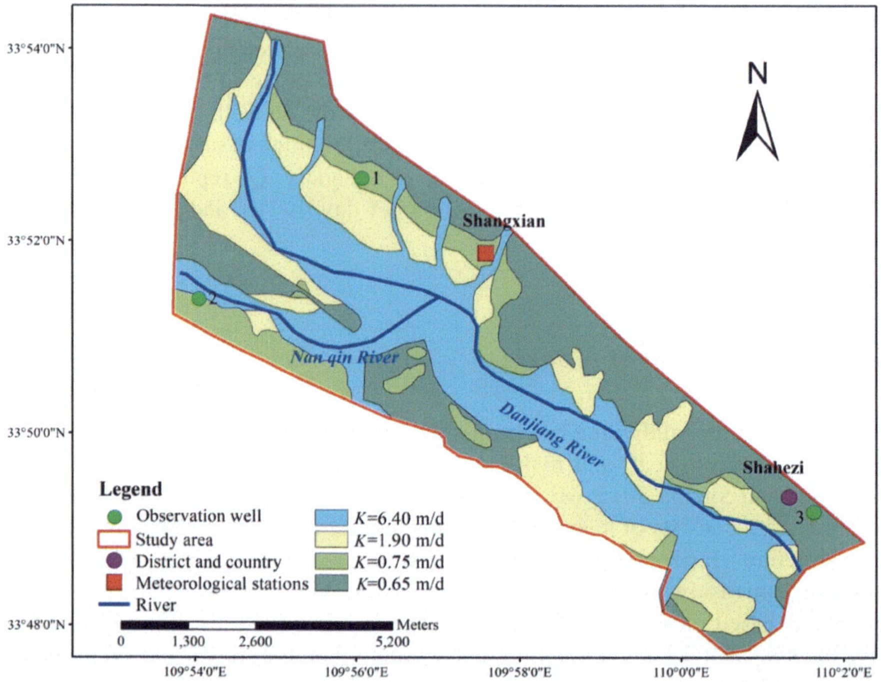

Fig. 4 Hydrogeological parameter zoning map of the study area

According to the hydrogeological survey in the study area, groundwater recharge in the study area includes mainly rainfall infiltration recharge, agricultural recharge, lateral runoff recharge and river infiltration recharge, while groundwater discharge is mainly evaporation, lateral runoff discharge, river discharge and artificial exploitation. Rainfall infiltration and evaporation can be calculated using the coefficient method based on the monthly rainfall and evaporation data; the lateral inflow and outflow are controlled by boundary conditions, and can be calculated as per the Darcy's law. The river recharge and discharge are added into the model by the river module and the artificial pumping water is added into the model in the form of groundwater pumping wells.

3.6 Model Calibration and Validation

In the groundwater flow model, model calibration is an essential process. In this process, the parameters are calibrated by iteration (Omar et al. 2020; Surinaidu et al. 2014). In this study, the model is calibrated by adjustment of aquifer parameters to fit the groundwater level data from January to December 2018 of the three observation wells. The calibrated hydraulic conductivity ranges from 0.65 to 6.4 m/d in the four units. The fitting curve between the calculated groundwater level and the measured value is shown in Fig. 5a. The water levels of the observation well are shown in Table 1. Figure 5a shows that the calculated water level of the model is in the 95% confidence interval of the observed water level, the RME is 0.416 m, and the normalized RMS is 0.649%. Therefore, the calibrated parameters are within the acceptable range, and the identified hydrogeological parameters are suitable (Table 2).

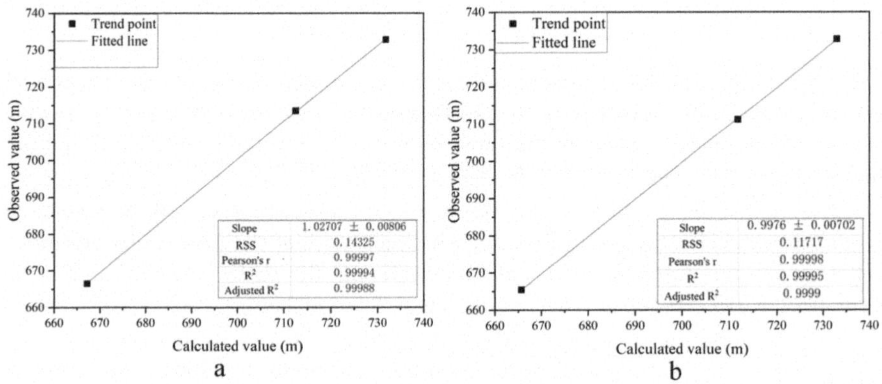

Fig. 5 Comparison diagram of calculated water level and measured water level of observation well

Table 1 Relative error between calculated water level and measured water level of observation wells in different periods

Period	Well	Observed value (m)	Calculated value (m)	Relative error (m)	R^2
Calibration period	Well 1	731.24	731.52	−0.28	0.9999
	Well 2	713.35	712.93	−0.42	
	Well 3	667.65	667.18	0.47	
Validation period	Well 1	733.02	732.78	0.24	0.9999
	Well 2	711.88	711.23	0.65	
	Well 3	665.63	665.45	0.18	

Table 2 Hydrogeological parameters in different hydrogeological units

Hydrogeological unit	Hydraulic conductivity (m/d)	Specific yield	Rainfall recharge coefficient
I	6.40	0.20	0.36
II	1.90	0.16	0.33
III	0.75	0.12	0.30
IV	0.65	0.10	0.09

Table 3 Groundwater budget calculation results ($\times 10^4$ m^3/a)

Recharge	Amount	Discharge	Amount
Rainfall infiltration	988.08	Evaporation	178.62
Irrigation water infiltration	405.76	Lateral outflow	49.10
Lateral outflow	712.42	River discharge	2087.80
River percolation	481.80	Artificial extraction	404.00
Total recharge	2588.06	Total discharge	2719.52

The verification is also an indispensable part of checking the accuracy of the model (Aliyari et al. 2019; Bailey et al. 2020). During the verification period, the hydrogeological parameter should be the same with those obtained from the calibration. The groundwater level data from January to December 2020 were selected for the verification. The fitting curve of the observed groundwater level and the measured groundwater level is shown in Fig. 5b, which shows that the observed water level of the model is within the 95% confidence interval, the RME is 0.346 m, and the normalized RMS is 0.505%. This indicates that the model constructed and calibrated is reliable in groundwater flow modeling, and can be used for further study.

The water budget calculated using the model is shown in Table 3. As shown in Table 3, the main recharge source in the basin is the rainfall infiltration followed by lateral outflow through the boundary, river percolation and irrigation water infiltration, and the total groundwater recharge is 2588.06×10^4 m^3/a. Approximately, the total amount of groundwater discharge by rivers, evaporation, groundwater outflow though the boundary and groundwater extraction is 2719.52×10^4 m^3/a. The relative difference between groundwater recharge and discharge is −2.48%, which indicates that groundwater in the study area is generally in a balanced state. This indicates that the model has converged. Therefore, it is considered that the model established in this study is reasonable and can be used in subsequent studies.

3.7 Groundwater Modeling Scenarios

The degree of groundwater exploitation and utilization in the study area is low, and groundwater potential is greater than its current exploitation amount. To meet increasing water demand in this area, more groundwater extraction may be expected in the near future. To determine the maximum permissible extraction amount in the area and to avoid over-exploitation, it is necessary to design different extraction scenarios. According to hydrogeological survey, the current annual groundwater extraction rate is 404×10^4 m^3/a. Based on this, in this study, 5 extraction scenarios were designed (Table 4). Groundwater modeling was conducted to predict the changes in groundwater level and groundwater flow in the nest ten year till 2030 under these 5 different extraction scenarios. The pumping well locations in the model are designed according to the field investigation. The maximum allowable

Table 4 Extraction scenarios for groundwater modeling

Extraction scenarios	Description of the scenarios
Scenario 1	Extracting with current water consumption rate (404 × 10^4 m^3/a)
Scenario 2	Double the amount of exploitation rate of Scenario 1 (808 × 10^4 m^3/a)
Scenario 3	Triple the amount of exploitation rate of Scenario 1 (1212 × 10^4 m^3/a)
Scenario 4	Quadruple the amount of exploitation rate of Scenario 1 (1616 × 10^4 m^3/a)
Scenario 5	Five times of the amount of exploitation rate of Scenario 1 (2020 × 10^4 m^3/a)

drawdown is 7.5 m in this study area. If the groundwater modeling results indicate a drawdown higher than 7.5 m in 2030, then the groundwater extraction rate is considered unsuitable.

4 Results and Discussion

4.1 Groundwater Allowable Extraction Rate Based on Modeling Results

Through the simulation of different extraction scenarios, the groundwater drawdown map is shown in Fig. 6. From the figure, it can be seen that under the extraction conditions of Scenario 1, the groundwater level will drop less than 1 m by 2030, and the maximum depth of descent is only 0.71 m, which is much lower than the maximum permissible depth of descent of 7.5 m in the study area, and at this time the gradual decline of the groundwater level will have a minimal impact on the groundwater flow field in the area. This indicates that the current extraction has not yet reached the maximum allowable extraction, suggesting that a moderate increase in extraction is possible.

Scenario 2 doubles the extraction volume to 808 × 10^4 m^3/year from Scenario 1. According to the model predictions, the maximum decrease in groundwater level reaches 3.42 m in the seventh year (Fig. 6) and remains relatively stable in the next three years. Compared to Scenario 1, the average rate of groundwater level decline for Scenario 2 is 0.34 m/year. Since the maximum permissible decline in groundwater levels has not yet been reached and there are no negative impacts on the surrounding environment, it can be inferred that there is potential for further groundwater extraction.

Scenario 3 further extends the extraction volume to 1212 × 10^4 m^3/year. During the simulation period, the maximum decrease in groundwater level reaches 6.34 m (Fig. 6), which is close to the maximum allowable decrease in groundwater level. The groundwater level stabilized in the ninth year and did not continue to decline.

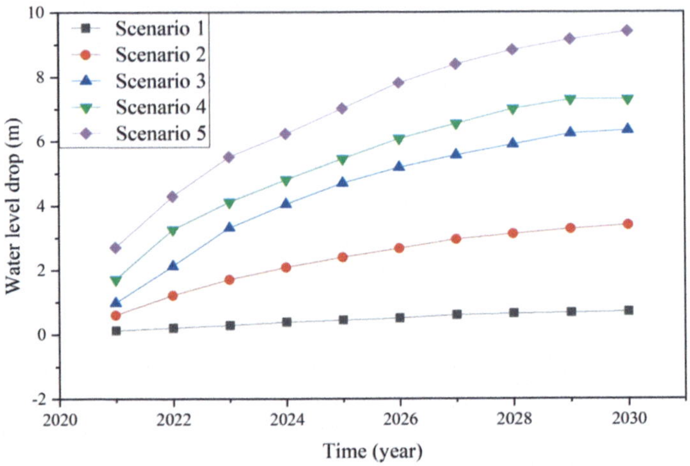

Fig. 6 Groundwater drawdown under different scenarios

This indicates that the maximum allowable extraction has not yet been reached and the extraction rate may be further increased.

Analysis of the predictive results for Scenarios 4 and 5 reveals that when the artificial extraction volume reaches 1616 $\times$ 10^4 m^3/year (Scenario 4), the groundwater level decrease reaches 7.30 m (Fig. 6), which is quite close to the maximum allowable drawdown in this area, and the groundwater level stabilizes in 2030 without continuous groundwater level decline. It can be assumed that the maximum allowable decrease has been reached. If the extraction volume continues to increase beyond this point, as in Scenario 5, the groundwater level decrease will exceed the maximum allowable decrease. The rate of groundwater level decrease in Scenario 5 is significantly faster than the other scenarios. Long-term groundwater extraction under Scenario 5 could result in a continuous decline in the groundwater level, leading to ecological and environmental problems such as water quality deterioration, well depletion, and vegetation reduction or degradation.

After comparing the groundwater level contours (Fig. 7) for these five extraction scenarios, it is found that with the incremental increase of the extraction volume, the water level shows a clear downward trend. This is reflected by the fact that larger groundwater abstraction procures higher drawdown. It is also worth noting that the water level at the boundary of the study area decreases more rapidly compared to the middle area. This is because near the surrounding boundary, the difference in landforms and lithology of surrounding bedrocks and alluvial sediments within the basin produce large hydraulic gradient. In addition, the surrounding bedrock is less abundant in groundwater resources than the alluvial sediments in the basin. Affected by landforms, lithology and water abundance potential all together make the regions along the boundary sensitive to groundwater extraction.

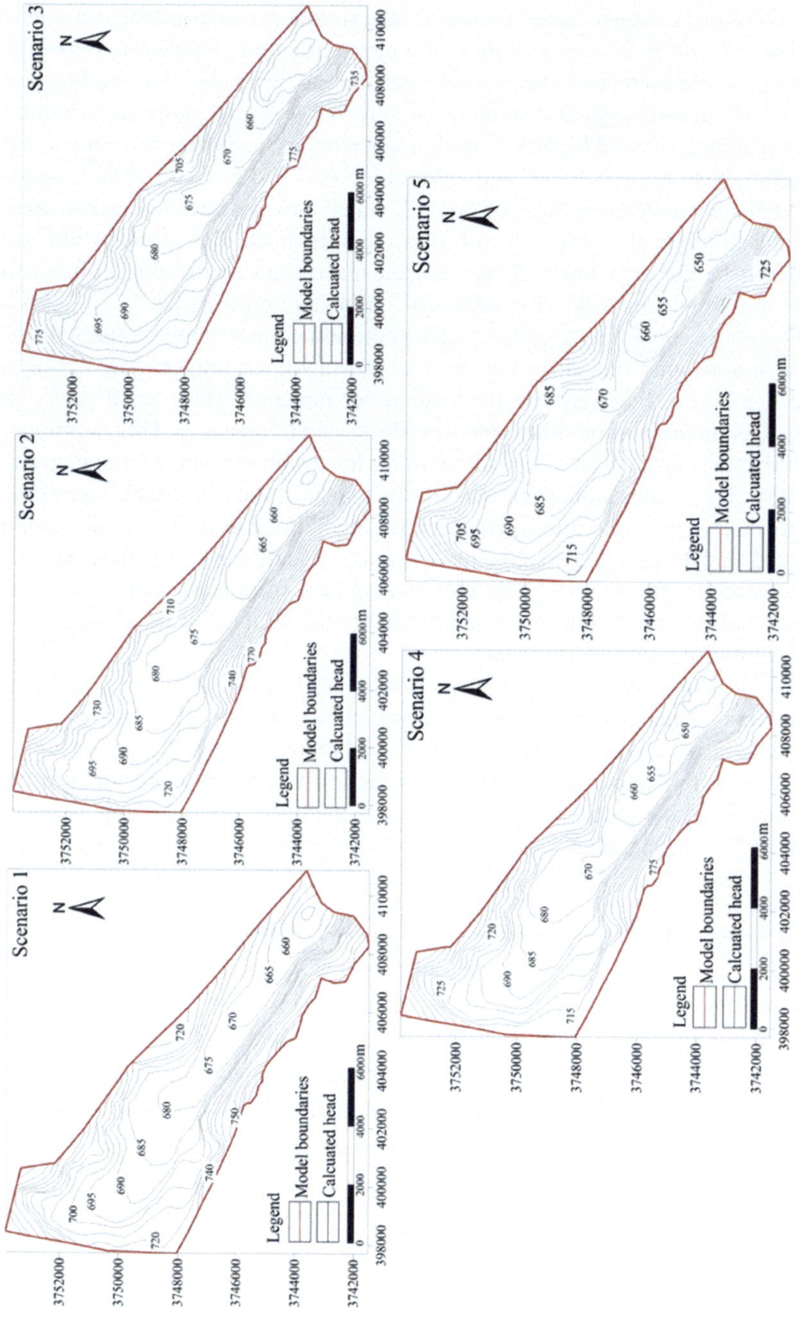

Fig. 7 Groundwater levels under different scenarios in 2030

A new water budget based on the modeling results of Scenario 4 is shown in Table 5. The comparison of Tables 3, 4 and 5 suggests that the natural groundwater discharge conditions change when the artificial extraction is expanded, and a new dynamic equilibrium is established under the new extraction conditions. When the extraction volume is increased, the groundwater storage decreases, thus leading to a decrease in natural discharge and an increase in recharge. In the study area, rainfall infiltration recharge showed highest variation when the groundwater is abstracted at the maximum extractable volume, with a variation of 336.12×10^4 m^3/a. Considering the inter-annual heterogeneity of rainfall, which may not be able to adequately replenish the groundwater in dry years, it is recommended that groundwater use quotas should be set at a lower level than the maximum extractable volume to coordinate the groundwater development, utilization, and protection.

Based on above modeling results, it can be determined that the maximum allowable extraction volume that does not produce a drawdown bigger than 7.5 m is 1616×10^4 m^3/year. For easy use for local water manager, 1600×10^4 m^3/year is suggested as the maximum allowable extraction rate in this area. This maximum extraction volume can be used as a benchmark for local water resource managers and policymakers. It provides a quantitative limit that ensures the sustainable use of the groundwater resource while minimizing environmental impacts. This value should be integrated into local water management strategies, extraction permissions, and long-term planning. It can also guide the development of policies and regulations related to groundwater extraction, helping to balance the needs of water consumption with the health of the local ecosystem.

However, it should be noted that these results are based on current conditions and assumptions. Changes in climate, land use, population, or other factors could alter the dynamics of groundwater flow and availability, potentially necessitating a reevaluation of this maximum allowable extraction volume. Regular monitoring and reevaluation of the groundwater situation are, therefore, crucial components of sustainable water management.

Table 5 Groundwater budget results based on modeling results of scenario 4

Recharge	($\times 10^4$ m^3/a)	Discharge	($\times 10^4$ m^3/a)
Rainfall infiltration	1324.20	Evaporation	110.95
Irrigation water infiltration	421.42	Lateral outflow	43.28
Lateral outflow	896.20	River discharge	1721.35
River percolation	445.98	Artificial extraction	1616.83
Total recharge	3087.80	Total discharge	3492.41

4.2 *Suggestions on Water Resources Development and Utilization*

The Shangzhou Basin's economy is relatively underdeveloped, with agriculture and forestry dominating its industrial structure which creates substantial demand for water resources. However, the current per capita water resources in the Shangzhou Basin fall below the internationally recognized absolute water shortage line of 500 m^3 (Esterhuizen et al. 2019), indicating a severe water resource deficit. Addressing this water shortage is a pressing issue.

In fact, Shangzhou Basin has a large total volume of water resources, but their development and utilization degree are low, thereby constraining local economic growth. Rational use and development of these water resources are crucial for local economic development and ecological protection (Yan et al. 2016). Hence, the following recommendations are proposed for the development and utilization of water resources in the Shangzhou Basin:

Increase Groundwater Resources Extraction: Despite the large total volume of water resources in the Shangzhou Basin, the current per capita water resources for local residents are limited. The actual groundwater development in the study area is low, indicating great potential for further exploitation. Therefore, a suitable expansion of groundwater extraction should be considered. Alongside increasing water supply from groundwater, efforts should be made to increase water use efficiency. Given the significant agricultural water demand in the Shangzhou Basin, the implementation of water-saving irrigation and changes in irrigation methods should be promoted to reduce agricultural water consumption (Wu et al. 2018). Moreover, with rainfall recharge accounting for 38.2% of total recharge, improving the rainwater collection system and utilization rate could increase water resource availability (Li et al. 2018, 2024).

Strengthen Groundwater Resource Monitoring: Regular groundwater monitoring system should be strengthened by setting up more groundwater monitoring wells and determining suitable monitoring frequencies. Especially, groundwater level monitoring frequency should be determined by considering the climate variations within a year, because rainfall is the primary source of groundwater recharge in the study area, groundwater levels fluctuate between wet season and the dry season. It is recommended to set a warning line, and groundwater extraction should be reduced or suspended when water levels in the dry seasons approach or reach the warning line. This will prevent over-exploitation of groundwater in the dry seasons.

Improve Groundwater Monitoring Infrastructure: While the groundwater flow model established in this study provides some insight into the groundwater conditions in the study area, its accuracy is limited by the availability of data from only three water level monitoring wells. Future work should consider setting up additional groundwater monitoring stations in areas with dense population distribution and high groundwater extraction. This will provide more comprehensive and detailed groundwater data, offering a more scientific and reliable basis for future water resources planning and management.

5 Conclusions

In this chapter, a numerical model was set up to determine the maximum allowable groundwater extraction rate in the Shangzhou Basin, aiming at providing scientific basis for sustainable groundwater management in this area. The following conclusions can be summarized:

(1) The model results demonstrate that the total groundwater recharge in the study area amounts to 2588.06×10^4 m^3/year, while the total discharge is 2719.52×10^4 m^3/year. This implies that the groundwater in the study area is essentially in a balanced state. Artificial extraction accounts for only 15.65% of the total discharge, with rivers being the primary form of groundwater discharge, accounting for 76.8%. These statistics reflect the low groundwater utilization rate in the Shangzhou Basin area.
(2) In alignment with the local actual situation, the prediction results of five extraction schemes indicate that the study area reaches the maximum allowable extraction amount when the extraction rate is set at 1614×10^4 m^3/year. At this point, the groundwater level almost reaches the maximum allowable drawdown. This suggests that the available extraction amount of groundwater in the study area can be set at 1600×10^4 m^3/year.
(3) The current groundwater development degree in the Shangzhou Basin is low, thus a suitable amount of groundwater, which does not exceed the maximum allowable groundwater extraction rate, can be used to meet the water demand in the Shangzhou Basin. However, a regular groundwater monitoring system should be enhanced by setting up more monitoring wells and suitable monitoring frequencies, and fully taking into account the climate variations within a year.

Acknowledgements This study received financial support from the National Natural Science Foundation of China (42272302 and 41761144059), the National Key Research and Development Program of China (2023YFC3706901) and the Qinchuangyuan "Scientist + Engineer" Team Development Program of the Shaanxi Provincial Department of Science and Technology (2022KXJ-005). The authors would like to express their gratitude towards the help rendered by Dr. Xiaodong He and Prof. Peiyue Li during field investigation and data analysis.

References

Ahmed I, Umar R (2009) Groundwater flow modelling of Yamuna-Krishni interstream, a part of central Ganga Plain Uttar Pradesh. J Earth Syst Sci 118:507–523

Akter A, Ahmed S (2021) Modeling of groundwater level changes in an urban area. Sustain Water Resour Manag 7:1–20

Aliyari F, Bailey RT, Tasdighi A, Dozier A, Arabi M, Zeiler K (2019) Coupled SWAT-MODFLOW model for large-scale mixed agro-urban river basins. Environ Model Softw 115:200–210

Bailey RT, Park S, Bieger K, Arnold JG, Allen PM (2020) Enhancing SWAT+ simulation of groundwater flow and groundwater-surface water interactions using MODFLOW routines. Environ Model Softw 126:104660

Berehanu B, Ayenew T, Azagegn T (2017) Challenges of groundwater flow model calibration using MODFLOW in Ethiopia: with particular emphasis to the Upper Awash River Basin. J Geosci Environ Protect 5:50–66

Bierkens MFP, Wada Y (2019) Non-renewable groundwater use and groundwater depletion: a review. Environ Res Lett 14:063002. https://doi.org/10.1088/1748-9326/ab1a5f

Biswas H, Melesse AM (2016) Regional scale groundwater flow modeling for Wakel River Basin: a case study of Southern Rajasthan. In: Melesse AM, Abtew W (eds) Landscape dynamics, soils and hydrological processes in varied climates. Springer International Publishing, Cham, pp 385–409

Cedrick MM, Alexander A, Nobert J, Mbudi CNUD (2021) Modeling groundwater flow under chaotic urbanization constraints in Kinshasa Capital Region (D.R. Congo). Phys Chem Earth, Parts A/B/C 124:102985. https://doi.org/10.1016/j.pce.2021.102985

Chakraborty S, Maity PK, Das S (2019) Investigation, simulation, identification and prediction of groundwater levels in coastal areas of Purba Midnapur, India, using MODFLOW. Environ Dev Sustain 22:3805–3837. https://doi.org/10.1007/s10668-019-00344-1

Colins J, Sashikkumar MC, Anas PA, Kirubakaran M (2016) GIS-based assessment of aquifer vulnerability using DRASTIC model: a case study on Kodaganar basin. Earth Sci Res J 20:1–8. https://doi.org/10.15446/esrj.v20n1.52469

De Graaf I, Van Beek L, Wada Y, Bierkens M (2014) Dynamic attribution of global water demand to surface water and groundwater resources: effects of abstractions and return flows on river discharges. Adv Water Resour 64:21–33

Devarajan K, Sindhu G (2015) Application of numerical and empirical models for groundwater level forecasting. Int J Res Eng Technol 4:127–133

Döll P, Müller Schmied H, Schuh C, Portmann FT, Eicker A (2014) Global-scale assessment of groundwater depletion and related groundwater abstractions: combining hydrological modeling with information from well observations and GRACE satellites. Water Resour Res 50:5698–5720. https://doi.org/10.1002/2014wr015595

Esterhuizen M, de Jager L, Jezewski WA (2019) The All Towns Reconciliation Strategies a partnership between water resource management and water services. Water Supply 19:79–87. https://doi.org/10.2166/ws.2018.054

Gebere A, Kawo NS, Karuppannan S, Hordofa AT, Paron P (2021) Numerical modeling of groundwater flow system in the Modjo River catchment, Central Ethiopia. Model Earth Syst Environ 7:2501–2515

Gobinath M, Sahu SK, Ram P, Jhariya DC (2022) Groundwater flow modeling study to assess the sustainability of groundwater resource in and around Bemetara Block, Chhattisgarh, India. J Geol Soc India 98:712–719. https://doi.org/10.1007/s12594-022-2048-4

Gropius M, Dahabiyeh M, Al Hyari M, Brückner F, Lindenmaier F, Vassolo S (2022) Estimation of unrecorded groundwater abstractions in Jordan through regional groundwater modelling. Hydrogeol J 30:1769–1787. https://doi.org/10.1007/s10040-022-02523-3

Hanasaki N, Yoshikawa S, Pokhrel Y, Kanae S (2018) A global hydrological simulation to specify the sources of water used by humans. Hydrol Earth Syst Sci 22:789–817. https://doi.org/10.5194/hess-22-789-2018

Jesudhas CJ, Chidambaram SM, Muniraj K (2017) Isolation of wells contaminated by tannery industries using principal component analysis. Arab J Geosci 10:8. https://doi.org/10.1007/s12517-017-3092-z

Jia X, O'Connor D, Hou D, Jin Y, Li G, Zheng C, Ok YS, Tsang DCW, Luo J (2019) Groundwater depletion and contamination: spatial distribution of groundwater resources sustainability in China. Sci Total Environ 672:551–562. https://doi.org/10.1016/j.scitotenv.2019.03.457

Kumar C (2013) Numerical modelling of ground water flow using MODFLOW. Indian J Sci 2:86–92
Lachaal F, Gana S (2016) Groundwater flow modeling for impact assessment of port dredging works on coastal hydrogeology in the area of Al-Wakrah (Qatar). Model Earth Syst Environ 2:1–15
Lavigne M-A, Nastev M, Lefebvre R (2010) Numerical simulation of groundwater flow in the Chateauguay River aquifers. Can Water Resour J 35:469–486. https://doi.org/10.4296/cwrj3504469
Li P, Qian H (2018) Water resources research to support a sustainable China. Int J Water Resour Dev 34:327–336. https://doi.org/10.1080/07900627.2018.1452723
Li P, He S, Yang N, Xiang G (2018) Groundwater quality assessment for domestic and agricultural purposes in Yan'an City, northwest China: implications to sustainable groundwater quality management on the Loess Plateau. Environ Earth Sci 77(23):775. https://doi.org/10.1007/s12665-018-7968-3
Li CS, Xing LT, Dong YA, Peng YM, Xing XR, Li CL, Zhao ZH (2022) Numerical simulation and protection of the dynamic change of Jinan karst spring based on coupling of seepage and conduit flow. Heliyon 8:13. https://doi.org/10.1016/j.heliyon.2022.e10428
Li P, Kou X, Wang Y, Niu L (2024) Building a more sustainable Chinese Loess Plateau. J Earth Sci 35(1):283–287. https://doi.org/10.1007/s12583-024-1970-3
Maity DK, Mandal S (2019) Identification of groundwater potential zones of the Kumari River basin, India: an RS & GIS based semi-quantitative approach. Environ Dev Sustain 21:1013–1034
Mirlas V, Kulagin V, Ismagulova A, Anker Y (2022) MODFLOW and HYDRUS modeling of groundwater supply prospect assessment for distant pastures in the Aksu River middle reaches. Sustainability 14:16783. https://doi.org/10.3390/su142416783
Mohamed MM, Al-Suwaidi N, Ebraheem A, Al Mulla M (2016) Groundwater modeling as a precursor tool for water resources sustainability in Khatt area, UAE. Environ Earth Sci 75:1–18
Mohanty S, Jha MK, Kumar A, Panda D (2013) Comparative evaluation of numerical model and artificial neural network for simulating groundwater flow in Kathajodi-Surua Inter-basin of Odisha, India. J Hydrol 495:38–51
Narula KK, Gosain A (2013) Modeling hydrology, groundwater recharge and non-point nitrate loadings in the Himalayan Upper Yamuna basin. Sci Total Environ 468:S102–S116
Niswonger RG, Panday S, Ibaraki M (2011) MODFLOW-NWT, a Newton formulation for MODFLOW-2005. US Geological Survey Techniques and Methods 6-A37, 44
Omar PJ, Gaur S, Dwivedi S, Dikshit P (2020) A modular three-dimensional scenario-based numerical modelling of groundwater flow. Water Resour Manag 34:1913–1932
Qu X, Shi L, Qu X, Qiu M, Gao W, Wang J (2021) Evaluation of groundwater resources and exploitation potential: a case from Weifang City of Shandong Province in China. ACS Omega 6:10592–10606
Rezaei M, Sargazi A (2009) The effects of running artificial recharge on the aquifer of Goharkooh Plain. J Earth Sci 19:99–106
Richey AS, Thomas BF, Lo MH, Famiglietti JS, Swenson S, Rodell M (2015) Uncertainty in global groundwater storage estimates in a total groundwater stress framework. Water Resour Res 51:5198–5216. https://doi.org/10.1002/2015wr017351
Sathe SS, Mahanta C (2019) Groundwater flow and arsenic contamination transport modeling for a multi aquifer terrain: assessment and mitigation strategies. J Environ Mmanag 231:166–181
Sathish S, Mohamed M, Klammler H (2018) Regional groundwater flow model for Abu Dhabi Emirate: scenario-based investigation. Environ Earth Sci 77:26. https://doi.org/10.1007/s12665-018-7544-x
Shen S-L, Xu Y-S (2011) Numerical evaluation of land subsidence induced by groundwater pumping in Shanghai. Can Geotech J 48:1378–1392
Singh RM, Shukla P (2016) Groundwater system simulation and management using visual MODFLOW and Arc-SWAT. J Water Resour Hydraul Eng 5:29–35
Surinaidu L, Rao VG, Rao NS, Srinu S (2014) Hydrogeological and groundwater modeling studies to estimate the groundwater inflows into the coal Mines at different mine development stages using MODFLOW, Andhra Pradesh, India. Water Resour Ind 7:49–65

Van Asselen S, Karssenberg D, Stouthamer E (2011) Contribution of peat compaction to relative sea-level rise within Holocene deltas. Geophys Res Lett 38:L24401. https://doi.org/10.1029/2011GL049835

Wada Y, van Beek LPH, van Kempen CM, Reckman JWTM, Vasak S, Bierkens MFP (2010) Global depletion of groundwater resources. Geophys Res Lett 37:L20402. https://doi.org/10.1029/2010gl044571

Wada Y, van Beek LPH, Bierkens MFP (2012) Nonsustainable groundwater sustaining irrigation: a global assessment. Water Resour Res 48:W00L06. https://doi.org/10.1029/2011wr010562

Wada Y, Flörke M, Hanasaki N, Eisner S, Fischer G, Tramberend S, Satoh Y, van Vliet MTH, Yillia P, Ringler C, Burek P, Wiberg D (2016) Modeling global water use for the 21st century: the water futures and solutions (WFaS) initiative and its approaches. Geosci Model Dev 9:175–222. https://doi.org/10.5194/gmd-9-175-2016

Walraevens K, Gebreyohannes Tewolde T, Amare K, Hussein A, Berhane G, Baert R, Ronsse S, Kebede S, Van Hulle L, Deckers J, Martens K, Van Camp M (2015) Water balance components for sustainability assessment of groundwater-dependent agriculture: example of the Mendae Plain (Tigray, Ethiopia). Land Degrad Dev 26:725–736. https://doi.org/10.1002/ldr.2377

Wang Y-Q, Wang Z-F, Cheng W-C (2019) A review on land subsidence caused by groundwater withdrawal in Xi'an, China. Bull Eng Geol Env 78:2851–2863

Wu C, Déry S, Wu W, Liu X, Xiong J, Gao W (2018) A review of water resources utilization and protection in Southwest China. Sci Cold Arid Reg 7:736–746

Yan L, Ma J, He LC, Wang F, Chen T (2016) International comparison of water resources utilization efficiency in China. In: International conference on water resource and hydraulic engineering. Springer International Publishing Ag, Hebei Univ Engn, Handan, PEOPLES R CHINA, pp 41–55. https://doi.org/10.1007/978-3-319-61630-85

Zareian MJ, Eslamian S (2021) Groundwater withdrawal adjustment based on changes in groundwater balance components (a case study: an arid region in central Iran). Arab J Geosci 14:1822. https://doi.org/10.1007/s12517-021-07983-7

Zhang Q, Qian H, Xu P, Liu R, Ke X, Furman A, Shang J (2023) A numerical assessment and prediction for meeting the demand for agricultural water and sustainable development in irrigation area. Remote Sens 15(3):571

Utilization of Green Chemistry Nano-Emulsion in Remediation of Real Contaminated Coastal Sediments by HOCs and Pb to Protect Shallow Aquifer

Moustafa Gamal Snousy, Mohamed Bakry Masod, Ahmed M. Rashad, Esam El-Sayed, Ashraf R. Abouelmagd, Abeer El Shahawy, Mahmoud F. Zawrah, and Jianhua Wu

Abstract The current study synthesized nano-emulsion as a safe environmental rehabilitation alternative for a site contaminated by hydrophobic organic compounds (HOCs) and lead (Pb) contaminants. The proposed nano-emulsion was prepared by a high-energy sonicator using a ternary combination of nonionic surfactant (Tween

M. G. Snousy
Egyptian Petroleum Sector, Petrotrade Co, 1 Anwar Al Moftty St.-Abbas El Akkad St.-Nasr City, Cairo, Egypt

M. B. Masod · A. M. Rashad
Analysis and Evaluation Department, Egyptian Petroleum Research Institute, Ahmed El-Zomer, Nasr City, PO Box 11727, Cairo, Egypt

E. El-Sayed
Faculty of Science, Geology Department, Minia University, PO Box 61519, El-Minia, Egypt

A. R. Abouelmagd
Egyptian Petroleum Sector, Nasr City, 85 Nasr Road, 1st District, Cairo, Egypt

A. E. Shahawy
Department of Civil Engineering, Faculty of Engineering, Suez Canal University, Ismailia, Egypt

M. F. Zawrah
Refractory, Ceramics and Building Materials Department, National Research Center, Dokki, Cairo 12622, Egypt

J. Wu (✉)
School of Water and Environment, Chang'an University, No. 126 Yanta Road, Xi'an 710054, Shaanxi, China
e-mail: wujianhua@chd.edu.cn; wjh2005xy@126.com

Key Laboratory of Subsurface Hydrology and Ecological Effects in Arid Region of the Ministry of Education, Chang'an University, No. 126 Yanta Road, Xi'an 710054, Shaanxi, China

Key Laboratory of Eco-Hydrology and Water Security in Arid and Semi-Arid Regions of the Ministry of Water Resources, Chang'an University, No. 126 Yanta Road, Xi'an 710054, Shaanxi, China

P. Li et al. (eds.), *Sustainable Groundwater and Environment: Challenges and Solutions*, Springer Hydrogeology, https://doi.org/10.1007/978-3-031-82194-3_12

80), sunflower oil, and water. Given that colloidal particles can transport harmful and helpful chemicals, they can be used to remediate polluted areas. A representative sediment sample of 25 kg was collected randomly from a refinery site in northwestern Suez Gulf, Egypt and utilized for the experiments. In this respect, a laboratory model was designed to inject the synthesized nano-emulsion with 1 kg of sediments polluted by HOCs and Pb. One cycle of nano-emulsion flushing (Fc) was applied to the contaminated soil and then rinsed with water six times (Wc). Dynamic light scattering (DLS) was used to analyze the typical nano-emulsion droplet size, while the total petroleum hydrocarbons (TPHs) were determined by the soxhlet extraction process. The total paraffines were determined by gas chromatography (GC) analysis. The results revealed that nearly all nano-emulsion droplets (98.7%) were smaller than 100 nm. The TPHs decreased from 4.54 to 1.86 wt.% after one flushing and sixth water washing run, with TPH removal efficiency (E) equal to 59%. These results indicated that the removal efficiency of Pb was 100%, i.e. the initial concentration of Pb (399.95 mg/kg) became 0.00 mg/kg after one flushing and sixth water washing cycles. The research is potentially meaningful for sediment contamination remediation, thus protecting groundwater from contamination from polluted sediments.

Keywords HOCs · TPHs · Pb · Adsorption · Treatment train · Nano-emulsion

1 Introduction

Soil/sediments and groundwater contamination are a serious problem that can cause many environmental problems for all organisms due to contaminants (Harvey 2011; Malik 2022). One of these contaminants is the oil spill or crude oil pollution. It harshly contaminates the environmental compartments and seriously threatens the ecosystems (Wang et al. 2010). There is a high probability that the oil can seep into bodies of standing water or percolate into the water table. Moreover, petroleum refining is a key industry that produces hydrophobic organic compounds (HOCs) during the refining process (Mohan and Pittman 2006; Bakry et al. 2024). The carcinogenic polycyclic aromatic- or polynuclear aromatic-hydrocarbons (PAHs) are always coupled with stable hydrocarbons during aquatic transport (Notar et al. 2001).

Toxic heavy metals like lead (Pb) coexist as contaminants with HOCs in oil refineries (Wang and Chen 2006). Lead is one of the most common hazardous heavy metals that contaminate the ecosystems and threaten for both humans and animals (Eick et al. 1999; Gardea-Torresdey et al. 2005). Due to its ability to accumulate in biomass and freely move between organisms in the food chain, Pb is dangerous even at low concentrations because it serves no essential biological functions (Mavropoulos et al. 2004). Because of how slowly HOCs and heavy metals biodegrade, they can linger in the environment for a long time.

In Egypt, the northwestern Suez Gulf area is of significant economic and investment, particularly in the northern region, which comprises petrochemical and oil refineries as shown in Fig. 1. The groundwater as the primary water supply in this region is being over-pumped from many wells sunk by locals and investors (Dabash 2008). Faults are the dominant subsurface structural feature in this region. Structural patterns and intensive use of groundwater through many wells are considered important in the dispersion of pollutants in this region. Thus, numerous studies have focused on the environmental monitoring of these regions' pollutants like HOCs and Pb. For example, although the degree of HOCs contamination in the northwest region of Suez Gulf has appeared to increase in the north direction, Ibrahim (Ibrahim 2004) found high concentrations of organic contaminants are ranged from 3487.25 to 5516.52 mg kg^{-1} wet weight at different stations far away from 10 to 60 km southern oil refineries. Northern industrial activities such as oil refining and petrochemical production are likely major anthropogenic causes of oil spills and leakage. This could indicate that significant and ongoing inputs of these pollutants have led to acute HOCs contamination in the sediments. PAHs, which are the carcinogenic constituents of HOCs, have been detected by Ali et al. (2014). They found that PAHs concentrations, especially from six-ring members, are alarmingly high enough to cause lethal toxicity for fish samples collected along the El-Sokhna area. The total concentrations of 16-PAHs ranged from 399.616 to 67631.779 μg/g. Belal (2001) reported that several marine creatures in Suez Bay water recorded the highest levels of heavy metals, such as Pb. Furthermore, Pb concentrations in water, sediments, gastropods, and green algae collected from the western shore (northern Gulf of Suez) were measured by El-Moselhy and Gabal (2004). The places that contain the highest amounts were those near several sources of pollution, like industrial drains. The Pb concentration in sediments was 13.90–28.34 μg/g. The Pb levels in sediments in the northern part of the Suez Gulf were determined by Abd El-Azim (1996) and EL-Moselhy et al. (1999). The recorded values were 18.0–29.2 μg/g and 21.69–91.16 μg/g dry wt, respectively. El-Moselhy and Gabal (2004) recorded Pb concentration in water as 1.84–2.57 μg/l. A variety of Pb concentrations have been recorded in coastal water of the northern part of Suez Gulf; for example, Hamed (1992) 3.00–6.70 μg/l, El-Moselhy (1993) 0.43–3.10 μg/l, Abd El-Azim (1996) 0.10–22.6 μg/l and EL-Moselhy et al. (1999) 1.133–3.448 μg/l. El-Moselhy and Gabal (2004) determined the concentration of Pb in marine organisms; they recorded the value of 6.59–0.49 μg/g wet wt in the digestive gland, head foot, and whole animal gastropod collected from the Suez Bay marine environment. Abd El-Azim (1996) recorded the concentration of Pb as 0.83–1.37 μg Pb/g on a wet wt. basis in the total soft tissues of gastropods collected from Suez Bay. Furthermore, the Pb concentration in green algae collected from the northern part of Suez Gulf was 3.87–23.29 and 2.03–29.17 μg/g dry wt. as detected by El-Moselhy and Gabal (2004) and Abd El-Azim (1996), respectively. Higher Pb concentration has been found in the northwest region of the Suez Gulf groundwater basin, according to the hydrogeochemistry study conducted by Mohamed (2010). The available contents of Pb were in the range of 0.05–0.32 ppm, which are upper the limit (nearly seven times higher) of drinking water standers in groundwater drilled wells, according to the Arab Republic of Egypt (ARE), World Health Organization (WHO), and the United States Environmental Protection Agency (EPA) standards. Hence, Zawrah et al. (2014) and Snousy et al. (2015, 2017, 2018) illustrated the relationship between HOCs and Pb and their dispersion in this region through many models.

It is well known that traditional methods for pollution remediation are ineffective in many situations. Hence, success in pollution remediation requires the development of novel and complex sets of materials and devices. Many potentially dangerous environmental agents can be eliminated at a low cost with the help of nanotechnology. Thus, this work aims to forecast the fate of HOCs and Pb pollutants in soils and subsoils by utilizing smart nano-colloidal as a new remediation method that has not been studied extensively.

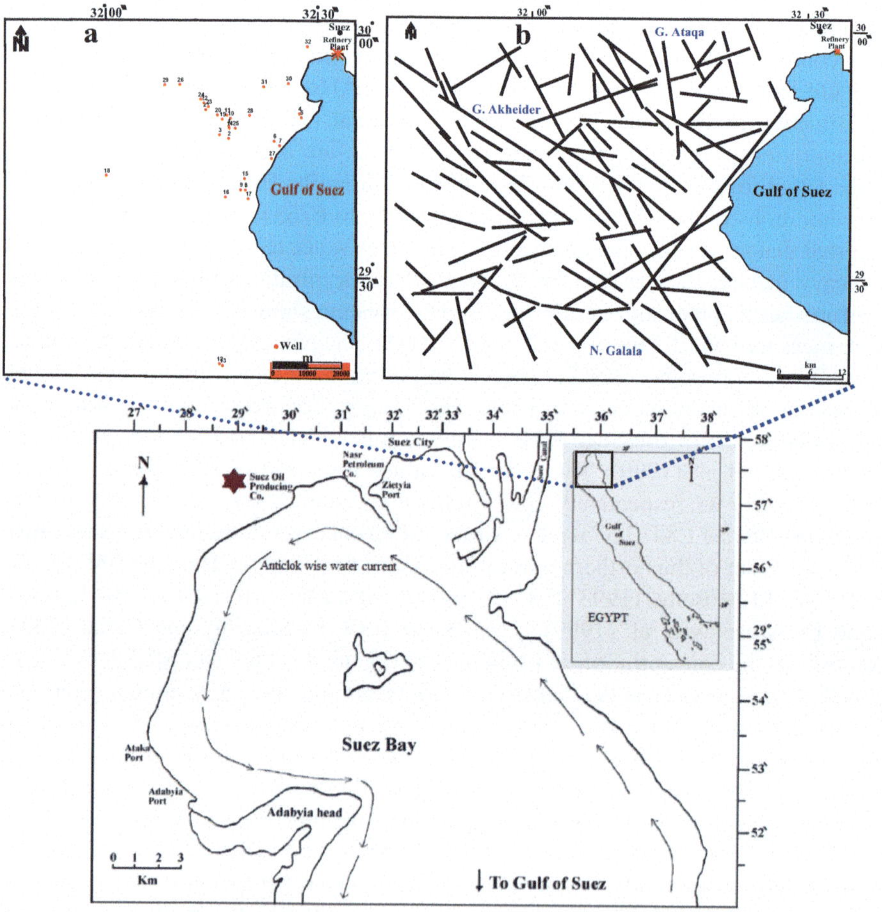

Fig. 1 Location of the petroleum refinery in northwestern Suez Gulf from which soil samples have been extracted (adapted from El-Moselhy and Gabal 2004), **a** configuration of a number of wells in the research site, and **b** lineament map of the area (Usama 2001)

Fig. 2 Sampling processes of tested soil

2 Materials and Methods

2.1 *Materials*

Benzene, methanol, hydrochloric acid, and Tween 80 (Polyoxyethylene (POE) sorbitan monooleate, $C_{24}H_{44}O_6$) chemicals were purchased from El Nasr Pharmaceutical Chemicals Co. (ADWIC), Cairo, Egypt, used without treatment. Vegetable oil, such as sunflower oil, was acquired from a store in Cairo, Egypt.

2.2 *Sampling*

To study the potential toxicity of the contaminants in soil sediments, an American auger was used to collect 25 kg of polluted sediments from various locations covering 2500 m^2 at a depth of 70 cm from a contaminated coastal refinery industrial plant in the northwest corner of the Gulf of Suez in Egypt as shown in Fig. 2. The samples were mixed, homogenized, air-dried in the laboratory and then stored at 25 °C until use (Dong et al. 2010). A representative sample (1 kg) of polluted sediments was taken for analysis and treatment. The particle size distribution of the soil sediment sample was determined by mechanical sieving.

2.3 *Sample Treatment and Analysis*

2.3.1 Preparation of Nano-Emulsion

The nano emulsion was prepared by mixing nonionic surfactant (Tween 80), sunflower oil and water. Water and surfactant were first combined then the oil was added later to prevent spontaneous emulsification (Anton and Vandamme 2009). For good results, the addition occurred by gentle agitation using a magnetic stirrer at approximately 300 rpm for 10 min. The emulsification temperature of this ternary system (tween 80/sunflower oil/water) was kept at room temperature of 25 °C. The concentration of oil and surfactant in the emulsion was kept constant at 5%. Finally, the emulsion was subjected to ultrasound sonicator model VCX ultrasonic processors, American Sonics Inc., at 30% power for 3 min to obtain milky nano-emulsion droplet diameters <100 nm. Dynamic light scattering (DLS) was used to measure the nano emulsion's mean droplet size and dispersion. A Malvern NanoZS instrument photon correlation spectrometer, Malvern, UK) was utilized. A variable-intensity argon laser ($\lambda = 633$ nm) covered the involved size range. At 25 °C and 173° scattering angles, the measurements were taken. The sample was measured by DLS one month after Preparation. This is a very satisfying time for Preparation and transfer to the polluted site. The proposed manufacturing process of nano-emulsion is simple and low-cost.

2.3.2 Treatment Model for Soil Sediment

A lab model was designed to remediate the field samples. The model consisted of a plastic bottle having a dimension of 37 cm (22 cm + 15 cm) height × 15 cm width, as shown in Fig. 3. A porous membrane was located below the model's upper part, allowing the eluent to wash out. The contaminated soil (1 kg) was packed in the model at 6 cm height. The contaminated soil sample was flushed by nano-emulsion for 3 min, followed by successive water washing cycles (six cycles −1 min for each); this step was enhanced by air flushing. The compressor injected the air at a constant pressure of 1 bar with an airflow rate of 6 L min^{-1}.

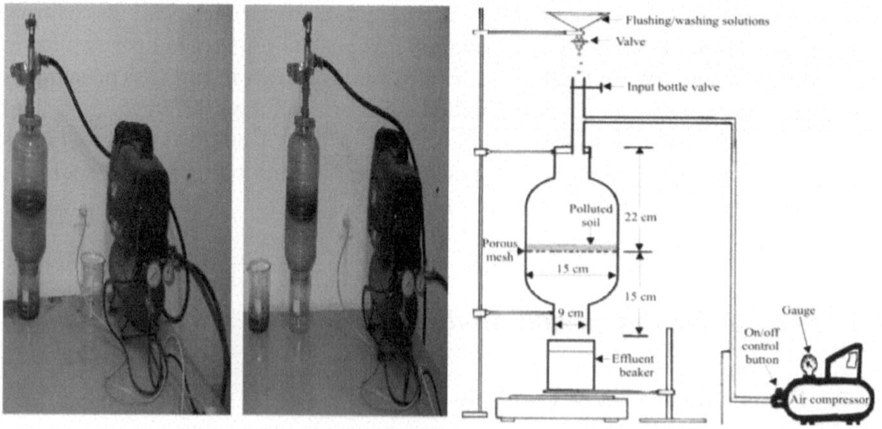

Fig. 3 The treatment laboratory experimental model

2.3.3 Determination of Total Petroleum Hydrocarbons in the Soil Before/After Treatment

The sediment samples were processed in a Soxhlet with benzene and methanol (70:30 vol./vol.) to isolate total petroleum hydrocarbons (TPHs). The extraction was carried out until the solvent in the soxhlet's collecting arm became colourless. The TPHs were reclaimed via distillation and then weighed in a beaker until a steady weight was achieved; the weight differences were used to determine the TPHs.

The removal efficiency (E) of TPH from the treated soil sediments at different stages of flushing by nano-emulsion and washing by water was calculated by the following equation:

$$E = \frac{C_i - C_f}{C_i} \times 100 \tag{1}$$

C_i and C_f are the initial and final concentrations of TPH before and after treatment.

2.3.4 Determination of Total Paraffin Hydrocarbons in Soil Sediment Before Treatment

The paraffin hydrocarbons (aliphatic hydrocarbons) in the sediment were analyzed via gas chromatography (GC) system. Hewlett-Packard's (HP-7890 plus) gas chromatograph equipped with a flame ionization detector (FID) was applied. A chromatograph ChemStation was utilized for data acquisition and processing. The samples were injected by the split mode, the injector and detector temperatures were kept at 300–320 °C, respectively. Nitrogen was utilized as the carrier gas at 2 ml min^{-1}. Fused silica capillary HP-5 (30 m, 0.32 mm o.d., with 250 μ film thickness) was employed to elute the samples by temperature programming fluctuating from 120 to 300 °C at 3 °C min^{-1} as shown in Table 1.

2.3.5 Determination of Lead Before/After Sediments Treatment

Before/after treatment of the samples by nano-emulsion, they were dissolved in 80 ml of 1:1 hydrochloric acid in a Teflon beaker. The solution was poured into a volumetric flask of 100 mL and brought up to volume with deionized water. A blank solution was prepared. The solutions were examined by flame atomic absorption spectrometer (ZEEnit, Analytikjena Co, Germany). The relationship between absorbance and Pb concentration was used to create a calibration curve. The concentrations of the metal were calculated using the obtained calibration curve. The removal efficiency (E) of Pb was calculated by Eq. 1.

Table 1 Paraffin concentrations in the investigated sample

Components	Carbon no	Carbon content wt.%
Pentadecanes	C_{15}	0.009
Hexadecanes	C_{16}	0.051
Heptadecanes	C_{17}	0.098
Octadecanes	C_{18}	0.112
Nonadecanes	C_{19}	0.170
Eicosanes	C_{20}	0.756
Eneicosanes	C_{21}	2.880
Dodeicosanes	C_{22}	5.092
Tricosanes	C_{23}	8.619
Tetraicosanes	C_{24}	8.690
Petaicosanes	C_{25}	8.863
Hexaicosanes	C_{26}	11.617
Heptaicosanes	C_{27}	8.313
Octaicosanes	C_{28}	8.200
Nonaicosanes	C_{29}	7.682
Tricontanes	C_{30}	6.535
Entricontanes	C_{31}	5.805
Dodetricontanes	C_{32}	4.298
Tritricontanes	C_{33}	3.392
Tetratricontanes	C_{34}	3.096
Pentatricontanes	C_{35}	2.318
Hexatricontanes	C_{36}	1.097
Hepatricontanes	C_{37}	0.752
Octatricontanes	C_{38}	0.682
Nonatricontanes	C_{39}	0.502
Tetracontanes	C_{40}	0.37
Total Paraffines %		9.0

3 Results and Discussion

3.1 Characteristics of Soil Sediments

It is well known that soil characteristics influence the remediation processes. Treating soil sediment under successive flushing/washing cycles enhanced by pulsing air injection requires knowledge about the particle size distribution and soil texture (i.e., gravel, sand, silt, and clay contents), as stated by Yang et al. (2013), Melanie et al. (2020) and Singh et al. (2022). Pollutants that may be readily adsorbed onto the surface of soil aggregates are not significantly altered. The large pores in the abundant coarse sand allow the petroleum hydrocarbon mixture to move quickly and contact soil components. Figure 4 shows the particle size distribution and the types of particle present in the polluted soil sediment under investigation as examined by mechanical sieving. It appears from Fig. 4a that this sample consists of gravel (18.72 wt.%), sand (76.18 wt.%), silt, and clay (5.10 wt.%).

On the other hand, the sieve analysis of sand fraction as shown in Fig. 4 indicates higher percentages of medium sand (37 wt.%) and fine sand (27 wt.%) with few fractions from coarse and very fine sand. From these results, it can be concluded that the sediment texture is sandy sediment. Abdel-Moghny et al. (2012) claimed that pollutants' pressurized liquid extraction technology is appropriate for sandy clay, sandy loam, loamy sand, and sand. The concentration of contaminants after being extracted from contaminated soil highly depends on the soil's heterogeneity.

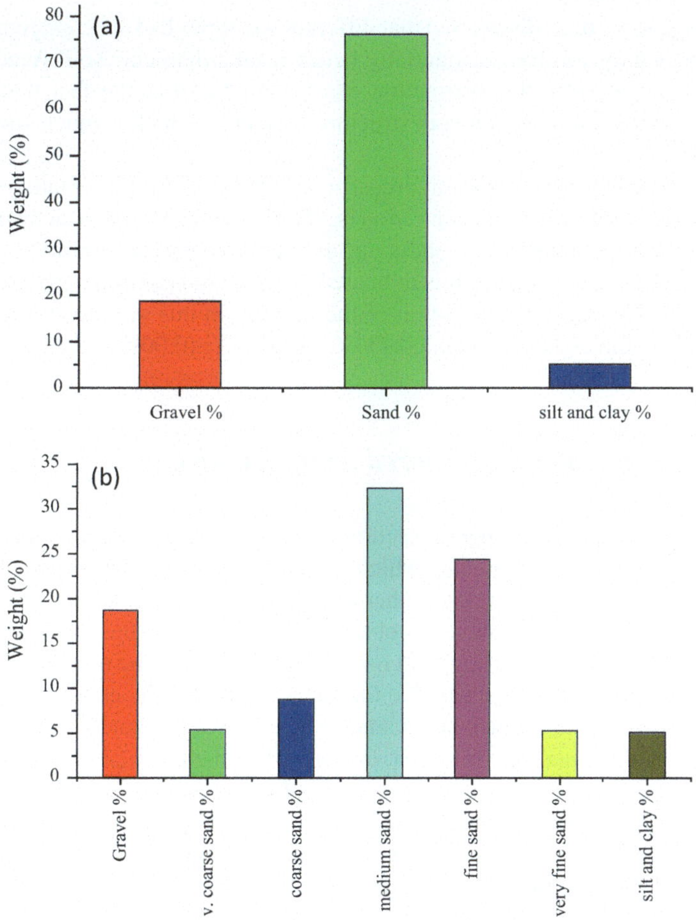

Fig. 4 Particle size distribution of **a** the soil sediment and **b** sand fraction

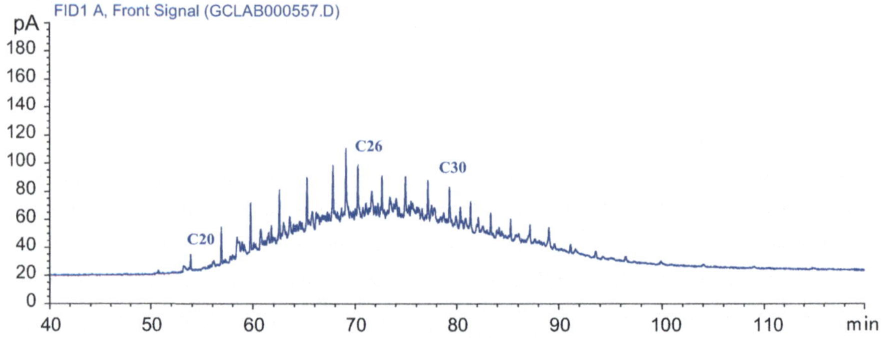

Fig. 5 GC Chromatogram of soil sediment before treatment

Additionally, it has been discovered that different soil types had different remediation efficiencies; sandy soil has substantially better results than clay soil. According to Kujawski et al. (2009), this distinction can be attributed to the fact that the soil particles in sandy soil have a looser structure than clay, which is much stickier and more pliable.

The measured total petroleum-hydrocarbons (TPH) in the representative soil sediment is 4.54%, while the Pb concentration is 399.95 mg/kg. The paraffin hydrocarbon content in the reprehensive soil sediment as measured by gas chromatography is presented in Table 1 and Fig. 5. It is indicated that the total paraffin percentage is 9% of total TPH. The carbon distribution of the paraffin in this sample is between C_{21} and C_{36}. C_{26} is the highest percentage (11.6), while C_{15} (0.009) is the lowest.

3.2 Characteristics of Prepared Nano-Emulsion

As mentioned in the experimental section, the nano-emulsion was prepared using the ultrasound sonicator technique, which produced intense and disruptive forces by ultrasound generators. Due to the higher energy input, nano-emulsion with smaller droplet size and milky appearance is obtained. The nature of the surfactant used to form the emulsion has a major impact on the droplets' size. The nonionic surfactant is usually considered the most suitable for forming fine droplet emulsions (Ostertag et al. 2012). Nonionic surfactant micelles can enhance the rate of dissolution of hydrophobic contaminants, which can, in turn, affect their availability for biodegradation. Tween 80 surfactant was chosen to clean soils contaminated with petroleum hydrocarbons, according to guidelines for surfactant selection (e.g., biodegradability, lower critical micelle concentration (CMC), turbidity, low toxicity, sorption to the soil, and solubility in groundwater. etc.) (Liu et al. 2006). The ideal surfactant system must exhibit high hardness tolerance (low sensitivity to electrolysis), low toxicity, quick biodegradation, low adsorption, increased solubilizing ability for pollutants,

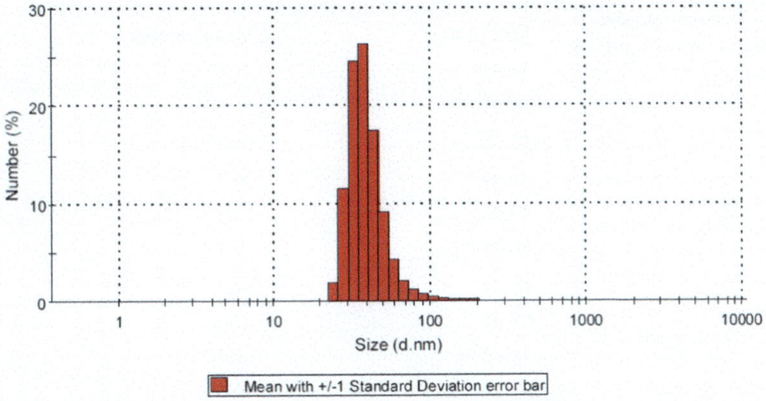

Fig. 6 Particle size distribution of prepared nano-emulsion as measured by dynamic light scattering

and reduced sorption to the soil. These requirements must be met in the aquifer under real circumstances, not only in a lab.

Figure 6 and Table 2 illustrate the particle size distribution of prepared nano-emulsions as measured by dynamic light scattering. It is indicated that 98.7% of droplets' sizes are below 100 nm; the sizes of droplets are extended between 24.36 and 91.28 nm. Only 1.2% of droplets sizes are above 100 nm, and the size of droplets extended from 105.7 nm to 220.2 nm with 0.4% and 0.1%, respectively. Generally, more than half of the droplet sizes (50.9%) lie between 32.67 nm (24.6%) and 37.84 nm (26.3%). This means that the prepared nano-emulsion has small particles and good stability against gravity separation or particle aggregation and creaming, i.e. the diffusion of molecules from small to big droplets as a result of pressure differences is low, and the primary breakdown mechanism of nano-emulsions is Ostwald ripening (Solans and Solé 2012). This is attributed to the steric impact caused by the surfactant molecules (Tween 80), which are adsorbed at the water/oil contact.

3.3 *Effect of Nano-Emulsion on HOCs and Pb Removal*

This part presents the effect of nano-emulsion that contains different extracting agents to desorb and solubilize TPH (4.54%; the amount of TPH in the soil sediments before treatment) and on removing of Pb (399.95 mg/kg; the amount of Pb in the soil sediments before treatment) from the contaminated soil. The laboratory treatment strategies consist of multi-stage extraction, i.e. one flushing cycle by nano-emulsion (Fc) followed by six repeated water washing cycles (Wc), as shown in Fig. 7. This process treats the soil contaminated by petroleum hydrocarbons and heavy metals.

Table 2 Size particle size distribution of nano emulsion

Size (d.nm)	Mean number %
24.36	1.8
28.21	11.5
32.67	24.6
37.84	26.3
43.82	17.5
50.75	9.1
58.77	4.2
68.06	2.0
78.82	1.1
91.28	0.6
105.7	0.4
122.4	0.3
141.8	0.2
164.2	0.1
190.1	0.1
220.2	0.1

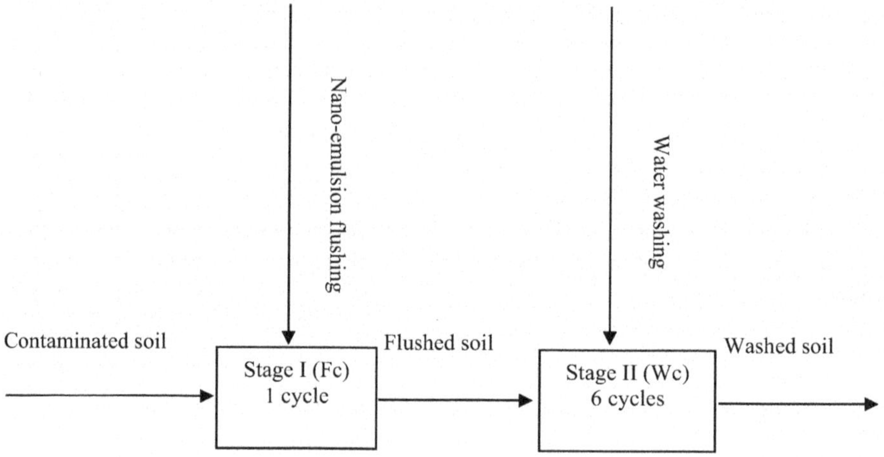

Fig. 7 Flushing/Washing cycles of soil sediments

3.4 *Effect of Nano-Emulsion on TPHs Removal from Contaminated Soil Sediment*

After the injection period of nano emulsion in the soil sediment, the water flood or water drive helps transport it across the target zone (contaminated sediments) and complete the recovery of the injected emulsion with a contaminant (i.e., polishing stage). This means that the nano-emulsion is used as a flusher for contaminated soil treatment while the water is used for washing and polishing. Figure 8 and Table 3 illustrate the total recovered TPH from soil sediment after one flushing cycle by nano-emulsion followed by six washing cycles by water. As mentioned before, the initial concentration of TPH in the soil sediment equals 4.54 wt.%. After the sediment sample was subjected to one flushing cycle by nano emulsion, the TPH concentration was reduced to 3.27 wt.%, i.e. 28% removal efficiency is recorded. After six-water washing cycles, the concentration of TPH gradually declined from 3.13 wt.% after the first water washing run to 1.86 wt.% after the sixth water washing run. The removal efficiency of TPH is increased from 28% after one flushing cycle by nano-emulsion to 59% after the sixth water washing cycle. In a published study by ITRC (2009), it was reported that polishing highly polluted soil was greatly interesting using water; however, it has no efficacy when no solubilized oil is found. The obtained relatively low removal efficiency (59%) might be related to the extremely low solubility of TPH and its slow desorption in aqueous phases. Delgado-Balbuena et al. (2013) hypothesized that the removal effectiveness would be slow after 112 days for soil that has been contaminated with hydrophobic organic contaminants for so long time. Thus, the interaction of hydrophobic organic compounds, such as TPH, with soils involves an initial quick, reversible sorption process, followed by a longer period of interactions. As a result, pollutants become more resistant. The relatively low removal efficiency might be related to the diffusion of TPHs in three-dimensional micropores of soil particles and become deeply sobbed within the soil matrix (Bogan et al. 2003).

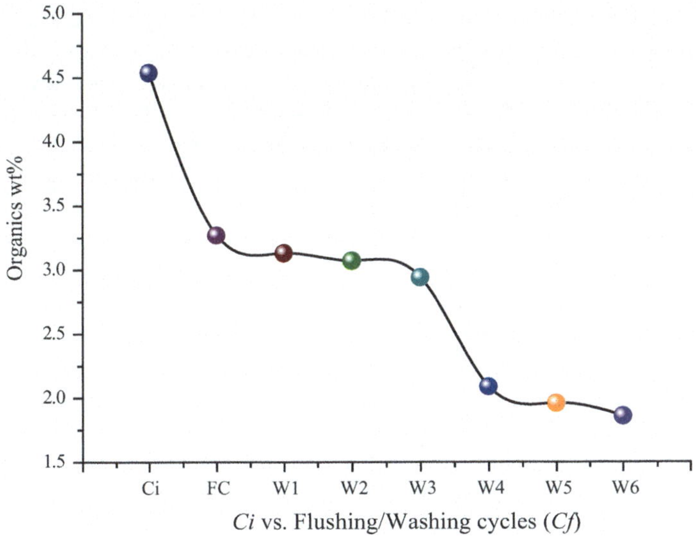

Fig. 8 The relation between treatment trains versus TPH removal

Table 3 The TPHs removal rate during the remediation processes

Organics wt.%	Initial Con. %	Nano-emulsion flushing	1st water washing cycle	2nd water washing cycle	3rd water washing cycle	4th water washing cycle	5th water washing cycle	6th water washing cycle
TPH %	4.54	3.27	3.13	3.07	2.94	2.09	1.96	1.86

Moreover, the micro-shells of sea creatures are considered components of soil grains which contain cracks, gaps, and caves; this leads to TPH detention, i.e. strong adsorption to soil and sediment, thus difficult to remediate. Pennell et al. (1994) found that the smaller soil particles with greater specific surface areas increase organics' solubilization. It has been reported that a larger soil aggregate leads to more release of TPH (Yang et al. 2013). This means that the TPH elimination is decreased with soil coarseness.

3.5 *Effect of Nano-Emulsion on Pb Removal*

A few works have reported the coexistence of hydrocarbons and metals in soil. Therefore, the extraction methods used to remove both pollutants should be selected carefully due to the dissimilarities between heavy metals and hydrocarbons. Agreeing with Samal et al. (2023) the proposed method used nano emulsion to eliminate Pb and TPH from the soil in the present study. Table 4 and Fig. 9 illustrate the change in Pb concentration during applying nano-emulsion. It is indicated that the initial concentration of pb is decreased from 399.95 to 111.95 mg/kg after extraction application by nano-emulsion. This means that 72% of Pb is removed after one flushing cycle by nano-emulsion. The removal efficiency is increased gradually during the later successive water washing cycles and reaches 100% (the Pb concentration is decreased from 111.95 mg/kg at the begging of the first water washing run to 0 at the end of 6th water washing run). This means that the soil rinsing or polishing stages (six water washing cycles) contribute only to the removal of 28% Pb. Mouton et al. (2009) reported that the removal of HOCs is typically attributed to the action of surfactants (ex. Tween 80), while the removal of metals can be accomplished utilizing various techniques. This means that the removal of Pb doesn't depend on the surfactant (Tween 80) addition but depends on the added oil to form the emulsion. Khodadoust et al. (2005) reported the ideal extraction conditions for heavy metals and organic contaminants. They claimed that using surfactants could not remove heavy elements like lead. Therefore, the solvent sunflower oil demonstrated more Pb scavenging capability than the surfactant (Liu et al. 1991; Doong et al. 1998). At low amphiphilic molecule concentrations, the surfactant only exists as monomers that facilitate the mobilization of organic and inorganic contaminants from soil particles. Pb can bind to micelles by complexation or be linked to the ligand (oil) dissolved in the micelles. The high specific area nano emulsion demonstrates high adsorption

Table 4 The change in Pb concentrations during the application of nano emulsion

Treatment cycles	Cycle no	Pb concentration, mg/kg
Initial concentration	C*i*	399.95
Nano-emulsion flushing	FC	111.95
Water washing cycles	W(1)	76.45
	W(2)	44.84
	W(3)	33.91
	W(4)	28.04
	W(5)	16.11
	W(6)	0.00

capacity (Ostertag et al. 2012; Sharma et al. 2013; Kumar et al. 2022). Similarly, the nano-emulsion provides an acidic medium that is appropriate for removing heavy metals since the vegetable oil constituents are oleic acid, simple phenols, phenolic acids, and dicarboxylic acid; these acids are readily biodegradable (Yap et al. 2010).

It is worth mentioning that the proposed concentration of prepared nano-emulsion (5% Tween 80 plus 5% sunflower oil) is suitable for remediation of the soil contaminated by HOCs and Pb owing to the surfactant solubility and the dissolution by sunflower oil. Higher surfactant concentration results in more surfactant sorption on the soil; this leads to a reduction of contaminant removal by introducing competing sites for the pollutant and by obstructing pathways in the soil matrix (Deitsch and Smith 1995; Sun et al. 1995; Bogan et al. 2003; He et al. 2022). The ideal nano emulsion concentration may vary depending on the soil system.

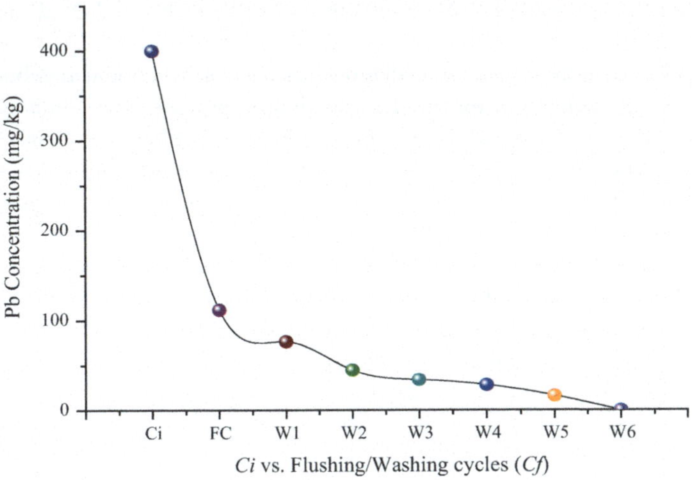

Fig. 9 The relation between treatment trains versus Pb removal

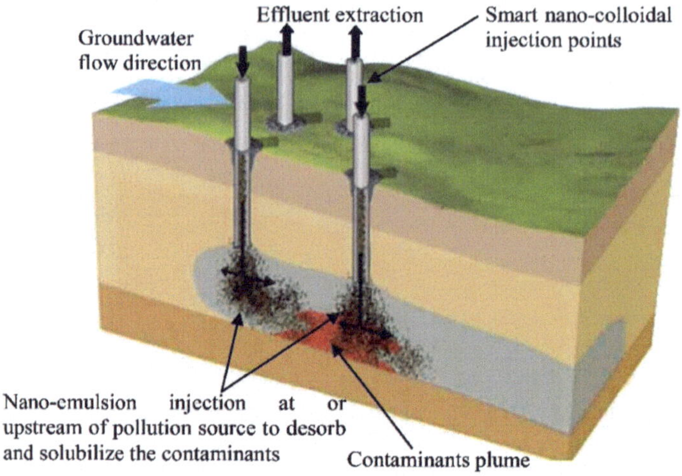

Fig. 10 The proposed scheme of nano emulsion injection in the contaminated soil/sediments

3.6 Mechanism of Nano-Emulsion Injection and Its Reaction with Contaminates

Figure 10 shows the proposed concept of nano-emulsion injected into the soil. Soil contaminated with petroleum hydrocarbons and heavy metals can be effectively remedied using the cleanup technique involving the injection of nano-emulsion into the soil followed by water rinsing for different cycles, i.e. multi-step process using many extracting solutions. Finally, extraction of effluent is conducted under force pressure.

Figure 11 shows the conceptual diagram of the soil treatment mechanism and the contaminants' behaviour during the remediation process. The contaminants are adsorbed on the soil grains and fill the cracks. On the other hand, the nano-emulsion micelles are composed of oil droplets surrounded by water molecules. After flushing the nano-emulsion and washing with water for various cycles, the contaminants are adsorbed and form swelling micelles. As a result of this desorption process, the concentration and flow of the aqueous phase increased during the subsequent water washing. This makes the contaminants released by creating cavities and opening the air passages closed by contaminants lumps (Sun et al. 2013).

In the present study, one flushing cycle by nano-emulsion did not remove most of the TPH and Pb from the contaminated soil. Consequently, additional water-rising cycles are proposed, followed by PLE cycles to increase mechanical separation and contamination clearance. The technique mentioned above appears to be effective in removing heavy metals and organic hydrocarbons.

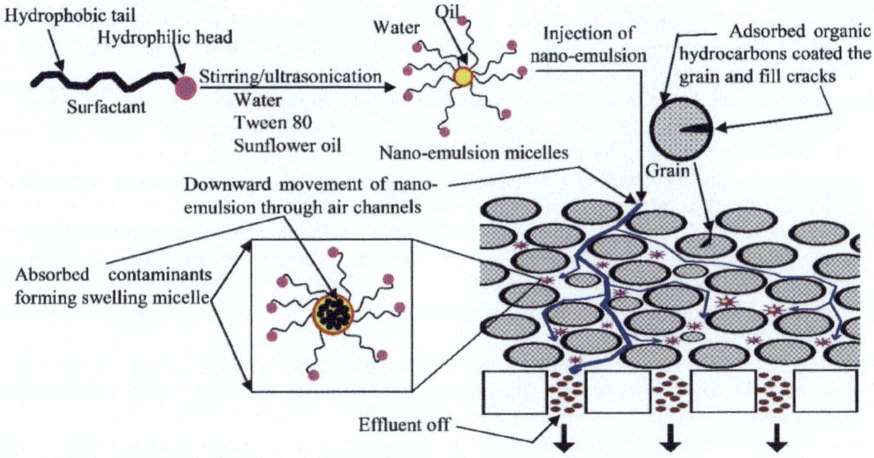

Fig. 11 Conceptual diagram of soil treatment mechanism and the contaminants' behaviour during the remediation process

4 Conclusions

It is important to find suitable techniques rather than traditional remediation methods to remediate the soil from heavy metals and organic hydrocarbons. In the present work, novel treatment sequences were developed to remediate the soil contaminated by hydrocarbons and Pb. Nano-emulsion was used as a desorption and chelating agent to remove the hydrocarbons and Pb from the contaminated soil. The mean droplet size of the prepared nano-emulsion lies below 100 nm for 98.7% of the droplets. The injected nano emulsion reduced the TPHs from 4.54 to 1.86 wt.% with an efficiency equal to 59% after six successive water washing.

On the other hand, the nano-emulsion was successfully used to remove 288 mg/kg Pb from 399.95 mg/kg (initial concentration of Pb) with an efficiency of 72%. The efficiency of Pb removing was reached 100% after the sixth times' water washing. The proposed method of remediation is economical and cost-effective.

Acknowledgements The authors gratefully acknowledge the researchers of the Analysis and Evaluation Department, Egyptian Petroleum Research Institute (EPRI), and the Faculty of Science, Geology Department, Minia University.

References

Abd El-Azim H (1996) Lead dynamics in Suez Bay. M.Sc. Thesis, Faculty of Science, Helwan University

Abdel-Moghny T, Mohamed RSA, El-Sayed E et al (2012) Effect of soil texture on remediation of hydrocarbons-contaminated soil at El-Minia District, Upper Egypt. ISRN Chem Eng 2012:1–13. https://doi.org/10.5402/2012/406598

Ali NA, Ahmed OE, Doheim MM (2014) Evaluation of poly-aromatic hydrocarbons (PAHs) in the aquatic species of Suez Gulf water along El-Sokhna area to the Suez refineries. Environ Monit Assess 186:1261–1269. https://doi.org/10.1007/s10661-013-3455-1

Anton N, Vandamme TF (2009) The universality of low-energy nano-emulsification. Int J Pharm 377:142–147. https://doi.org/10.1016/j.ijpharm.2009.05.014

Bakry M, El-Azab WI, El-Fiqi A, Ebiad MA (2024) Eco-friendly removal of methyl tert-butyl ether from contaminated water using steam and CO2-activated carbons. Int J Environ Sci Technol. https://doi.org/10.1007/s13762-023-05431-5

Belal A (2001) Ecological studies on the polychaetes in the Suez Bay. Ph.D. Thesis, Faculty of Science, Zagazig University

Bogan BW, Trbovic V, Paterek JR (2003) Inclusion of vegetable oils in Fenton's chemistry for remediation of PAH-contaminated soils. Chemosphere 50:15–21. https://doi.org/10.1016/S0045-6535(02)00490-3

Dabash M (2008) Hydrogeological studies on the Nubian sandstone aquifer in El-Bahariya Oasis, Western Desert, Egypt. Ph.D. Thesis, Fac. of Sci., Assiut University

Deitsch JJ, Smith JA (1995) Effect of Triton X-100 on the rate of trichloroethene desorption from soil to water. Environ Sci Technol 29:1069–1080. https://doi.org/10.1021/es00004a029

Delgado-Balbuena L, Romero-Tepal EM, Luna-Guido ML et al (2013) Removal of anthracene from recently contaminated and aged soils. Water Air Soil Pollut 224:1–8. https://doi.org/10.1007/s11270-012-1420-1

Dong L, Zhu Z, Qiu Y, Zhao J (2010) Removal of lead from aqueous solution by hydroxyapatite/magnetite composite adsorbent. Chem Eng J 165:827–834. https://doi.org/10.1016/j.cej.2010.10.027

Doong RA, Wu YW, Lei WG (1998) Surfactant enhanced remediation of cadmium contaminated soils. Water Sci Technol 37:65–71. https://doi.org/10.1016/S0273-1223(98)00235-2

Eick MJ, Peak JD, Pesek JD, Brady PV (1999) Kinetics of lead adsorption/desorption on goethite: residence time effect. Soil Sci. https://doi.org/10.1097/00010694-199901000-00005

El-Moselhy K (1993) Studies on the heavy metals level in some economic fishes in the Suez Gulf. M.Sc. Thesis, Faculty of Science, El-Mansoura University

El-Moselhy KM, Gabal MN (2004) Trace metals in water, sediments and marine organisms from the northern part of the Gulf of Suez, Red Sea. J Mar Syst 46:39–46. https://doi.org/10.1016/j.jmarsys.2003.11.014

EL-Moselhy M, Diab AA, Tolba M, Mohamadein L (1999) Levels of some heavy metals in coastal water, sediment and the limpet Patella Sp. From the Northern Part of the Gulf of Suez (Suez Bay). Egypt J Aquat Biol Fish 3:69–84. https://doi.org/10.21608/ejabf.1999.3418

Gardea-Torresdey JL, Peralta-Videa JR, De La Rosa G, Parsons JG (2005) Phytoremediation of heavy metals and study of the metal coordination by X-ray absorption spectroscopy. Coord Chem Rev 249:1797–1810. https://doi.org/10.1016/j.ccr.2005.01.001

Hamed M (1992) Seawater quality at the northern part of the Gulf of Suez and the nearby area of the Suez Canal. M.Sc. Thesis, Faculty of Science, El-Mansoura University

Harvey RG (2011) Polycyclic aromatic hydrocarbons chemistry and carcinogenicity. Cambridge University Press

He M, Li Z, Chen C, Mei P (2022) Impact of soil types and root exudates on cadmium and petroleum hydrocarbon phytoremediation by Sorghum sudanense, Festuca arundinace, and Lolium perenne. Front Ecol Evol 10:1–14. https://doi.org/10.3389/fevo.2022.1036765

Ibrahim MBM (2004) Levels and sources of polycyclic aromatic hydrocarbons in sediments from the Gulf of Suez. Mar Pollut Bull 49:356–360. https://doi.org/10.1016/j.marpolbul.2004.04.010

ITRC (Interstate Technology & Regulatory Council) (2009) Evaluating natural source zone depletion at sites with LNAPL. LNAPL-1. Interstate Technology & Regulatory Council, LNAPLs Team, Washington, DC

Khodadoust AP, Reddy KR, Maturi K (2005) Effect of different extraction agents on metal and organic contaminant removal from a field soil. J Hazard Mater 117:15–24. https://doi.org/10.1016/j.jhazmat.2004.05.021

Kujawski W, Koter I, Koter S (2009) Membrane-assisted removal of hydrocarbons from contaminated soils-laboratory test results. Desalination 241:218–226. https://doi.org/10.1016/j.desal.2008.02.032

Kumar M, Bolan N, Jasemizad T et al (2022) Mobilization of contaminants: Potential for soil remediation and unintended consequences. Sci Total Environ 839:156373. https://doi.org/10.1016/j.scitotenv.2022.156373

Liu Z, Laha S, Luthy RG (1991) Surfactant solubilization of polycyclic aromatic hydrocarbon compounds in soil-water suspensions. Water Sci Technol 23:475–485. https://doi.org/10.2166/wst.1991.0447

Liu W, Sun D, Li C et al (2006) Formation and stability of paraffin oil-in-water nano-emulsions prepared by the emulsion inversion point method. J Colloid Interface Sci 303:557–563. https://doi.org/10.1016/j.jcis.2006.07.055

Malik JA (2022) Advances in bioremediation and phytoremediation for sustainable soil management principles, monitoring and remediation. Springer, Cham

Mavropoulos E, Rocha NCC, Moreira JC et al (2004) Characterization of phase evolution during lead immobilization by synthetic hydroxyapatite. Mater Charact 53:71–78. https://doi.org/10.1016/j.matchar.2004.08.002

Melanie M, Kosasih FY, Kasmara H et al (2020) Antifeedant activity of Lantana Camara nano suspension prepared by reverse emulsion of ethyl acetate active fraction at various surfactant organic-phase ratio. Biocatal Agric Biotechnol 29:101805. https://doi.org/10.1016/j.bcab.2020.101805

Mohamed H (2010) Hydrochemistry of groundwater and assessment of treated water quality along the Western Side of Gulf Suez—Egypt. M.Sc. Thesis, Fac. Sci., Sohag University

Mohan D, Pittman CU (2006) Activated carbons and low cost adsorbents for remediation of tri- and hexavalent chromium from water. J Hazard Mater 137:762–811. https://doi.org/10.1016/j.jhazmat.2006.06.060

Mouton J, Mercier G, Drogui P, Blais JF (2009) Experimental assessment of an innovative process for simultaneous PAHs and Pb removal from polluted soils. Sci Total Environ 407:5402–5410. https://doi.org/10.1016/j.scitotenv.2009.06.042

Notar M, Leskovšek H, Faganeli J (2001) Composition, distribution and sources of polycyclic aromatic hydrocarbons in sediments of the Gulf of Trieste, Northern Adriatic Sea. Mar Pollut Bull 42:36–44. https://doi.org/10.1016/S0025-326X(00)00092-8

Ostertag F, Weiss J, McClements DJ (2012) Low-energy formation of edible nanoemulsions: factors influencing droplet size produced by emulsion phase inversion. J Colloid Interface Sci 388:95–102. https://doi.org/10.1016/j.jcis.2012.07.089

Pennell KD, Jin M, Abriola LM, Pope GA (1994) Surfactant enhanced remediation of soil columns contaminated by residual tetrachloroethylene. J Contam Hydrol 16:35–53. https://doi.org/10.1016/0169-7722(94)90071-X

Samal SK, Datta SP, Dwivedi BS et al (2023) Phytoextraction of nickel, lead, and chromium from contaminated soil using sunflower, marigold, and spinach: comparison of efficiency and fractionation study. Environ Sci Pollut Res 30:50847–50863. https://doi.org/10.1007/s11356-023-25806-y

Sharma M, Vyas RK, Singh K (2013) A review on reactive adsorption for potential environmental applications. Adsorption 19:161–188. https://doi.org/10.1007/s10450-012-9436-9

Singh V, Waris Z, Saha S et al (2022) Chapter 22—Role of biosurfactants on microbial degradation of oil-contaminated soils. In: Malik JA (ed) Microbes and microbial biotechnology for green remediation. Elsevier, pp 423–441

Snousy MG, Rashad AM, Ebiad MA, Helmy HM, Abd El Bassier MA (2018) Lead and associated micropollutant propagations in the North Suez Gulf, Egypt. Int J Environ Res 12:357–371. https://doi.org/10.1007/s41742-018-0094-y

Snousy MG, Zawrah MF, Abdel-Moghny T, Ebiad MA, Rashad AM, Khalil MM, Abu El Ella EM, El-Sayed E, Tantawy MA (2017) Mobility and fate of pollutants in the aquifer system of the Northwestern Suez Gulf, Egypt. In: de Voogt, P. (eds) Reviews of environmental contamination and toxicology Volume 240, Springer, Cham. https://doi.org/10.1007/398_2016_5

Snousy MG, Zawrah MF, Rashad AM, Ebiad MA, El-Sayed E, Tantawy MA (2015) HPLC evaluation of PAHS polluted soil in coastal petroleum refinery site Northwestern Suez Gulf, Egypt. Res J Environ Toxicol 9(5):251–260.

Solans C, Solé I (2012) Nano-emulsions: formation by low-energy methods. Curr Opin Colloid Interface Sci 17:246–254. https://doi.org/10.1016/j.cocis.2012.07.003

Sun S, Inskeep WP, Boyd SA (1995) Sorption of nonionic organic compounds in soil-water systems containing a micelle-forming surfactant. Environ Sci Technol 29:903–913. https://doi.org/10.1021/es00004a010

Sun M, Luo Y, Teng Y et al (2013) Remediation of polycyclic aromatic hydrocarbon and metal-contaminated soil by successive methyl-β-cyclodextrin-enhanced soil washing-microbial augmentation: a laboratory evaluation. Environ Sci Pollut Res 20:976–986. https://doi.org/10.1007/s11356-012-1064-0

Usama A (2001) Geophysical studies on Wadi Hagoul-wadi Badaa area, North of Ain Sukhna, Gulf of Suez. Ph.D. Thesis, Damietta Fac. Sci., Mansoura Univ

Wang J, Chen C (2006) Biosorption of heavy metals by Saccharomyces cerevisiae: a review. Biotechnol Adv 24:427–451. https://doi.org/10.1016/j.biotechadv.2006.03.001

Wang G, Sun Q, Zhang Y et al (2010) Sorption and regeneration of magnetic exfoliated graphite as a new sorbent for oil pollution. Desalination 263:183–188. https://doi.org/10.1016/j.desal.2010.06.056

Yang M, Yang YS, Du X et al (2013) Fate and transport of petroleum hydrocarbons in vadose zone: compound-specific natural attenuation. Water Air Soil Pollut 224. https://doi.org/10.1007/s11270-013-1439-y

Yap CL, Gan S, Ng HK (2010) Application of vegetable oils in the treatment of polycyclic aromatic hydrocarbons-contaminated soils. J Hazard Mater 177:28–41. https://doi.org/10.1016/j.jhazmat.2009.11.078

Zawrah MF, Ebiad MA, Rashad AM, El-Sayed E, Snousy MG, Tantawy MA (2014) GC estimation of organic hydrocarbons that threaten shallow quaternary sandy aquifer Northwestern Gulf of Suez, Egypt. Environ Monit Assess 186:7579–7591. https://doi.org/10.1007/s10661-014-3949-5

Magnetic Induction on pH, Oxidation-Reduction Potential, Conductivity and Dissolved Oxygen Level of Water and Its Properties: A State-of-the-Art Review

Augustine Chung Wei Yap, Hwang Sheng Lee, and Joo Ling Loo

Abstract For over a century, the technique of using magnetic induction or 'magnetic water treatment' (MWT) has been largely employed in overcoming problematic scaling. MWT has become an alternative water treatment method over conventional electrochemical and chemical methods which are often costly and less eco-friendly. Different effects of magnetic induction on pH, oxidation-reduction potential (ORP), conductivity and dissolved oxygen (DO) concentration have been reported in recent literature. While these effects were regarded as advantageous, the scientific understanding and systematic review of the magnetic induction on water quality improvement are still inadequate in literature. Hence, this article aims to present a comprehensive overview of the MWT and its enhancing effects on water parameters and properties. Firstly, the important principles of MWT which led to water parameters and properties enhancement were summarized. Then, the effects of magnetic induction on pH, ORP, conductivity and DO concentration were systematically reviewed and discussed. This is followed by the analysis of the impacts of electron on water clustering, hydrogen bonding and thermal properties, as well as their correlation with magnetic induction effects on these water properties. Lastly, the applications of MWT and the potential contributions of this technology with greater emphasis on the contamination control, microbes inhibition, fish and plant growth as well as healthcare were highlighted.

A. C. W. Yap · H. S. Lee (✉)
Department of Mechanical and Material Engineering, Lee Kong Chian Faculty of Engineering and Science, Universiti Tunku Abdul Rahman, Jalan Sungai Long, Bandar Sungai Long, 43000 Cheras, Kajang, Selangor, Malaysia
e-mail: leehs@utar.edu.my

H. S. Lee
Centre for Advanced and Sustainable Materials Research, Universiti Tunku Abdul Rahman, Jalan Sungai Long, Bandar Sungai Long, 43000 Cheras, Kajang, Selangor, Malaysia

J. L. Loo
Department of Mechatronics and Biomedical Engineering, Lee Kong Chian Faculty of Engineering and Science, Universiti Tunku Abdul Rahman, Jalan Sungai Long, Bandar Sungai Long, 43000 Cheras, Kajang, Selangor, Malaysia

P. Li et al. (eds.), *Sustainable Groundwater and Environment: Challenges and Solutions*, Springer Hydrogeology, https://doi.org/10.1007/978-3-031-82194-3_13

Keywords Electron in water · Dissolved oxygen concentration · Magnetic water treatment · Water quality parameters · Water properties

1 Introduction

Water, including groundwater and surface water, plays an indispensable role in many aspects of life on earth. Many industries and sectors of the economy, for examples, food processing industries, manufacturing industries, electricity power production plants and agriculture require clean water to operate smoothly (Brini et al. 2017). Clean water is sustained by good water quality. Good water quality is then monitored through many water parameters. The most common water parameters that are enlisted in major water quality control and monitoring system can be categorized as physical, chemical and biological indicators. The physical and chemical indicators include pH, turbidity, dissolved oxygen, biochemical oxygen demand, total dissolved solids and suspended solids (Uddin et al. 2021) whereas bioindicators include microbiological parameters of coliforms, *Escherichia coli* and other pathogens. Table 1 shows that these parameters are routinely regulated for wastewater and drinking water treatments by international and local organizations such as the World Health Organization, the Ministry of Health Malaysia and the Department of Environmental Malaysia (World Health Organization 2011; Ministry of Health Malaysia 2016; Department of Environment Malaysia 2019). Among the water quality parameters, pH, conductivity, dissolved oxygen (DO) concentration and oxidation-reduction potential (ORP) are prominent in water quality and environmental monitoring. In water and wastewater treatment, these water parameters are the indicators of water quality and treatment efficiency.

pH and ORP involve in important water and wastewater treatment processes, for examples, acid-base and neutralization reaction, coagulation, disinfection and corrosion control (American Public Health Association et al. 2017). Conductivity measurement of water provides insight into the concentration of ions and the ability

Table 1 Drinking water quality parameters and their standard limits as regulated by Malaysian authorities and World Health Organization (WHO)

Parameters	Standard limits	
	Malaysia	WHO
pH	6.5–9.0	6.5–8.5
Turbidity	5 NTU	Not stated
Total coliform	Not detected/100 mL sample	Not detected/100 mL sample
Escherichia coli or Thermotolerant coliform	Absent/100 mL sample	Not detected/100 mL sample
Conductivity	1000 μS/cm	Not stated
Dissolved oxygen	7 mg/L	Not stated

of water to carry electric current due to the mobility of these ions, which is a useful indicator of water's purity (American Public Health Association et al. 2017; Omer 2020). Then, the extent of water pollution caused by chemical and biological contaminants can be revealed by chemical and biochemical oxygen demands, which are calculated based on DO concentration. This shows that fundamentally, the overall DO level of water is influenced by the physical, chemical and biological activities in water, by which they are all inter-correlated. (American Public Health Association et al. 2017).

Regrettably, rapid global population growth, as well as the urbanization and industrialization in this modern era have impacted water quality. Following the rise in water usage which amounts to a maximum of 70% of available freshwater, a large amount of inadequately treated water and wastewater are generated at the same time (Santhosh et al. 2016). This increases the demand for good quality water sources at all times. Conventionally, water quality parameters such as pH, conductivity, ORP and DO concentration can be improved using electrolysis and ionizing radiation methods (Santhosh et al. 2016). In electrolysis, electron exchange redox reactions induce changes to pH, ORP and DO concentration through reduction of O_2 and oxidation of H_2 at the cathode and anode, respectively based on the reactions expressed in Eqs. (1) and (2) (Cheng and Jiang 2015):

$$O_2 + 4H^+ + 4e^- \rightarrow 2H_2O \tag{1}$$

$$H_2 \rightarrow 2H^+ + 2e^- \tag{2}$$

On the contrary, the ionizing radiation method employs radioactive irradiation on water. Generally, it generates primary electron (e^-) upon ionization and through subsequent hydration, aqueous electrons and H_3O^+ are the end products (Eq. 3) (Garrett et al. 2005; Horne et al. 2017). Aqueous electron in water would later form H_2 and OH^- as shown by Ali et al. (2021) (Eq. 4).

$$H_2O \xrightarrow[\text{ionization}]{\text{radiation,}} H_2O^+ + e^- \xrightarrow{\text{hydration}} H_3O^+,\ e^-(aq) \tag{3}$$

$$e^-(aq) + 2H_2O \rightarrow H_2 + 2OH^- \tag{4}$$

However, these techniques use electricity and chemicals that can be costly at times. Additionally, usage of the chemicals could possess environmental and health risks if the wastes are not properly managed and disposed of. Therefore, this could lead to serious environmental drawbacks. In conserving and protecting the environment, there has been a trend towards implementing applications that are simple and non-chemical. Since the last decade, permanent magnets and electromagnetic induction are reportedly used in the remediation and improvement of water quality which includes pH, conductivity, ORP and DO concentration (Yin et al. 2011; Kotb 2013; Hassan and Rahman 2016; Lee et al. 2019; Yap et al. 2019, 2021). In most cases, water was exposed to the magnetic field by flowing in between magnet pairs or electromagnets. This technique is ideal for a facile, safe, environmental-friendly and economical alternative as it does not involve a lot of power and electricity consumption (Tu et al. 2015; Esmaeilnezhad et al. 2017).

With this, several applications in the water industry have incorporated the use of magnetic inductions in water quality control. For example, in scale inhibition, pH could be controlled to the desired level by magnetic induction without alkali pre-treatment and thus reducing the scale precipitation by 48% (Parsons et al. 1997). Hassan and Rahman (2016) also demonstrated that by flowing water through magnetic field, the highest pH and DO increments were from 7.11 to 7.42 and 3.84 to 4.51 mg/L, respectively under 0.1 Tesla with 4 magnetization cycles. Meanwhile, the enhancements of these water parameters proved to be effective on the hatching rate of *Artemia salina* (0.15 Tesla, 4 magnetization cycles) as compared to normal water. These promising outcomes suggest that the magnetic induction method can be considered as a green and sustainable alternative approach to water quality improvement. A thorough comprehension of the magnetic induction on water quality improvement should be established. Thus, in this article, the background and development of the water magnetization technique are first introduced. Then, the key concepts of water magnetization which include its principles, technical considerations, as well as effects on pH, ORP, conductivity and DO are systematically reviewed and discussed. This is followed by the analysis of the magnetic induction effect on water properties and a summary of its potential applications. Lastly, implications and prospects arising from this facile technology are recommended, which can spark innovative ideas and improvements in the near future, especially in the healthcare and medical fields.

2 Background and Development of Water Magnetization

Water magnetization is defined as the technique by which water or fluid is exposed to the magnetic field. It can be achieved by either flowing the water through a channel in between magnetic field or letting water be stagnant within the magnetic field. It is more commonly known as 'magnetic water treatment' (MWT) by researchers specifically in reporting the inhibition of scale formation (Busch and Busch 1997; Gabrielli et al. 2001; Fathi et al. 2006; Mosin and Ignatov 2014; Latva et al. 2016), despite some of the latest works which researched on other aspects such as the impact of magnetic field on water parameters as well as plant and fish growth (Abhay and Chandran 2018; Ahmed and NB Abdul El-Hamed 2020). Therefore, the term MWT can be used to refer particularly to the technique and methodology of water magnetization.

MWT has been adopted to improve the quality of water and wastewater in the industries since the 1950s (Ozair and Gutierrez 2010). In 1894, a device was invented and patented by Harris (1894) which incorporated magnetic induction to separate diamagnetic impurities in water. As opposed to separating paramagnetic substances via attracting to magnetic field, the impurities that were settled in water would be repelled off by magnetic induction and as a result, purified water would be passed out from the device. This invention could be the pioneer in magnetic water softening. Water softening refers to the elimination of scums forming substances in water, such as calcium and magnesium which can contribute to water hardness, either physically or chemically (Akram and Rehman 2018). It is a very significant contribution to the water industry as these scums will potentially lead to salt encrustation and scaling which can decline the efficiencies of pipelines, equipment and machinery by clogging (Gabrielli et al. 2006). About half a century later, another patent was filed in the 1940s (Vermeiren 1948) to treat scaling problems. It involved a magnetization device consisting of permanent magnets in between a pipe to which water flowed. The magnetic field generated by the magnets was perpendicular to the water flow. Subsequent invention such as the one patented by Harcourt and collaborators (1997) further developed the magnetic water softening device using electromagnetic coils instead of permanent magnets. In particular, they also mentioned that their invention can soften water through the improvement of several water parameters such as total hardness reduction and decreasing biological encrustation and algae growth in pipes (Harcourt and Harcourt 1997). Despite the availability of these inventions and patents, evident data are not well-established in these documents which can further substantiate the improvement of water quality by water softening.

In more scholarly literature, other common conventional physical and chemical water softening methods were reported, for examples, ion-exchange and electrically induced (electrochemical) precipitation. Ion-exchange method generally uses resins to substitute dissolved Ca^{2+} and Mg^{2+} cations with Na^{+} cations. The Na^{+} cations are originated from NaCl salt or brine from resin regeneration (Rocha et al. 2021). One disadvantage of this method is the leaching of high amount of sodium into the environment which can pose adverse health effects (Rocha et al. 2021). Meanwhile, electrochemical water softening method is a technique to induce precipitation of scaling salts such as calcium carbonate through electrolysis. Typically, it involves an electrolyser which consists of a pair of electrodes for electrical current flow. Then, calcium carbonate salts are deposited on the cathodic plates which can be eluted, leaving the purified water as the final product. This is more environmentally friendly than the ion exchange method due to lesser chemical additives but requires periodical maintenance (Gabrielli et al. 2006). In Table 2, these methods of water softening are compared with MWT. Fox et al. (2011) reported that electrically induced precipitation and MWT achieved a 50% reduction of scaling compared to those without treatment. It is still evident that MWT is considered cost-effective and economical in terms of capital costs, average annual operation and maintenance costs and average 10-year life cycle costs based on Table 2 (Fox et al. 2011; Overbo and Heger 2019). This is owing to the features of MWT which is simple, facile and does not require costly chemicals, materials and electricity (US Department of Energy 1998). Furthermore, magnetic induction can be achieved using simple magnet assembly or solenoids (Esmaeilnezhad et al. 2017). Therefore, MWT is also considered a sustainable green technology.

Table 2 Average capital, operation, maintenance and life-cycle costs of water treatment technologies (Fox et al. 2011; Overbo and Heger 2019)

Treatment methods	Capital costs (Approximate USD)	Average annual operation and maintenance costs (Approximate USD)	Average 10-year life cycle cost (Approximate USD)
Magnetic water treatment	840	12	945
Ion exchange	2246	186	3844
Electrically induced precipitation	2625	214	4588

As mentioned earlier, MWT was largely practiced in the water and wastewater treatment industry for scale reduction, especially from the 1940s to the early 2000s. Several works have reported its applications in preventing encrustation in boilers, industrial pipelines, and heat exchangers (Vermeiren 1948; Wang et al. 1997; Gabrielli et al. 2001). The effectiveness of magnetic induction in such applications is often attributed to the alteration of important indicator water parameters, such as pH and conductivity which can influence the scaling process. In a study by Parsons et al. (1997), a 48% reduction of scaling was recorded when pH of media was controlled by magnetic effect in the range of 7.5–8.0. Besides, magnetic effect tended to change pH from a higher range of >9.0 to approximately 8.5 (Alabi et al. 2015). When pH was lower than 9.0, calcium carbonate precipitation was enhanced. This showed that in the case of scaling, magnetic induction played an essential role in pH regulation of the media. Meanwhile, Jiang et al. (2015) reported that the conductivity value of artificial hard water containing $CaCl_2$ and Na_2CO_3 was changed from approximately 980–1090 μS/cm by magnetic field induction. From the finding, the authors confirmed that in the presence of magnetic field, lesser calcium carbonate precipitation was formed which agreed with the increased conductivity values. Additionally, the study could suggest that magnetic induction alters the nucleation process between $CaCl_2$ and Na_2CO_3 during precipitation. In this case, when calcium carbonate precipitation rate in the solution (homogenous nucleation) is higher, the heterogenous nucleation process, which refers to the formation of scales on pipe walls, is lower, thus scaling effect is weaker (Crabtree et al. 1999).

Therefore, considering these MWT potentials, subsequent investigations of magnetic induction on water quality improvement have been extended to other applications in the recent decade from the 2010s to 2020s, for examples, biological wastewater treatment and water quality improvement, apart from the classical anti-scaling studies. The study by Mohammed et al. (2014) demonstrated that magnetic induction combined with conventional absorption by activated carbon in a reactor further increased the removal efficiency of colour, total suspended solids and chemical oxygen demand of palm oil mill effluent by 79%, 98% and 99%, respectively compared to 57, 61 and 68%, without magnetic induction. Wu and Brant (2020) provided an insight into the fundamental investigation of magnetic induction on hard water's physical parameters, such as pH and conductivity. In this case, municipal tap water represented as hard water showed an increase in pH from 7.00 to 7.22 and an increase in conductivity from 100 to 250 μS/cm. The authors related this to the increased hydrogen bonding between water molecules, free protons and hydronium ions. The weakening of hydration shells around ions which increased the internal electrical field of the water eventually caused conductivity increment.

3 Principles of MWT

3.1 Magnetohydrodynamic and the Lorentz Force

Fundamentally, the principle of MWT is attributed to the concept of 'magnetohydrodynamic'. 'Magnetohydrodynamic' theory suggests that when a fluid containing charged particles or ions, such as a dilute electrolyte, flows through a transverse magnetic field, current will be generated with respect to velocity, v and magnetic induction, B. The magnetic field is always orthogonal to the flow of fluid (Al-Habahbeh et al. 2016). Similarly, the Lorentz force theory can be used to express magnetohydrodynamic, which is commonly referenced to explain the magnetic effect on liquid characteristics. As suggested by Silva et al. (2015), Lorentz force must involve charge of ions, electric field, magnetic field perpendicular to flow, and velocity. An expression of Lorentz Force, F is expressed in Eq. (5):

$$\vec{F} = q\left(\vec{E} + v\vec{B}\right) \tag{5}$$

where q = charge of particle, E = electric field vector, v = velocity and B = magnetic field vector.

When charged particles in a solution flow through a magnetic dipole, the Lorentz force will act on the charged ions, promoting ion polarization. This will suppress ions and counterions to bind together, altering the molecular orientation of ions and precipitation reaction. One of the fundamental examples is the application of scale suppression. For Lorentz force to be valid, there must also be flowing motion, otherwise, the effect is not pronounced when the solution is stagnant in the magnetic field (Silva et al. 2015). Besides, the higher the velocity of charged particles, the greater the Lorentz force acting on them (Zaidi et al. 2014). In short, both 'magnetohydrodynamic' and Lorentz Force theories express the fact that a flowing motion of charged particles through the magnetic effect at a perpendicular angle will result in current generation due to the displacement of charged particles.

When the Lorentz theory is applied, water flow rate or velocity is influential on the magnetization effect of water, which is represented by the force, F acting on water particles based on Eq. (5). Sronsri et al. (2021) verified that water flowing under 6000 G strength at 3.18 m/s yielded optimum improvements in water properties such as electrical conductivity, heat capacity and infrared absorption. In this study, electrical conductivity and infrared absorption increased by 4.12% and 9.75%, respectively while heat capacity decreased by 2.89%. Meanwhile, when circulation time was gradually increasing at a fixed flow rate of 3.18 m/s, the infrared absorption continued to increase from 0.81 arbitrary unit (0 min) up to 0.89 arbitrary unit at 60 min of circulation time when it reached saturation. These findings gave clear evidence that there was a shortened exposure time of water to magnetic induction when flow velocity increased beyond 3.18 m/s, whereby the time for water to absorb the magnetic effect had decreased. Subsequently, this lowered the retention time between water and magnetic induction, which resulted in a lower magnetization efficiency. Therefore, it demonstrated that induction or retention time remains an essential factor in many magnetic induction studies. In another investigation, Jouzdani and Reisi (2020) found that tap water magnetized under a lower flow rate (9 L/min), instead of higher flow rates such as 18 and 27 L/min, contributed to concrete quality improvement, which includes the highest reduction in consumption of superplasticizer (a self-compacting concrete additive to produce high strength concrete) and the highest increment in compressive strength as a whole. Therefore, these findings offered an insight into the improvement of water properties by using sufficiently lower flow rates rather than higher ones when flowing water through magnetic induction.

3.2 Reversed Electric Motor Principle

In the Lorentz Force equation, the correlation for velocity of charged particle, electric field and magnetic field are expressed. This provides insight into the influence of charge velocity, which must be present only when there is a force or motion, on the electric field within the magnetic induction. Fundamentally, the principle that defines this statement is Faraday's Law of Electromagnetic Induction which is founded by Michael Faraday in 1832 (Faraday 1832; Klempner and Kerszenbaum 2018). The law states that current flowing in a conductor produces magnetic field.

In a recent study, a 'reversed electric motor' principle has been postulated by Lee et al. (2019) to explain the changes in water parameters under magnetic induction. It is derived from Faraday's 'electric motor' principle (Faraday 1832; Al-Khalili 2015) which states the working principle of a classic electric motor such that within a magnetic field, a conductor coil will experience a torque (force) generated by the electric current that passes through it, causing the coil to rotate. The 'reversed electric motor' principle is precisely the reversed process of the 'electric motor' principle.

In terms of water parameter changes due to MWT, Lee et al. (2019) proposed that the important components in the reversed electric motor principle include magnetic field, dynamic flow of water, and electron generation. In their work, distilled water's pH, conductivity and DO concentration increased whereas ORP decreased. The findings were then attributed to the reactions that have occurred according to Eqs. (6) and (7) (Perlt et al. 2017):

$$H_2O \leftrightharpoons H^+ + OH^- \tag{6}$$

$$4OH^- \leftrightharpoons 2H_2O + O_2 + 4e^- \quad (7)$$

Water ionizes to form H^+ and OH^- in equilibrium Eq. (6). When the flow of water is perpendicular to magnetic field, current will be induced which indicates electron flow. Electron will be attached to its counterion, H^+ ions via metastable bonding. Consequently, the OH^- concentration will be higher than H^+ concentration, which will then cause pH to increase. In this case, the equilibrium is disturbed. To restore chemical equilibrium, OH^- ions will further ionize to form O_2, electron and water (Eq. (7)) according to Le Chaterlier's principle (Le Chatelier 1884; Dasmeh et al. 2009). Both the well-known Lorentz force theory and reversed electric motor principle must involve motion (flowing) of water to produce the magnetization effect. The study of Lee et al. (2019) particularly delivered the mechanism of generating electrons and increasing electron density in water alongside pH, ORP and DO improvements.

4 Technical Considerations in MWT

Magnetic induction is commonly achieved by using permanent magnets and electromagnets. Based on previous studies reported in the literature, both methods of magnetic induction are still being used. Whichever method is preferred, the magnetic field must be consistent within the passage whereby the flowing fluid can have maximum exposure to the magnetic effect. Thus, the most effective magnetization can then be achieved. In general, two modes of water magnetizations can be classified, which are static and dynamic. They refer to water without flowing, and water flowing within magnetic induction, respectively (Alabi et al. 2015; Lin et al. 2020). For flowing experiments, flow rate plays an important role in achieving optimum magnetization changes on water parameters. In general, most literature sources showed that a lower flow rate corresponded to a longer exposure time of water with magnetic induction, prolonging the retention time of magnetic water treatment. For example, in the recent investigation by Sronsri et al. (2021), its key finding showed that the optimum flow rate of 3.18 m/s corresponded to the highest improvement of water properties and parameters such as infrared absorption property and electrical conductivity increased by 9.75% and 4.12%, respectively. Then, when water circulated at flow velocities beyond 3.18 m/s, which was between 3.5 and 4.0 m/s, the changes of water property and parameter declined. Electrical conductivity decreased by approximately 0.25% while infrared absorbance decreased by approximately 2.22%. A prior study by Kadhim and Al-Rufaye (2018) also attested that the variation percentages of water parameters subjected to magnetic induction were greater when the lowest flow rate of 0.1 L/s was used compared to the highest flow rate of 1.0 L/s. At 0.1 L/s flow rate, pH, ORP and electrical conductivity differed at least by 0.94%, 18.2% and 3.0%, respectively. Conversely, at 1.0 L/s flow rate, the differences descended drastically to 0.11%, 1.18% and 0.70%, respectively.

There are criteria to be taken into consideration when using permanent magnets or electromagnet techniques as they can influence the magnetic induction on water. For instance, when using the electromagnet technique, the thickness of coils, the number of coiling or turns and the power of electric source used can affect the strength of magnetic field (Alabi et al. 2015). On the contrary, the magnetic strength and economic perspective of permanent magnets depend largely on their synthesis material, for examples, ferrite material and rare-earth material permanent magnets (Coey 2020). Furthermore, magnetic field patterns generated by different arrangements of permanent magnets (Gabrielli et al. 2001) as well as geometries and shapes (Yap et al. 2019) can also influence its effect on water. In terms of magnets' arrangement, there are two types, namely inverted and non-inverted (Gabrielli et al. 2001). They are also termed reversed and non-reversed polarity magnets arrangement (Yap et al. 2021). Inverted or reversed polarity means magnets are arranged in such a way that the poles are opposite to each other as shown schematically in Fig. 1, whereas non-inverted or non-reversed polarity refers to magnets arranged with the same poles adjacent to each other (Fig. 2).

Based on Gabrielli et al. (2001)'s explanation, the inverted or reversed polarity arrangement produced less homogenous and uniform magnetic field waveforms in contrast with the opposite form. It is crucial that to achieve a successful and effective magnetization, the direction of water flow and magnetic field lines must be perpendicular to each other.

Fig. 1 Schematic diagram of magnets arranged in the inverted polarity orientation. N denotes North Pole while S denotes South Pole

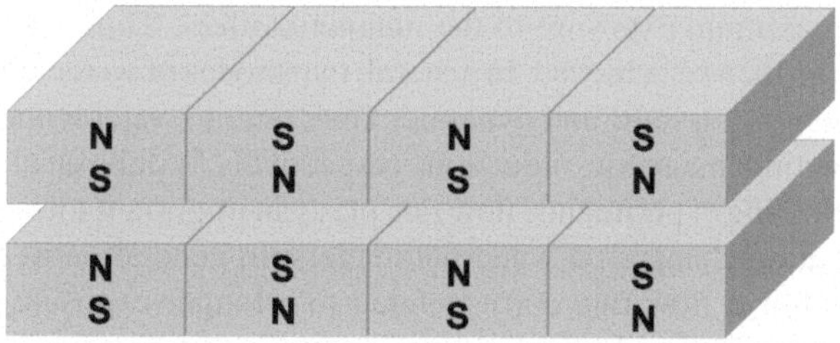

Fig. 2 Schematic diagram of magnets arranged in the non-inverted polarity orientation. N denotes North Pole while S denotes South Pole

5 Effects of Magnetic Induction on Water Parameters

5.1 PH and ORP

pH reflects directly the measurement of the hydrogen ion, H^+ concentration, which is often expressed mathematically as $-\log_{10}[H^+]$. pH value <7, =7 and >7 indicates that water is acidic, neutral and alkaline, respectively. ORP parameter is the measurement of the oxidation and reduction potential. Commonly, it indicates the oxidising and reducing strength of a liquid, especially when it contains dissolved chemical substances which can either be oxidising or reducing agents (Petrov et al. 2021).

The effects of different magnetic induction conditions on pH of water are summarized in Table 3. Recent studies revealed that pH of water could change via magnetic effect. Sales et al. (2015) investigated the magnetic effect on drinking water samples using permanent magnet with 2000 G magnetic strength. Based on their findings, the pH of drinking water increased by 0.3–0.4 after being exposed to magnetic effect. For 60 min, the pH of magnetized drinking water seemed to stabilize within a minor difference of approximately 0.05–0.1. On the contrary, the pH of unmagnetized drinking water experienced huge fluctuations approximately within 0.18–0.2. This phenomenon might suggest that magnetic effect could help in prolonging the pH of water over a certain period while at the same time stabilizing it.

Kotb (2013) investigated the effect of recirculating water through a commercial magnetic water conditioner with 1700 G magnetic strength. They found that pH of water increased by 1.15 (from 7.40 to 8.55) over 820 min until a saturation point was reached whereby the pH of magnetized water was maintained for 10.5 h after magnetization. In another investigation (Hasaani et al. 2015), tap water was flowed through a 28-cm length of magnetic region created by permanent magnets with 2 min retention time. The finding showed that pH of water increased exponentially from 7.1 to approximately 7.9 (by 0.8) when the magnetic strength was 1000–2000 G. Then, pH of water remained plateau when magnetic strength beyond 2000 G up to 6500 G was used. The authors attributed the rise of pH to the formation of OH^- ions due to calcium bicarbonate formation and the presence of alkaline material. In comparison, Kotb (2013)'s and Hasaani et al. (2015)'s findings for the magnetic induction on water's pH were agreeable with each other, suggesting that a relatively low gauss strength permanent magnets within 1000–2000 G is sufficient to raise pH of tap water by 1.0. In the most recent study, Mghaiouini et al. (2020) experimented the magnetic induction generated by an electromagnetic device on several water parameters, including pH. In this case, the electromagnetic field of 40 nT was generated using an Aqua 4 D system operating at an electric voltage of 220 V. Water was flowed through the electromagnetic field continuously between 5 to 20 min under the flow rate between 0.18 and 0.6 L/s. The result showed that pH of tap water increased from 7.69 to 8.25 (by 0.56).

The counterpart of pH is ORP. Both pH and ORP are correlated according to the potentiometry principle (Copeland and Lytle 2014). Just as pH is influenced by the activity of hydrogen ions, ORP is influenced by the electron activity. ORP is

Table 3 Improvement of pH, ORP, conductivity and DO concentration through magnetic induction

Water parameter	Source of magnetic induction	Mode of magnetic induction	Magnetic strength	Experimental conditions	Key results	References
pH	Rare-earth permanent magnet	Flowing	2000 G	60 min exposure time	• pH of magnetized drinking water was 0.3–0.4 units higher than that of unmagnetized drinking water • pH of magnetized drinking water was stabilized over 60 min	Sales et al. (2015)
	Commercial magnetic water conditioner	Flowing	1700 G	820 min exposure time with recirculation	• pH increment by 1.15 (from 7.40 to 8.55); maintained up to 10 h after magnetic effect removal	Kotb (2013)
	Permanent magnets	Flowing	1000–2000 G	2 min contact time; distance of magnetic region = 28 cm	• pH of tap water increased by 0.8 (from 7.1 to 7.9)	Hasaani et al. (2015)
	Electromagnetic device Aqua 4D	Flowing	40 nT	5, 10 and 20 min exposure times; 0.18 L/s and 0.6 L/s	• pH of tap water increased from 7.69 to 8.25 (by 0.56)	Mghaiouini et al. (2020)
	Nd–Fe–B magnetic bead	Stagnant	5000 G	12 h exposure time	• pH of magneztied water increased by 0.5–1.0 compared to untreated water	Yin et al. (2011)
	Permanent magnets	Flowing	1000–1500 G	Non-reversed polarity magnets arrangement; 0.1–0.5 mL/s	• pH of DI water increased from 5.40–5.42 to 5.58–5.62 (+3.5%)	Yap et al. (2021)

(continued)

Table 3 (continued)

Water parameter	Source of magnetic induction	Mode of magnetic induction	Magnetic strength	Experimental conditions	Key results	References
ORP	Nd–Fe–B magnetic bead	Stagnant	2800, 3200 and 5000 G	Between 12 and 24 h of exposure time	• ORP of mineral water, tap water and distilled water decreased by 6 and 7 mV, respectively upon magnetic exposure to 2800 G after 24 h • When exposed to 5000 magnetic effect, the values decreased by 12, 17 and 26 mV, respectively	Yin et al. (2011)
	Permanent magnets	Flowing	1000–1500 G	Non-reversed polarity magnets arrangement; 0.1–0.5 mL/s	• ORP of DI water decreased from 392 to 365 mV (−6.9%). ORP decrement corresponded to 8.0×10^{13} number of electron generated in water	Yap et al. (2021)
Conductivity	Electromagnet system	Static	1030, 1529, 2030 and 2470 G	Between 0 and 120 min of exposure time	• Within the first 30 min, conductivity of tap water increased for all magnetic strengths • The highest change of conductivity achieved (+60 μS/cm; 160–220 μS/cm) when 1529 G magnetic strength was used	Soni et al. (2018)
	Electromagnetic system	Static	800 G	24 h exposure time	• Conductivity of partially degassed DI water increased from 0.62 to 0.68 μS/cm (difference of 0.06 μS/cm)	Lee et al. (2013)
DO concentration	Superconducting magnets	Static	Maximum 10 T	1 h exposure time; 15.0 ± 0.1 °C	• Dissolution of oxygen was enhanced when water was exposed to magnetic field • The optimum result showed that within 1 h of exposure time, DO increased from 12.89 to 17.67 mg/L (4 Tesla)	Hirota et al. (2000)

(continued)

Table 3 (continued)

Water parameter	Source of magnetic induction	Mode of magnetic induction	Magnetic strength	Experimental conditions	Key results	References
	Magnet device	Flowing	Between 1000 and 2000 G	Circulation intervals ranging from 1 to 4 times; 5 h each interval	• DO concentration of water increased by 0.67 mg/L (3.84–4.51 mg/L) after 4 magnetization times (intervals)	Hassan and Rahman (2016)
	Permanent magnets	Flowing	1000–1500 G	Non-reversed polarity magnets arrangement; 0.1–0.5 mL/s	• DO concentration of DI water increased from 7.30 to 7.71 mg/L (+5.6%)	Yap et al. (2021)

conventionally measured using an electrochemical cell in which a redox reaction can occur based on Eq. (8):

$$\text{Oxidant} + ne^{-} \leftrightharpoons \text{Reductant} \tag{8}$$

where, n = number of mole and e^{-} = electron. In the cell, a potential difference is induced from electron exchange between the inert platinum (Pt) electrode and the reference electrode. This potential difference creates a current across the circuit and it is measured as the E value, which is also the ORP value. The ORP electrode performs similarly based on the concept of the electrochemical cell. The measured ORP by the electrode is due to the potential difference created across the two electrodes. When the solution that is being measured has fewer electrons, the ORP electrode loses electron to the solution. This creates an increasing (positive) potential. In contrast, when the solution has more electrons, it donates the electrons to the electrode. This favours a decreasing (negative) potential which was picked up by the electrode sensor. As a result, the ORP value decreases which indicates that the solution is a reducing agent with more electron (Copeland and Lytle 2014).

Yin et al. (2011) used Nd–Fe–B magnetic beads of 2800, 3200, and 5000 G to magnetize distilled water, mineral water and tap water by immersing the magnetic beads into the water of a fixed volume and observing the water parameters' changes up to 12 h. Their results showed that after 24 h of magnetic exposure, the ORP of three types of water decreased with increasing magnetic strength (Table 3). One interesting observation was that the ORP changes of distilled water was the greatest followed by tap water and mineral water. For example, as magnetic strength increased in the order of 0, 2800, 3200 and 5000 G, the ORP values of distilled water decreased which showed 177, 170, 166 and 151 mV, respectively. The ORP range, in this case, was 26 mV. Similarly, the ORP range of mineral water and tap water was 12 mV and 17 mV, respectively. This could provide an insight into the influence of water's purity on the effectiveness of magnetization whereby water with less impurity showed more pronounced ORP change. Concurrently, the pH of three different water after magnetic exposure for up to 12 h was measured. The results concluded that upon magnetic exposure of 5000 G, the pH was generally higher than that without magnetic induction by approximately 0.5–1.0 unit. This observation aligned with the trends of pH and ORP of water after being subjected to magnetic induction as reported by Copeland and Lytle (2014).

The ORP measurement in magnetized water is particularly important as it indicates the presence of electron in water (Mohri 2002; Esmaeilnezhad et al. 2017; Lee et al. 2019). A better correlation of pH, ORP and electron generation was revealed in the work of Yap et al. (2021). DI water was flowed through magnetic induction produced by permanent magnets of 1000–1500 G, arranged in non-reversed orientation and flow rate of 0.1–0.5 mL/s, whereby the magnetization improvement of water's pH and ORP were found to be 3.5% (from 5.40–5.42 to 5.58–5.62) and −6.9% (from 392 to 365 mV), respectively. The ORP decrement also corresponded to an overall increment of 8.0×10^{13} number of electron in water after magnetization.

5.2 *Conductivity*

Conductivity measurement in water indicates the movement of ions and their concentrations as these dissolved ions or particles carry electric charge in aqueous form (Petrov et al. 2021). In Table 3, the conductivity changes due to magnetic induction were summarized. Soni et al. (2018) observed that the conductivity of normal tap water increased with increasing magnetic strengths of 1030, 1529, 2030 and 2470 G generated using an electromagnet system to which the water was exposed at stagnant conditions. Within 30 min of magnetic effect exposure, the conductivity of tap water increased for all magnetic strengths. Out of the four magnetic strengths, the conductivity of tap water increased by the largest magnitude when exposed to 1529 G magnetic strength, from 160 to 220 μS/cm (+37.5%). On the other hand, when water was exposed to 1030, 2030 and 2470 G magnetic induction, the conductivity only changed by +20 to +30 μS/cm (+13 to +18%). In another study by Lee et al. (2013), results indicated that conductivity of non-degassed deionized (DI) water increased from 0.62 to 0.68 μS/cm (difference of 0.06 μS/cm) after exposure to pulsating 800 G electromagnetic field. Conductivity does not precisely indicate the flow of electrons but rather indicates current in the water due to the activity of charged particles. The changes in water's conductivity due to magnetic induction could be attributed to the hydration of ions and hydrogen bonding characteristic of the water-ion interface. Water molecules tend to create hydration shells around ions. Magnetic induction weakens these hydration shells which delocalizes the water molecules with ions while forming stronger hydrogen bonds with other water molecules instead (Chang and Weng 2008; Chibowski et al. 2018). This increases the mobility of ions which explains the conductivity increment after magnetization.

5.3 *DO Concentration*

DO concentration refers to the amount of dissolved molecular oxygen (O_2) in water. In aquatic environments, DO is essential as it is utilized by aquatic lifeforms such as fish, plants and invertebrates. The level of DO in water is also indicative of contamination. For instance, low DO in water body may suggest the presence of chemical pollutants or microorganisms which metabolise the oxygen in water. Several works had reported the influence of magnetic induction on the DO concentration in water, which were tabulated in Table 3. In Hirota et al. (2000)'s work, the authors corroborated a mechanism to discuss the enhancement of DO concentration in degassed water using a superconducting magnet up to 10 Tesla. By supplying pure oxygen to the water sample exposed to magnetic field in an inert environment at 15 °C, the authors deduced that magnetic induction would induce convection in bulk water due to the magnetic force acting on paramagnetic oxygen of magnetized water. The convection induced by magnetic induction was similar to a stirring force and thus enhanced the dissolution of pure oxygen into the water. The optimum result showed that within 1 h of magnetic field exposure time, DO concentration in water increased from 12.89 mg/L (0 G) to 17.67 mg/L (4 Tesla).

Hassan and Rahman (2016) also found that water exposed to 1000 G magnetic induction intensity could raise DO concentration by 0.67 mg/L (from 3.84 to 4.51 mg/L) after four magnetization intervals, which could further improve the hatchability of *Artemia salina* eggs. Meanwhile, they observed the pH of water was also raised by 0.31 using the same intensity. DO improvement in DI water was also reported by Yap et al. (2021), whereby DO concentration was raised from 7.30 to 7.71 mg/L (+0.41 mg/L or +5.6%) under 1000–1500 G magnetic strength when water was flowed through permanent magnets arranged in non-reversed polarity configuration. The DO improvement was correlated with electron increment in water through the oxidation of OH^- ions which was explained based on reversed electric motor principle.

6 Effects of Magnetic Induction on Water Properties

Apart from imparting changes to the water parameters, magnetic induction can also alter water properties. Conventionally, literature in the past mostly attributed these changes to the hydrogen bonding, water clustering behaviours and thermal properties of magnetized water which can be characterized by various analytical techniques. The alteration of these water properties has a correlation with electron density in water. Since one of the key revelations of magnetic induction on water is electron generation (Lee et al. 2019), there could be a correlation between water properties changes and the magnetic induction effect. Therefore, in this section of the review, the effects of magnetic induction on water properties will be discussed accordingly, with emphasis on associating them with electron density in water.

6.1 Influence of Electron on Water Properties

Water is a polar molecule. Thus, at room temperature and thermodynamic equilibrium, water molecules can interact with each other by hydrogen bonding and cluster formation (Su and Wu 2003). Each cluster can consist of approximately 100 water molecules at room temperature (Su and Wu 2003; Ghorbani et al. 2019).

In the earlier section of this article, we have mentioned that electron can be generated in water by ionizing radiation and electrolysis. These methods show that when electron is formed in water by using electricity or high-energy radiation source, water parameters such as pH and oxygen level are changed. Apart from that, electron can influence water properties such as molecular structure characteristics and hydrogen bonding. Chen et al. (2016) precisely worked on the production of electron-doped DI water with reduced hydrogen bonds using gold nanoparticles (AuNPs) that were resonantly illuminated. These electrons that were generated from excited AuNPs were captured by water molecules. As a result, the hydrogen bonds of water were broken and the electrons would then merge with the smaller and fragmented water at a metastable state, which could last for around 8 days as determined via zeta potential measurement. The findings also validated that the produced electron-doped water exhibited several unique properties including reduced heat capacity, faster freezing time, a higher reductive and electron-donating activity. Their findings were summarized in Table 4. The authors later inferred that the electrons generated by activated AuNPs served as energy sources which lowered the energy gap. Thus, several reaction rates of water properties such as thermodynamics, freezing time and reducing properties were increased and enhanced.

On top of that, several investigations were also reported on water clusters fragmentation using electron via the electron-impact ionization (EI) process (Lengyel et al. 2014). Typically, they found that EI of water cluster, $(H_2O)_n$, where $n =$ size range between 10 and 150, yielded fragmented protonated water clusters, in the reported forms of $H^+(H_2O)_n/H_3O^+(H_2O)_n$, and aqueous electron ($e^-_{(aq)}$). EI also led to large fragmentation of water clusters when electron energies are close to the ionization threshold, which was 13 eV (Table 4). In another investigation of EI enhancement via magnetic induction, Qi et al. (2020) showed that 150 G magnetic field could enhance the electron intensity in the EI process due to the Lorentz Force effect, with the optimum electron intensity reached under 150 G magnetic strength. To demonstrate the relationship between magnetic field and electron intensity, electron beam was

Table 4 Effects of electron-doping and electron-impact ionization on water properties

Effects	Types of characterization	Key observations	References
Electron-doping from activated AuNPs	Boiling point and heat capacity	• Decreased boiling point (94.1 °C) compared to normal water (97.3 °C) • Reduced specific heat capacity from 1.000 to 0.945 ± 0.0012 cal/g/°C between 25 and 40 °C	Chen et al. (2016)
	Freezing time	• Shortened freezing time, which was 2.4 min, compared to that of normal water which was 2.6 min	
	Chemical activity of water	• Accelerated reduction of $AgNO_3$ by treated water in the presence of sodium citrate under 80 °C • Chemical reaction with $AgNO_3$ occurred after 1 h, as opposed to normal water that did not show any changes to the solution	
Electron-impact ionization	Water clustering	• Fragmented protonated water clusters, as well as aqueous electron were produced • At 13 eV ionization threshold, large fragmentation was observed	Lengyel et al. (2014)

applied in a chamber under different magnetic fields. They found that the electron beam intensity increased from 0.025 nA (0 G) and eventually reached the peak value of 0.035 nA under 150 G, which gradually declined as magnetic intensity increased beyond 150 G up to 900 G. Thus, this finding could suggest that magnetic induction and its intensity exhibit a certain influence on the electron generation in water.

6.2 Influence of Magnetic Induction on Water Clustering and Hydrogen Bonding

Since Lorentz Force acts on charged particles in aqueous solution due to interaction with magnetic field, it changes the clustering of ions or charged particles, which in turn weakens the solute/solvent interaction and increases ions mobility (Guo et al. 2011; Chibowski et al. 2018). The conductivity of the solution increases as ions' mobility increases. Similarly, when the ion-water interaction (hydration shell thickness around the ion) weakens, water clusters also weaken. This favours the formation of smaller water cluster groups with stronger hydrogen bonds (Al Helal et al. 2018; Chibowski et al. 2018). Besides, Toledo et al. (2008) also demonstrated that magnetic induction changed the viscosity and surface tension of water. They inferred that when water was exposed to an external magnetic field, the 'intra clusters' hydrogen bonds were weakened. As a result, larger clusters are broken, promoting the formation of smaller 'inter clusters' of hydrogen-bonded water that are more stable. This inference was attributed to the increase in viscosity and surface tension of water, which indicated that some of the water activities were enhanced due to the existence of smaller water molecules (Barham et al. 2021).

The –OH asymmetric and symmetric stretching of water which is typically found in the range between 2900 and 3700 cm^{-1} wavenumber is usually analyzed (Durickovic 2016). Pang and Deng (2008) performed Raman scattering analysis on pure water and magnetized water. Magnetized water was prepared by exposing 200 mL of pure water to 4000 G magnetic induction for 30 min. Overall, the absorption strength of magnetized water was 3.5-fold higher than unmagnetized pure water. Besides, a new Raman scattering peak was also observed at 7900 cm^{-1} wavenumber in the 7000–9000 cm^{-1} region. As concluded by the authors, magnetic effect changed the polarization state of water molecules while maintaining the molecular state of water as H_2O. Besides, Raman characterization of water can give insights into the stretching and vibration characteristics of water molecule's OH bond. Pang (2014) deduced that the intensification of Raman bands for magnetized water was attributed to the displacements and excitations of atoms and transition of electron in the atomic scale of water (Dueñas et al. 2020). This could suggest that magnetic effect has altered water molecular structures as well as the electron motions in water when the molecules were excited by the Raman laser. Similarly, a higher Raman OH stretching vibration intensity of magnetized DI water (magnetized under 1000–1500 G, non-reversed polarity magnets arrangement, 0.1–0.5 mL/s flow rate) within the range of 2600–4000 cm^{-1} wavenumber was reported (Yap et al. 2021). These literature aligned with each other in deducing the influence of electron on the Raman spectra of water.

Apart from these characterizations, thermal properties of magnetized water such as specific heat capacity and boiling point were also reported. Specific heat capacity of water is defined as the amount of heat required to raise the temperature of 1 kg of water by 1 Kelvin (Cengel and Boles 2005). Pang et al. (2012) performed a specific heat study on magnetized water (pure water magnetized with 4400 G magnetic induction for 40 min) using a calorimeter. The results showed that on average, the specific heat capacity of magnetized water was lower than the control. Interestingly, when water was magnetized dynamically (flowing), the average specific heat capacity was 0.833 cal/g °C contrary to water magnetized statically (without flowing), which recorded 0.876 cal/g °C. The finding had proven and substantiated the effectiveness of flowing in MWT as stated by both Lorentz Force and reversed electric motor principle. Agreeable results were also reported by Wang et al. (2018) in which tap water magnetized via flowing through 3000 G magnetic induction had in general, the lowest specific heat capacity compared to unmagnetized tap water. For example, at room tempature of 25 °C, the specific heat capacity of water was approximately 4.04 J/g K, followed by 4.06, 4.09 and 4.13 J/g K when magnetized under magnetic strengths of 3000, 2000, 4000 and 1000 G, respectively. Specific heat capacity of unmagnetized tap water at 25 °C was 4.16 J/g K. Meanwhile, the boiling point of unmagnetized and magnetized tap water was analyzed, and results indicated that magnetized water (3000 G) had the lowest boiling point of 97.23 °C compared to that of unmagnetized water, which was 99.29 °C. Most up-to-date work reported by Sronsri et al. (2021) also observed similar results in terms of specific heat capacity, such that the specific heat capacity of magnetized DI water (6000 G) was <4.05 J/g K, compared to unmagnetized water which was near 4.20 J/ g K. By comparison, a relatively high magnetic strength of 6000 G was used in the work of Sronsri et al. (2021) which yielded almost similar range of reading as reported by Wang et al. (2018), in terms of specific heat capacity of <4.05 J/g K. This may be attributed to the differences in experimental methodologies, design and the type of water used in both studies. Nevertheless, based on the thermal properties studies by Pang et al. (2012), Wang et al. (2018) and Sronsi et al. (2021), we could postulate that the decrement of specific heat capacity and boiling point of magnetized water might be contributed by the storage of internal energy, which in this case could be electron induced from magnetic induction. Since electron carries energy, magnetized water with higher electron density tends to require a lower amount of heat energy to raise its temperature, hence leading to lower specific heat capacity (Toledo et al. 2008; Zhou et al. 2010; Chen et al. 2016).

7 Applications of Magnetic Water Treatment

7.1 Contamination Control

The presence of emerging pollutants in water, including trace metals and heavy metal elements is an ongoing hazard to the environment and ecosystem (Kamaruzaman et al. 2017; Razak et al. 2018). Examples of these contaminants are alumnium, copper and zinc. They tend to be persistent in the environment as they are not easily degraded (Selvi et al. 2019). Recent studies disclosed that these contaminants can be removed via MWT. Khater and Ibraheim (2015) investigated the magnetic induction on running tap water through the use of a magnetic liquid modifier. Heavy metals analysis showed that the concentrations of Aluminum (Al), Zinc (Zn), Copper (Cu) and Cobalt (Co) in tap water decreased significantly ($p < 0.05$) by 44.6%, 68.0%, 100% and 40%, respectively. In a separate study (Khater and Ibraheim 2015), the concentrations of Al, Zn, Cu and Co in tap water were determined after subjecting to magnetization of 18 G, with respect to time. The findings showed that the concentrations of Al, Zn, Cu and Co reduced significantly ($p < 0.05$) by 48.8%, 43.1%, 42.9% and 50%, respectively after 7-day exposure to magnetic effect generated from stacked magnets. Several literature works have suggested that the magnetic effect on trace elements removal from water is often attributed to the Lorentz force effect which alters the polarization of the ions in water. In this case, magnetic effect weakens the hydration shell between water molecules and ions. Thus, counterions particles can bind. As a result, precipitation and settling of these particles will be enhanced. For this reason, water can be purified when these particles are eliminated with the water flowing out from the magnetic region (Wahid et al. 2001; Putro and Endarko 2016; Vaskina et al. 2020).

7.2 Microbes Inhibition

One potential application of MWT is the inhibition of pathogenic microorganisms. This is evident in several reported studies. For example, Goyal et al. (2017) investigated the impact of mouth rinse produced from magnetized water on oral microorganism, for an example, *Streptococcus mutans (S. mutans)*. *S. mutans* is a particular species of normal oral flora that can cause cavities (Kaspar et al. 2021). In their 2-week-study, 10 mL of magnetized water (reverse osmosis water kept over magnets for 72 h) was rinsed in the volunteers' mouths for 3 min, twice a day. Results showed that the mean of *S. mutans* in the post-rinsed saliva significantly decreased after each interval week, from 2.07 initially to 1.47 in Week 1 and eventually, 1.13 in Week 2. Bali and Gueddari (2018) also found that magnetic induction on water flowing at 8 L/s through 2 pairs of permanent magnets arranged in non-inverted configurations with each magnetic strength of 0.8 Tesla was able to reduce total coliforms, *Escherichia coli* and fecal streptococci concentration by 2.02, 0.95 and 1.06 log CFU/

100 mL, respectively. This was attributed to the concurrent study of scale reduction in hard water by the authors. The reduction of scale's adherence on the wall surface by magnetic induction is agreeable to the microbial removal since microbes' attachment to pipe scales are often associated with each other (Makris et al. 2014).

Another aspect that is worth noting is the control of pathogens in drinking water via magnetic induction. Selamat et al. (2021) researched the impact of low-frequency electromagnetic field on *E. coli* in river water. An electromagnetic system was designed comprising externally wrapped coils around a column where water would flow through, with an adjustable magnetic field from 2 to 10 mT. A promising result was shown in which after 6 h of magnetic exposure (8 mT), 100% removal of *E.coli* was achieved. The outcome of this study can herald prospective innovations in the future, especially for water treatment industries that face challenges in disinfecting opportunistic and emerging pathogens (Zhang et al. 2021) as well as chlorine-resistant microorganisms which can survive their way into the drinking water treatment process and distribution (Al-Berfkani et al. 2014).

7.3 *Fish and Plant Growth Improvement*

The majority of the studies in the past that involved the applications of magnetized water for the field of agriculture revealed beneficial effects on plants, fishes and livestock. This section particularly highlights the positive effects of using magnetized water on the growth of fish and plants since water is the most essential element that sustains the lives of these two groups of living things. Hassan et al. (2018) investigated the magnetic induction on water quality parameters and the impact on the growth condition of African catfish (*Clarias batrachus*). They found that magnetized water significantly improved the egg hatching rate of *C. batrachus* by 22.59% when exposed to water magnetized under 2000 G field intensity. Furthermore, the water quality parameters, for examples, DO concentration, pH and ammonium (NH_4–N) concentration improved significantly ($p < 0.05$) after being subjected to magnetic effect of up to 2000 G intensity. The pH and DO concentration increased from 8.03 to 8.19; and from 5.92 to 6.33 mg/L, respectively while NH_4–N concentration decreased from 0.20 to 0.16 mg/L. These data were summarized in Table 5. To further substantiate the magnetic induction on fish growth, the experimental findings showed an enhanced hatching rate of fish from 51.68% (Control) to 66.76% (2000 G). Similarly, Irhayyim et al. (2020) investigated a similar technique (0.8 mT or 8 G electromagnetic field, 28-day-recirculating aquaculture system) on the water quality and growth performance of common carp. At such a relatively low gauss field, overall means of water parameters such as pH, DO, ammoniacal nitrogen, nitrate and nitrite did not differ significantly ($p > 0.05$) between magnetized water and control group (before magnetized) (Table 5). But notwithstanding this, fishes in the magnetized water group had a higher specific growth rate per day (1.82%) compared to the control group (1.61%). The finding differs from Hassan et al. (2018)'s work whereby the pH and ammonium were significantly changed after magnetization, and could be

Table 5 Summary of aquaculture water quality data after subjecting to magnetic induction

Source	Hassan et al. (2018)		
Parameters	pH	DO concentration (mg/L)	Ammonium concentration (mg/L)
Before magnetization	8.03	5.92	0.20
After magnetization	8.19	6.33	0.16
Source	Irhayyim et al. (2020)		
Before magnetization	7.28	7.97	0.35
After magnetization	7.29	7.96	0.38

attributed to the difference in magnetic strength used. Therefore, this could probably suggest that a higher magnetic strength, for example, between 1000 and 2000 G is rather feasible in improving water quality parameters in a large scale recirculating aquaculture system.

Per contra, magnetized water could remediate the growth of several plant species under extreme conditions. For example, Sadeghipour (2015) reported that magnetized water produced by magnetizing ordinary water using a pair of magnets (15 cm × 5 cm × 5 cm) could remediate drought injuries experienced by mung beans (*Vigna radiata* L. Wilczek). This study highlighted that magnetized water could further enhance the antioxidant enzymes production in plants that suffered oxidative stress due to overproduction of reactive oxygen species (ROS) which was resulted from drought and water stress. In cell biology, ROS refers to oxidant molecules that have cell damaging properties (Sies et al. 2022). As a result, the antioxidant defense systems of plant were activated and ROS could be scavenged by the increased antioxidant enzymes. In another investigation performed by Abhay and Chandran (2018), the results showed that by irrigating with magnetized water (magnetized using five magnetic rings of approximately 1000 G each), the germination of several seeds was enhanced as compared to that using normal water. Among them, *Vigna radiata* (mung bean) exhibited better growth performance by at least 50% in terms of shoot length, root length and leaf size upon 72 h of irrigation with magnetized water compared to normal water. Meanwhile, the investigation also showed that *V. radiata* seedlings were able to absorb more water when irrigated with magnetized water over time compared to normal water. This suggested that magnetized water improved the penetration and hydration process of the seedlings, which had a direct effect on the overall growth performance of the plant.

7.4 Potential Contributions in Healthcare and Medicine Fields

As far as electron generation and oxygen increment in water due to magnetic induction is concerned, MWT can have promising contributions to the fields of healthcare and medicine. Electron can serve as an antioxidant in treating oxidative stress damage of cells caused by free radicals. Therefore, this technology unveils the potential of naturally generated electron which can be prospective in health sciences and medical applications, particularly in cancer therapy. Most recent reviews have revealed that in radiation cancer cell therapy, low-energy auger electrons are emitted from the decay of radionuclides near targeted cancer cells. These auger electrons are cytotoxic to cancer cells by inducing DNA and cell damage (Rezaee et al. 2017; Ku et al. 2019). This discovery has shed light on the potential application of electron in cancer treatment. Furthermore, cancer development can also be associated with hypoxia, which refers to a state of low oxygen concentration in body tissues (López-Lázaro 2009). One way to overcome this is by supplying adequate oxygen to tissues and cells, which can be suggested by the use of magnetized water with improved electron density and DO level. On top of that, free radical oxidative damage in the body can be potentially caused by the novel Coronavirus SARS-CoV-2 which is responsible for the Coronavirus pandemic since 2020 (Wu 2020; Laforge et al. 2020). Therefore, the use of antioxidant drug such as uridine; and molecular hydrogen (H_2) was proposed to combat the adverse effects of the virus on tissues and organs (Qin et al. 2020; Malaysian Health Technology Assessment Section 2020). As an alternative, naturally generated electron in water through magnetic induction can also act as a natural antioxidant and contribute to the curing of these free radical-induced infectious diseases.

8 Conclusion and Future Perspectives

In the current review, the state-of-the-art concepts of MWT and its major effects and enhancements of water parameters pH, ORP, conductivity and DO concentration were summarized and discussed. MWT can be explained through several principles such as magnetohydrodynamic, Lorentz Force and the reversed electric motor principle. In particular, the reversed electric motor principle explains the generation of electron in water due to the collective improvement of pH, ORP, conductivity and DO concentration. This must be achieved through the action of water flowing through magnetic induction. In terms of magnetic induction on water parameters, literature sources suggested that pH, ORP, conductivity and DO concentration could be influenced by magnetic induction using permanent magnets and electromagnets. The typical magnetic field intensity was between 1000 and 5000 G, where 1000–2000 G magnetic field intensity was commonly used. Furthermore, the effect of magnetic induction on other water properties such as water clustering, hydrogen bonding characteristics and thermal properties were evident, which could be highly attributed to the increased electron density in water. This facile and eco-friendly technology was also acquired in several applications that led to positive impacts on the environment and health, for examples, trace elements removal in water, pathogens disinfection as well as the enhancement of fish and plant growth. As a future recommendation, this technology can herald prospective contributions to the healthcare and medicine fields, particularly in the treatment of free-radical induced diseases. Nevertheless, future advanced research and studies such as kinetic studies, electron profiling analysis and technology upscaling in the industry are very much necessary to better elucidate the complete mechanisms of water quality improvement in the scientific and molecular aspects by magnetic induction.

Acknowledgements The authors acknowledge the funding by the Ministry of Higher Education (MoHE) Malaysia through Fundamental Research Grant Scheme under the grant number FRGS/1/2017/TK10/UTAR/02/5. The authors are also grateful to Universiti Tunku Abdul Rahman (UTAR) and the Institute of Postgraduate Studies and Research (IPSR) for allocating the UTAR Research Fund (UTARRF) Top Up Scheme under Vote Number 6235/L06.

References

Abhay KVK, Chandran K (2018) Comparative efficiency of magnetized (biotron) water and normal tap water on various seeds for their germination and growth. Chem Sci Rev Lett 7:301–309. https://chesci.com/wp-content/uploads/2018/07/V7i25_45_CS042049012_Krishna_301-309.pdf

H Ahmed N, NB Abdul El-Hamed N (2020) Impact of magnetic water treatment technology on water parameters, growth performance and blood parameters of the Nile Tilapia (Oreochromis niloticus). Egypt J Aquat Biol 24:645–655. https://doi.org/10.21608/ejabf.2020.126947

Akram S, Rehman F (2018) Hardness in drinking-water, its sources, its effects on humans and its household treatment. J Chem Appl 4:1–4. https://www.avensonline.org/fulltextarticles/JCAP-2380-5021-04-0009.html

Al Helal A, Soames A, Gubner R, Iglauer S, Barifcani A (2018) Influence of magnetic fields on calcium carbonate scaling in aqueous solutions at 150 °C and 1 bar. J Colloid Interface Science 509:472–484. https://doi.org/10.1016/j.jcis.2017.09.028

Alabi A, Chiesa M, Garlisi C, Palmisano G (2015) Advances in anti-scale magnetic water treatment. Environ Sci Water Res 1:408–425. https://doi.org/10.1039/C5EW00052A

Al-Berfkani MI, Zubair AI, Bayazed H (2014) Assessment of chlorine resistant bacteria and their susceptibility to antibiotic from water distribution system in Duhok province. J Appl Biol Biotechnol 2:10–13. https://doi.org/10.7324/JABB.2014.2603

Al-Habahbeh OM, Al-Saqqa M, Safi M, Khater TA (2016) Review of magnetohydrodynamic pump applications. Alex Eng J 44:1347–1358. https://doi.org/10.1016/j.aej.2016.03.001

Ali I, Imanova GT, Garibov AA, Agayev TN, Jabarov SH, Almalki AS, Alsubaie A (2021) Gamma rays mediated water splitting on nano-ZrO_2 surface: kinetics of molecular hydrogen formation. Radiat Phys Chem 183:109431. https://doi.org/10.1016/j.radphyschem.2021.109431

Al-Khalili J (2015) The birth of the electric machines: a commentary on Faraday (1832) 'Experimental researches in electricity.' Philos Trans R Soc A 373:20140208. https://doi.org/10.1098/rsta.2014.0208

American Public Health Association, American Water Works Association (2017) Water Environment Federation, Standard methods for the examination of water and wastewater

Bali M, Gueddari M (2018) The effect of magnetic treatment on the physico-chemical and microbiological characteristics of hard waters. Sep Sci Technol 53:1405–1411. https://doi.org/10.1080/01496395.2018.1444640

Barham WS, Albiss B, Lataifeh O (2021) Influence of magnetic field treated water on the compressive strength and bond strength of concrete containing silica fume. J Build Eng 33:101544. https://doi.org/10.1016/j.jobe.2020.101544

Brini E, Fennel CJ, Fernandez-Serra M, Hribar-Lee B, Luksic M, Dill KA (2017) How water's properties are encoded in its molecular structure and energies. Chem Rev 117:12385–12414. https://doi.org/10.1021/acs.chemrev.7b00259

Busch KW, Busch MA (1997) Laboratory studies on magnetic water treatment and their relationship to a possible mechanism for scale reduction. Desalination 109:131–148. https://doi.org/10.1016/S0011-9164(97)00059-3

Cengel YA, Boles MA (2005) Thermodynamics: an engineering approach, 5th edn. McGraw-Hill, New York

Chang KT, Weng CI (2008) An investigation into the structure of aqueous NaCl electrolyte solutions under magnetic fields. Comput Mater Sci 43:1048–1055. https://doi.org/10.1016/j.commatsci.2008.02.020

Chen HC, Mai FD, Hwang BJ, Lee MJ, Chen CH, Wang SH, Tsai HY, Yang CP, Liu YC (2016) Creation of electron-doping liquid water with reduced hydrogen bonds. Sci Rep-UK 6:22166. https://doi.org/10.1038/srep22166

Cheng Y, Jiang SP (2015) Advances in electrocatalysts for oxygen evolution reaction of water electrolysis-from metal oxides to carbon nanotubes. Prog Nat Sci-Mater 25:545–553. https://doi.org/10.1016/j.pnsc.2015.11.008

Chibowski E, Szcześ A, Hołysz L (2018) Influence of magnetic field on evaporation rate and surface tension of water. Colloid Interfaces 2:68. https://doi.org/10.3390/colloids2040068

Coey JMD (2020) Perspective and prospects for rare earth permanent magnets. Engineering 6:119–131. https://doi.org/10.1016/j.eng.2018.11.034

Copeland A, Lytle DA (2014) Measuring the oxidation-reduction potential of important oxidants in drinking water. J AWWA 106:E10–E20. https://doi.org/10.5942/jawwa.2014.106.0002

Crabtree M, Eslinger D, Fletcher P, Johnson A, King G (1999) Fighting scale—Removal and prevention. https://www.slb.com/-/media/files/oilfield-review/fighting. Accessed 2 May 2022

Dasmeh P, Searles DJ, Ajloo D, Evans DJ, Williams SR (2009) On violations of Le Chatelier's principle for a temperature change in small systems observed for short times. J Chem Phys 131:12B602. https://doi.org/10.1063/1.3261849

Department of Environment Malaysia (2019) National Water Quality Standards for Malaysia. https://www.doe.gov.my/portalv1/wp-content/uploads/2019/05/Standard-Kualiti-Air-Kebangsaan.pdf. Accessed 25 Jan 2022

Dueñas JA, Weiland C, Núñez MA, Ruiz-Rodriguez FJ (2020) Effect of low intensity static magnetic field on purified water in stationary condition: ultraviolet absorbance and contact angle experimental studies. J Appl Phys 127:133907. https://doi.org/10.1063/1.5132420

Durickovic I (2016) Using Raman spectroscopy for characterization of aqueous media and quantification of species in aqueous solution. In: Stauffer MT (ed) Applications of molecular spectroscopy to current research in the chemical and biological sciences. IntechOpen Limited, London, pp 406–427. https://www.intechopen.com/chapters/51969

Esmaeilnezhad E, Choi HJ, Schaffie M, Gholizadeh M, Ranjbar M (2017) Characteristics and applications of magnetized water as a green technology. J Clean Prod 161:908–921. https://doi.org/10.1016/j.jclepro.2017.05.166

Faraday M (1832) V. Experimental researches in electricity. Philos Trans R Soc (London) 125–162. https://doi.org/10.1098/rstl.1832.0006

Fathi A, Mohamed T, Claude G, Maurin G, Mohamed BA (2006) Effect of a magnetic water treatment on homogeneous and heterogeneous precipitation of calcium carbonate. Wat Res 40:1941–1950. https://doi.org/10.1016/j.watres.2006.03.013

Fox P, Wiest M, Thomure T, Lee W (2011) Evaluation of alternatives to domestic ion exchange water softeners. https://www.geyserenergy.co.uk/wp-content/uploads/Evaluation-of-Alternative-to-Domestic-Ion-Exchange.pdf. Accessed 20 Apr 2022

Gabrielli C, Jaouhari R, Maurin G, Keddam M (2001) Magnetic water treatment for scale prevention. Wat Res 35:3249–3259. https://doi.org/10.1016/S0043-1354(01)00010-0

Gabrielli C, Maurin G, Francy-Chausson H, Thery P, Tran TTM, Tlili M (2006) Electrochemical water softening: principle and application. Desalination 201:150–163. https://doi.org/10.1016/j.desal.2006.02.012

Garrett BC, Dixon DA, Camaioni DM, Chipman DM, Johnson MA, Jonah CD, Kimmel GA, Miller JH, Rescigno TN, Rossky PJ, Xantheas SS (2005) Role of water in electron-initiated processes and radical chemistry: issues and scientific advances. Chem Rev 105:355–390. https://doi.org/10.1021/cr030453x

Ghorbani S, Ghorbani S, Tao Z, De Brito J, Tavakkolizadeh M (2019) Effect of magnetized water on foam stability and compressive strength of foam concrete. Constr Build Mater 197:280–290. https://doi.org/10.1016/j.conbuildmat.2018.11.160

Goyal AK, Rathore AS, Garg M, Mathur R, Sharma M, Khairwa A (2017) Effect of magnetized water mouthrinse on Streptococcus mutans in plaque and saliva in children: an in vivo study. Int J Clin Pediat Dent 10:335. https://doi.org/10.5005/jp-journals-10005-1461

Guo B, Han HB, Chai F (2011) Influence of magnetic field on microstructural and dynamic properties of sodium, magnesium and calcium ions. Trans Nonferrous Metals Soc of China 21:s494–s498. https://doi.org/10.1016/S1003-6326(11)61631-2

Harcourt GA, Harcourt DL (1997) Method and apparatus for magnetically treating a fluid. U.S. Pat. 5,683,586. https://patentimages.storage.googleapis.com/7d/16/17/ffa13a1a9f5160/US5683586.pdf

Harris JT (1894) Method of and apparatus for purifying liquids. U.S. Pat. 531,183A. https://patentimages.storage.googleapis.com/fd/6f/83/5c416055004e74/US531183.pdf

Hasaani AS, Haidi ZL, Rasheed KA (2015) Experimental study of the interaction of magnetic fields with flowing water. Int J Basic Appl Sci 3:1–8. https://www.researchgate.net/profile/Zaid-Hadi-3/publication/311714420_Experimental_Study_of_the_Interaction_of_Magnetic_Fields_with_Flowing_Water/links/5e9db09b4585150839ec4701/Experimental-Study-of-the-Interaction-of-Magnetic-Fields-with-Flowing-Water.pdf

Hassan SM, Rahman RA (2016) Effects of exposure to magnetic field on water properties and hatchability of Artemia salina. ARPN. J Agric Biol Sci 11:416–423. http://www.arpnjournals.org/jabs/research_papers/rp_2016/jabs_1116_829.pdf

Hassan SM, Rahman RA, Kamaruddin RH, Madlul SN (2018) Effect of exposure of African catfish (*Clarias batrachus*) to magnetic field on water properties and egg hatching. Borneo J Marine Sci Aquac 2:54–59. https://doi.org/10.51200/bjomsa.v2i0.1553

Hirota N, Ikezoe Y, Uetake H, Nakagawa J, Kitazawa K (2000) Magnetic field effect on the kinetics of oxygen dissolution into water. Mater Trans JIM 41:970–980. https://doi.org/10.2320/matertrans1989.41.976

Horne GP, Pimblott SM, LaVerne JA (2017) Inhibition of radiolytic molecular hydrogen formation by quenching of excited state water. J Phys Chem B 121:5385–5390. https://doi.org/10.1021/acs.jpcb.7b02775

Irhayyim T, Beliczky G, Havasi M, Bercsényi M (2020) Impacts of magnetic water treatment on water quality, feeding efficiency and growth performance of common carp in integrated recirculating aquaculture systems. J. Central Eur Agric 21:246–255. https://doi.org/10.5513/JCEA01/21.2.2672

Jiang L, Zhang J, Li D (2015) Effects of permanent magnetic field on calcium carbonate scaling of circulating water. Desalin Water Treat 53:1275–1285. https://doi.org/10.1080/19443994.2013.850450

Jouzdani BE, Reisi M (2020) Effect of magnetized water characteristics on fresh and hardened properties of self-compacting concrete. Constr Build Mater 242:118196. https://doi.org/10.1016/j.conbuildmat.2020.118196

Kadhim KN, Al-Rufaye AHR (2018) Experimental study of magnetization effect on ground water properties. Jordan J Civil Eng 12:254–262. https://jjce.just.edu.jo/issues/paper.php?p=4248.pdf

Kamaruzaman S, Aris NF, Yahaya N, Hong LS, Razak MR (2017) Removal of Cu (II) and Cd (II) ions from environmental water samples by using cellulose acetate membrane. J Environ Anal Chem 4:4172. https://doi.org/10.4172/2380-2391.1000220

Kaspar JR, Lee K, Richard B, Walker AR, Burne RA (2021) Direct interactions with commensal streptococci modify intercellular communication behaviors of *Streptococcus mutans*. ISME J 15:473–488. https://doi.org/10.1038/s41396-020-00789-7

Khater Z, Ibraheim MH (2015) Ecological studies on the effect of magnetic field on water. Int J Curr Res Acad Rev 3:262–279. https://www.ijcrar.com/vol-3-8/Zeinab%20Z.%20K.%20Khater%20and%20Mona%20H.%20Ibraheim.pdf

Klempner G, Kerszenbaum I (2018) Principles of operation of synchronous machines. In: El-Hawary ME (ed) Handbook of large turbo-generator operation and maintenance. Wiley-IEEE Press, New Jersey, pp 1–52

Kotb A (2013) Magnetized water and memory meter. Energy Power Eng 5:422–426. http://www.scirp.org/journal/PaperInformation.aspx?PaperID=35920

Ku A, Facca VJ, Cai Z, Reilly RM (2019) Auger electrons for cancer therapy—A review. EJNMMI Radiopharm Chem 4:1–36. https://doi.org/10.1186/s41181-019-0075-2

Laforge M, Elbim C, Frère C, Hémadi M, Massaad C, Nuss P, Benoliel JJ, Becker C (2020) Tissue damage from neutrophil-induced oxidative stress in COVID-19. Nat Rev Immunol 20:515–516. https://doi.org/10.1038/s41577-020-0407-1

Latva M, Inkinen J, Rämö J, Kaunisto T, Mäkinen R, Ahonen M, Matilainen J, Pehkonen S (2016) Studies on the magnetic water treatment in new pilot scale drinking water system and in old existing real-life water system. J Water Process Eng 9:215–224. https://doi.org/10.1016/j.jwpe.2016.01.009

Le Chatelier HL (1884) On a general statement of the laws of chemical equilibrium. Comput Rendus 99:786–789 [in French]

Lee SH, Jeon SI, Kim YS, Lee SK (2013) Changes in the electrical conductivity, infrared absorption, and surface tension of partially-degassed and magnetically-treated water. J Mol Liquids 187:230–237. https://doi.org/10.1016/j.molliq.2013.07.017

Lee HS, Yap ACW, Ng CC, Mohd NS, Loo JL (2019) Increased electron density and dissolved oxygen level in water through magnetic effect. IOP Conf Ser Earth Enviorn (IOP Publishing). https://doi.org/10.1088/1755-1315/257/1/012010

Lengyel J, Pysanenko A, Poterya V, Kočišek J, Fárník M (2014) Extensive water cluster fragmentation after low energy electron ionization. Chem Phys Lett 612:256–261. https://doi.org/10.1016/j.cplett.2014.08.038

Lin L, Jiang W, Xu X, Xu P (2020) A critical review of the application of electromagnetic fields for scaling control in water systems: mechanisms, characterization, and operation. NPJ Clean Water 3:1–19. https://doi.org/10.1038/s41545-020-0071-9

López-Lázaro M (2009) Role of oxygen in cancer: looking beyond hypoxia. Anti-Cancer Agents Med Chem 9:517–525. https://doi.org/10.2174/187152009788451806

Makris KC, Andra SS, Botsaris G (2014) Pipe scales and biofilms in drinking-water distribution systems: undermining finished water quality. Crit Rev Environ Sci Technol 44:1477–1523. https://doi.org/10.1080/10643389.2013.790746

Malaysian Health Technology Assessment Section (MaHTAS) (2020) Molecular hydrogen therapy for management of COVID-19. https://covid-19.moh.gov.my/kajian-dan-penyelidikan/mahtas-covid-19-rapid-evidence-updates/10_Molecular_Hydrogen_Therapy_For_Management_Of_COVID-19_.pdf. Accessed 2 Mar 2022

Mghaiouini R, Elmlouky A, El Moznine R, Monkade M, El Bouari A (2020) The influence of the electromagnetic field on the electric properties of water. Mediterr J Chem 10:507–515. https://doi.org/10.13171/mjc10502005181406rm

Ministry of Health Malaysia (2016) Drinking water quality standard. http://engineering.moh.gov.my/v4/download.php?i=43. Accessed 22 Jan 2022

Mohammed RR, Ketabchi MR, McKay G (2014) Combined magnetic field and adsorption process for treatment of biologically treated palm oil mill effluent (POME). Chem Eng J 243:31–42. https://doi.org/10.1016/j.cej.2013.12.084

Mohri K (2002) Gradual decreasing characteristics and temperature stability of electrical resistivity in water triggered with milligauss AC field. IEEE Trans Magn 38:3353–3355. https://doi.org/10.1109/TMAG.2002.802307
Mosin O, Ignatov I (2014) Basic concepts of magnetic water treatment. Eur J Mol Biotechnol 4:72–85. http://www.medicalbiophysics.bg/en/1408289997.html
Omer NH (2020) Water Quality Parameters. In: Summers K (ed) Water quality—Science, assessments and policy. IntechOpen Limited, London, pp 1–18
Overbo A, Heger S (2019) Costs and benefits of household water softening: a review. https://www.wrc.umn.edu/sites/wrc.umn.edu/files/overbo_watersofteningcostbenefit_jan2019.pdf. Accessed 20 Apr 2022
Ozair G, Gutierrez JT (2010) An overview of magnetic water treatment system & further course of study. https://www.academia.edu/18068484/An_Overview_of_Magnetic_Water_Treatment_System_and_Further_Course_of_Study. Accessed 25 Jan 2022
Pang XF, Deng B (2008) The changes of macroscopic features and microscopic structures of water under influence of magnetic field. Phys B 403:3571–3577. https://doi.org/10.1016/j.physb.2008.05.032
Pang XF, Deng B, Tang B (2012) Influences of magnetic field on macroscopic properties of water. Mod Phys Lett B 26:1250069. https://doi.org/10.1142/S0217984912500698
Pang XF (2014) The magnetic field effects on water and its magnetization. In: Yu S (ed) Water molecular structure and properties. World Scientific Publishing, Singapore, pp 1–88
Parsons SA, Wang BL, Judd SJ, Stephenson T (1997) Magnetic treatment of calcium carbonate scale—Effect of pH control. Water Res 31:339–342. https://doi.org/10.1016/S0043-1354(96)00238-2
Perlt E, von Domaros M, Kirchner B, Ludwig R, Weinhold F (2017) Predicting the ionic product of water. Sci Rep-UK 7:1–10. https://doi.org/10.1038/s41598-017-10156-w
Petrov D, Taron KF, Hilleringmann U, Joubert TH (2021) Low-cost sensor system for on-the-field water quality analysis. In: 2021 smart systems integration (SSI), pp 27–29. https://doi.org/10.1109/SSI52265.2021.9466956
Putro T, Endarko (2016) The influence of electron discharge and magnetic field on calcium carbonate ($CaCO_3$) precipitation. AIP Conf Proc 1725:020067 (AIP Publishing LLC). https://doi.org/10.1063/1.4945521
Qi G, Tian D, Gao G, Liu G, Qiu C, Long T (2020) Numerical simulations and the design of magnetic field-enhanced electron impact ion source with hollow cylinder structure. J Anal Methods Chem 2020:2809485. https://doi.org/10.1155/2020/2809485
Qin M, Cao Z, Wen J, Yu Q, Liu C, Wang F, Zhang J, Yang F, Li Y, Fishbein G, Yan S (2020) An antioxidant enzyme therapeutic for COVID-19. Adv Mater 32:2004901. https://doi.org/10.1002/adma.202004901
Razak MR, Yusof NA, Haron MJ, Ibrahim N, Mohammad F, Kamaruzaman S, Al-Lohedan HA (2018) Iminodiacetic acid modified kenaf fiber for waste water treatment. Int J Biol Macromol 112:754–760. https://doi.org/10.1016/j.ijbiomac.2018.02.035
Rezaee M, Hill RP, Jaffray DA (2017) The exploitation of low-energy electrons in cancer treatment. Radiat Res 182:123–143. https://doi.org/10.1667/RR14727.1
Rocha LS, Nogueira J, Daniel-da-Silva AL, Marques P, Fateixa S, Pereira E, Trindade T (2021) Water softening using graphene oxide/biopolymer hybrid nanomaterials. J Environ Chem Eng 9:105045. https://doi.org/10.1016/j.jece.2021.105045
Sadeghipour O (2015) Magnetized water alleviates drought damages by reducing oxidative stress and proline accumulation in mung bean (*Vigna radiate* L. Wilczek). Bull Environ Pharmacol Life Sci 4:62–69. http://www.bepls.com/july_2015/11f.pdf
Sales FHS, dos Santos LG, Ferreira KS (2015) Effects of magnetic field generated by permanent magnet in water pH. https://www.researchgate.net/publication/311269438_EFFECTS_OF_MAGNETIC_FIELD_GENETARED_BY_PERMANENT_MAGNET_IN_WATER_pH. Accessed 1 Feb 2022

Santhosh C et al (2016) Role of nanomaterials in water treatment applications: a review. Chem Eng J 306:1116–1137. https://doi.org/10.1016/j.cej.2016.08.053

Selamat R, Ariffin A, Shaarani ZS, Adam HAM, Jamil R (2021) Low-frequency electromagnetic field exposure on Escherichia coli growth in river water. Int J Eng Adv Res 3:16–25. https://myjms.mohe.gov.my/index.php/ijear/article/view/16454

Selvi A, Rajasekar A, Theerthagiri J, Ananthaselvam A, Sathishkumar K, Madhavan J, Rahman PK (2019) Integrated remediation processes toward heavy metal removal/recovery from various environments-a review. Front Environ Sci 7:66. https://doi.org/10.3389/fenvs.2019.00066

Sies H, Belousov VV, Chandel NS, Davies MJ, Jones DP, Mann GE, Murphy MP, Yamamoto M, Winterbourn C (2022) Defining roles of specific reactive oxygen species (ROS) in cell biology and physiology. Nat Rev Mol Cell Bio 1–17. https://doi.org/10.1038/s41580-022-00456-z

Silva IB, Queiroz Neto JC, Petri DFS (2015) The effect of magnetic field on ion hydration and sulfate scale formation. Colloids Surf A 465:175–183. https://doi.org/10.1016/j.colsurfa.2014.10.054

Soni A, Sharma K, Verma SS (2018) Effect of magnetic field on some physical properties of tap water. Glob J Sci Front Res 18:35–48. https://journalofscience.org/index.php/GJSFR/article/view/2268

Sronsri C, Kongpop U, Sittipol W (2021) Analyses of vibrational spectroscopy, thermal property and salt solubility of magnetized water. J Mol Liquids 323:114613. https://doi.org/10.1016/j.molliq.2020.114613

Su N, Wu CF (2003) Effect of magnetic field treated water on mortar and concrete containing fly ash. Cement Concrete Compos 25:681–688. https://doi.org/10.1016/S0958-9465(02)00098-7

Toledo EJL, Ramalho TC, Magriotis ZM (2008) Influence of magnetic field on physical-chemical properties of the liquid water: insights from experimental and theoretical models. J Mol Struct 888:409–415. https://doi.org/10.1016/j.molstruc.2008.01.010

Tu R, Jin W, Xi T, Yang Q, Han S, Abomohra AE (2015) Effect of static magnetic field on the oxygen production of Scenedesmus obliquus cultivated in municipal wastewater. Water Res 86:132–138. https://doi.org/10.1016/j.watres.2015.07.039

Uddin MG, Nash S, Olbert AI (2021) A review of water quality index models and their use for assessing surface water quality. Ecol Indic 122:107218. https://doi.org/10.1016/j.ecolind.2020.107218

US Department of Energy (1998) Non-chemical technologies for scale and hardness control. https://www.gmxinternational.com/downloads/pdf/Federal_Technology_Alert.pdf. Accessed 20 Apr 2022

Vaskina I, Roi I, Plyatsuk L, Vaskin R, Yakhnenko O (2020) Study of the magnetic water treatment mechanism. J Ecol Eng 21:251–260. https://doi.org/10.12911/22998993/116341

Vermeiren TIS (1948) Magnetic treatment device for liquids. U.S. Pat. 2,652,925. https://patentimages.storage.googleapis.com/03/1c/80/1cfc9de7524f43/US2652925.pdf

Wahid ZA, Othman F, Sohaili J (2001) Electromagnetic technology on sewage treatment. Malays J Civil Eng 13:11–21. https://doi.org/10.11113/mjce.v13.15638

Wang Y, Babchin JA, Chernyi LT, Chow RS, Sawatzky RP (1997) Rapid onset of calcium carbonate crystallization under the influence of a magnetic field. Water Res 31:346–350. https://doi.org/10.1016/S0043-1354(96)00243-6

Wang Y, Wei H, Li Z (2018) Effect of magnetic field on the physical properties of water. Results Phys 8:262–267. https://doi.org/10.1016/j.rinp.2017.12.022

World Health Organization (2011) Guidelines for drinking water quality. https://apps.who.int/iris/bitstream/handle/10665/44584/9789241548151_eng.pdf. Accessed 25 Jan 2022

Wu J (2020) Tackle the free radicals damage in COVID-19. Nitric Oxide 102:39–41. https://doi.org/10.1016/j.niox.2020.06.002

Wu T, Brant JA (2020) Magnetic field effects on pH and electrical conductivity: implications for water and wastewater treatment. Environ Eng Sci 37:717–727. https://doi.org/10.1089/ees.2020.0182

Yap ACW, Lee HS, Loo JL, Mohd NS (2019) Investigation of the magnetic effects on water properties using permanent magnets. AIP Conf Proc 2157:020016 (AIP Publishing LLC). https://doi.org/10.1063/1.5126551

Yap ACW, Lee HS, Loo JL, Mohd NS (2021) Electron generation in water induced by magnetic effect and its impact on dissolved oxygen concentration. Sustain Environ Res 31:7. https://doi.org/10.1186/s42834-021-00080-0

Yin J, Zhang JK, Wu L, Li X (2011) Influence of water physical and chemical performance by magnetizing. Adv Mater Res 281:223–227. https://doi.org/10.4028/www.scientific.net/AMR.281.223

Zaidi NS, Sohaili J, Muda K, Sillanpää M (2014) Magnetic field application and its potential in water and wastewater treatment systems. Sep Purif Rev 43:206–240. https://doi.org/10.1080/15422119.2013.794148

Zhang C, Struewing I, Mistry JH, Wahman DG, Pressman J, Lu J (2021) Legionella and other opportunistic pathogens in full-scale chloraminated municipal drinking water distribution systems. Water Res 205:117571. https://doi.org/10.1016/j.watres.2021.117571

Zhou LP, Wang BX, Peng XF, Du XZ, Yang YP (2010) On the specific heat capacity of CuO nanofluid. Adv Mech Eng 2:172085. https://doi.org/10.1155/2010/172085

The Deficit Libres-Oriental Aquifer (Balsas Region) of Eastern-Central Mexico Under Global Warming: Hydrochemistry, Moisture Recycling and Rock-Water Interaction

Priyadarsi D. Roy, Oscar Agesandro García-Arriola, Sekar Selvam, and José Luis Sánchez-Zavala

Abstract This chapter evaluated physico-chemical characteristics as well as stable isotope compositions of groundwater in four different seasons of a year from the deficit Libres-Oriental aquifer of the Balsas hydrological administrative region at eastern-central Mexico in the context of global warming and its effect on recurrent droughts. Irrespective of the season, the groundwater mostly showed homogeneity with Ca–Mg–HCO_3 facies, and comparable values of the stable isotope fractions and d-excess. This similarity reflected possible decadal scale mixing of different moisture sources and comparable recycling mechanisms. Autumn samples with the highest variations in $\delta^{18}O$ (ca. 4.51‰), δ^2H (ca. 27.66‰) and d-excess (ca. 8.92‰) suggested additional recycling processes in the wetter months. Even though the ion exchange between water and aquifer materials was crucial for the water chemistry, the observation of persistent simple dissolution of carbonates and evaporites (salinity) in all the seasons is precarious in the framework of depleted precipitation and aquifer overextraction. Thus, the sustainability of groundwater quality of this aquifer should be considered as one of the immediate priorities.

Keywords Groundwater · Seasonal chemistry · Chemical facies · Stable isotope composition · d-excess · México

P. D. Roy (✉) · J. L. Sánchez-Zavala
Instituto de Geología, Universidad Nacional Autónoma de México, Del. Coyoacán, 04510 Ciudad de México, CP, Mexico
e-mail: roy@geologia.unam.mx

O. A. García-Arriola
Posgrado en Ciencias del Mar y Limnología, Universidad Nacional Autónoma de México, Del. Coyoacán, 04510 Ciudad de México, CP, Mexico

S. Selvam
Department of Geology, V.O. Chidambaram College, Tuticorin, Tamil Nadu 628008, India

P. Li et al. (eds.), *Sustainable Groundwater and Environment: Challenges and Solutions*, Springer Hydrogeology, https://doi.org/10.1007/978-3-031-82194-3_14

1 Introduction

The global warming, triggered from an amplified concentration of anthropogenically sourced greenhouse gases in the atmosphere, has already led to an ascent in the average global temperature and evapotranspiration (Masson-Delmotte et al. 2021). Additionally, the associated forcings of the monsoonal system and Atlantic Meridional Overturning Circulation (AMOC) weakening as well as the increase in frequency of the El Niño Southern Oscillation-related extreme weather events due to the warm sea have reduced the annual precipitation and the groundwater recharge in several regions (Cai et al. 2015; Cheng et al. 2016; Masson-Delmotte et al. 2021). Thus, the growing requirements for human and industrial consumptions as well as the irrigational needs have steered the expansion of overexploited aquifers (Famiglietti 2014). Cherlet et al. (2018) have observed that the rate of groundwater diminution has doubled globally over the last six decades, mostly in the arid and semi-arid areas. Under the estimation of Intergovernmental Panel on Climate Change (IPCC) that suggests up to 4 °C warmer temperature towards the end of twenty-first century under the scenario of SSP3-7.0, the drylands that presently represent about 41% of the global surface would continue to expand further and contribute to the upsurge in water-stress in some of the developing countries (Feng and Fu 2013; Cherlet et al. 2018; Masson-Delmotte et al. 2021; Roy et al. 2021, 2024).

Mexico, with a total population of about 127 million, is a developing country in North America and it is located at south of USA and north of Guatemala and Belize, between 14 and 32°N latitudes. Its surface area of about 1.97 million km^2 is divided into 32 states, including the Mexico City. The Sierra Madre Oriental, Sierra Madre Occidental and Sierra Madre Sur mountains and as well as the Trans Mexican Volcanic Belts comprising of active and dormant volcanoes with elevations reaching more than 3000 m asl marked the variable geomorphology. About two-third of its territory receives less than 500 mm of annual precipitation, and it hosts 77% of the total population and contributes to 82% of the national GDP. The rest one-third territory receives more than 2000 mm of total precipitation and host 23% of the population. A total of 653 aquifers of this country are grouped into 13 different hydrological administrative regions. Nine of them with the average per capita renewable water of 1528 m^3/hab./yr represent the semi-arid and arid regions (<500 mm/yr precipitation) in the central part and to the north. The drought monitoring shows more than 80% of the national territory has been under some degree of drought in the last four years (Source: National Meteorological Service of Mexico) with the drought areas comparable in extension to the 2011 mega drought that declined the agricultural productivity by 40% and led to the necessity of food distribution to hard hit indigenous Tarahumara community in the dry regions of Chihuahua state (Allen et al. 2012; Stahle et al. 2016). The groundwater presently contributes about 39.4% of the total volume of water used in Mexico and about 71% of this water resource is presently used for the irrigation (CONAGUA 2022). The consistent droughts, however, have increased the number of overexploited aquifers to 275 and the forecasts of reduced precipitation under the global warming scenario might continue to aggravate the impact on aquifers across Mexico (Cavazos et al. 2013; Arreguín-Cortés et al. 2016; CONAGUA 2022). The drought recurrence might further decline the hydrological resources and induce more than 6 million adult Mexicans to emigrate across the international borders between Mexico-USA by 2080 (Feng et al. 2010; Hunter et al. 2013; Nawrotzki et al. 2015).

Under the persistent droughts, the impact of climate change on groundwater can be identified by deciphering the seasonal evolution through the estimation of chemical facies, evaluation of rock-water interactions, demarcation of recharge regions, and identification of the moisture recycling through the physical parameters, major ion chemistry, and the stable oxygen and hydrogen isotope compositions (e.g., Cejudo et al. 2020; Roy et al. 2021; Torres-Martínez et al. 2021). The d-excess values additionally would help to constrain the moisture sources for the aquifer recharge (e.g. Risi et al. 2013; Tian et al. 2018; Roy et al. 2022, 2024). This chapter evaluated the groundwater from the deficit Libres-Oriental aquifer of the Balsas hydrological administrative region in eastern-central Mexico by evaluating the seasonal changes in physicochemical parameters as well as $\delta^{18}O$ and $\delta^{2}H$ compositions and provided new data for drought mitigation in a part of Mexico that is getting increasingly water scarce. The oxygen and hydrogen isotope compositions, and deuterium-excess values were compared with local precipitations and global meteoric water line to identify the hydrological processes and moisture sources. Similarly, the distributions of ion chemistry in Piper, modified Gibbs and Durov diagrams demarcated their chemical facies and evolutions in terms of different rock-water interactions.

2 Libres-Oriental Aquifer and Geology

The hydrological administrative region Balsas covers an area of ca. 116,500 km^2 in central Mexico. It has water renewability of 23,446 hm^3/year (or 1915 m^3/hab./yr, Fig. 1), comprises of several aquifers serving to about 9.5% of the national population and contributes to 6% of the GDP. The Libres-Oriental aquifer is located at its north-eastern margin (Fig. 1). It is a deficit aquifer with annual extraction (157.9 hm^3) surpassing the annual recharge (137.4 hm^3) (CONAGUA 2023). The static levels (5–160 m) and hydraulic transmissivity (0.01–13.70 $\times$ 10^{-3} m^2/s) are variable due to over-extraction and reduced precipitation in the previous decades. The static level mostly remains between 20 and 30 m, and it has been decreasing at rates of 0.3–0.6 m/year. A total of 16% of the extraction is presently used for rural-urban consumptions through the public supply to the five surrounding municipalities and about 81.4% is used for agricultural needs, including the irrigation of industrial-scale agricultural fields (Fig. 2; CONAGUA 2023). About 11–12% of the habitants in this region directly depend on the agriculture (source: www.datamexico.org). The alluvial-lacustrine deposits are used for the cultivation of corn, bean, wheat, barley, potato, apple, plum, pear, peach and other vegetables (INEGI 2000). Data of 1993/94 showed that the corn, wheat, and bean cultivations in the Libres covered about 126,000 ha, 13,000 ha and 10,000 ha, respectively (INEGI 2000).

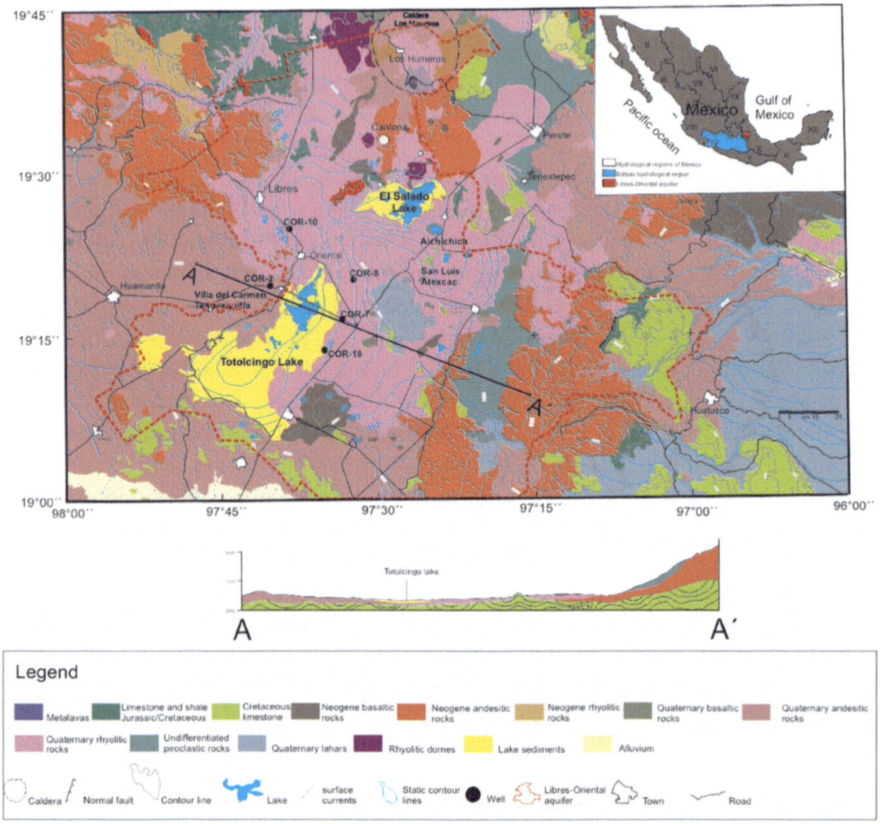

Fig. 1 Demarcation of the déficit Libres-Oriental aquifer (Balsas region) in a geological map of eastern-central Mexico. The location of sampled groundwater wells and static levels are also shown in the map

This unconfined to semiconfined, heterogenous, and anisotropic aquifer covers surface area of ca. 3995 km^2. It is comprised of Pliocene-to-Quaternary unconsolidated alluvial and lacustrine deposits as well as volcanic rocks in the upper part and the Cretaceous limestone in the lower part (CONAGUA 2023, Fig. 1). Geomorphologically, this endorheic basin has radial centrifugal and subdendritic and rectangular drainages, and it is limited by the volcanic and sedimentary mountain ranges. Upper Tamaulipas Formation (Albian-Cenomanian) is the oldest sedimentary with outcrops in the northern, northeastern, south-central and southeastern portions of the basin. It is composed of stratified limestone with bands of black chert and shale. It is transitionally covered by the Agua Nueva and San Felipe formations (Turionian-Santonian), made up of limestone, shale and bentonite horizons. Andesita Teziutlán Formation (Pliocene) represents andesite and andesitic to dacitic breccia to the west are rocks from the magmatic events of the Trans-Mexican Volcanic Belt. The San Antonio Formation (Pliocene-early Pleistocene) is made up mainly of spills and

Fig. 2 A drone's-eye view of the industrial scale agricultural fields at western margin of the study area in eastern-central Mexico

basaltic breccia, products of the eruptions of monogenetic cones, representing the activity of the Los Humeros volcanic field to the north. The late Pleistocene and Holocene lithologies are represented by dome structures of rhyolitic-dacitic composition, weathered rhyolitic tuff, basaltic scoria and basalt spills from the eruption of monogenetic cones and formation of maar-lakes (e.g., Alchichica) (Vergara-Martínez et al. 1997; Carrasco-Nuñez et al. 2017; CONAGUA 2023). Two of the lake systems (i.e., Lake El Saldo and Lake Totolcingo) have become ephemeral and saline due to the recurrent droughts over the last decade and over-exploitation of the aquifer.

3 Seasonality in Climatic Conditions

Seasonality of this semi-arid temperate region is demarcated by well-defined dry winter–spring and the wetter and warmer summer–autumn (Fig. 3). Tropical convective activity related to the dynamics in the Inter Tropical Convergence Zone and easterly trade winds during the summer and tropical cyclones in autumn contribute most of the precipitation. This wetter phase is temporarily interrupted by the relative minimum in precipitation, known as the mid-summer drought (Magaña et al. 1999, 2003). Similarly, the El Niño conditions reduces the number of tropical cyclones over the Intra Americas Seas (Magaña et al. 2003). National Meteorological Service of Mexico data between 1981 and 2010 from the stations at Villa de El Carmen (west) and San Luis Atexcac (east) characterized the seasonality (Fig. 3; CONAGUA 2023). The station at Villa de El Carmen registered average annual precipitations of 454 mm in 56 days of rainfall with 72% of those occurring in the summer (June–August) and

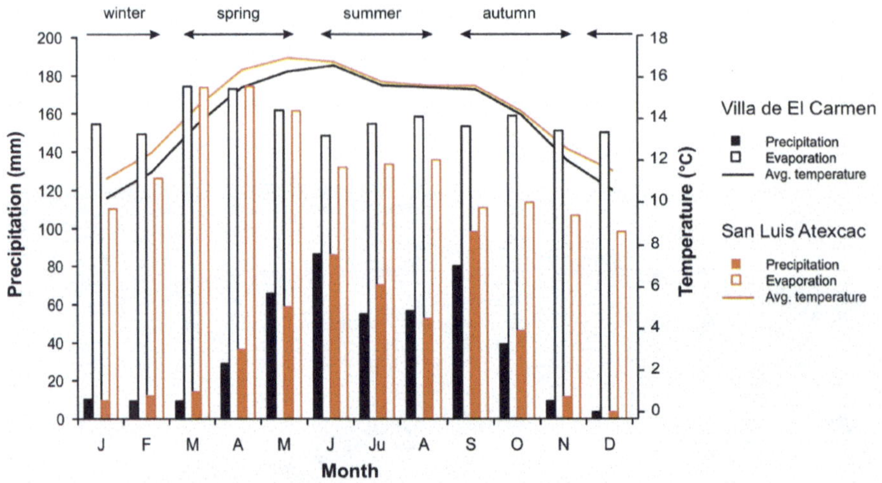

Fig. 3 Seasonal and monthly distributions of average precipitation, evaporation and temperature in the study area between 1981 and 2010 from the meteorological stations at Villa de El Carmen and San Luis Atexcac

autumn (September–November). Even though the evaporation in both wetter seasons was comparable (24.5%), the precipitation remained much higher in the warmer (average temperature: 15.5–16.6 °C) summer (198 mm, 43.5%) compared to the autumn (129 mm, 28.5%). The total annual evaporation was fourfold higher than the precipitation with 1889 mm, and it was the highest representing 27% (510 mm) of the total in spring (March–May: average temperature: 13.7–16.3 °C). The winter (December-February) was characterized by minimal precipitation (5%) and evaporation (24%) with cooler conditions, represented by the average temperature of 10.2–11.5 °C.

The data from San Luis Atexcac showed 500 mm of annual precipitation in an average 78 days/yr and almost 3-folds higher evaporation of 1580 mm (source: National Meteorological Service of Mexico or SMN). The seasonality is again marked by the changes in precipitation and evaporation as well as the average temperatures. Both the summer (June–August: 209 mm, 42%) and autumn (September–November: 156 mm, 31%) seasons received more precipitation compared to the winter (December-February: 26 mm, 5%) and spring (March–May: 110 mm, 22%). Evaporation was the highest in spring (511 mm, 32.5%), and it was the lowest in autumn and winter (335 mm, 21%). In terms of seasonality, the summer and autumn represented the wetter interval with more precipitation and relatively lower evaporation. Spring was the driest interval with less precipitation and the highest evaporation, reflecting the highest average temperature in May (16–17 °C).

4 Methodology

A total of 20 groundwater samples (5 samples/season) from five borewells with static levels between 10 and 30 m were collected in the months of February (winter), May (spring), August (summer) and November (autumn) during 2023–2024 in representation of the four seasons (Fig. 1). This sampling had the intention of documenting the overall temporal variations in terms of average seasonal changes. The samples, along with a blank, were transported in duplicate to laboratory in pre-cleaned Teflon bottles under <4 °C for the analysis of major cation and anion chemistry as well as compositions of $\delta^{18}O$ and $\delta^{2}H$ after measuring the physicochemical parameters such as electrical conductivity (EC), total dissolved solid (TDS), and Hydrogen ion concentration (pH) in-situ with a portable field Hanna instrument (HI 98130). The cation (Na^+, Ca^{2+}, Mg^{2+} and K^+) and anion (HCO_3^-, Cl^- and SO_4^{2-}) were measured in Waters corporation liquid chromatograph (Model 1525). Compositions of $\delta^{18}O$ and $\delta^{2}H$ were estimated in a Liquid Water Isotope Analyzer (Los Gatos Research model) and the results were expressed after normalization to V-SMOW. The deuterium excess was calculated as $d = \delta^{2}H - 8\delta^{18}O$ (Dansgaard 1964; Rozanski et al. 1993). Water-minerals saturation indices (SI) for some minerals (calcite, gypsum, thenardite etc.) and $\text{Log}pCO_2$ were estimated using the WATEQ4F software. The negative SI values suggested mineral undersaturation, positive values indicated mineral supersaturation and SI between −0.5 and 0.5 indicated the equilibrium of water samples with a mineral phase. $\text{Log}pCO_2$ characterized the partial pressure of CO_2 in groundwater with respect to the atmosphere (−3.40).

5 Results

5.1 *Physical and Chemical Characteristics*

The groundwater remained acidic to alkaline with pH between 6.6 and 9.6 and the EC varied between 346 and 2080 μS/cm (Table 1). TDS (171–1030 mg/L) demarcated all the samples, except for one well as fresh (TDS < 1000 mg/L) and the sample from one well remaining slightly brackish (TDS:1000–10,000 mg/L (e.g., Freeze and Cherry 1979). The chemical composition also reflected their homogeneity irrespective of the seasons (i.e., Ca–Mg–HCO_3, Fig. 4). One of the autumn samples reflected Na–HCO_3–Cl facies. The following section describes physical parameters and ion chemistry of the seasonal samples.

Winter: Samples were slightly acidic to neutral (pH: 6.6–7.6) with EC between 414 and 2060 μS/cm and TDS between 206 and 1030 mg/L. They contained more Na (25.6–110.7 mg/L) compared to K (2.1–9.2 mg/L) among the alkali metals and more Ca (33.6–188.5 mg/L) compared to Mg (8.1–122.7 mg/L) among the alkaline earths (Table 1). Among the anions, HCO_3 was more dominant (154.3–832.9 mg/L) compared to SO_4 (10.8–186 mg/L) and Cl (2.8–159.1 mg/L). Piper diagram

Table 1 Physical and chemical characteristics as well as the oxygen and hydrogen isotope compositions ((minimum–maximum (average)) of groundwater from the Libres-Oriental aquifer (Balsas region) of eastern-central Mexico

	Winter	Spring	Summer	Autumn
pH	6.6–7.6 (7.1)	7–7.8 (7.6)	7–8.2 (7.6)	7.3–9.6 (8)
EC (μS/cm)	414–2060 (1070)	433–1995 (1085)	426–2080 (1114.8)	346–2020 (1035)
TDS (mg/L)	206–1030 (532.8)	220–992 (541.4)	212–1030 (553.8)	171–1000 (515.2)
Na (mg/L)	25.6–110.7 (57.5)	25.5–100 (55.8)	24.3–92 (52.9)	28.4–130.1 (69.4)
K (mg/L)	2.1–9.2 (6.2)	2.2–8.1 (5.4)	3–7.1 (5.3)	2–9.1 (6.3)
Ca (mg/L)	33.6–188.5 (97.3)	34–169.4 (91.8)	28.3–191.7 (90.8)	11.2–105.2 (54.4)
Mg (mg/L)	8.1–122.7 (55.1)	9.4–115.3 (54.9)	8–108.2 (51.1)	9.6–118.5 (53.6)
HCO_3 (mg/L)	154.3–832.9 (438.1)	169.4–1074.4 (575.8)	162–797.7 (466.1)	145.5–1017.1 (511.1)
SO_4 (mg/L)	10.8–186 (53.8)	10.3–127.3 (50.5)	2.8–146.5 (41.1)	10.5–164.8 (60.9)
Cl (mg/L)	2.8–159.1 (70.9)	5.9–140.5 (61.6)	4.8–142.3 (62.4)	4.7–144.2 (65.5)
δ^2H (‰)	−94.41 to −78.20 (−87.32)	−91.84 to −76.01(−85.37)	−96.27 to −78.60 (−89.19)	−95.89 to −68.23 (−85.71)
$\delta^{18}O$ (‰)	−13.08 to −10.71 (−11.90)	−12.99 to −10.67 (−11.98)	−13.32 to −11.23 (−12.38)	−13.55 to −9.04 (−11.80)
d-excess (‰)	4.29–10.83 (7.91)	8.19–13.27 (10.44)	6.67–12.36 (10.53)	4.05–12.97 (8.68)

(Piper 1944) evaluated the relationship between dissolved ions and the dominant water chemistry (Fig. 4). The major cations represented no-dominant type and major anions indicated HCO_3 type, and both projected the samples into Ca–Mg–HCO_3 facies. The Gibbs diagram modified by Marandi and Shand (2018) reflected variations of Na/(Na + Ca) versus TDS and it indicated variable rock-water interaction of freshwater recharge with mostly carbonates and some along the trend showing evaporite dissolution of Sahib et al. (2016). The interaction with relatively more silicates was restricted to one well (Fig. 5). The modified Durov (1948) diagram of Al-Bassam and Khalil (2012) displayed major ion concentrations with respect to TDS and pH (Fig. 6). It helped to understand the hydro-chemical and ion exchange processes of the groundwater system (e.g., Tiwari et al. 2019; Brindha et al. 2020). Distributions in this plot indicated that cation chemistry varied between Ca and Mg dominances and the anion chemistry between HCO_3 and SO_4 abundances (Fig. 6). The samples in classes of 1, 2, 4 and 5 represented ion exchange between the water and aquifer material, dissolution of sulfate minerals, such as gypsum and/or anhydrite, and the simple dissolution of carbonate minerals.

Spring: Groundwater of this season remained neutral to slightly alkaline (pH: 7–7.8) with variable EC (433–1995 μS/cm) and TDS (220–992 mg/L). Na (25.5–100 mg/L) was more compared to K (2.2–8.1 mg/L) among the alkali metals, and Ca (34–169.4 mg/L) was higher compared to Mg (9.4–115.3 mg/L) among the

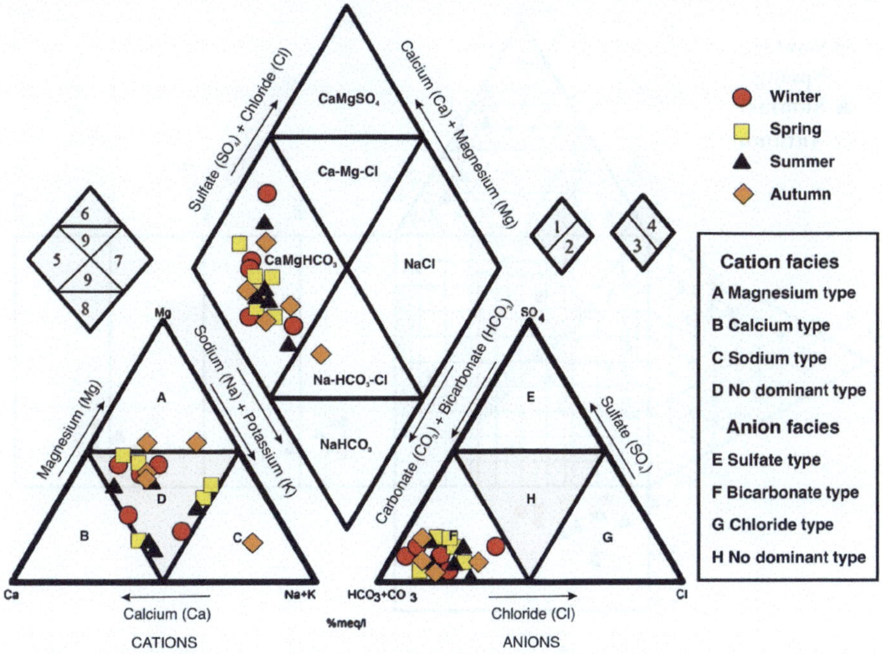

Fig. 4 Hydrochemical facies of seasonal groundwater from the deficit Libres-Oriental aquifer (Balsas region) at eastern-central Mexico in the trilinear Piper diagram

Fig. 5 Groundwater chemistry of deficit Libres-Oriental aquifer (eastern-central Mexico) evaluated in the modified Gibbs plot of Marandi and Shand (2018)

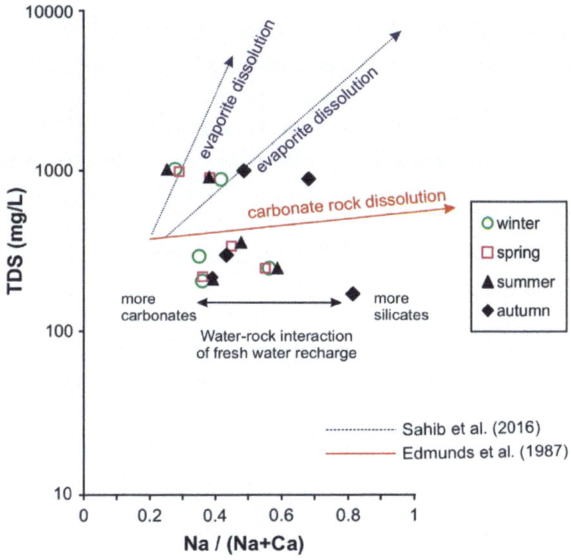

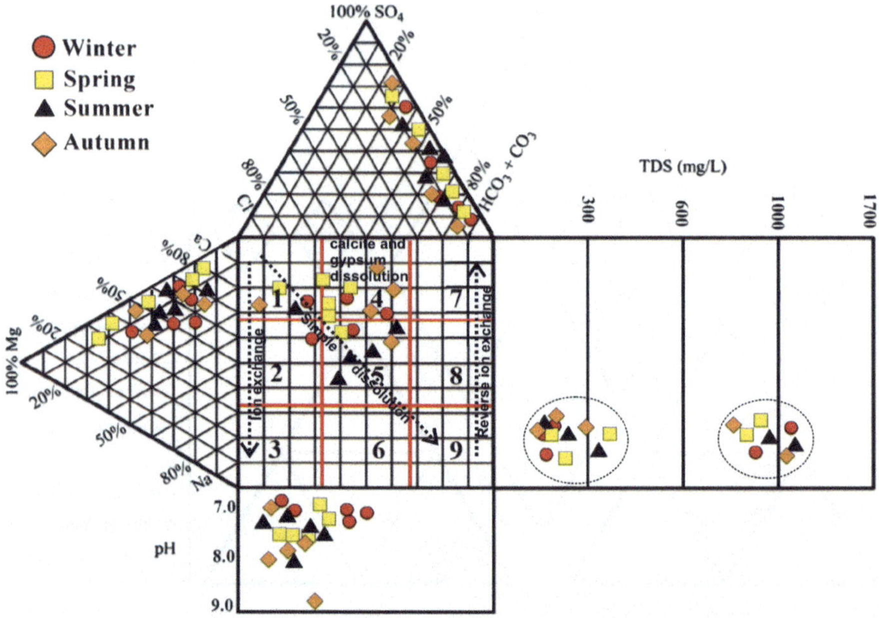

Fig. 6 Major ion concentrations of seasonal groundwater samples from deficit Libres-Oriental aquifer (eastern-central Mexico) with respect to TDS and pH in modified Durov (1948) diagram of Al-Bassam and Khalil (2012)

alkaline earths (Table 1). Among the anions, the concentration of HCO_3 (169.4–1074.4 mg/L) increased compared to the winter and it was much higher compared to SO_4 (10.3–127.3 mg/L) and Cl (5.9–140.5 mg/L). Piper diagram grouped the samples in no-dominant cation and dominant HCO_3-type anion fields, and the Ca–Mg–HCO_3 facies (Fig. 4). Both the carbonate as well the evaporite dissolutions dominated the rock-water interaction with the silicate weathering of freshwater discharge remaining minimal (Fig. 5). The distribution mostly in class 4 of the modified Durov (1948) diagram of Al-Bassam and Khalil (2012) represented homogenous gypsum dissolution (Fig. 6).

Summer: Samples were slightly neutral to alkaline (pH: 7–8.2) with EC of 426–2080 μS/cm and TDS between 212–1030 mg/L. They contained more Na (24.3–92 mg/L) compared to K (3–7.1 mg/L), and more Ca (28.3–191.7 mg/L) compared to Mg (8–108.2 mg/L) (Table 1). HCO_3 (162–797.7 mg/L) was more than SO_4 (2.8–146.5 mg/L) and Cl (4.8–142.3 mg/L). The classification in Piper diagram grouped them in Ca–Mg–HCO_3 facies with no-dominant cation and HCO_3-type dominant anion (Fig. 4). Carbonate dissolution, evaporite dissolution as well as silicate weathering reflected the heterogeneous rock-water interactions (Fig. 5). The Durov diagram suggested (class 1 and 5) ion exchange between the water and aquifer materials as well as the simple dissolution (Fig. 6).

Autumn: Samples were mostly alkaline (pH: 7.3–9.6) with EC (346–2020 μS/cm) and TDS (171–1000 mg/L) remaining the most heterogenous among the seasons (Table 1). Na (28.4–130.1 mg/L) and K (2–9.1 mg/L) also remained the most heterogeneous with some samples showing the highest among all the seasons. Ca (11.2–105.2 mg/L) remained minimal among all the seasons but higher compared to Mg (9.6–118.5 mg/L) among the alkaline earths (Table 1). Among the anions, HCO_3 (145.5–1017.8 mg/L) was most heterogenous among the seasons but remained more compared to SO_4 (10.8–186 mg/L) and Cl (2.8–159.1 mg/L). In the Piper diagram (Piper 1944, Fig. 4), the dominant major cations (Mg-type, no-dominant and Na + K type) were heterogenous and major anions remained homogeneous (HCO_3 type). Both projected all the samples, except for one into Ca–Mg–HCO_3 and classified one sample as Na–HCO_3–Cl facies. The modified Gibbs diagram of Marandi and Shand (2018) also mirrored the variable rock-water interaction of freshwater recharge with some along the trend showing more carbonate dissolution and some showing more interaction with silicates (Fig. 5). Samples in class 4 and 5 of the modified Durov (1948) diagram indicated both the gypsum/anhydrite and simple dissolutions on water chemistry (Fig. 6).

5.2 *Water-Mineral Saturation and* pCO_2

Saturation indices (SI) of calcite ($CaCO_3$), magnesite ($MgCO_3$), dolomite ($CaMg(CO_3)_2$), gypsum ($CaSO_4$. $2H_2O$), anhydrite ($CaSO_4$), thenardite (Na_2SO_4) and mirabilite (Na_2SO_4. $10H_2O$) predicted the seasonal changes in water-mineral equilibrium (Table 2). Irrespective of the season, the SI of all the sulphate minerals remained below −0.50 and reflected their undersaturation nature. SI of gypsum varied between −3.89 and −1.19, anhydrite between −4.08 and −1.39, thenardite between −10.83 and −5.78 and mirabilite between −10.29 and −5.18 (Table 2). All carbonate minerals (calcite, magnesite and dolomite) with variable SI indicated undersaturation, equilibrium and supersaturation states. The partial pressure of CO_2, estimated as logpCO_2 (g), was more compared to the atmosphere (−3.40) and it varied between −3.15 and −1.68 in different seasons (Table 2).

Winter: Samples remained undersaturated with respect to magnesite (−0.99 to −0.72), dolomite (−1.16 to −0.86), gypsum (−3.65 to −1.74), anhydrite (−3.83 to −1.91), thenardite (−10.83 to −8.14) and mirabilite (−10.29 to −7.62) and in equilibrium with calcite (−3.03 to 0.82). The partial pressure of CO_2 (logpCO_2) varied between −2.90 and −1.77 (Table 2).

Spring: Samples continued to remain undersaturated with respect to gypsum (−3.41 to −1.19), anhydrite (−3.59 to −1.39), thenardite (−8.94 to −5.78) and mirabilite (−8.46 to −5.18). The saturation states varied between undersaturation and equilibrium with respect to magnesite (−1.14 to 0.02) and dolomite (−1.15 to 0.09) and in equilibrium with calcite (−0.45 to −0.01). LogpCO_2 varied between −3.15 and −1.73, with some samples showing lower CO_2 partial pressure compared to the winter (Table 2).

Table 2 Water-minerals saturation indices for different carbonate minerals and evaporites as well as partial pressure of CO_2 (logpCO_2) in groundwater from the Libres-Oriental aquifer (Balsas region) of eastern-central Mexico (minimum-to-maximum (average))

	Winter	Spring	Summer	Autumn
Calcite	−0.34 to −0.10 (−0.23)	−0.45 to −0.01 (−0.16)	−0.16 to 0.55 (0.18)	−0.28 to −0.07 (−0.16)
Magnesite	−0.99 to −0.72 (−0.83)	−1.14 to 0.02 (−0.39)	−0.23 to 0.52 (0.07)	−0.54 to −0.34 (−0.45)
Dolomite	−1.16 to −0.86 (−0.98)	−1.15 to 0.09 (−0.47)	−0.32 to 1.13 (0.32)	−0.69 to −0.44 (−0.53)
Gypsum	−3.65 to −1.74 (−2.87)	−3.41 to −1.19 (−2.31)	−3.02 to −2.21 (−2.62)	−3.89 to −1.24 (−2.51)
Anhydrite	−3.83 to −1.91 (−3.04)	−3.59 to −1.39 (−2.49)	−3.21 to −2.40 (−2.81)	−4.08 to −1.42 (−2.69)
Thenardite	−10.83 to −8.14 (−9.24)	−8.94 to −5.78 (−7.93)	−7.98 to −6.53 (−7.50)	−10.56 to −6.52 (−7.79)
Mirabilite	−10.29 to −7.62 (−8.74)	−8.46 to −5.18 (−7.34)	−7.31 to −5.92 (−6.85)	−9.98 to −5.98 (−7.23)
Log pCO_2	−2.90 to −1.77 (−2.27)	−3.15 to −1.73 (−2.58)	−2.33 to −1.89 (−2.10)	−2.84 to −1.68 (−2.10)

Summer: Samples remained undersaturated with respect to gypsum (−3.02 to −2.21), anhydrite (−3.21 to −2.40), thenardite (−7.98 to −6.53) and mirabilite (−7.31 to −5.92). Both calcite (−0.16 to 0.55) and magnesite (−0.23 to 0.52) were mostly in equilibrium, and dolomite (−0.32 to 1.13) varied between equilibrium to saturation states. The partial pressure of CO_2 increased compared to the spring and it remained the most homogeneous among all the seasons (LogpCO_2: −2.33 to −1.89).

Autumn: Samples were undersaturated with respect to gypsum (−3.89 to −1.24), anhydrite (−4.08 to −1.42), thenardite (−10.56 to −6.52) and mirabilite (−9.98 to −5.98). Both calcite (−0.28 to −0.07) and magnesite (−0.54 to −0.34) remained in equilibrium states, and dolomite (−0.69 to −0.44) varied between undersaturation and equilibrium. The partial pressure of CO_2 remained comparable to the summer with logpCO_2 between −2.84 to −1.68 (Table 2).

5.3 *Stable Oxygen and Hydrogen Isotopes*

In general, the $\delta^{18}O$ (−13.55 to −9.04‰) and $\delta^{2}H$ (−96.27 and −68.23‰) compositions of most of the groundwater remained within the linear regression equations of the global meteoric water line (GMWL, Craig 1961) and the local meteoric water line (LMWL, Sánchez-Murillo et al. 2023) (Table 1, Fig. 7). Isotopic compositios ($\delta^2H = 7.9\delta^{18}O + 11.2$) of precipitation of 3 years (2018–2021) from the coast of

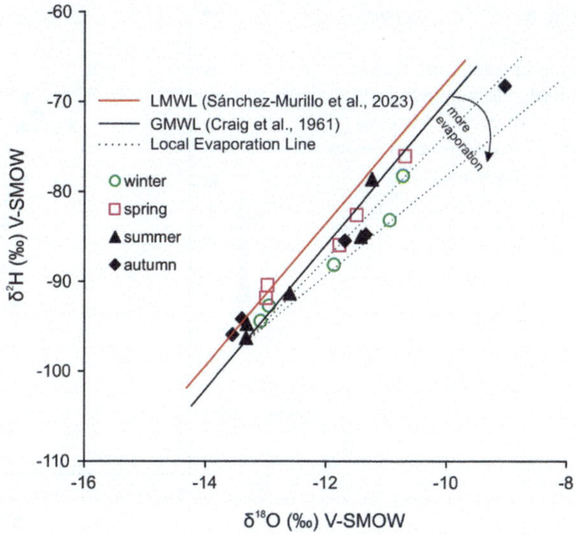

Fig. 7 Distributions of $\delta^{18}O_{H2O}$ versus $\delta^{2}H_{H2O}$ in groundwater from deficit Libres-Oriental aquifer (eastern-central Mexico) with respect to the local meteoric water line (LMWL; Sánchez-Murillo et al. 2023) and global meteoric water line (GMWL; Craig 1961)

Gulf of Mexico (ca. 155 km east) were considered here as the reference for local precipitation. The samples deviating from tendencies of the meteoric water lines with enriched $\delta^{18}O$, irrespective of the season, defined different degrees of evaporation at the recharge areas (Fig. 7). Deuterium excess (d-excess) fluctuated between 4.05 and 13.27‰ with the d-excess values decreasing with increasing $\delta^{18}O$ (Fig. 8). Samples with the lowest $\delta^{18}O$ showed the highest d-excess and samples with the highest ^{18}O showed the lowest value of d-excess. In TDS vs d-excess distribution, the samples showed a tendency of variable degrees of moisture recycling and mineral dissolution (e.g., Huang and Pang 2012; Fig. 9). Irrespective of the season, the samples from two wells with enriched salinity and less variable d-excess reflected more mineral dissolution and the samples from three wells with more variable d-excess and low TDS reflected more moisture recycling.

Winter: $\delta^{18}O$ varied between −13.08‰ and −10.71‰, and $\delta^{2}H$ fluctuated between −94.41 and −78.20‰. The d-excess values (4.29–10.83‰) remained comparable to global average precipitation (+10‰; Clark & Fritz 1997) and the local precipitation (+11.2‰; Sánchez-Murillo et al. 2023) in two wells and below these values in three wells.

Spring: The compositions of $\delta^{18}O$ between −12.99 and −10.67‰, and $\delta^{2}H$ between −91.84 and −76.01‰ were comparable to the winter samples. The d-excess values (8.19–13.27‰) with an average of 10.44‰ remained comparable to global and local precipitation.

Summer: $\delta^{18}O$ between −13.32 and −11.23‰, and $\delta^{2}H$ between −96.27 and −78.60‰ were the most homogeneous among the seasons. Except for one well, the d-excess values (6.67–12.36‰) were comparable to global average precipitation and local precipitation.

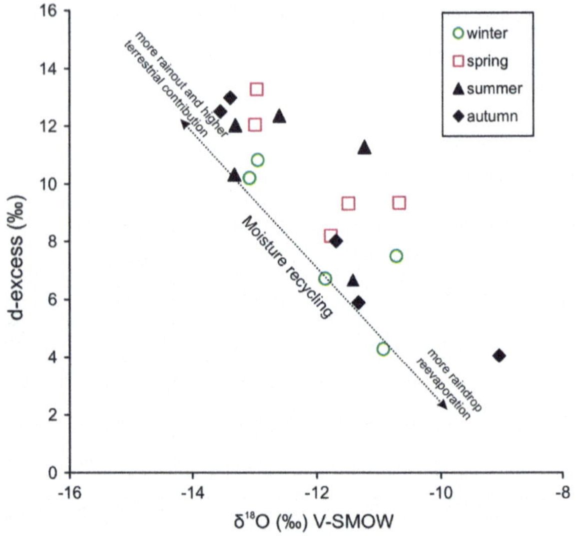

Fig. 8 $\delta^{18}O_{H2O}$ versus d-excess plot for groundwater from deficit Libres-Oriental aquifer (eastern-central Mexico) showing different processes of moisture recycling

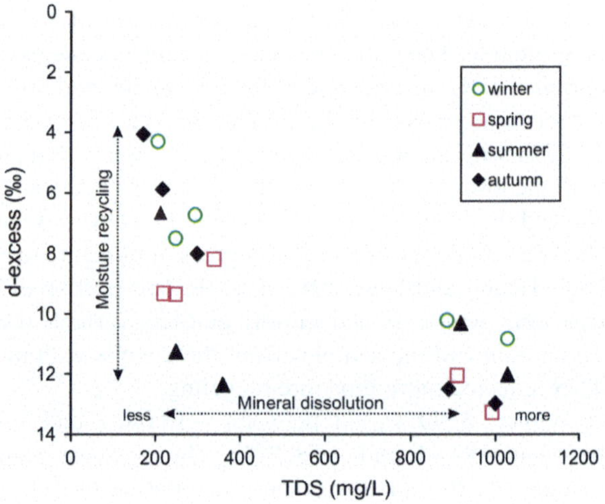

Fig. 9 Seasonal groundwater samples from deficit Libres-Oriental aquifer (eastern-central Mexico) in d-excess versus TDS (mg/L) plot showing the influences of mineral dissolution and moisture recycling

Autumn: Groundwater of this season were the most heterogenous with $\delta^{18}O$ between −13.55 and −9.04‰, and $\delta^{2}H$ between −95.89 and −68.23‰. The d-excess values (4.05–12.97‰) also showed the heterogeneity with an average of 8.68‰, lower comparable to the global and local precipitations.

6 Discussion

6.1 *Seasonality of Hydrogeochemical Evolution*

Data from both the meteorological stations characterize seasonality with the highest evaporations in spring, with 23% and 32.5% of the total. The amount of precipitation remained lower than the summer-autumn seasons (22–23% of annual) and the average temperature reached the highest values of 16.3 and 17 °C, respectively. Similarly, the summer (42–43.5% of annual) and autumn (28.5–31% of annual) are relatively wetter with more precipitations and lower evaporation (21–25.5%). The winter received about 5% of annual rainfall and showed the lowest average annual temperature (11.2–12.4 °C) among all the seasons (Fig. 3). The effect of seasonality on groundwater was unclear with mostly homogeneous facies in the trilinear Piper plot (Fig. 4). Except for one, all the samples maintained uniform facies of Ca–Mg–HCO_3 in all the seasons. The facies of one sample (Na–HCO_3–Cl) represented heterogeneity in wetter autumn season and it also showed the highest Na content (130.1 mg/L). The heterogeneity of autumn samples was also mirrored by the highest variations in pH (7.3–9.6), EC (346–2020 μS/cm) and TDS (171–1000 mg/L) values (Table 1). Apart from the evaporite dissolution, some of the autumn samples showed more carbonate dissolution and others indicated more interaction with the silicates. In comparison, the groundwater of three other seasons with homogeneous facies indicated mostly carbonate and evaporite dissolutions in the modified Gibbs diagram (Fig. 5). The wetter autumn samples continued to remain heterogenous with respect to the stable isotopes showing the highest variations in $\delta^{18}O$ (ca. 4.51‰), $\delta^{2}H$ (ca. 27.66‰) and d-excess (ca. 8.92‰) compared to all other seasons. In contrast, the drier spring samples mostly remained homogenous with the lowest variations in $\delta^{2}H$ (ca. 15.83‰) and d-excess (ca. 5.08‰), and a variability of ca. 2.32‰ in $\delta^{18}O$.

6.2 *Moisture Recycling of Recharge*

The variabilities in isotopic characteristics of groundwater, although comparable among seasons, highlight moisture source conditions and mositure recycling (e.g., Pfahl and Sodemann 2014).The global and local meteoric water lines, reflecting the average precipitations, have mean d-excess of 10‰ and 11.2‰, respectively (Clark and Fritz 1997; Sánchez-Murillo et al. 2023). The average d-excess in spring (10.44‰) and summer (10.53‰) remained similar to the average precipitations, and the average values of autumn (8.68‰) and winter (7.91‰) remained lower compared to the global and regional rainfalls. The terrestrial evapotranspiration, sub-cloud raindrop re-evaporation, net rainout, altitude of the recharge area, evaporation, and conditions of the evaporating moisture such as the sea-surface temperature (SST) at the source ocean cause the moisture recycling and lead to the large-scale d-excess variations in groundwater (Clark and Fritz 1997; Pfahl and Sodemann 2014; Cui et al.

2017; Xia and Winnick 2021; Natali et al. 2022). The autumn precipitation in northern hemisphere shows higher d-excess compared to the spring rainfall as the water vapor sourced from the ocean with warmer SST has higher d-excess (Pfahl and Sodemann 2014). Rainfall events at higher altitudes also show higher d-excess values compared to the rainfall at lower altitudes. Natali et al. (2022) reported a mean gradient of + 0.73‰/100 m, indicating the influence of sub-cloud evaporation in elevated terrains of the Apuan Alps (Italy). Terrestrial moisture through evaporation from the canopy storage and the net rainout enhance the d-excess values (Gat and Matsui 1991; Putman et al. 2019; Xia and Winnick 2021). In constrast, the downwind sub-cloud raindrop re-evaporation and evaporation at recharge regions lower the d-excess values of rainfall.

The d-excess variabilities were influenced by moisture recycling along with the evaporation. Irrespective of the season, the groundwater samples with a tendency similar to the GMWL and LMWL represented the local precipitation and the samples deviating away with enriched oxygen isotope composition supported the evaporation at recharge zones (Fig. 7). The d-excess values reflected different moisture sources, i.e., tropical storm in autumn and deep-convective system in summer (e.g., Adams and Comrie 1997; Aggarwal et al. 2016; Pérez-Quezadas et al. 2021). The Western Hemisphere warm pool of summer-autumn might have also increased d-excess of the sourced moisture (e.g., Wang and Enfield 2001; Pfahl and Sodemann 2014). The moisture recycling from terrestrial evapotranspiration and net rainout, and the more altitude of recharge areas might have led to higher d-excess (e.g., Liebminger et al. 2006; Xia and Winnick 2021; Putman et al. 2019; Natali et al. 2022). Higher d-excess along with depleted $\delta^{18}O$ groundwater reflected the influences of tropical storms and more net rainout (Xia and Winnick 2021; Pérez-Quezadas et al. 2021). Similarly, the enhanced d-excess values might be indicative of contribution from terrestrial evapotranspiration as well as the moisture sourced from warmer ocean. The recharge zones at higher altitudes might have also contributed to the enriched d-excess values in groundwater. In contrast, the groundwater samples with depleted d-excess showed the effects of raindrop re-evaporation and evaporation at recharge areas, generally associated with warm and dry areas, as well as the moisture sourced from the ocean with lower SST (e.g., Putman et al. 2019). The moisture recycling is evident in samples with less influence of mineral dissolution (Fig. 9). Stable isotopes of oxygen and hydrogen of shallow groundwater in Mexico represent the long-term decadal-scale infiltration inputs through the warm season precipitations (e.g., Wassenaar et al. 2009; Roy et al. 2021, 2024). Hence the groundwater of this study represented infiltrations and moisture recycling in decadal-scale (e.g., Pérez-Quezadas et al. 2021; Xia and Winnick 2021; Natali et al. 2022). The comparability in stable isotope characteristic in different seasons reflected decadal-scale mixing of different moisture sources and their recycling. However, the autumn samples with highest $\delta^{18}O$ (ca. 4.51‰), $\delta^{2}H$ (ca. 27.66‰) and d-excess (ca.8.92‰) variabilities suggested additional recycling mechanisms in the wetter season.

6.3 Rock-Water Interaction

Modified Gibbs diagram and Durov plot helped to evaluate the influences of rock-water interactions on groundwater. The aquifer lithologies is represented by abundant limestone and volcanic deposits of different compositions ranging from basalt to rhyolite. In the Gibbs diagram of Marandi and Shand (2018), and the Durov plot showed the influences of the silicate weathering, and the dissolution of both carbonate and evaporites (Fig. 5). Their distribution in TDS vs d-excess plot showed different degrees of mineral dissolution (Fig. 9). Among the samples with nearly uniform d-excess, the wells with lower TDS showed less influences of mineral dissolution, and the wells with higher TDS showed more influence of mineral dissolution. In the samples with lower TDS, the variable d-excess indicated the influences of moisture recycling (discussed in Sect. 6.2). Different mineral dissolution is also reflected in the Durov plot in samples with TDS around 1000 mg/L showing more evaporite/carbonate dissolution compared to the samples with TDS around 300 mg/L. The saturation indices of evaporites reflected the different equilibrium states. The sulphate minerals reflected undersaturation nature in groundwater of all the seasons (Table 2). For example, the SI of gypsum varied between −3.89 and −1.19 (Table 2). The carbonate minerals, however, remained undersaturated, in equilibrium and supersaturated. For example, the calcite with SI between −0.45 and 0.55 mostly remained in equilibrium state, magnesite with SI between −1.14 and 0.52 varied between undersaturation and equilibrium and the dolomite with SI between −1.16 and 1.13 indicated undersaturation to saturation, especially saturation in some samples of the summer season. Higher pCO_2 compared to the atmosphere (−3.15 to 1.68) facilitated the dissolutions of carbonate and sulphate minerals. Relatively higher CO_2 partial pressure in the autumn season (pCO_2: up to −1.68) caused more dissolution of carbonates and sulphates with all these minerals reflecting their undersaturation and equilibrium states.

Higher partial pressure in some autumn samples might be indicating additional CO_2 in the watershed topsoil from the degradation of organic material. Their scatterings in the modified Durov (1948) diagram also indicated dissolutions of gypsum/anhydrite and carbonates (Fig. 6). Even though the Durov diagram did not suggest ion-exchange in autumn, the influence of silicate weathering on water chemistry of autumn was shown in the Na/(Na + Ca) versus TDS scatter plot of Marandi and Shand (2018). One of the studied wells indicated higher Na/(Na + Ca) values without significant change in TDS (Fig. 5). The groundwater from summer and winter seasons was influenced by ion exchange between the water and aquifer materials as well as the simple dissolution. Even with dominant limestone in the subsurface, the ion exchange between Ca in groundwater, and Na and K from the volcanic deposits suggested that the ion-exchange remained one of the dominant geochemical process for the groundwater. The drier spring samples, however, was influenced only by the gypsum dissolution (Fig. 6). In context of the global warming and depleted precipitation in the region, the aquifer salinity would continue to intensify through more carbonate and evaporite dissolutions. Thus, the sustainability of groundwater quality from this aquifer should be considered as one of the immediate priorities.

7 Conclusions

The Libres-Oriental aquifer in hydrological administrative region Balsas of Mexico is a deficit aquifer with annual extraction (157.9 hm^3) surpassing the annual recharge (137.4 hm^3), and variable static levels (5–160 m) and hydraulic transmissivities (0.01–13.70 $\times$ 10^{-3} m^2/s). Under the impact of global warming and its possible effects, the evaluation of groundwater from this aquifer for the seasonal evolution, moisture sources and their recycling, and the rock-water interactions resulted the following conclusions.

The effect of seasonality was unclear with mostly homogeneous facies of Ca–Mg–HCO_3 in the trilinear Piper plot. Facies of one sample (Na–HCO_3–Cl) represented the heterogeneity in autumn. The heterogeneity in autumn was also mirrored in the highest variations of pH (7.3–9.6), EC (346–2020 μS/cm) and TDS (171–1000 mg/L).

The d-excess variabilities were influenced by different moisture sources (tropical storm vs. deep-convective system) and moisture recycling. Irrespective of the season, the groundwater samples deviating away from the meteoric water lines supported evaporation at the recharge zones. The enhanced d-excess might be indicative of contribution from the terrestrial evapotranspiration, net rainout, moisture sourced from warmer ocean, and recharge zones at higher altitudes. In contrast, the depleted d-excess showed the effects of raindrop re-evaporation and evaporation at recharge areas. The comparability in stable isotope characteristic in different seasons reflected decadal-scale mixing. However, the autumn samples with highest $\delta^{18}O$ (ca. 4.51‰), δ^2H (ca. 27.66‰) and d-excess (ca. 8.92‰) variabilities suggested additional recycling mechanisms in the wetter months.

Modified Gibbs diagram and Durov plot showed the influences of silicate weathering, and the dissolutions of both carbonate and evaporites on water chemistry. Among the samples with nearly uniform d-excess, the wells with lower TDS showed less influence of mineral dissolution, and the wells with higher TDS showed more influence of mineral dissolution. The saturation indices reflected undersaturation nature of sulphate minerals and equilibrium states of calcite in all the seasons. Higher pCO_2 compared to the atmosphere (between -3.15 and -1.68) facilitated the dissolutions of evaporites. Even with abundant limestone in the subsurface, the exchange between Ca in groundwater and Na–K from the volcanic deposits suggested that the ion-exchange remained one of the dominant geochemical process for the groundwater chemistry during the summer and winter seasons along with the simple dissolutions of carbonate and gypsum. Gypsum dissolution dominantly influenced groundwater of the drier spring, and it is a matter of concern in context of the global warming and depleted precipitation with the potential to further elevate the aquifer salinity in the near future.

Acknowledgements Financial support from the Support Program for Research and Technological Innovation Projects of DGAPA-UNAM (proposal PAPIIT-IN101124) and technical helps from Olivia Zamora Martínez and Francisco Otero (LANGEM, UNAM) are thankfully acknowledged.

References

Adams DK, Comrie AC (1997) The north American monsoon. Bull Am Meteor Soc 78(10):2197–2214

Aggarwal PK, Romatschke U, Araguas-Araguas L, Belachew D, Longstaffe FJ, Berg P et al (2016) Proportions of convective and stratiform precipitation revealed in water isotope ratios. Nat Geosci 9(8):624–629

Al-Bassam AM, Khalil AR (2012) DurovPwin: a new version to plot the expanded Durov diagram for hydro-chemical data analysis. Comput Geosci 42:1–6

Allen J, Assaf A, Small J, Riebeek H (2012) Drought causes Mexico food shortages. NASA Earth Observatory Images

Arreguín-Cortés FI, López-Pérez M, Ortega-Gaucin D, Ibañez-Hernández Ó (2016) La política pública contra la sequía en México: avances, necesidades y perspectivas. Tecnología y Ciencias Del Agua 7(5):63–76

Brindha K, Paul R, Walter J, Tan ML, Singh MK (2020) Trace metals contamination in groundwater and implications on human health: comprehensive assessment using hydrogeochemical and geostatistical methods. Environ Geochem Health 42:3819–3839

Cai W, Santoso A, Wang G, Yeh SW, An SI, Cobb KM et al (2015) ENSO and greenhouse warming. Nat Clim Change 5(9):849–859

Carrasco-Nuñez G, Hernández J, De León L, Dávila G, Norini J, Bernal JP, Jicha B, Navarro M, López-Quiroz P (2017) Geologic Map of Los Humeros volcanic complex and geothermal field, eastern. Trans-Mexican Volcanic Belt 2:1–11

Cavazos T, Salinas JA, Martínez B, Colorado G, de Grau P, Prieto-González R, Bravo ME (2013) Actualización de escenarios de cambio climático para México como parte de los productos de la Quinta Comunicación Nacional. informe, México, Instituto Nacional de Ecología y Cambio Climático

Cejudo E, Acosta-Gonzalez G, Leal-Bautista RM, Estrada-Medina H (2020) Water stable isotopes (δ2H and δ18O) in the Peninsula of Yucatan, Mexico. Hydrol Earth Syst Sci Discuss 2020:1–20

Cheng J, Liu Z, Zhang S, Liu W, Dong L, Liu P, Li H (2016) Reduced interdecadal variability of Atlantic Meridional Overturning Circulation under global warming. Proc Natl Acad Sci 113(12):3175–3178

Clark ID, Fritz P (1997) Environmental isotopes in hydrogeology. CRC Press/Lewis Publishers, Boca Raton, Florida, pp 1–342

CONAGUA (2022) Estadísticas del Agua en México 2021. Comisión Nacional del Agua, Ciudad de México (In spanish). https://sinav30.conagua.gob.mx:8080/PDF/EAM_2021.pdf

CONAGUA (2023) Actualización de la disponibilidad media anual de agua en el acuífero Libres-Oriental (2102), Estado de Puebla. Subdirección General Técnica Genercia de Aguas Subterráneas. Comisión Nacional del Agua, Ciudad de México (In spanish)

Cherlet M, Hutchinson C, Reynolds J, Hill J, Sommer S, Von Maltitz G (2018) World atlas of desertification, Publications Office of the European Union, Luxembourg. ISBN 978-92-79-75350-3, https://doi.org/10.2760/9205, JRC111155

Craig H (1961) Isotopic variations in meteoric waters. Science 133:1702–1703

Cui J, Tian L, Biggs TW, Wen R (2017) Deuterium-excess determination of evaporation to inflow ratios of an alpine lake: implications for water balance and modeling. Hydrol Process 31(5):1034–1046

Dansgaard W (1964) Stable isotopes in precipitation. Tellus 16(4):436–468

Durov SA (1948) Natural waters and graphic representation of their composition. Dokl Akad Nauk SSSR 59(3):87–90

Edmunds WM, Cook JM, Darling WG, Kinniburgh DG, Miles DL, Bath AH, ... Andrews JN (1987) Baseline geochemical conditions in the Chalk aquifer, Berkshire, UK: a basis for groundwater quality management. Appl Geochem 2(3):251–274

Famiglietti JS (2014) The global groundwater crisis. Nat Clim Chang 4(11):945–948

Feng S, Fu Q (2013) Expansion of global drylands under a warming climate. Atmos Chem Phys 13(19):10081–10094

Feng S, Krueger AB, Oppenheimer M (2010) Linkages among climate change, crop yields and Mexico–US cross-border migration. Proc Natl Acad Sci 107(32):14257–14262

Freeze RA, Cherry JA (1979) Groundwater. Prentice Hall, Englewood Cliffs

Gat JR, Matsui E (1991) Atmospheric water balance in the Amazon Basin: an isotopic evapotranspiration model. J Geophys Res: Atmos 96(D7):13179–13188

Huang T, Pang Z (2012) The role of deuterium excess in determining the water salinisation mechanism: a case study of the arid Tarim River Basin, NW China. Appl Geochem 27:2382–2388

Hunter LM, Murray S, Riosmena F (2013) Rainfall patterns and US migration from rural Mexico. Int Migr Rev 47(4):874–909

INEGI (2000) Agricultura, Sintesis Geografica del Estado de Puebla. Instituto Nacional de Estadistuca Geografica e Informatica, Aguascaliente, Mexico, pp 85–90

Liebminger A, Haberhauer G, Papesch W, Heiss G (2006) Correlation of the isotopic composition in precipitation with local conditions in alpine regions. J Geophys Res: Atmos 111(D5)

Magaña V, Amador JA, Medina S (1999) The midsummer drought over Mexico and Central America. J Clim 12(6):1577–1588

Magaña VO, Vázquez JL, Pérez JL, Pérez JB (2003) Impact of El Niño on precipitation in Mexico. Geofísica Internacional 42(3):313–330

Marandi A, Shand P (2018) Groundwater chemistry and the Gibbs Diagram. Appl Geochem 97:209–212

Masson-Delmotte V, Zhai P, Pirani A, Connors SL, Péan C, Berger S et al (2021) Climate change 2021: the physical science basis. Contribution of working group I to the sixth assessment report of the intergovernmental panel on climate change

Natali S, Doveri M, Giannecchini R, Baneschi I, Zanchetta G (2022) Is the deuterium excess in precipitation a reliable tracer of moisture sources and water resources fate in the western Mediterranean? New insights from Apuan Alps (Italy). J Hydrol 614:128497

Nawrotzki RJ, Hunter LM, Runfola DM, Riosmena F (2015) Climate change as a migration driver from rural and urban Mexico. Environ Res Lett 10(11):114023

Pérez-Quezadas J, Adams D, Murillo RS, Lagunes AJ, Castañeda JLR (2021) Isotopic variability (δ18O, δ2H and d-excess) during rainfall events of the north American monsoon across the Sonora River Basin, Mexico. J S Am Earth Sci 105:102928

Pfahl S, Sodemann H (2014) What controls deuterium excess in global precipitation? Clim Past 10(2):771–781

Piper AM (1944) A graphic procedure in the geochemical interpretation of water-analyses. EOS Trans Am Geophys Union 25:914–928

Putman AL, Fiorella RP, Bowen GJ, Cai Z (2019) A global perspective on local meteoric water lines: meta-analytic insight into fundamental controls and practical constraints. Water Resour Res 55(8):6896–6910

Risi C, Landais A, Winkler R, Vimeux F (2013) Can we determine what controls the spatio-temporal distribution of d-excess and 17O-excess in precipitation using the LMDZ general circulation model? Clim Past 9:2173–2193

Roy PD, Selvam S, Venkatramanan S, Logesh N, Lakshumanan C, Sánchez-Zavala JL (2021) Identification of sources and groundwater recharge zones from hydrochemistry and stable isotopes of an agriculture-based paleo-lacustrine basin of drought-prone northeast Mexico. Geochemistry 81(2):125742

Roy PD, Selvam S, Gopinath S, Logesh N, Sánchez-Zavala JL, Lakshumanan C (2022) Geochemical evolution and seasonality of groundwater recharge at water-scarce southeast margin of the Chihuahuan Desert in Mexico. Environ Res 203:111847

Roy PD, García-Arriola OA, Selvam S (2024) Seasonality of hydrogeochemical evolutions and isotopic variabilities (δ18O, δ2H and d-excess) in the surface water as well as groundwater from tropical central-south Mexico. Environ Res 118529. https://doi.org/10.1016/j.envres.2024.118529

Rozanski K, Araguás-Araguás L, Gonfiantini R (1993) Isotopic patterns in modern global precipitation. Clim Change Cont Isot Rec 78:1–36

Sahib LY, Marandi A, Schüth C (2016) Strontium isotopes as an indicator for groundwater salinity sources in the Kirkuk region, Iraq. Sci Total Environ 562:935–945

Sánchez-Murillo R, González-Hita L, Mejía-González MA, Carteño-Martinez B, Aparicio-González JC, Mañón-Flores D et al (2023) Tracing isotope precipitation patterns across Mexico. Plos Water 2(10):e0000136. https://doi.org/10.1371/journal.pwat.0000136

Stahle DW, Cook ER, Burnette DJ, Villanueva J, Cerano J, Burns JN et al (2016) The Mexican Drought Atlas: tree-ring reconstructions of the soil moisture balance during the late pre-Hispanic, colonial, and modern eras. Quat Sci Rev 149:34–60

Tian C, Wang L, Kaseke KF, Bird BW (2018) Stable isotope compositions (δ2H, δ18O and δ17O) of rainfall and snowfall in the central United States. Sci Rep 8:1–15

Tiwari AK, Orioli S, De Maio M (2019) Assessment of groundwater geochemistry and diffusion of hexavalent chromium contamination in an industrial town of Italy. J Contam Hydrol 225:103503

Torres-Martínez JA, Mora A, Mahlknecht J, Daesslé LW, Cervantes-Avilés PA, Ledesma-Ruiz R (2021) Estimation of nitrate pollution sources and transformations in groundwater of an intensive livestock-agricultural area (Comarca Lagunera), combining major ions, stable isotopes and MixSIAR model. Environ Pollut 269:115445. https://doi.org/10.1016/j.envpol.2020.115445

Vergara-Martínez A, Zárate-López J, Jiménez-Hernández A (1997) Carta Geológica Minera de Veracruz E14-3, Servicio Geológico Mexicano

Wang C, Enfield DB (2001) The tropical Western Hemisphere warm pool. Geophys Res Lett 28(8):1635–1638

Wassenaar LI, Van Wilgenburg SL, Larson K, Hobson KA (2009) A groundwater isoscape (δD, δ18O) for Mexico. J Geochem Explor 102:123–136

Xia Z, Winnick MJ (2021) The competing effects of terrestrial evapotranspiration and raindrop re-evaporation on the deuterium excess of continental precipitation. Earth Planet Sci Lett 572:117120

Radionuclides as Environmental Contaminants of Concern: Threats to Public Health Through Soil and Groundwater

Johnbosco C. Egbueri, Johnson C. Agbasi, Chiedozie C. Aralu, Mahamuda Abu, Henry C. Uwajingba, and Sani I. Abba

Abstract Radionuclides, both naturally occurring and anthropogenic, pose significant challenges as environmental contaminants, with far-reaching implications for public health, soil and groundwater management, and ecosystem quality. This study examined the distribution, behavior, and health impacts of radionuclides in Nigeria, contextualizing the findings within a global perspective. It analyzed the sources, pathways, and risks associated with radionuclide contamination. The investigation revealed significant spatial variability in radionuclide concentrations across Nigeria. Notably, areas such as the Jos Plateau exhibited higher levels of natural radioactivity, with activity concentrations of ^{232}Th, ^{238}U, and ^{40}K reaching 436.08, 168.83, and 346.1 Bq/kg, respectively, substantially exceeding global averages. In contrast, some southwestern regions generally showed lower concentrations, with mean activity levels of ^{232}Th, ^{226}Ra, and ^{40}K at 23.23, 12.14, and 270.14 Bq/kg, respectively. Groundwater investigations revealed considerable variations, with some areas showing elevated radon (^{222}Rn) concentrations up to 24.16 Bq/L. The mobility

J. C. Egbueri (✉) · J. C. Agbasi
Department of Geology, Chukwuemeka Odumegwu Ojukwu University, Uli, Nigeria
e-mail: johnboscoegbueri@gmail.com; jc.egbueri@coou.edu.ng

J. C. Egbueri
Research Management Office (RMO), Chukwuemeka Odumegwu Ojukwu University, Anambra, Nigeria

C. C. Aralu
Department of Pure and Industrial Chemistry, Nnamdi Azikiwe University, Awka, Anambra State, Nigeria

M. Abu
Department of Geological Engineering, University for Development Studies, Tamale, Ghana

H. C. Uwajingba
Department of Surveying and Geoinformatics, Federal Polytechnic, Ugep, Cross River State, Nigeria

S. I. Abba
Department of Civil Engineering, Prince Mohammad Bin Fahd University, Al Khobar, Saudi Arabia

P. Li et al. (eds.), *Sustainable Groundwater and Environment: Challenges and Solutions*, Springer Hydrogeology, https://doi.org/10.1007/978-3-031-82194-3_15

and bioavailability of radionuclides were found to be governed by complex interactions between the soil properties, climatic conditions, and anthropogenic factors. Public health risk assessments indicated that while many areas fall within acceptable limits, localized hotspots pose potential long-term health risks. The study offer some key recommendations for future studies, including enhancement of standardization of measurement techniques, expansion of geographical coverage of investigations, and developing integrated research approaches that link environmental measurements with epidemiological studies.

Keywords Groundwater pollution · Human health risks · Nigeria · Radioactive contaminants · Radioactive elements · Radionuclides in water

1 Introduction

Radionuclides have garnered significant global attention due to their potential impacts on public health (Elaboudi et al. 2024), soil quality (Chernysh et al. 2024), and groundwater resources (Schubert et al. 2024). These unstable atoms, which emit ionizing radiation through radioactive decay, pose unique challenges in environmental management and human health protection. The presence of radionuclides in the environment, whether naturally occurring or anthropogenic, necessitates a comprehensive understanding of their sources, distribution, and potential risks to develop effective mitigation strategies. The global concern surrounding radionuclide contamination stems from the ubiquitous nature of these elements in the Earth's crust and their potential to enter various environmental compartments. Natural radionuclides, such as uranium, thorium, and their decay products, are present in varying concentrations in rocks, soils, and water bodies worldwide (Jha et al. 2024). In addition, human activities, including mining, nuclear energy production, and industrial processes, can significantly alter the natural distribution and concentration of these elements, potentially exacerbating environmental and health risks (Elaboudi et al. 2024; Atipo et al. 2020).

Groundwater, a critical resource for domestic, agricultural, and industrial use in many regions globally, is particularly vulnerable to radionuclide contamination. The mobility of radionuclides in subsurface environments depends on complex geochemical processes, which are influenced by factors such as soil properties, hydrogeological conditions, and climate (Ribeiro et al. 2018; Saint-Fort 2018; Iurian et al. 2015). Understanding these processes is crucial for predicting the fate and transport of radionuclides in groundwater systems and assessing potential exposure pathways to humans and ecosystems. Recent studies have highlighted the global variability in radionuclide occurrence and distribution in groundwater. For instance, Manawi et al. (2024) conducted a comprehensive review of global groundwater radon concentrations, revealing significant spatial heterogeneity and identifying regions with elevated levels that may pose health risks. The public health implications of radionuclide exposure through contaminated groundwater are of paramount concern. Chronic exposure to low-level radiation from radionuclides in drinking water has been associated with increased cancer risk and other adverse health effects (Schubert et al. 2024). A study by Martinelli et al. (2023) in the Rio Grande do Norte State, Brazil, found a significant spatial association between cancer occurrence and areas with high natural radioactivity, particularly in regions with pegmatite outcrops rich in uranium. Their findings revealed an odds ratio of 1.21 (95% CI 1.20; 1.23) for cancer cases in high-intensity natural radioactivity areas, emphasizing the urgent need for comprehensive monitoring, risk assessment, and health management strategies in regions with elevated levels of naturally occurring radioactive materials (NORMs).

In the context of soil management, radionuclides present unique challenges due to their persistence and potential for bioaccumulation in food chains (Fan et al. 2023). Agricultural practices, particularly the use of phosphate fertilizers, can inadvertently introduce radionuclides into soils (Szaciłowski 2024), potentially affecting crop quality and human health through dietary exposure. Recent research by Kumar and Choudhury (2024) has explored innovative phytoremediation techniques for radionuclide-contaminated soils, offering promising avenues for sustainable soil management in affected areas. The global interest in radionuclide contamination extends beyond immediate health and environmental concerns to encompass broader issues of environmental justice (Höffken and Ramana 2024) and sustainable development (Wimmers et al. 2024). Disparities in exposure risks and remediation efforts between developed and developing nations have been increasingly recognized (Doyi et al. 2017), prompting calls for international collaboration and knowledge sharing. A study by Orellana et al. (2021) in Latin America and the Caribbean highlights the importance of capacity building, international support, and public–private partnerships in addressing nuclear-related challenges.

In Nigeria, a nation characterized by various geological formations and rich mineral resources, the issue of radionuclide contamination takes on particular significance. The country's diverse geology, encompassing Precambrian crystalline rocks, sedimentary basins, and alluvial deposits (Tijani 2023), provides a unique setting for studying the natural occurrence and distribution of radionuclides. Understanding the geologic context is crucial for assessing potential hotspots of radionuclide contamination and developing targeted management strategies. The interaction between the geologic units and human activities, such as mining and oil exploration, creates a complex landscape of potential radionuclide sources and pathways. This necessitates a multidisciplinary approach to radionuclide management, which could integrate geosciences, environmental health, and policy studies. Recent advancements in geospatial analysis and machine learning offer promising tools for mapping and predicting radionuclide distribution across diverse terrains. For instance, Rezaie et al. (2023) demonstrated the effectiveness of enhanced machine learning algorithms, specifically a group method of data handling (GMDH) integrated with metaheuristic optimization, in modeling indoor radon distribution in South Korea. Such innovative approaches, which incorporate soil properties, geogenic factors, and topography, could potentially be adapted to map radionuclide concentrations in other regions, including Nigeria. By leveraging these advanced modeling techniques, researchers and policymakers could gain valuable insights into radionuclide hotspots, enabling more targeted and cost-effective management strategies across varied landscapes. As global awareness of radionuclide contamination grows, there is an increasing recognition of the need for comprehensive, science-based policies and regulations. The development of such frameworks requires an enhanced understanding of the sources, pathways, and risks associated with radionuclides in different environmental contexts (IAEA 2024).

This chapter aims to provide a comprehensive analysis of radionuclides as environmental contaminants, examining their occurrence, distribution, and potential impacts on soil and groundwater resources from a global perspective. While maintaining a broad international scope, the study uses Nigeria as a focal point to illustrate challenges and opportunities in radionuclide management within developing nations. Through this approach, the work explores the public health implications of radionuclide contamination and evaluates management strategies applicable across diverse geological and socioeconomic contexts. The specific objectives of this study are as follows: (1) To synthesize current knowledge on radionuclide presence and levels in soil and groundwater resources globally, with detailed insights from Nigeria as a case study; (2) To evaluate potential public health risks associated with radionuclide exposure through various pathways, considering both universal and region-specific factors; and (3) To identify emerging issues and research needs in the field of radionuclide contamination and management, contributing to the global scientific discourse on this topic. The novelty of this study lies in its comprehensive approach to radionuclide contamination in the world, with a focus on Nigeria, integrating geological, environmental, and public health perspectives. By synthesizing current knowledge and identifying critical gaps, this work aims to provide a holistic understanding of radionuclide issues specific to Nigeria's unique geological and environmental contexts. This approach not only contributes to the national discourse on environmental management but also offers valuable insights for global efforts in addressing radionuclide contamination in diverse settings.

This comprehensive analysis of radionuclides as environmental contaminants makes several significant contributions to the field. Firstly, it bridges the gap between global perspectives on radionuclide contamination and specific insights from a geologically diverse developing nation. By using Nigeria as a focal point while maintaining a broad international scope, this study offers a unique synthesis of broad and localized knowledge, enhancing our understanding of radionuclide issues in varied contexts. Secondly, the chapter contributes to the global body of knowledge on radionuclide management by providing an in-depth analysis of challenges and potential solutions. These insights are applicable to countries with similar geological settings or developmental stages, thus extending the study's relevance beyond its primary focus area. Lastly, by addressing both universal and region-specific aspects of radionuclide contamination, this work promotes international collaboration and knowledge exchange. It fosters a more holistic approach to this pervasive environmental challenge, encouraging a global perspective on local issues.

2 Naturally Occurring Radionuclides

2.1 *Overview and Sources*

NORMs are ubiquitous components of the Earth's crust, contributing significantly to background radiation levels worldwide. These materials have been present since the planet's formation. The distribution of NORMs varies globally, influenced by geological formations, tectonic activity, and mineralization processes (Seow et al. 2024). On a global scale, areas of elevated natural radioactivity have been identified in several regions. Notable examples include the monazite sand areas of Brazil (Larijani et al. 2017) and India (Natarajan et al. 2024), the high background radiation areas in Ramsar, Iran (Fathabadi et al. 2017), and certain granitic regions in Finland (Kallio et al. 2023) and China (Yang and Sun 2022). These areas often exhibit radiation levels significantly above the global average, sometimes reaching levels comparable to or exceeding those in some nuclear facilities. In Africa, the continent's diverse geology contributes to a varied distribution of NORMs. The East African Rift System, for instance, is known for its elevated levels of radon due to the presence of volcanic rocks and hot springs (Muga 2017; Davidson et al. 2016). In South Africa, the gold mining regions of the Witwatersrand Basin have been subjects of extensive radiological studies due to the presence of uranium in the gold-bearing reefs (Tibane and Mamba 2024; Tutu et al. 2009; Wade and Coetzee 2008). Focusing on Nigeria, the country's complex geology plays a crucial role in the distribution of NORMs (Fig. 1). The Nigerian landmass is composed of two main geological components: the Basement Complex and the Sedimentary Basins. The Basement Complex, which covers about half of the country, is particularly significant in terms of natural radioactivity (Olatunji et al. 2022). The North-Central region of Nigeria, especially the Jos Plateau, has garnered significant attention due to its high background radiation levels. Some areas in this region exhibit radioactivity values comparable to known high background radiation areas worldwide, such as the Odisha coast in India (Kumar et al. 2024; Olatunji et al. 2022). These elevated levels are primarily attributed to the presence of uraniferous and thoriferous minerals, often exposed through mining activities (Olatunji et al. 2022).

2.2 *Radiological Properties and Behavior*

The radiological properties of naturally occurring radionuclides are fundamental to understanding their environmental impact and potential health effects. The most significant naturally occurring radionuclides are Potassium-40 (^{40}K), and the isotopes and decay products of uranium and thorium. These radionuclides exhibit a wide range of half-lives, from billions of years for the primordial radionuclides to mere seconds for some of their decay products (Fig. 2). Uranium-238 (^{238}U), with a half-life of approximately 4.5 billion years, is the progenitor of a complex decay series that

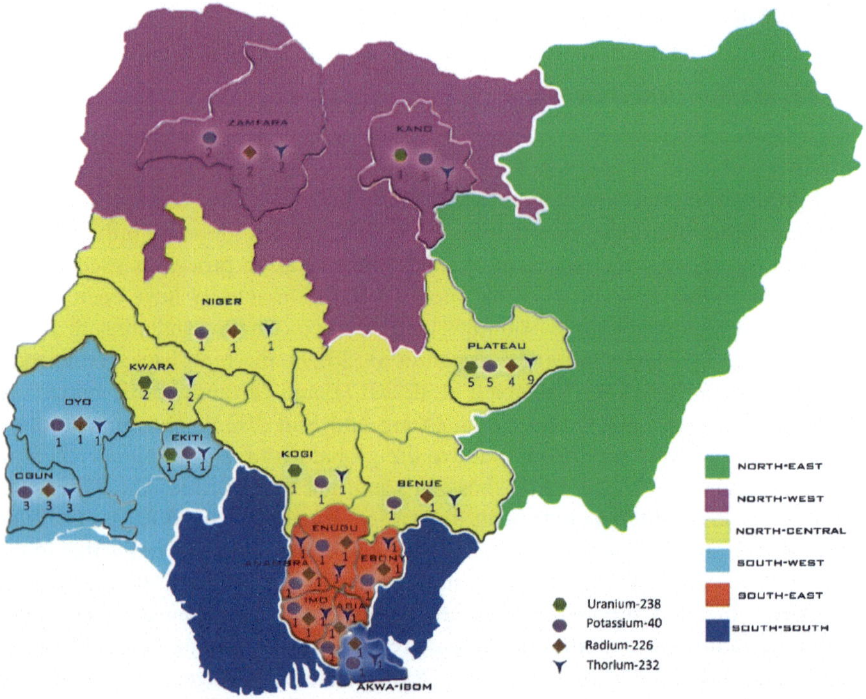

Fig. 1 A detailed map of Nigeria illustrating the variation in NORMs across different geopolitical zones (Olatunji et al. 2022)

includes radium-226 (^{226}Ra) and radon-222 (^{222}Ra) (Fig. 2). This series, known as the uranium series, consists of 14 radionuclides with half-lives ranging from microseconds to billions of years. Thorium-232 (^{232}Th), with a half-life of about 14 billion years, heads another important decay series comprising 10 radionuclides. ^{40}K, despite its relatively low isotopic abundance of 0.0117% (Kuhnel et al. 2021), contributes significantly to natural background radiation due to its ubiquity in the environment and its long half-life of 1.25 billion years (Storm 2020).

The behavior of these radionuclides in the environment is governed by their physical half-lives and chemical properties. Uranium, for instance, exists primarily in two oxidation states in nature: U(IV) and U(VI). Under oxidizing conditions, uranium tends to form soluble complexes, particularly with carbonate ions (Alloway 2013; Aly and Hamza 2013), enhancing its mobility in the environment. Conversely, under reducing conditions, uranium tends to precipitate as insoluble U(IV) compounds, limiting its movement (Kim et al. 2024). Thorium, on the other hand, exists predominantly in the +4 oxidation state and is generally less mobile due to its lower solubility and strong tendency to hydrolyze and form colloidal particles. However, in the presence of certain organic ligands, thorium mobility can be enhanced (Saha et al. 2023).

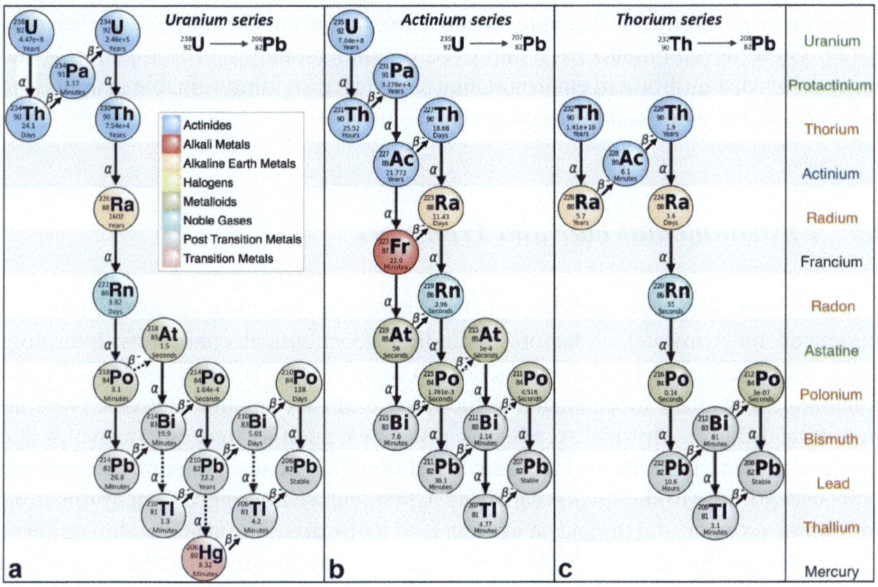

Fig. 2 A comprehensive decay series diagram for **a** uranium-238 (^{238}U), **b** uranium-235 (^{235}U), and **c** thorium-232 (^{232}Th) (Tan 2016)

Radium, a decay product of both uranium and thorium, exhibits behavior similar to other alkaline earth elements. It can be relatively mobile in the environment, especially in chloride-rich waters (Walters 1995). The mobility of radium is often controlled by co-precipitation with or adsorption onto metal oxides and clay minerals (Chen and Kocar 2018). Radon, being a noble gas, presents unique challenges in environmental radiological protection. It can easily diffuse through soil and rock (Tan et al. 2016), potentially accumulating in buildings and posing health risks. The behavior of radon is largely governed by physical processes such as diffusion and advection, rather than chemical reactions (Elmaghraby et al. 2021). The decay of these radionuclides results in the emission of various types of ionizing radiation, including alpha particles, beta particles, and gamma rays. Alpha particles, while highly ionizing, have limited penetrating power and are primarily of concern when radionuclides are internalized. Beta particles have moderate penetrating power, while gamma rays are highly penetrating and can pose external exposure risks.

Understanding the concept of radioactive equilibrium is crucial when studying naturally occurring radionuclides. In undisturbed systems, the decay series may reach a state of secular equilibrium, where the activity concentrations of all members of the decay chain are equal (Ivanovich 1994). However, environmental processes can disrupt this equilibrium, leading to disequilibrium conditions that can significantly affect the distribution and behavior of radionuclides in the environment (Ivanovich 1994). Recent advances in analytical techniques, such as alpha spectrometry coupled with machine learning algorithms, have enhanced the ability to characterize these radiological properties with unprecedented precision (Jinia et al. 2024; Dragović 2022). These developments have improved the understanding of isotope ratios and decay series disequilibria in environmental samples, providing valuable insights into radionuclide behavior and transport mechanisms.

2.3 Environmental Fate and Transport

The environmental fate and transport of naturally occurring radionuclides are influenced by a myriad of factors, including geochemical conditions, hydrological processes, and biological interactions (Landa 2007). Weathering of rocks and minerals is a primary mechanism for the release of radionuclides into the environment (Gadd 2007). Physical weathering processes, such as freeze-thaw cycles and abrasion, can increase the surface area of radionuclide-bearing minerals, enhancing their susceptibility to chemical weathering. Chemical weathering, driven by reactions with water, oxygen, and organic acids, can lead to the dissolution and mobilization of radionuclides (Tawfic et al. 2021). Once released, radionuclides can be transported through various environmental compartments. In aquatic systems, radionuclides can exist in dissolved forms, associated with suspended particles, or incorporated into bottom sediments (Hirose and Povinec 2022). The partitioning between these phases is governed by factors such as pH, redox conditions, and the presence of complexing agents. For instance, uranium tends to form stable carbonate complexes in alkaline waters, enhancing its mobility, while in reducing environments, it may precipitate as insoluble minerals (Kim et al. 2024; Alloway 2013; Aly and Hamza 2013).

Atmospheric transport of radionuclides can occur through the suspension of radioactive dust particles (Abed et al. 2024) and the emanation of radon gas from soil and rocks (Kaminski et al. 2024). Radon, being a noble gas, can diffuse through soil pores and enter the atmosphere. Its short-lived decay products can attach to aerosol particles, leading to their deposition on surfaces or inhalation by organisms (Vaupotič 2024). In terrestrial ecosystems, the uptake of radionuclides by plants represents a significant pathway for their entry into the food chain. Factors influencing plant uptake include soil properties, plant species, and the chemical form of the radionuclide (Hegazy et al. 2013). Some plants have been found to accumulate certain radionuclides preferentially (e.g., epiphytic plants; Manigandan 2009), a property that has been exploited in phytoremediation strategies (Wang and Chen 2024). Microbial activities play a crucial role in the environmental fate of radionuclides. Microorganisms can alter the oxidation state of radionuclides, affecting their solubility and mobility. For example, certain bacteria can reduce soluble U(VI) to insoluble U(IV), potentially immobilizing uranium in contaminated environments (Yu et al. 2024). Conversely, microbial oxidation of reduced species can enhance radionuclide mobility (Wilkins et al. 2007).

The transport of radionuclides in groundwater systems is of particular concern due to the potential for contamination of drinking water sources. Advection, dispersion, and diffusion are the primary physical processes governing radionuclide transport

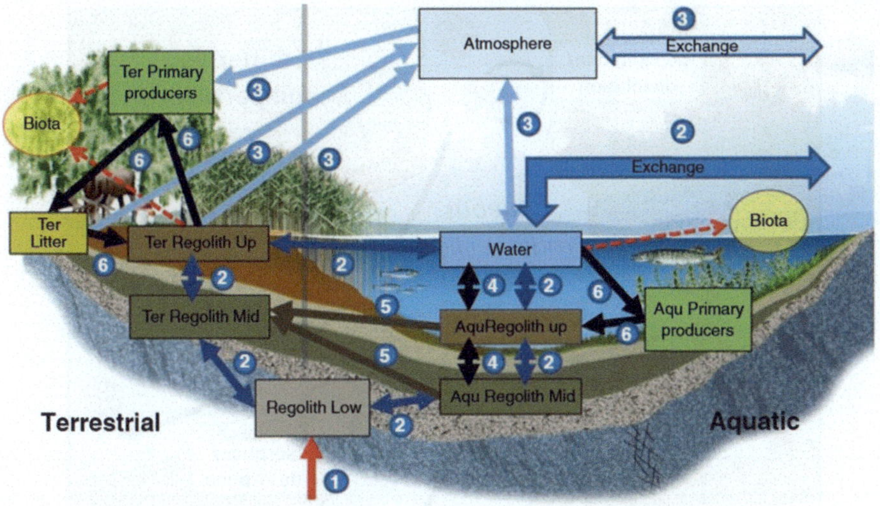

Fig. 3 A conceptual model illustrating the major pathways of radionuclide transport in the environment (Avila et al. 2013)

in porous media (Elmaghraby et al. 2021). However, chemical processes such as sorption, precipitation, and complexation can significantly affect transport rates and patterns (Zhang et al. 2022). Climate and hydrological conditions play a significant role in radionuclide transport. Precipitation patterns influence weathering rates and the flushing of radionuclides from soils. In arid regions, evaporation can lead to the concentration of radionuclides in surface waters and the formation of secondary minerals (Rentería-Villalobos et al. 2017).

Recent advances in modeling techniques have improved the ability to predict radionuclide fate and transport in complex environmental systems (Dragović 2022). Coupled hydrogeochemical models, incorporating both physical transport processes and chemical reactions, provide valuable tools for assessing long-term behavior and potential risks associated with naturally occurring radionuclides (Borrego-Alonso et al. 2023) (Fig. 3).

2.4 Potential Interactions Between Radionuclides in Soil and Groundwater

The interface between soil and groundwater represents a critical zone for radionuclide interactions, with implications for both environmental quality and human health. The behavior of radionuclides at this interface is governed by a complex interplay of physical, chemical, and biological processes (Fig. 4). Sorption processes play a crucial role in controlling radionuclide mobility at the soil-groundwater interface (Sun et al. 2022). Adsorption onto mineral surfaces, particularly clay minerals and

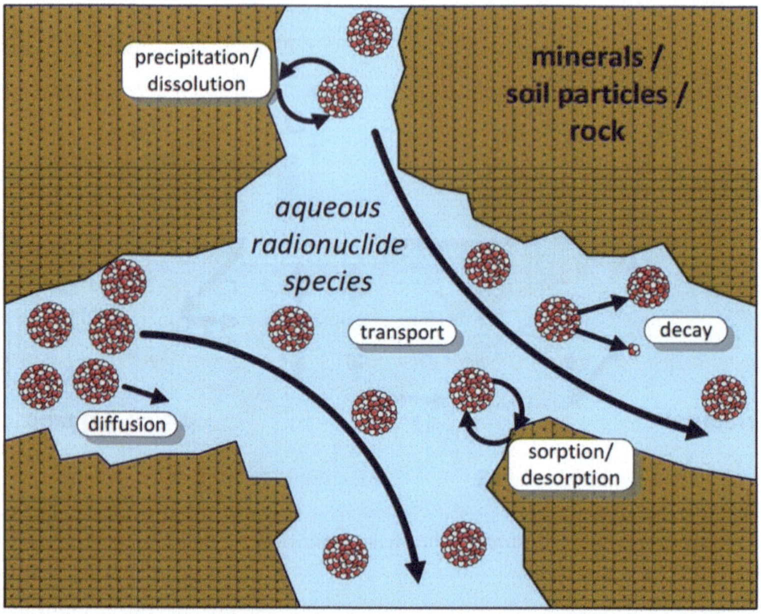

Fig. 4 Soil-groundwater radionuclide interactions (Heuel-Fabianek 2014)

metal oxides, can significantly retard radionuclide transport. The extent of sorption is influenced by factors such as pH, ionic strength, and the presence of competing ions. For instance, uranium sorption tends to increase with pH, reaching a maximum in the near-neutral range, before decreasing at higher pH values due to the formation of soluble carbonate complexes (Katsoyiannis et al. 2006).

Ion exchange processes, particularly important for cationic radionuclides like radium and cesium, can affect the partitioning of radionuclides between the solid and aqueous phases (Heuel-Fabianek 2014). The cation exchange capacity of soils, largely determined by clay content and organic matter, influences the extent of these interactions (Iurian et al. 2015). Precipitation and co-precipitation reactions can lead to the immobilization of radionuclides at the soil-groundwater interface (Zafar et al. 2021). Changes in geochemical conditions, such as redox potential or pH, can trigger the formation of secondary minerals incorporating radionuclides. For example, the precipitation of iron and manganese oxides can scavenge uranium and other radionuclides from solution (de Moraes and Ladeira 2021; Abdou et al. 2018).

Complexation with organic matter can significantly influence radionuclide behavior. Dissolved organic matter can form stable complexes with radionuclides, potentially enhancing their mobility (Matsunaga et al. 2004). Conversely, sorption to solid-phase organic matter can retard radionuclide transport (Stockdale et al. 2013). The nature and concentration of organic matter thus play a crucial role in determining radionuclide fate. Colloidal transport has gained increasing recognition as an important mechanism for radionuclide migration (Malkovsky et al. 2023). Radionuclides associated with mobile colloidal particles, such as clay minerals, organic matter, or metal oxides, can be transported over greater distances than predicted by considering only dissolved species (Salbu 2024).

Microbial activities at the soil-groundwater interface can significantly influence radionuclide behavior. Biofilms can alter the physical and chemical properties of mineral surfaces, affecting sorption processes. Microbial reduction or oxidation of radionuclides can change their solubility and mobility. For instance, microbial reduction of U(VI) to U(IV) can lead to the precipitation of insoluble uraninite, potentially immobilizing uranium (Zhang et al. 2021). The presence of co-occurring contaminants can significantly influence radionuclide behavior at the soil-groundwater interface (Blake et al. 2019). Competition for sorption sites, formation of mixed complexes, and alterations in redox conditions can all affect radionuclide mobility. For example, the presence of phosphate can lead to the formation of insoluble uranium phosphate minerals, reducing uranium mobility (Knox et al. 2008). Hydrological factors, such as groundwater flow rates and fluctuations in water table levels, can influence the transport and distribution of radionuclides at the soil-groundwater interface (Cao et al. 2017). Seasonal variations in precipitation and evapotranspiration can lead to cyclic patterns of radionuclide mobilization and immobilization (Payne and Edis 2012).

Recent advances in spectroscopic and microscopic techniques have provided new insights into radionuclide speciation and behavior at the molecular scale. Techniques such as X-ray absorption spectroscopy and electron microscopy have revealed the complex nature of radionuclide interactions with mineral surfaces and microbial cells (Ohnuki et al. 2020).

3 Radionuclides in the Environment

3.1 Distribution and Levels of Radionuclides in Soil and Groundwater

The presence and distribution of radionuclides in soil and groundwater are of significant concern due to their potential impact on human health and the environment. Soil serves as a primary repository for various environmental contaminants, including radionuclides. The behavior and distribution of radionuclides in soil are influenced by a complex interplay of factors, including soil physicochemical properties, parent rock types, historical geochemical processes, and anthropogenic activities (Ogunyele et al. 2020; Jasaitis et al. 2020; Adebiyi et al. 2021). These factors collectively determine the extent of radionuclide absorption and retention in soil matrices.

In Nigeria, numerous studies have been conducted to assess the levels of radionuclides in soil across different regions (Table 1). A comprehensive review of these studies reveals varying concentrations of key radionuclides such as ^{226}Ra, ^{232}Th, ^{238}U, and ^{40}K. Many investigations reported higher activity concentrations of ^{40}K compared to other radionuclides (Akpan et al. 2020; Ademola 2021; Oluyide et al. 2019; Adedokun et al. 2019; Ibikunle et al. 2019; Popoola et al. 2019). This observation aligns with the assertion that environmental media with high absorption activities for ^{40}K typically exhibit lower absorption activities for ^{226}Ra and vice versa (Adedokun et al. 2019). However, some studies reported contrasting findings, with relatively lower activity levels of ^{40}K compared to other radionuclides (Adesiji and Ademola 2019; Adebiyi and Ore 2020; Atipo et al. 2020; Ogundele et al. 2020). This variability accentuates the heterogeneous nature of radionuclide distribution in Nigerian soils, which can be attributed to the diverse geological formations, soil types, and human activities across the country.

The global context of radionuclide distribution in soil presents a similarly complex picture. Studies conducted in various countries have reported wide-ranging concentrations of radionuclides. For instance, research in China has shown that the distribution of radionuclides in soil is strongly influenced by geological factors, with higher concentrations often associated with specific rock types (Chen et al. 2022). In Europe, a comprehensive study across multiple countries revealed significant spatial variability in radionuclide concentrations, with some areas exhibiting elevated levels due to both natural and anthropogenic factors (Rühm et al. 2020). In the United States, the U.S. Geological Survey has conducted extensive mapping of soil radioactivity, identifying regions with naturally high concentrations of uranium and thorium, particularly in the western states (Duval et al. 1990).

3.1.1 Radionuclide Concentrations in Groundwater

The presence of radionuclides in groundwater is a critical environmental and public health concern, as groundwater serves as a primary source of drinking water for many communities worldwide. In Nigeria, several studies have investigated the concentrations of radionuclides in groundwater across different regions. A review of recent studies in Nigeria reveals varying levels of radionuclides in groundwater (Table 2). For instance, Abiodun et al. (2019) reported concentrations of ^{238}U, ^{232}Th, and ^{40}K in groundwater samples from Osogbo, Osun State, with values of 8.38 Bq/L, 6.45 Bq/L, and 202.70 Bq/L, respectively. These levels are notably higher than those reported in some other regions of the country. To illustrate, a study conducted in Ebonyi State by Ononugbo and Nwaka (2017) found lower concentrations of ^{232}Th (4.17 Bq/L), ^{40}K (9.82 Bq/L), and ^{226}Ra (1.22 Bq/L) in groundwater samples. The variability in radionuclide concentrations across different regions of Nigeria can be attributed to factors such as local geology, hydrogeological conditions, and potentially, anthropogenic influences. Of particular interest is the presence of radon (^{222}Rn) in groundwater, which has been the focus of several studies due to its potential health impacts. Ajibola et al. (2021) reported a mean ^{222}Rn concentration of 24.16 Bq/L in groundwater samples from Kwara State, while Olise et al. (2016) found an average concentration of 23.036 Bq/L in southwestern Nigeria. These values are within the range of global averages but highlight the need for continued monitoring and assessment of radon in Nigerian groundwater resources.

Table 1 Levels of radionuclides in Nigerian soil (Bq/kg)

Location	Region	^{226}Ra	^{232}Th	^{238}U	^{40}K	References
Ogun State	SW	28.69	45.86	–	481.22	Ademola (2021)
Asa, Kwara State	NC	–	18.15	42.86	570.91	Orosun et al. (2021)
Lagos State	SW	11.2	10.99		19.38	Adedokun et al. (2020)
Akure, Ondo State	SW	–	0.89	132.13	51.52	Adebiyi and Ore (2020)
Akwa Ibom	SS	–	36	23	145	Akpan et al. (2020)
Jos, Plateau State	NC	–	168.83	436.08	346.1	Atipo et al. (2020)
Mowe, Ogun state	SW	–	11.21	67.28	1071.85	Egunjobi et al. (2020)
Benue State	NC	29.44	39.10	–	113.02	Kungur et al. (2020)
Ondo State	SW	–	19.8	171.8	146.2	Ogundele et al. (2020)
Osun State	SW	17.17	106.03	–	131.11	Olalekan and Adebiyi (2020)
Jos, Plateau State	NC	–	1776.08	242.13	374.01	Adesiji and Ademola (2019)
Abeokuta, Ogun state	SW	47.10	30.87	–	269.29	Ekhaguere et al. (2019)
South Western Cities	SW	76.79	52.91	–	393.73	Ibikunle et al. (2019)
Lagos	SW	–	10.44	11.47	403.07	Ilori and Alausa (2019)
Ile Ife, Osun State	SW	–	23.23	12.14	270.14	Oluyide et al. (2019)
Delta State	SS	561.67	54.43	–	413.64	Ononugbo and Amah (2019)
Edo State	SS	–	6.89	2.07	57.80	Popoola et al. (2019)
South western cities	SW	–	61.12	21.53	554.2	Ajayi et al. (2018)
Ogun State	SW	–	103.19	30.51	350.75	Gbadamosi et al. (2018)
Ewekoro, Ogun State	SW	–	51.13	1.95	285.34	Usikalu et al. (2018)
South western cities	SW	–	17.37	8.27	151.72	Ajayi et al. (2017)
Elere, Oyo State	SW	–	29.05	36.55	537.3	Alusa et al. (2017)
Aramoko, Ekiti State	SW	–	34.74	22.62	316.72	Ajiboye et al. (2016)

SW South-West; ***NC*** North-Central; ***SS*** South-South

From a global perspective, radionuclide concentrations in groundwater exhibit considerable variation. In the United States (U.S.), a study in Maine by Yang et al. (2014) revealed that 4.8% of private bedrock wells in greater Augusta exceeded the U.S. Environmental Protection Agency's (US-EPA) Maximum Contaminant Level (MCL) for uranium (30 μg/L). Moreover, 29% of sampled wells showed radon concentrations above 4,000 pCi/L, the EPA's Alternative Maximum Contaminant Level. Notably, elevated levels of both radionuclides were associated with proximity

Table 2 Levels of radionuclides in groundwater (Bq/L)

Location	Region	^{238}U	^{232}Th	^{40}K	^{222}Rn	^{226}Ra	References
South western regions	SW	–	–	–	23.036	–	Olise et al. (2016)
Osogbo, Osun State	SW	8.38	6.45	202.70	–	–	Abiodun et al. (2019)
Ilorin, Kwara State	NC	–	–	–	–	3.7	Nwankwo (2013)
South Western Nigeria	SW	–	–	–	–	80.3	Olatunde et al. (2018)
Ebonyi State	SE	–	4.17	9.82	–	1.22	Ononugbo and Nwaka (2017)
Rivers State	SS	14.86	31.94	33.7	–	–	Ononugbo and Tutumeni (2016)
Ondo state	SW	1.77	1.41	33.03	–	–	Abey et al. (2017)
Kwara State	NC	–	–	–	24.16	–	Ajibola et al. (2021)
Ogbomoso, Oyo State	SW	–	–	–	1.86	–	Oni and Adagunodo (2019)
River State	SS	10.81	13.49	170.19	–	–	Ononugbo et al. (2017)

SW South-West; ***NC*** North-Central; ***SE*** South-East; ***SS*** South-South

to granitic intrusions. This geological correlation is further supported by findings in other regions of the U.S., such as the Nambe region of northern New Mexico, where over 50% of wells exceed the EPA's recommended drinking water standard, with concentrations reaching as high as 1200 ng/mL (Balaram et al. 2022). The issue of uranium contamination extends beyond the U.S., affecting various parts of the world. In India, several states including Punjab, Rajasthan, Karnataka, Telangana, and Madhya Pradesh have reported uranium concentrations in groundwater ranging from 0 to 1443 ng/mL, significantly exceeding the World Health Organization (WHO) guideline of 30 ng/mL in many locations (Balaram et al. 2022). This contamination pattern is mirrored in neighboring Pakistan, particularly in the Sindh and Punjab regions, which are geologically connected to the affected areas in India. Other countries facing similar challenges include Canada, Mongolia, Burundi, Zambia, South Korea, Jordan, Afghanistan, China, and Myanmar, with some areas reporting uranium concentrations up to 1400 ng/mL (Balaram et al. 2022). In Europe, countries like Finland have observed higher radionuclide concentrations in groundwater associated with uranium-rich granite bedrock (Pohjolainen 2015). While some regions have naturally elevated levels due to their underlying geology, others face challenges from anthropogenic sources or water-rock interactions that mobilize radionuclides into groundwater systems. For instance, in South Korea, a study of 4,140 wells found that 160 wells, primarily in plutonic bedrock regions, exceeded the WHO limit for uranium (Shin et al. 2016). The presence of uranium in these groundwaters was attributed to desorption and re-dissolution processes, with uranyl carbonates being the predominant uranium species (Shin et al. 2016). Interestingly, the relationship between geology and uranium concentration is not always straightforward. In Switzerland, while high uranium concentrations in groundwater are mainly found in alpine regions and can be traced to bedrock geology, there's no direct correlation with other trace elements (Stalder et al. 2012). This highlights the complex nature of radionuclide mobility in groundwater systems.

Redox conditions play a crucial role in uranium mobility. In Central Italy, oxidizing conditions in cold waters favor the formation of soluble uranium species, while anoxic conditions in thermal waters promote relatively insoluble species (Cinti et al. 2015). Conversely, in countries like Kazakhstan and Kyrgyzstan, where groundwater is predominantly anoxic, uranium concentrations in drinking water are generally low. However, surface waters in these countries can have extremely high uranium concentrations (up to 1920 ng/mL) due to leaching from ore bodies (Uralbekov et al. 2011). The global prevalence of uranium contamination in groundwater may be more extensive than currently recognized. For example, groundwater in a mountainous region near Rio de Janeiro, Brazil, has shown uranium concentrations up to 930 ng/mL, 30 times higher than the WHO limit (Godoy et al. 2019). This is particularly concerning given the area's popularity for tourism and water-related industries.

3.2 *Factors Influencing Radionuclide Mobility and Bioavailability*

The capacity of the soil itself to immobilize radionuclides is the main factor controlling activity concentrations available to biota, and it operates in conjunction with numerous external factors. Environmental factors such as soil redox potential, clay-sized soil fraction, degree of weathering, concentration of contaminants, organic matter content, pH, temperature fluxes, and precipitation have been identified as elements that controls the bioavailability, uptake, and transfer of these radionuclides in soils and different food chains (Ribeiro et al. 2018; Saint-Fort 2018).

3.2.1 Soil Organic Matter

Climate, soil type, vegetation, and soil organisms all have an impact on the amount of organic matter that accumulates in soil because most organic material in shallow soil horizons in tropical environments decomposes quickly (Iurian et al. 2015). Because of the organic acids, lipids, and humic acids that make up its composition, radionuclides can interact and react in a variety of ways with these substances (Iurian et al. 2015). A key mechanism of this interaction is the creation of chelates and mineral–organic complexes. These combinations' mobility is determined by the size and composition of the soil, while their stability is determined by the bioclimatic conditions, soil pH, cation concentration in the soil, functional group, and degree of saturation of the possible sorption sites (Ruedig and Johnson 2015).

3.2.2 Redox Potential (E_h) and the pH

The chemical behavior of radionuclides in soil is determined by their level of oxidation (Timothy and Robert 2012). It is known that the redox potential (Eh) and pH of the soil are important variables influencing the chemical speciation of radionuclides (Ribeiro et al. 2018). Water is necessary for the occurrence of all redox reactions in soil (Timothy and Robert 2012). Since the redox potential and pH have a negative correlation, a reduction in E_h will cause the pH to rise. Soil redox potential can be influenced by a number of variables, including biological activity, soil type, and the presence of a water-impermeable soil horizon (Ribeiro et al. 2018).

It is conceivable for radionuclide mobility to occur in different types of soil and is linked to particular minerals (Timothy and Robert 2012). Different adsorbing components are present in silt and clay soil fractions; the most significant ones include minerals like smectite, illite, imogolite, and allophane, as well as the oxides and hydroxides of silica, aluminum, iron, and manganese (Korobova et al. 2014; Timothy and Robert 2012). The adsorption capacity of cations is caused by the clay minerals' permanent negative charge on their surface, while the adsorption of anions or cations is caused by the clay minerals' variable charge, which can be either positive or negative and is based on the pH of the soil as well as the degree of protonation of functional Si, Al, Fe, or C groups ((Korobova et al. 2014).

3.2.3 Rainfall

The primary regulator of gas exchange between the soil and the atmosphere is the water content of the soil, which directly affects radionuclide speciation and redox potential (Ribeiro et al. 2018). While the sluggish entry of water along macropores promotes the chemical interaction between the soil solution and matrix, the water flow along macropores can carry on largely unaltered particles (Ribeiro et al. 2018). Strong rainfalls may accelerate the migration of radionuclides into deeper soil layers. Furthermore, Thørring et al. (2014) suggested that the composition of precipitation might affect radiocaesium mobility in the early post-deposition phase, likely prior to considerable time-dependent fixation.

3.2.4 Soil Structure and Texture

Radionuclide adsorption in natural environments seemed to be caused by silt (0.010 mm) to clay (0.001 mm) particle size fractions in soils. According to Korobova et al. (2014), radionuclide retention in soil is caused by tiny particle agglomeration that naturally occurs and is resistant to water (Timothy and Robert 2012). On the other hand, a soil layer containing a significant amount of coarse fractions and aggregated fine particles may smooth the distribution of radionuclides in the various mass fractions (Rong et al. 2022). It should be noted that radioactive sorption is known to be spatially and in-soil depth specific in response to the particle size and composition of the soils (Cardi et al. 2017). Because of its interactions with rocks and soils, groundwater may include radioactive materials at quantities that can be detected (Rong et al. 2022). The surrounding environment and the characteristics of the water affect the spread of these radionuclides (Cardi et al. 2017).

3.2.5 Climate Change and Soil Management

It's possible that past soil management practices and climate variations have left us with different conditions now. Direct alterations to temperature and water balance may also occur, as well as indirect changes to plant growth conditions and soil organism activity (Adebiyi et al. 2021). Significant changes in land use and the amount and quality of organic matter in the soil take years or decades, but soil parameters like temperature, redox potential, and water content can change in a matter of days or weeks. Increased biological availability and a mobilization of integrated radionuclides may arise from a faster turnover of organic matter (Adebiyi et al. 2021). Erosion by wind or water may increase in both humid and arid environments, which could lead to a greater spatial dispersion of radionuclides in polluted soil (Timothy and Robert 2012).

Radionuclide mobility would be favored by changing the characteristics of the soil and increasing soil deterioration when farmland was converted to waste land (Iurian et al. 2015). Elevated temperatures and greater precipitation levels impact not only the long-term behavior of radionuclides in soil but also accelerate all other soil processes (Timothy and Robert 2012). Raising temperatures in conjunction with growing aridity lead to slower weathering processes and less organic matter breakdown, which in turn reduces radionuclide mobilization (Iurian et al. 2015). However, organic matter decomposition is inhibited and radionuclide absorption in the organic soil component may be boosted at low temperatures and high humidity (Pontedeiro et al. 2010). Additionally, increasing soil acidity (pH value tends to fall) can increase radioactive mobility in the soil, resulting in increased potential root uptake or radionuclide loss to deeper soil layers (Rachkova et al. 2010).

3.3 Case Studies and Site-Specific Investigations

3.3.1 Nigerian Case Studies of Soil Radionuclide Contamination

Building upon the overview provided in Sect. 3.1, several site-specific investigations in Nigeria have revealed notable variations in radionuclide concentrations across different regions. These studies provide valuable insights into the factors influencing radionuclide distribution in Nigerian soils. A study by Atipo et al. (2020) in Jos, Plateau State, an area known for tin mining activities, reported significantly elevated levels of radionuclides in soil samples. The activity concentrations of ^{232}Th, ^{238}U, and ^{40}K were found to be 436.08, 168.83, and 346.1 Bq/kg, respectively. These values substantially exceed the global averages and can be attributed to the mineral composition of the soils in this region, particularly the presence of radioactive minerals associated with tin deposits. In contrast, investigations in southwestern Nigeria have generally reported lower radionuclide concentrations. For instance, Oluyide et al. (2019) found mean activity concentrations of ^{232}Th, ^{226}Ra, and ^{40}K in soils to be 23.23, 12.14, and 270.14 Bq/kg, respectively. These values are below the worldwide averages, suggesting a lower radiological risk in this region. However, the authors emphasized the importance of monitoring occupational exposure to prevent long-term radiation buildup in the soil matrix. The influence of anthropogenic activities on radionuclide distribution was highlighted in a study by Akpan et al. (2020) in Akwa Ibom State. The researchers observed slightly higher concentrations of ^{40}K in locations with increased human activities, indicating that land use patterns can impact radionuclide distribution in soils.

3.3.2 Nigerian Case Studies of Groundwater Radionuclide Contamination

Groundwater investigations in Nigeria have revealed significant variations in radionuclide concentrations across different geological settings. A study by Olise et al. (2020) in a solid-minerals hosted area of southwestern Nigeria reported a mean ^{222}Rn concentration of 23.036 Bq/L in groundwater samples. The authors attributed the elevated radon levels to the mica-granite geological structure of the area, emphasizing the role of bedrock composition in determining radionuclide concentrations in groundwater. In Osogbo, Osun State, Abiodun et al. (2019) reported notably high activity concentrations of ^{238}U, ^{232}Th, and ^{40}K in groundwater, with values of 8.38, 6.45, and 202.70 Bq/L, respectively. The exceptionally high levels of ^{40}K observed in this study underscore the importance of considering local geological formations when assessing radiological risks associated with groundwater consumption. Conversely, a study by Ononugbo and Nwaka (2017) in Ebonyi State found relatively lower radionuclide concentrations in borehole water, with mean activity levels of ^{232}Th, ^{40}K, and ^{226}Ra at 4.17, 9.82, and 1.22 Bq/L, respectively. These values were within safe limits for radiological health risk parameters, illustrating the spatial variability of radionuclide contamination across different regions of Nigeria. An investigation by Abey et al. (2017) in a bitumen deposit area of Ondo State provided insights into the potential impact of natural hydrocarbon deposits on groundwater radionuclide levels. The study reported mean activity concentrations of ^{238}U, ^{232}Th, and ^{40}K at 1.77, 1.41, and 33.03 Bq/L, respectively. The authors suggested that the observed radioactivity levels were indicative of contamination from the bitumen deposits, highlighting the complex interactions between geological formations and groundwater quality.

4 Public Health Implications of Radionuclides

4.1 Exposure Pathways and Routes

Radionuclides pose significant public health concerns globally, with their presence and impacts extensively studied across various regions, including Africa, Europe, North America, and Asia (Adedokun et al. 2020). These radioactive elements are ubiquitous in the environment, found in soils, water bodies, air, and geological formations (Ramola et al. 2008). The distribution and concentration of radionuclides such as ^{40}U, ^{228}Th, and ^{232}Ra vary significantly across different geographical locations, influenced by factors such as local geology, anthropogenic activities, and atmospheric transport. Human exposure to radionuclides occurs through multiple pathways, with the food chain being recognized as one of the most critical routes (Tsai et al. 2008; International Commission on Radiological Protection 2012; El-Gamal et al. 2019). This pathway involves the uptake of radionuclides by food crops from contaminated soils, subsequently entering the human body through ingestion. Recent studies have highlighted the importance of understanding the soil-to-plant transfer factors for various radionuclides, as these can significantly influence human exposure levels.

While ingestion remains a primary exposure route, inhalation and direct external exposure also contribute significantly to overall human exposure (Egbueri et al. 2024). Inhalation of radon gas and its progeny, particularly in areas with high natural background radiation or in poorly ventilated buildings, has gained increased attention due to its association with lung cancer. Direct external exposure, especially in regions with elevated natural background radiation or near nuclear facilities, continues to be a concern for public health officials and radiation protection authorities. Emerging research has also begun to focus on the potential for radionuclide exposure through dermal absorption, particularly in occupational settings or areas with high levels of contamination. This exposure route, while less significant than ingestion or inhalation, highlights the need for comprehensive risk assessments and protective measures in radiation-contaminated environments.

4.2 Health Effects and Risks Associated with Radionuclide Exposure

The health effects of radionuclide exposure are primarily associated with long-term, chronic exposure rather than acute, short-term incidents. This characteristic makes the assessment and management of radionuclide-related health risks particularly challenging, as effects may not manifest for years or even decades after initial exposure (Ryan 2012; WHO 2016). Chronic exposure to ionizing radiation from radionuclides can lead to a spectrum of health effects, ranging from cellular damage to the development of various cancers. Recent epidemiological studies have strengthened the evidence linking prolonged radionuclide exposure to an increased risk of leukemia, thyroid cancer, and other solid tumors (Kitahara and Schneider 2022; Rühm et al. 2022). The inhalation of radon and its decay products remains a significant public health concern, with recent global estimates suggesting that it may be responsible for several lung cancer cases worldwide, depending on the regional radon levels and smoking prevalence (Riudavets et al. 2022; de Bernardo Gisbert et al. 2025).

Beyond cancer, emerging research has highlighted potential non-cancer effects of chronic low-dose radiation exposure. Cardiovascular diseases have been increasingly linked to radiation exposure, with recent studies suggesting that even low doses may contribute to an elevated risk of heart disease and stroke (Little et al. 2023; Jahng et al. 2024). Additionally, there is growing evidence of potential neurological effects, including cognitive decline and an increased risk of neurodegenerative diseases, associated with long-term exposure to low-dose radiation (Jayan et al. 2022). The field of radiobiology has made significant advances in understanding the mechanisms by which ionizing radiation affects human health at the molecular and cellular levels. Recent studies have enlightened the role of oxidative stress, DNA damage, and epigenetic alterations in radiation-induced health effects (Danforth et al. 2022; Ruprecht et al. 2024). These insights are crucial for developing more accurate risk models and potential interventions to mitigate the health impacts of radionuclide exposure.

An emerging area of concern is the potential for transgenerational effects of radionuclide exposure. While still a subject of ongoing research, some studies suggest that parental exposure to ionizing radiation may have health implications for offspring, possibly through epigenetic mechanisms (Belli and Tabocchini 2020; Chadha 2021). This highlights the need for long-term, multigenerational studies to fully understand the extent of radionuclide-related health risks. The global scientific community continues to refine its understanding of dose-response relationships for various radionuclides and exposure scenarios. Recent advancements in dosimetry and biomarkers of radiation exposure are enhancing the ability to accurately assess individual doses and potential health risks (Ainsbury et al. 2022; Blakely et al. 2024). These developments are crucial for improving radiation protection standards and public health interventions in areas affected by elevated radionuclide levels.

As global awareness of radionuclide-related health risks grows, there is an increasing emphasis on developing and implementing effective risk communication strategies. Public perception and understanding of radiation risks can significantly impact the success of mitigation efforts and health interventions. Thus, tailoring risk communication to different cultural contexts and addressing public concerns about long-term health effects are becoming integral components of comprehensive radionuclide management strategies worldwide.

4.3 *Risk Assessment and Dose Estimation*

In the evaluation and health risk assessment of radionuclides, some equations and indices have been used over the years. For instance, the specific activities (Bq/kg) for soil samples (Eq. 1) and activity concentrations (Bq/l) for water samples (Eq. 2) (Shanthi et al. 2010; Jahan et al. 2016). Also, the radiological hazard indices are estimated for ^{40}U, ^{228}Th, and ^{232}Ra in soils that are used for activities e.g. cultivation of food crops. Radionuclide indices like internal hazard index (H_{in}), representative level index (RLI) (Amana 2017), absorbed dose rate (D), Radium equivalent (Ra_{eq}), excess lifetime cancer risk (ELCR), and external hazard index (H_{ex}) (Adedokun et al., 2020). The following equations have been used:

$$A(Bq/kg) = C/(t \times m \times \varepsilon \times PY) \quad (1)$$

$$A(Bq/l) = C/(t \times V \times \varepsilon \times PY) \quad (2)$$

With A as the specific concentration of radionuclide in soil (Eq. 1) and water (Eq. 2), m (kg) is the mass of the soil, V is the volume of water (litters), t is the counting life (seconds), C is the net count of the sample in peak range, while ε is the detector energy dependent on efficiency, and then P_Y is the gamma-ray yield per disintegration of the nuclides (Shanthi et al. 2010; Jahan et al. 2016; Adedokun et al. 2020).

H_{in} estimates the level of health effect on the internal organs (especially the respiratory organs) of humans on long-term exposure to radionuclides (radon) and is evaluated using Eq. 3 (Harb et al. 2014).

$$H_{in} = (A_{Ra}/185\,Bq/kg) + (A_{Th}/259\,Bq/kg) + (A_K/4810\,Bq/kg) \quad (3)$$

where A_{Ra}, A_{Th}, and A_K are specific activities of ^{226}Ra, ^{232}Th, and ^{40}K, respectively.

Hex which is based on the exposure to the aforestated radionuclides following the use of farmland soils is estimated using Eq. 4.

$$Hex = (A_{Ra}/370\,Bq/kg) + (A_{Th}/259\,Bq/kg) + (A_K/4810\,Bq/kg) \quad (4)$$

The values of H_{in} and H_{ex} must be less or equal to 1 for the radiation hazard to be insignificant (Adedokun et al. 2020).

Ra_{eq} is one of the equations used in radionuclides hazard assessment and it is a single parameter that is often used to generalize human exposure to the radionuclides stated above. The estimation of this (Ra_{eq}) from the soil is based on the assumption that 10 Bq/kg of ^{226}Ra, 7 Bq/kg of ^{232}Th, and 130 Bq/kg of ^{40}K produce equal gamma dose (Eq. 5) (Arafat et al. 2017).

$$Ra_{eq} = A_{Ra} + 1.43A_{Th} + 0.077A_K \quad (5)$$

Also, to evaluate the level of radionuclide radiations associated with the varied concentrations of ^{226}Ra, ^{232}Th, and ^{40}K, the RLI is used (Eq. 6).

$$RLI = (A_{Ra}/150\,Bq/kg) + (A_{Th}/100\,Bq/kg) + (A_K/1500\,Bq/kg) \quad (6)$$

For the safety use of any material, the RLI need not exceed 1. There is also a direct link between terrestrial gamma rays and radionuclide concentrations in the soil. To assess the dose rate of external gamma rays coming from 1.0 m above the soil, the absorbed dose rate (Eq. 7) (Sharma et al. 2017) is usually used. It is assumed that the concentration of ^{226}Ra, ^{232}Th, and ^{40}K is uniformly distributed at 1.0 m above the ground. The dose rate per unit (nGy/h)/(Bq/kg) concentration of ^{40}K, ^{232}Th, and ^{226}Ra is 0.043, 0.621, and 0.462, respectively according to UNSCEAR (2000).

$$D\ (\mathrm{nGy/h}) = 0.043A_{K} + 0.623A_{Th} + 0.427A_{Ra} \tag{7}$$

The excess lifetime cancer risk (ELCR) (Eq. 8) is the probable carcinogenic risk assessment parameter associated with long-term exposure to radionuclides associated with the inhalation of air from radionuclides-contaminated soils of irrigated farmlands (Thabayneh and Jazzar 2012).

$$\mathrm{ELCR} = E_{out} + \mathrm{RF} + \mathrm{DL} \tag{8}$$

where E_{out} is the annual effective dose (outdoor), DL is the duration of life (70 years globally) and RF is the fatal cancer risk factor per Sievert which is 0.05 for the public. However, the DL for Nigerians according to Olabamiji et al. (2023) is 55 years and 2 months (55.2 years).

5 Conclusions

The comprehensive assessment of radionuclides as environmental contaminants in Nigeria indicated a complex landscape of challenges and opportunities. This study has highlighted significant spatial variability in radionuclide concentrations across the country, with areas such as the Jos Plateau exhibiting notably elevated levels of natural radioactivity. The interplay between geological formations, soil properties, and anthropogenic activities was identified as a crucial determinant of radionuclide distribution and mobility. Research progress in Nigeria has been substantial, with numerous studies providing valuable insights into radionuclide levels in soil and groundwater across various regions. These investigations have contributed to a growing database of radionuclide concentrations, enabling more accurate risk assessments and informed decision-making. However, several shortcomings in the current research are evident. There is a notable lack of standardization in sampling and analytical methodologies across studies, potentially hindering direct comparisons and trend analyses. Additionally, the geographical coverage of investigations remains uneven, with some regions being understudied. The public health implications of radionuclide exposure in Nigeria warrant continued attention. While many studies report radionuclide levels below international safety thresholds, localized hotspots of elevated concentrations underscore the need for targeted interventions and ongoing monitoring. The development and implementation of region-specific remediation strategies remain a critical challenge, particularly given the resource constraints faced by many Nigerian communities. Looking forward, there is a pressing need for more integrated, multidisciplinary research approaches. Future studies should aim to bridge the gap between environmental measurements and epidemiological investigations to better understand the long-term health impacts of chronic low-dose exposure to radionuclides. Furthermore, the potential interactions between radionuclide behavior and climate change impacts in Nigeria represent an important area for future research. In conclusion, while significant strides have been made in understanding and addressing radionuclide contamination in Nigeria, continued efforts are needed to enhance monitoring capabilities, develop locally appropriate remediation technologies, and strengthen regulatory frameworks. International collaboration and knowledge exchange will be crucial in addressing this complex environmental and public health challenge.

References

Abdou AA, Nagar MS, Ghazala RA (2018) Removal of radium from uranium effluent by manganese oxide coated modified bentonite (Mn-NaB). Mediterr J Chem 7(2):105–114. https://doi.org/10.13171/mjc72/01807121310-abdou

Abed TK, Fayad MA, Alamiery AA, Wahhab HA, Mohammed JK, Majdi HS (2024) Radon gas emission from home appliances: understanding sources, implications, and mitigation strategies. Results Eng 22:102133. https://doi.org/10.1016/j.rineng.2024.102133

Abey JA, Adebiyi FM, Asubiojo OI, Tchokossa P (2017) Assessment of radioactivity and polycyclic aromatic hydrocarbon levels in ground waters within the bitumen deposit area of Ondo State, Nigeria. Energy Sour Part A 39(13):1443–1451. https://doi.org/10.1080/15567036.2017.1336820

Abiodun OA, Shola EA, Odunayo TO (2019) Investigation of radionuclide levels in groundwater around transmission company of Nigeria for environmental impact assessment. Nucl Sci 4(4):66–71. https://doi.org/10.11648/j.ns.20190404.15

Adebiyi FM, Ore OT (2020) Measurement of radioactivity level of oil contaminated soils around mechanic workshops for environmental impact assessment. Energy Sour Part A 42(17):2145–2154

Adebiyi FM, Ore OT, Adeola AO, Durodola SS, Akeremale OF, Olubodun KO, Akeremale OK (2021) Occurrence and remediation of naturally occurring radioactive materials in Nigeria: a review. Environ Chem Lett. https://doi.org/10.1007/s10311-021-01237-4

Adedokun MB, Aweda MA, Maleka PP, Obed RI, Ogungbemi KI, Ibitoye ZA (2019) Natural radioactivity contents in commonly consumed leafy vegetables cultivated through surface water irrigation in Lagos state. J Radiat Res Appl Sci 12(1):147–156. https://doi.org/10.1080/16878507.2019.1618084

Adedokun MB, Aweda MA, Maleka PP, Obed RI, Ibitoye AZ (2020) Evaluation of natural radionuclides and associated radiation hazard indices in soil and water from selected vegetable farmlands in Lagos, Nigeria. Environ Forensics. https://doi.org/10.1080/15275922.2020.1850557

Ademola AK (2021) Radiological risks from natural radionuclides in surface soil of Agbara industrial area, Ogun State, Nigeria. J Appl Sci Environ Manag 25(5):733–739

Adesiji NE, Ademola JA (2019) Soil-to-maize transfer factor of natural radionuclides in a tropical ecosystem of Nigeria. J Environ Radioact 201:1–4

Ainsbury EA, Moquet J, Sun M, Barnard S, Ellender M, Lloyd D (2022) The future of biological dosimetry in mass casualty radiation emergency response, personalized radiation risk estimation and space radiation protection. Int J Radiat Biol 98(3):421–427

Ajayi OS, Balogun KO, Dike CG (2017) Spatial distributions and dose assessment of natural radionuclides in rocks and soils of some selected sites in southwestern Nigeria. Hum Ecol Risk Assess Int J 23(6):1373–1388. https://doi.org/10.1080/10807039.2017.1312278

Ajayi OS, Dike CG, Balogun KO (2018) Elemental and radioactivity analysis of rocks and soils of some selected sites in southwestern Nigeria. Environ Forensics 19(2):87–98

Ajibola TB, Orosun MM, Lawal WA, Akinyose FC, Salawu NB (2021) Assessment of annual effective dose associated with radon in drinking water from gold and bismuth mining area of edu, Kwara, North-Central Nigeria. Pollution 7(1):231–240

Ajiboye Y, Badmus OG, Ojo OD, Isinkaye MO (2016) Measurement of Radon concentration and radioactivity in soil samples of Aramoko, Ekiti State, Nigeria. Int J Publ Health Res 4(5):37–41

Akpan AE, Ebong ED, Ekwok SE, Eyo JO (2020) Assessment of radionuclide distribution and associated radiological hazards for soils and beach sediments of Akwa Ibom Coastline, southern Nigeria. Arab J Geosci 13:753. https://doi.org/10.1007/s12517-020-05727-7

Alloway BJ (2013) Uranium. In: Alloway B (eds) Heavy metals in soils. Environmental pollution, vol 22. Springer, Dordrecht (pp 565–577). https://doi.org/10.1007/978-94-007-4470-7_26

Alusa SK, Odunaike K, Adeniji IA (2017) Transfer factor of radionuclides from soil-topalm oil produced from Elere palm tree plantation near Ibadan Oyo state, Nigeria. Niger J Pure Appl Phys 7(1):7–12

Aly MM, Hamza MF (2013) A review: studies on uranium removal using different techniques. Overview. J Dispers Sci Technol 34(2):182–213. https://doi.org/10.1080/01932691.2012.657954

Amana MS (2017) Radiation hazard index of common imported ceramic using for building materials in Iraq. Austral J Basic Appl Sci 11(10):94–102

Arafat AA, Salama MHM, El-Sayed SA, Elfeel AA (2017) Distribution of natural radionuclides and assessment of the associated hazards in the environment of Marsa Alam-Shalateen area, Red Sea Coast, Egypt. J Radiat Res Appl Sci 10:219–232

Atipo M, Olarinoye O, Awojoyogbe B (2020) Comparative analysis of NORM concentration in mineral soils and tailings from a tin-mine in Nigeria. Environ Earth Sci 79(16):1–17. https://doi.org/10.1007/s12665-020-09136-7

Avila R, Kautsky U, Ekström PA, Åstrand PG, Saetre P (2013) Model of the long-term transport and accumulation of radionuclides in future landscapes. Ambio 42:497–505. https://doi.org/10.1007/s13280-013-0402-x

Balaram V, Rani A, Rathore DP (2022) Uranium in groundwater in parts of India and world: a comprehensive review of sources, impact to the environment and human health, analytical techniques, and mitigation technologies. Geosyst Geoenviron 1(2):100043. https://doi.org/10.1016/j.geogeo.2022.100043

Belli M, Tabocchini MA (2020) Ionizing radiation-induced epigenetic modifications and their relevance to radiation protection. Int J Mol Sci 21(17):5993

Blake JM, Avasarala S, Ali AM, Spilde M, Lezama-Pacheco JS, Latta D, Artyushkova K, Ilgen AG, Shuey C, Nez C, Cerrato JM (2019) Reactivity of As and U co-occurring in Mine Wastes in northeastern Arizona. Chem Geol 522:26–37. https://doi.org/10.1016/j.chemgeo.2019.05.024

Blakely WF, Port M, Ostheim P, Abend M (2024) Radiation research society journal-based historical review of the use of biomarkers for radiation dose and injury assessment: acute health effects predictions. Radiat Res 202(2):185–204

Borrego-Alonso D, Quintana-Arnés B, Lozano JC (2023) Natural radionuclides behaviour in drinking groundwaters from Castilla y León (Spain); radiological implications. Water Res 245:120616. https://doi.org/10.1016/j.watres.2023.120616

Cao X, Hu L, Wang J, Wang J (2017) Radionuclide transport model for risk evaluation of high-level radioactive waste in Northwestern China. Hum Ecol Risk Assess Int J 23(8):2017–2032. https://doi.org/10.1080/10807039.2017.1361811

Caridi F, Messina M, D'Agostino M (2017) An investigation about natural radioactivity, hydrochemistry, and metal pollution in groundwater from Calabrian selected areas, southern Italy. Environ Earth Sci 76(19):668

Chadha VD (2021) Role of epigenetic mechanisms in propagating off-targeted effects following radiation-based therapies—A review. Mutat Res/Rev Mutat Res 787:108370

Chen MA, Kocar BD (2018) Radium sorption to iron (hydr) oxides, pyrite, and montmorillonite: implications for mobility. Environ Sci Technol 52(7):4023–4030. https://doi.org/10.1021/acs.est.7b05443

Chen Z, Wang S, Hou H, Chen K, Gao P, Zhang Z, Jin Q, Pan D, Guo Z, Wu W (2022) China's progress in radionuclide migration study over the past decade (2010–2021): sorption, transport and radioactive colloid. Chin Chem Lett 33(7):3405–3412. https://doi.org/10.1016/j.cclet.2022.02.054

Chernysh Y, Chubur V, Ablieieva I, Skvortsova P, Yakhnenko O, Skydanenko M, Plyatsuk L, Roubík H (2024) Soil contamination by heavy metals and radionuclides and related bioremediation techniques: a review. Soil Syst 8(2):36. https://doi.org/10.3390/soilsystems8020036

Cinti D, Poncia PP, Brusca L, Tassi F, Quattrocchi F, Vaselli O (2015) Spatial distribution of arsenic, uranium and vanadium in the volcanic-sedimentary aquifers of the Vicano-Cimino Volcanic District (Central Italy). J Geochem Explor 152:123–133

Danforth JM, Provencher L, Goodarzi AA (2022) Chromatin and the cellular response to particle radiation-induced oxidative and clustered DNA damage. Front Cell Dev Biol 10:910440

Davidson JR, Fairley J, Nicol A, Gravley D, Ring U (2016) The origin of radon anomalies along normal faults in an active rift and geothermal area. Geosphere 12(5):1656–1669. https://doi.org/10.1130/GES01321.1

de Bernardo Gisbert LM, Ríos MP, Lema LV, Dios JMB, Ravina AR (2025) Indoor radon as a public health problem. Available evidence on radon and its health effects. In: Communicating public health risk. Routledge, pp 11–28

de Moraes ML, Ladeira AC (2021) The role of iron in the rare earth elements and uranium scavenging by Fe–Al-precipitates in acid mine drainage. Chemosphere 277:130131. https://doi.org/10.1016/j.chemosphere.2021.130131

Doyi IN, Essumang DK, Dampare SB, Duah D, Ahwireng AF (2017) Evaluation of radionuclides and decay simulation in a terrestrial environment for health risk assessment. Sci Rep 7(1):16537. https://doi.org/10.1038/s41598-017-16659-w

Dragović S (2022) Artificial neural network modeling in environmental radioactivity studies—A review. Sci Total Environ 847:157526. https://doi.org/10.1016/j.scitotenv.2022.157526

Duval JS, Jones WJ, Riggle FR, Pitkin JA (1990) Potassium and thorium maps of the conterminous United States. United States Department of the Interior, Geological Survey. Open-File Report, 90–338. U.S. Geological Survey 927 National Center Reston, VA 22092

Egbueri JC, Agbasi JC, Ezugwu AL, Omeka ME, Ucheana IA, Aralu CC, Abugu HO (2024) Metal(loid)s, nitrate, polycyclic aromatic hydrocarbons, and radioactive contaminants in Nigerian water resources: state-of-the-art of their ecological and health risk assessments. Environ Dev Sustain. https://doi.org/10.1007/s10668-024-05423-6

Egunjobi KA, Adesina KE, Chime IP (2020) Radiological assessment of soil samples from selected dumpsites in Mowe, Ogun State Nigeria. CCU J Sci 1(1):72–80

Ekhaguere OA, Alatise OO, Oyeyemi KD (2019) Assessment of natural radionuclides in a fertilized farmland in Abeokuta, Nigeria: implications for environmental radioactivity evaluation and monitoring. J Phys Conf Ser 1299(1):012093. https://doi.org/10.1088/1742-6596/1299/1/012093

Elaboudi Z, Madinzi A, Saadi R, Laissaoui A, Kurniawan TA, Anouzla A, Souabi S (2024) A review of radionuclides impacts and remediation techniques. Euro-Mediterr J Environ Integr. https://doi.org/10.1007/s41207-024-00613-0

El-Gamal H, Hussien MT, Saleh EE (2019) Evaluation of natural radioactivity levels in soil and various foodstuffs from Delta Abyan, Yemen. J Radiat Res 12(1):226–233

Elmaghraby EK, Ataalla NN, Afifi MB, Salem E (2021) Radon exhalation and transfer processes in aqueous media. Eur Phys J Plus 136(12):1217. https://doi.org/10.1140/epjp/s13360-021-02231-z
Fan Y, Pan D, Yang M, Wang X (2023) Radiolabelling and in vivo radionuclide imaging tracking of emerging pollutants in environmental toxicology: a review. Sci Total Environ 866:161412. https://doi.org/10.1016/j.scitotenv.2023.161412
Fathabadi N, Salehi AA, Naddafi K, Kardan MR, Yunesian M, Nodehi RN, Deevband MR, Shooshtari MG, Hosseini SS, Karimi M (2017) Radioactivity levels in the mostly local foodstuff consumed by residents of the high level natural radiation areas of Ramsar, Iran. J Environ Radioact 169–170:209–213. https://doi.org/10.1016/j.jenvrad.2016.12.011
Gadd GM (2007) Roles of micro-organisms in the environmental fate of radionuclides. In: Lake JV, Bock GR, Cardew G (eds) Ciba foundation symposium 203—Health impacts of large releases of radionuclides. Wiley, Chichester, UK, pp 94–108. https://doi.org/10.1002/9780470515006.ch7
Gbadamosi MR, Afolabi TA, Banjoko OO, Ogunneye AL, Abudu KA, Ogunbanjo OO, Jegede DO (2018) Spatial distribution and lifetime cancer risk due to naturally occurring radionuclides in soils around tar-sand deposit area of Ogun State, southwest Nigeria. Chemosphere 193:1036–1048
Godoy JM, Ferreira PR, Souza EM, Silva LI, Bittencourt IC, Fraifeld F (2019) High uranium concentrations in the groundwater of the Rio de Janeiro State, Brazil, mountainous region. J Braz Chem Soc 30(2):224–233
Harb S, El-Kamel AH, Zahran AM, Abbady A, Ahmed FA (2014) Assessment of natural radioactivity in soil and water samples from Aden Governorate South of Yemen region. Int J Recent Res Phys Chem Sci 1(1):1–7
Hegazy AK, Afifi SY, Alatar AA, Alwathnani HA, Emam MH (2013) Soil characteristics influence the radionuclide uptake of different plant species. Chem Ecol 29(3):255–269. https://doi.org/10.1080/02757540.2012.744827
Heuel-Fabianek B (2014) Partition coefficients (Kd) for the modelling of transport processes of radionuclides in groundwater. Division of Safety and Radiation Protection (S), Berichte des Forschungszentrums Julich: 4375
Hirose K, Povinec PP (2022) Ten years of investigations of Fukushima radionuclides in the environment: a review on process studies in environmental compartments. J Environ Radioact 251–252:106929. https://doi.org/10.1016/j.jenvrad.2022.106929
Höffken J, Ramana MV (2024) Nuclear power and environmental injustice. Wiley Interdiscip Rev: Energy Environ 13(1):e498. https://doi.org/10.1002/wene.498
IAEA (International Atomic Energy Agency) (2024) Development and application of level 1 probabilistic safety assessment for nuclear power plants. IAEA Safety Standards, No. SSG-3 (Rev. 1). International Atomic Energy Agency, VIENNA. IAEAL 24-01658
Ibikunle SB, Arogunjo AM, Ajayi OS (2019) Characterization of radiation dose and soil-to-plant transfer factor of natural radionuclides in some cities from south-western Nigeria and its effect on man. Sci Afr 3:e00062
ICRP (International Commission on Radiological Protection) (2012) Compendium of dose coefficients based on ICRP Publication 60. ICRP Publication 119. Annals of ICRP, vol 41. Elsevier, Amsterdam, the Netherlands. ISBN 978-1-4557-5430-4
Ilori AO, Alausa SK (2019) Estimation of natural radionuclides in grasses, soils, and cattle-dungs from a cattle rearing-field at Mangoro-Agege, Lagos State, Nigeria. FUW Trends Sci Technol J 4(1):018–024
Iurian AR, Phaneuf MO, Mabit L (2015) Mobility and bioavailability of radionuclides in soils. In: Walther C, Gupta D (eds) Radionuclides in the environment. Springer, Cham, pp 37–59. https://doi.org/10.1007/978-3-319-22171-7_2
Ivanovich M (1994) Uranium series disequilibrium: concepts and applications. Radiochim Acta 64(2):81–94. https://doi.org/10.1524/ract.1994.64.2.81

Jahan I, Ali ML, Haydar MA, Ali MI, Paul D, Islam SMA (2016) Distribution of natural and probable artificial radioactivity in the sediment and water samples collected from low-lying areas of Savar industrial zone, Bangladesh. J Nucl Part Phys 6(2):25–34

Jahng JW, Little MP, No HJ, Loo Jr BW, Wu JC (2024) Consequences of ionizing radiation exposure to the cardiovascular system. Nat Rev Cardiol 1–19

Jasaitis D, Klima V, Pečiulienė M, Vasiliauskienė V, Konstantinova M (2020) Comparative assessment of radiation background due to natural and artificial radionuclides in soil in specific areas on the territories of state of Washington (USA) and Lithuania. Water Air Soil Pollut 231(7):1–10

Jayan J, Roshi H, Ashraf FFP, Nair PG, Vijayakumar A, Nair AS et al (2022) Effects of radiation exposure on brain health: a state of the art and new challenges. Environ Sci Pollut Res 29(58):87068–87081

Jha SK, Patra AC, Verma GP, Iyer IS, Aswal DK (2024) Natural radiation and environment. In: Aswal DK (ed) Handbook on radiation environment, vol 1. Springer, Singapore, pp 27–72. https://doi.org/10.1007/978-981-97-2795-7_2

Jinia AJ, Clarke SD, Moran JM, Pozzi SA (2024) Intelligent radiation: a review of machine learning applications in nuclear and radiological sciences. Ann Nucl Energy 201:110444. https://doi.org/10.1016/j.anucene.2024.110444

Kallio A, Leikoski N, Otaki M (2023) Natural radioactivity of residues from groundwater treatment facilities in Finland. J Radiol Prot 43(3):031517. https://doi.org/10.1088/1361-6498/acf8d1

Kaminski MD, Daiyega N, Magnuson M (2024) A review of the resuspension of radioactively contaminated particles by vehicle and pedestrian traffic—Current theory, practice, gaps, and needs. Health Phys 126(4):216–240. https://doi.org/10.1097/HP.0000000000001797

Katsoyiannis IA, Althoff HW, Bartel H, Jekel M (2006) The effect of groundwater composition on uranium (VI) sorption onto bacteriogenic iron oxides. Water Res 40(19):3646–3652. https://doi.org/10.1016/j.watres.2006.06.032

Kim S, Lee Y, Park M, Jeong HY (2024) Immobilization of hexavalent uranium U (VI) by hydroxyapatite under oxic conditions. J Nucl Mater 595:155059. https://doi.org/10.1016/j.jnucmat.2024.155059

Kitahara CM, Schneider AB (2022) Epidemiology of thyroid cancer. Cancer Epidemiol Biomark Prev 31(7):1284–1297

Knox AS, Brigmon RL, Kaplan DI, Paller MH (2008) Interactions among phosphate amendments, microbes and uranium mobility in contaminated sediments. Sci Total Environ 395(2–3):63–71. https://doi.org/10.1016/j.scitotenv.2008.01.061

Korobova EM, Linnik VG, Chizhikova NP, Alekseeva TN, Shkinev VM, Brown J, Dinu MI (2014) Granulometric and mineralogic investigation for explanation of radionuclide accumulation in different size fractions of the Yenisey floodplain soils. J Geochem Explor 142:49–59

Kuhnel WW, Jacobsen SB, Li Y, Ku Y, Petaev MI, Huang S, Wu Z, Wang K (2021) High-temperature inter-mineral potassium isotope fractionation: implications for K–Ca–Ar chronology. ACS Earth Space Chem 5(10):2740–2754. https://doi.org/10.1021/acsearthspacechem.1c00147

Kumar P, Choudhury D (2024) Optimized phytoremediation process for the sustainable management radionuclides. In: Kumar P, Srivastav AL, Chaudhary V, Hullebusch ED, Busquets R (Eds) Bioremediation of emerging contaminants from soils. Elsevier, pp 443–464. https://doi.org/10.1016/B978-0-443-13993-2.00020-7

Kumar NS, Raju MK, Ratnaraju M, Satyanarayana GV, VidyaSagar D, Das NL (2024) Assessment of radioactivity levels in soil of coastal dwellings in Odisha and Andhra Pradesh, India. J Radioanal Nucl Chem. https://doi.org/10.1007/s10967-024-09652-x

Kungur ST, Ige TA, Ikyo BA (2020) Analysis of natural radionuclides and evaluation of radiation hazard indices in soil samples from Benue state, Nigeria. Int J Innov Res Sci Eng 5(5):1765–1769

Landa ER (2007) Naturally occurring radionuclides from industrial sources: characteristics and fate in the environment. Radioact Environ 10:211–237. https://doi.org/10.1016/S1569-4860(06)10010-8

Larijani C, Schwendner P, Cockell C, Ivanov P, Russell B, Aitken-Smith P, Pearce AK, Regan PH (2017) Destructive and non-destructive measurements of NORM in monazite-rich sands of Brazil. Radiat Phys Chem 140:180–185. https://doi.org/10.1016/j.radphyschem.2017.01.010

Little MP, Azizova TV, Richardson DB, Tapio S, Bernier MO, Kreuzer M et al (2023) Ionising radiation and cardiovascular disease: systematic review and meta-analysis. BMJ 380:e072924. https://doi.org/10.1136/bmj-2022-072924

Malkovsky VI, Petrov VA, Yudintsev SV, Ojovan MI, Poluektov VV (2023) Influence of rock structure on migration of radioactive colloids from an underground repository of high-level radioactive waste. Sustainability 15(1):882. https://doi.org/10.3390/su15010882

Manawi Y, Hassan A, Atieh MA, Lawler J (2024) Overview of radon gas in groundwater around the world: health effects and treatment technologies. J Environ Manag 368:122176. https://doi.org/10.1016/j.jenvman.2024.122176

Manigandan PK (2009) Activity concentration of radionuclides in plants in the environment of Western Ghats. Iran J Radiat Res 7(2):85–90

Martinelli S, Medeiros AN, de Souza RF, Marconi AM, Navoni JA (2023) Analysis of the cancer occurrence related to natural radioactivity in the Rio Grande do Norte State, Brazil. Environ Sci Pollut Res 30(38):89140–89152. https://doi.org/10.1007/s11356-023-28708-1

Matsunaga T, Nagao S, Ueno T, Takeda S, Amano H, Tkachenko Y (2004) Association of dissolved radionuclides released by the Chernobyl accident with colloidal materials in surface water. Appl Geochem 19(10):1581–1599. https://doi.org/10.1016/j.apgeochem.2004.02.002

Muga J (2017) Geochemical assessment for morendat East geothermal prospect, using radon and carbondioxide concentration from soil, Kenya. J Geol Resour Eng 1:38–47. https://doi.org/10.17265/2328-2193/2017.01.004

Natarajan T, Sahoo SK, Inoue K, Arae H, Aono T, Fukushi M (2024) Natural radionuclides and radiological risk assessment in the stream and river sediments of a high background natural radiation area Kanyakumari, India. Environ Monit Assess 196(3):330. https://doi.org/10.1007/s10661-024-12456-x

Nwankwo LI (2013) Determination of natural radioactivity in groundwater in Tanke-Ilorin, Nigeria. West Afr J Appl Ecol 21(1):111–119

Ogundele LT, Ayeku PO, Inuyomi SO, Ogunsakin OM, Oladejo OF, Adejoro IA (2020) Assessment of naturally occurring 40K, 232Th and 238U and their associated radiological hazard indices in soils used for building in Ondo West Local Government Area, Southwestern, Nigeria. EQA Int J Environ Qual 37:11–21

Ogunyele AC, Obaje SO, Akingboye AS, Adeola AO, Babalola AO, Olufunmilayo AT (2020) Petrography and geochemistry of Neoproterozoic charnockite–granite association and metasedimentary rocks around Okpella, southwestern Nigeria. Arab J Geosci 13:780. https://doi.org/10.1007/s12517-020-05785-x

Ohnuki T, Ozaki T, Kozai N, Utsunomiya S (2020) Function of microbes on chemical species transformation of radionuclides. In: Kato K, Konoplev A, Kalmykov S (eds) Behavior of radionuclides in the environment I. Springer, Singapore, pp 67–92. https://doi.org/10.1007/978-981-15-0679-6_3

Olabamiji AO, Alausa SK, Adejumobi CA, Alabi TA (2023) Environmental impact assessment of radionuclide and metal contamination in heavily urbanized river, Ibadan, Nigeria. Nigerian J Phys 32(1)

Olalekan EO, Adebiyi FM (2020) Measurement of radioactivity levels in oil impacted soils around petroleum products retailing stations. Pet Sci Technol 38(1):53–63

Olatunde MO, Philip AA, Abraham AA (2018). Simulation of absorbed dose to human organs and tissues associated with radon in groundwater use in southwestern Nigeria. Radiat Phys Chem. https://doi.org/10.1016/j.radphyschem.2018.08.029

Olatunji MA, Khandaker MU, Nwankwo VU, Idris AM (2022) The status of natural radioactivity in Nigerian environments. Radiat Environ Biophys 61(4):597–608. https://doi.org/10.1007/s00411-022-00993-3

Olise FS, Akinnagbe DM, Olasogba OS, Speldewinde PC (2016) Radionuclides and radon levels in soil and ground water from solid minerals-hosted area, south-western Nigeria. Cogent Environ Sci 2(1). https://doi.org/10.1080/23311843.2016.1142344

Oluyide SO, Tchokossa P, Akinyose FC, Orosun MM (2019) Assessment of radioactivity levels and transfer factor of natural radionuclides around iron and steel smelting company located in Fashina village, Ile-ife, Osun state, Nigeria. Facta Universitatis, Series: Working and Living Environmental Protection, pp 241–256

Oni EA, Adagunodo TA (2019) Assessment of radon concentration in groundwater within Ogbomoso, SW Nigeria. J Phys: Conf. Ser 1299:012098 (IOP Publishing). https://doi.org/10.1088/1742-6596/1299/1/012098

Ononugbo CP, Amah OS (2019) Radioactivity concentrations and their radiological significance in sediments of some communities in Andoni, Rivers State, Nigeria. Asian J Adv Res Rep. https://doi.org/10.9734/ajarr/2019/v6i330151

Ononugbo CP, Avwiri GO, Agbalagba EO (2017) Radioactive pollution and excess lifetime cancer risk due to gamma exposure of soil and groundwater around open landfills in River State Nigeria. Can J Pure Appl Sci 11(1):4121–4130

Ononugbo CP, Nwaka BU (2017) Natural radioactivity and radiological risk estimation of drinking water from Okposi and Uburu salt lake area, Ebonyi State, Nigeria. Phys Sci Int J 15(3):1–15

Ononugbo CP, Tutumeni G (2016) Natural radioactivity and radiation dose estimation in various water samples in Abua/Odua Area, Rivers State, Nigeria. Phys Sci Int J 11(4):1–12

Orellana P, Mut F, Estrada E, Lette MM, Pellet O, Morozova O, El-Haj N, Bucheli JC, Pynda Y, Okolielova T, Cherit A (2021) Status of nuclear medicine in Latin America and the Caribbean: IAEA analysis of development in the past 6 years. J Nucl Med 62(06):23N–29N

Orosun MM, Usikalu MR, Oyewumi KJ, Oladapo OF (2021) Radiological hazard assessment of sharp-sand from Ilorin-East, Kwara State, Nigeria. J Phys Conf Ser 1734(1):012040

Payne TE, Edis R (2012) Mobility of radionuclides in tropical soils and groundwater. In: Twining JR (ed) Radioactivity in the environment, vol 18. Elsevier, pp 93–120. https://doi.org/10.1016/B978-0-08-045016-2.00003-5

Pohjolainen E (2015) Uranium deposits of Finland. In: Maier WD, Lahtinen R (eds) Mineral deposits of Finland. Elsevier, pp 659–683. https://doi.org/10.1016/B978-0-12-410438-9.00025-X

Pontedeiro EM, van Genuchten MT, Cotta RM, Simunek J (2010) The effects of preferential flow and soil texture on risk assessments of a NORMWaste Disposal Site. J Hazard Mater 174:648

Popoola FA, Fakeye OD, Basiru QB, Adesina DA, Sulola MA (2019) Assessment of radionuclide concentration in surface soil and human health risk associated with exposure in two higher institutions of Esan land, Edo State, Nigeria. J Appl Sci Environ Manag 23(12):22679–32284. https://doi.org/10.4314/jasem.v23i12.29

Rachkova NG, Shuktomova II, Taskaev AI (2010) The state of natural radionuclides of Uranium, Radium and Thorium in soils. Eurasian Soil Sci 43:651

Ramola RC, Gusain GS, Badoni M, Prasad Y, Prasad G, Ramachandran TV (2008) ^{226}Ra, ^{232}Th and ^{40}K contents in soil samples from Garhwal Himalaya, India, and its radiological implications. J Radiol Prot 28:379–385

Rentería-Villalobos M, Covarrubias-Muños A, Pinedo-Álvarez A, Manjon-Collado G (2017) Hydrochemistry and ^{222}Rn concentrations in spring waters in the arid zone El Granero, Chihuahua, Mexico. Geosciences 7(1):12. https://doi.org/10.3390/geosciences7010012

Rezaie F, Panahi M, Bateni SM, Kim S, Lee J, Lee J, Yoo J, Kim H, Kim SW, Lee S (2023) Spatial modeling of geogenic indoor radon distribution in Chungcheongnam-do, South Korea using enhanced machine learning algorithms. Environ Int 171:107724. https://doi.org/10.1016/j.envint.2022.107724

Ribeiro FCA, Lauria D, Rodrigues JI, Lima SE, Sobrinho NM, Pérez DV (2018) Baseline and quality reference values for natural radionuclides in soils of Rio de Janeiro State, Brazil Revisita Brasileira Ciencia do Solo 1–15

Riudavets M, Garcia de Herreros M, Besse B, Mezquita L (2022) Radon and lung cancer: current trends and future perspectives. Cancers 14(13):3142

Rong H, Steven L, Meng Z, Xiao-Dong Z, Hongmei L (2022) Environmental evaluation of radioactivity levels and associated radiation hazards in groundwater around the WIPP site. Ecotoxicol Environ Saf 242:113849. https://doi.org/10.1016/j.ecoenv.2022.113849
Ruedig E, Johnson TE (2015) An evaluation of health risk to the public as a consequence of in situ uranium mining in Wyoming, USA. J Environ Radioact 150:170–178
Rühm W, Ainsbury E, Breustedt B, Caresana M, Gilvin P, Knežević Ž, Rabus H, Stolarczyk L, Vargas A, Bottollier-Depois JF, Harrison R (2020) The European radiation dosimetry group—Review of recent scientific achievements. Radiat Phys Chem 168:108514
Rühm W, Laurier D, Wakeford R (2022) Cancer risk following low doses of ionising radiation–Current epidemiological evidence and implications for radiological protection. Mutat Res/Genet Toxicol Environ Mutagen 873:503436
Ruprecht NA, Singhal S, Schaefer K, Panda O, Sens D, Singhal SK (2024) A review: multi-omics approach to studying the association between ionizing radiation effects on biological aging. Biology 13(2):98
Ryan JL (2012) Ionizing radiation: the good, the bad, and the ugly. J Invest Dermatol 132(3 Pt 2):985–993. https://doi.org/10.1038/jid.2011.411
Saha A, Kumari K, Sharma S, Kumar R, Sahu M, Dumpala RM, PP MS, Deb SB, Saxena MK (2023) Selective extraction of Thorium (IV) from Uranium and rare earth elements using Tetraphenylethane-1, 2-diylbis (phosphoramidate). Inorg Chem 62(24):9391–9399. https://doi.org/10.1021/acs.inorgchem.3c00359
Saint-Fort R (2018) Understanding sorption behavior and properties of radionuclides in the environment. In: Principles and applications in nuclear engineering: radiation effects, thermal hydraulics, radionuclide migration in the environment, vol 121
Salbu B (2024) Release of radioactive particles to the environment. Radiat Res 202(2):260–272. https://doi.org/10.1667/RADE-24-00016.1
Schubert M, Lin M, Clark JF, Kralik M, Damatto S, Copia L, Terzer-Wassmuth S, Harjung A (2024) Short-lived natural radionuclides as tracers in hydrogeological studies—A review. Sci Total Environ 920:170800. https://doi.org/10.1016/j.scitotenv.2024.170800
Seow SQ, Mohan Viswanathan P, Dodge-Wan D (2024) Distribution of natural radioactivity in different geological formations and their environmental risk assessment in Malaysia. Environ Sci Pollut Res 31:43292–43308. https://doi.org/10.1007/s11356-024-33906-6
Shanthi G, Kumaran JTT, Raj GA, Maniyan CG (2010) Natural radionuclides in the South Indian foods and their annual dose. Nucl Inst Methods Phys Res A 619:436–440
Sharma BA, Singh NS, Devi PT, Basu H, Saha S, Singha RK (2017) Assessment of radioactivity in the soil samples from Imphal city, India, and its radiological implication. Radiat Prot Environ 40:149–153
Shin W, Oh J, Choung S, Cho B-W, Lee K-S, Yun U, Kim HK (2016) Distribution and potential health risk of groundwater uranium in Korea. Chemosphere 163:108–115. https://doi.org/10.1016/j.chemosphere.2016.08
Stalder E, Blanc A, Haldimann M, Dudler V (2012) Occurrence of uranium in Swiss drinking water. Chemosphere 86(6):672–679. https://doi.org/10.1016/j.chemosphere.2011.11
Stockdale A, Bryan ND (2013) The influence of natural organic matter on radionuclide mobility under conditions relevant to cementitious disposal of radioactive wastes: a review of direct evidence. Earth-Sci Rev 121:1–17
Storm DL (2020) Chemical monitoring of California's public drinking water sources: public exposures and health impacts. In: Wang RGM (ed) Water contamination and health. CRC Press, pp 67–124
Sun H, Zhou S, Jiang Y, Xi X, Tan Y, Zhang G, Jiang N, Zhou T, Yin X, Wang M, Gao B (2022) Fate and transport of engineered nanoparticles in soils and groundwater. In Gao B (ed) Emerging contaminants in soil and groundwater systems. Elsevier, pp 205–51. https://doi.org/10.1016/B978-0-12-824088-5.00003-3

Szaciłowski G (2024) Evaluation of natural radioactivity and assessment of dose increase due to the use of fertilizers produced in Poland. J Radioanal Nucl Chem 333(7):3425–3430. https://doi.org/10.1007/s10967-023-09241-4
Tan CL (2016) Big gaps and short bridges: a model for solving the discontinuity problem. Answ Res J 9(6):149–162
Tan W, Li Y, Tan K, Duan X, Liu D, Liu Z (2016) Fractal theory and field cover experiments: implications for the fractal characteristics and radon diffusion behavior of soils and rocks. Health Phys 111(6):506–512. https://doi.org/10.1097/HP.0000000000000569
Tawfic AF, Mira HI, Omar AM (2021) Chemical alteration processes for uranium migration and their significance on uranium isotopic ratio changes. Int J Environ Anal Chem 101(14):2379–2391. https://doi.org/10.1080/03067319.2019.1702172
Thabayneh KM, Jazzar MM (2012) Radioactivity levels in plant samples in Tulkarem district, Palestine and its impact on human health. Radiat Prot Dosimetry 153(4):467–474
Thørring H, Skuterud L, Steinnes E (2014) Influence of chemical composition of precipitation on migration of radioactive caesium in natural soils. J Environ Radioact 134:114–119
Tibane LV, Mamba D (2024) Environmental risk assessment, principal component analysis, tracking the source of toxic heavy metals of solid gold mine waste tailings, South Africa. Environ Forensics 25(4):254–270. https://doi.org/10.1080/15275922.2023.2172478
Tijani MN (2023) Geology of Nigeria. In: Faniran A, Jeje LK, Fashae OA, Olusola AO (eds) Landscapes and landforms of Nigeria. World geomorphological landscapes. Springer, Cham, pp 3–32. https://doi.org/10.1007/978-3-031-17972-3_1
Timothy EP, Robert E (2012) Mobility of radionuclides in tropical soils and groundwater. Radioact Environ 18:93–120. https://doi.org/10.1016/B978-0-08-045016-2.00003-5
Tsai T, Lin C, Wang T, Chu T (2008) Radioactivity concentrations and dose assessment for soil samples around nuclear power plant IV in Taiwan. J Radiol Protect: Off J Soc Radiol Protect 28:347–360
Tutu H, Cukrowska EM, McCarthy TS, Hart R, Chimuka L (2009) Radioactive disequilibrium and geochemical modelling as evidence of uranium leaching from gold tailings dumps in the Witwatersrand Basin. Int J Environ Anal Chem 89(8–12):687–703. https://doi.org/10.1080/03067310902968749
UNSCEAR (2000) Sources, effects and risks of ionizing radiation. United Nations Scientific Committee on the effects of Atomic Radiation. Exposures from natural sources, 2000 report to General Assembly, Annex B, New York
Uralbekov BM, Smodis B, Burkitbayev M (2011) Uranium in natural waters sampled within former uranium mining sites in Kazakhstan and Kyrgyzstan. J Radioanal Nucl Chem 289:805–810. https://doi.org/10.1007/s10967-011-1154-3
Usikalu MR, Oderinde A, Adagunodo ER, Akinpelu A (2018) Radioactivity concentration and dose assessment of soil samples in cement factory and environs in Ogun State, Nigeria. Int J Civil Eng Technol 9(9):1047–1059
Vaupotič J (2024) Radon and its short-lived products in indoor air: present status and perspectives. Sustainability 16(6):2424. https://doi.org/10.3390/su16062424
Wade P, Coetzee H (2008) Risk assessment of uranium in selected gold mining areas in South Africa. In: Merkel BJ, Hasche-Berger A (eds) Uranium, mining and hydrogeology. Springer, Berlin, Heidelberg, pp 141–150. https://doi.org/10.1007/978-3-540-87746-2_21
Walters MO (1995) Radium in coastal Sarasota County ground water. Groundw Monit Remed 15(4):114–118. https://doi.org/10.1111/j.1745-6592.1995.tb00558.x
Wang J, Chen C (2024) Phytoremediation of radium contaminated soils: recent advances and prospects. Radiochim Acta 112(9):583–604. https://doi.org/10.1515/ract-2023-0248
Wilkins MJ, Livens FR, Vaughan DJ, Beadle I, Lloyd JR (2007) The influence of microbial redox cycling on radionuclide mobility in the subsurface at a low-level radioactive waste storage site. Geobiology 5(3):293–301. https://doi.org/10.1111/j.1472-4669.2007.00101.x

Wimmers A, Böse F, Beppler J, Morawe P, Weber M, von Hirschhausen C (2024) (Re)integrating radioactive materials and waste into a global sustainable development context. Radiat Environ Biophys. https://doi.org/10.1007/s00411-024-01088-x

World Health Organization (WHO) (2016) Ionizing radiation, health effects and protective measures. World Health Organization

Yang J, Sun Y (2022) Natural radioactivity and dose assessment in surface soil from Guangdong, a high background radiation province in China. J Radiat Res Appl Sci 15(1):145–151. https://doi.org/10.1016/j.jrras.2022.01.019

Yang Q, Smitherman P, Hess CT, Culbertson CW, Marvinney RG, Smith AE, Zheng Y (2014) Uranium and radon in private bedrock well water in Maine: geospatial analysis at two scales. Environ Sci Technol 48(8):4298–4306. https://doi.org/10.1021/es405020k

Yu X, Xiong F, Zhou C, Luo Z, Zhou Z, Chen J, Sun K (2024) Uranium bioprecipitation mediated by a phosphate-solubilizing Enterobacter sp. N1-10 and remediation of uranium-contaminated soil. Sci Tot Environ 906:167688. https://doi.org/10.1016/j.scitotenv.2023.167688

Zafar AM, Javed MA, Hassan AA, Mohamed MM (2021) Groundwater remediation using zero-valent iron nanoparticles (nZVI). Groundw Sustain Dev 15:100694. https://doi.org/10.1016/j.gsd.2021.100694

Zhang L, Chen Y, Xia Q, Kemner KM, Shen Y, O'Loughlin EJ, Pan Z, Wang Q, Huang Y, Dong H, Boyanov MI (2021) Combined effects of Fe(III)-bearing clay minerals and organic ligands on U(VI) bioreduction and U(IV) speciation. Environ Sci Technol 55(9):5929–5938. https://doi.org/10.1021/acs.est.0c08645

Zhang X, Ma F, Dai Z, Wang J, Chen L, Ling H, Soltanian MR (2022) Radionuclide transport in multi-scale fractured rocks: a review. J Hazard Mater 424:127550. https://doi.org/10.1016/j.jhazmat.2021.127550

Future Research Imperatives in Hydrogeology

Rakesh Roshan Gantayat, Vetrimurugan Elumalai, and Peiyue Li

Abstract This chapter explores critical advancements in the field of hydrogeology and outlines essential priorities for future research by emphasizing on growing importance of groundwater management and sustainability in the face of climate change, population growth, and environmental degradation. It examines intricate relation between changes in precipitation patterns primarily caused by rising global temperatures and their impact groundwater recharge, quality, and surface-groundwater interactions. Additionally, it explores the potential of managed aquifer recharge as a sustainable solution for mitigating water scarcity, groundwater depletion, and enhancing aquifer replenishment. The chapter highlights the application and benefits of implementing emerging innovative technologies such as remote sensing, isotopic analysis, and machine learning for groundwater monitoring and management. The integration of advanced hydrogeochemical modelling with emerging tools such as machine learning are improving our traditional techniques and understanding of groundwater systems more accurately and would be highly facilitative of developing strategies for long-term sustainability. The chapter concludes by stressing the need for interdisciplinary research and technological innovations to address global water challenges.

Keywords Hydrogeology · Groundwater management · Climate change impacts · Managed aquifer recharge · Remote sensing · Machine Learning · Hydrochemical modelling

---

R. R. Gantayat · V. Elumalai (✉)
Department of Hydrology, University of Zululand, Kwa-Dlangezwa, Richards Bay, Durban 3886, South Africa
e-mail: ElumalaiV@unizulu.ac.za

P. Li
School of Water and Environment, Chang'an University, No. 126 Yanta Road, Xi'an 710054, Shaanxi, China

Key Laboratory of Subsurface Hydrology and Ecological Effects in Arid Region of the Ministry of Education, Chang'an University, No. 126 Yanta Road, Xi'an 710054, Shaanxi, China

Key Laboratory of Eco-Hydrology and Water Security in Arid and Semi-Arid Regions of the Ministry of Water Resources, Chang'an University, No. 126 Yanta Road, Xi'an 710054, Shaanxi, China

P. Li et al. (eds.), *Sustainable Groundwater and Environment: Challenges and Solutions*, Springer Hydrogeology, https://doi.org/10.1007/978-3-031-82194-3_16

1 Introduction

Hydrogeology, the scientific front of water related studies including its movement, distribution and quality within our planets natural systems. It primarily focuses on critical processes in the hydrological cycle including precipitation, evaporation, infiltration and ground/surface water flow while exploring water interaction with rocks, soil, atmosphere and various organisms in aquatic systems (Winter 1988). Surface water and groundwater systems are two primary components of focus in this field. Hydrologists focus on rivers, lakes, and streams in case of surface water whereas water stored in aquifers in case of groundwater. Key aspects such as understanding run-off, excess waterflow over land after rainfall towards river and lakes are major part of surface water hydrology (Kelbe and Germishuyse 2010). On the other hand, key aspect of groundwater includes the study recharge, replenishment and discharge of water stored in permeable rock or sediments in underground aquifers (Tanner and Hughes 2015). The understanding of such intricate natural dynamics helps hydrologists to predict changes due to climate change, land use and human interventions by providing critical insights for ensuring sustainable water usage. As a result, field of hydrology has gained a broad, international recognition from just branch of geosciences to a cornerstone discipline in within geosciences.

Historically, our water resources are dependent on the change in surrounding physical and environmental factors. The pressure on river, catchments and aquifers have increased over decades due to increase in human population and related activities. Factors such as climatic inputs towards water resources have been found very critical since 1980s, which previously assumed to be constant (Calow and Petts 2009). Various research conducted in the past have directly linked acceleration of hydrological cycle of the planet due to changes in climatic condition especially increase in temperature impacting the precipitation and evapotranspiration across continents (Pratap and Markonis 2022). Such increase in temperature, could cause erratic rainfall pattern impacting the river flow (Bajracharya et al. 2023; Sharma et al. 2019), including timing of high/low flow, which are the major reasons behind droughts and water scarcity (Jehanzaib et al. 2020). As per UN in SDG reports, 2.2 billion people around the world lack access to safe drinking water where 703 million people lack basic water service. In scenario of growing global economies along with population, water scarcity is becoming a more complex issue requiring a sustainable way for clean water supply to meet rising requirements (Singh 2022). Such scenarios are even worsened by the impact of land use changes that mainly increases water contamination. Land use changes such as increasing agriculture, industrialization and population demographics are contributing more than 2 million tons of sewage towards both surface and groundwater. Major water consuming sector is agriculture consisting of 70% of the water usage followed by industries (20%) and domestic usage standing at 10% worldwide (Brauman et al. 2016). In terms of usage USA is the largest consumer of freshwater followed by China whereas India is largest consumer of groundwater with 124% higher usage compared to both China and the USA (Singh 2018) (Fig. 1).

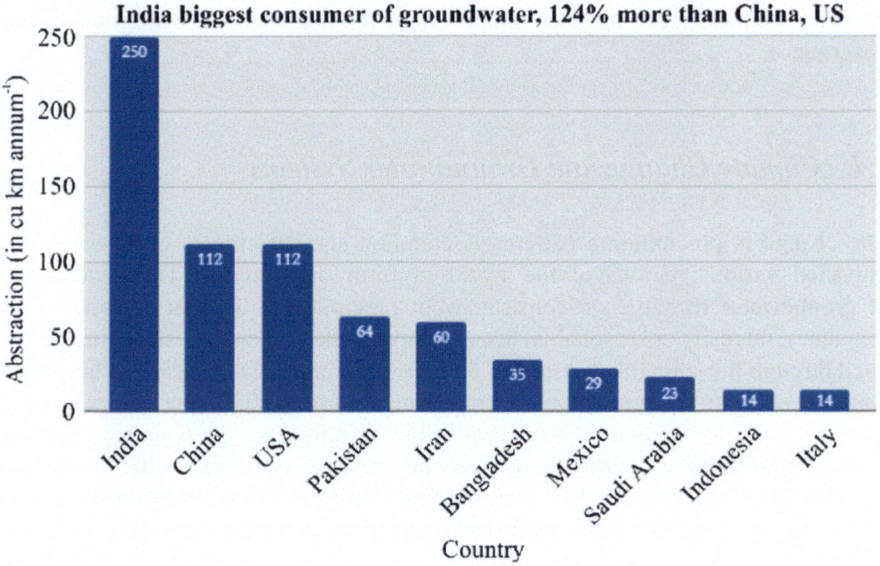

Fig. 1 Groundwater consumption by country (Singh 2018)

Keeping mentioned issues in priority, world's water resources in current state necessitates critical assessment especially due to increased pressure on water availability due to growing population, climate change and contamination. Continuous research is required to understand how climate change might impact water resources in future with continuous changes in precipitation pattern, snowmelt, extreme weather events such as flood and droughts. Research is required to evade water scarcity with the emerging technologies such as water harvesting, desalinisation and efficient water use in agriculture and industries. On the other hand, innovation in water quality improvement research including identification of contaminants and treatment technologies would provide an efficient water resource management for long term sustainability.

2 Emerging Themes in Hydrogeological Research

The study of groundwater and its interaction with interacting geological formations is essential for managing this critical resource sustainably especially during the rise of critical issues such as climate change, population growth and environmental degradation. As discussed before, groundwater provides much of the worlds drinking water and supports agriculture and various ecosystems, new research areas focusing on addressing the challenges related to hydrogeology is more necessary than ever in current century to the future water needs of humanity in sustainable manner. The major focus on such might include groundwater contamination, effects of climate change, development of advance monitoring technologies along with groundwater governance.

2.1 Climate Change and Groundwater Systems

The change is precipitation patterns is common sign of climate change and has impacted various regions of the world in form of droughts. The dependency of groundwater recharge on consistency of precipitation whether in form of its frequency, intensity or seasonality, heavily impacts the volume of water being percolated through the soil replenishing the groundwater table. Major factors behind such patterns are due to change in global average temperature at 0.08 °C per decade causing uneven spatial temperature distribution in higher altitudes than equatorial regions making high-altitude regions more wet while reverse effect in visible in dry and equatorial regions (USGRP 2018; Earman and Dettinger 2011). Additionally, excessive extraction, combined with inconsistent precipitation and recharge puts enormous effect on groundwater table especially in shallow and small aquifers, which are more vulnerable towards prolonged droughts (Dao et al. 2023). The example of such impact is well recorded in California Central Valley, where groundwater accounted for 60% of the water supply during the drought period in 2012–2016 putting enormous pressure on its aquifer forcing state to implement groundwater sustainability laws to combat over extraction (Faunt et al. 2024; Liu et al. 2022). Similar examples are reported around the world, such as reduced groundwater supply during "Millennium Drought" (1997–2009) in Murray-Darling Basin, Australia (Van Dijk et al. 2013), over extraction due to agriculture and erratic monsoon rains in south Asian region especially in Ganges basin of India and Bangladesh, Insufficient recharge and prolonged dry spell in Sub-Saharan countries like Kenya and Ethiopia etc. (Davamani et al. 2024). The impact of irregular rainfall is not only limited to reduced rainfall and impacts groundwater due to heavy pouring leading to higher run-off rather than efficient recharge in contrary effect favouring extreme weather event (Mileham et al. 2009). For example, in sub-Saharan region, continental surface temperatures have increased by 0.7 °C over the last century and expected to increase by 0.2–0.5 °C per decade with extensively higher precipitation, higher evaporative demand, higher run-off and low recharge (Hulme et al. 2001; Joubert and Hewitson 1997; Milly et al. 2005).

Groundwater quality is another factor, on which impact of rising temperature is enormous further complicating the water needs of the world. Several studies have confirmed that groundwater up to 100 m are highly susceptible to atmospheric temperature change in form of potential evaporation (Cui et al. 2020). The subsequent change of such effect could cause change organic composition and biochemical composition (McDonough et al. 2020; Lipczynska-Kochany et al. 2018) and increase mineral weathering over prolonged period, ultimately altering chemical composition of groundwater (Li et al. 2023; Swain et al. 2022). Such impact of regional surface temperature transfer to sub-surface groundwater in porous and unconfined aquifers is noted in Germany (Menberg et al. 2014). Studies consisting of 2196 aquifers in the United States by McDonough et al. (2020) suggested that 1 °C increase in ambient temperature could lead to approximately 3.4% dissolved organic matter increase in groundwater, simultaneously reducing the soil organic matter by over 10%. Such impact could result in changing pH, DO concentration groundwater along with release of toxic metals from upper soil or vadose zones Dong et al. 2022). Such impact is additionally compensated by active land use changes such as forest fire, application of CO_2 based fertilizers and modification in vegetation cover (Pitz 2016), further escalating the preceding positive shift surface temperature to subsurface resources. For example, agriculture dominated catchments could see increase in nutrient release into groundwater due to soil organic matter are known to be the source of nutrients for groundwater resources and could alter nitrogen cycle in the soil and groundwater. AT 5 °C higher temperature than usual could cause 45% higher nitrogen mineralisation in upper soil (Butler et al. 2012), which could dissolve under changing conditions to contaminate the groundwater. In case of arid and semi-arid regions, concentration of chloride and nitrate proses a greater threat than others due to rain-fed incentive agricultural practices and could cause anthropogenic salination of aquifers (Gurdak et al. 2007) (Fig. 2).

On the other hand, sea level rise is a direct result of global temperature rise and is actively influencing hydraulic gradient between freshwater and saltwater interface in groundwater systems causing seawater to penetrate deep into the coastal aquifers (Fig. 3). These scenarios are further accentuated by over extraction of coastal groundwater due to extensive farming, coastal aquafarms practices along with increased organization in the coast. Additional factors such as uneven precipitation patterns and higher evapotranspiration especially arid and semi-arid regions contributes to such salinization actively by further pushing back the freshwater boundaries in coastal

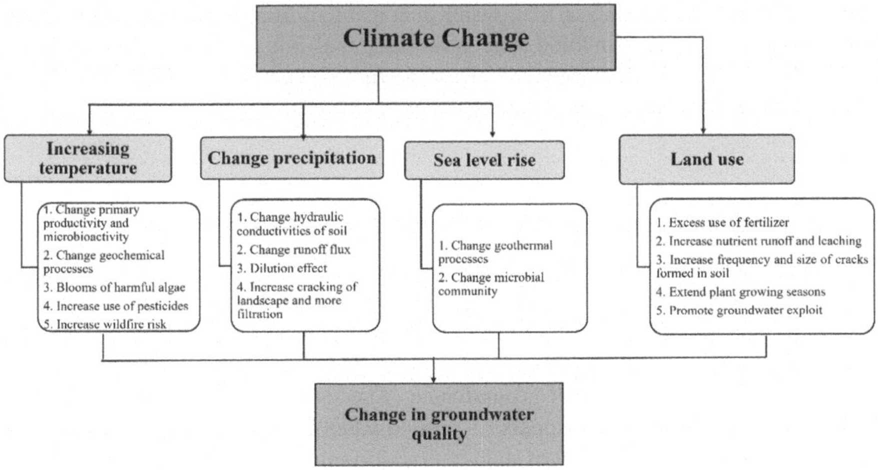

Fig. 2 Natural and anthropogenic impact of climate change on hydrogeology

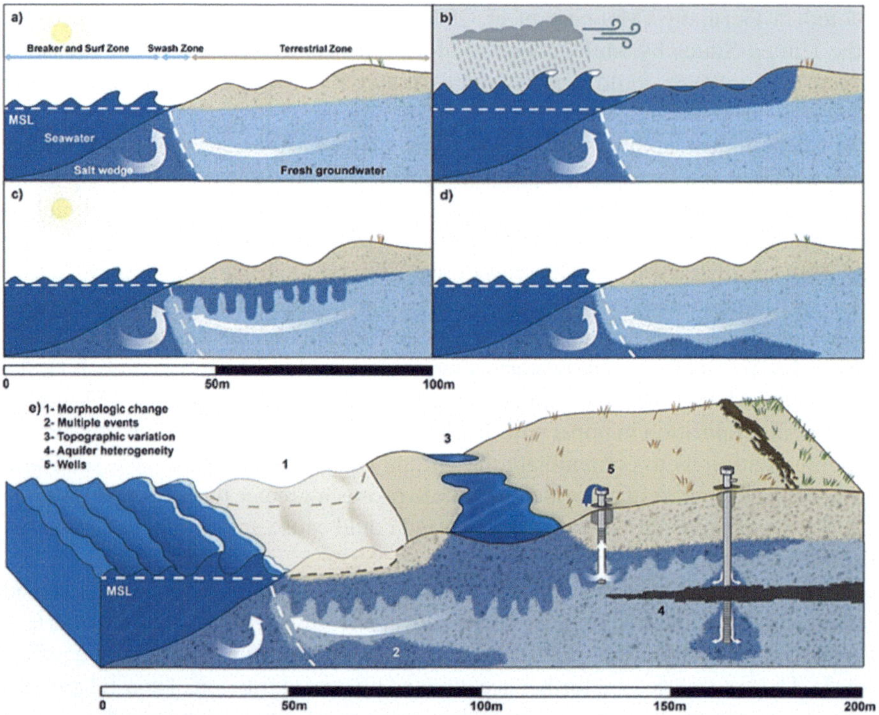

Fig. 3 Seawater influence mechanism on groundwater systems under changing global temperature (Cantelon et al. 2022)

aquifers (Telahigue et al. 2020; Alfarrah and Walraevens 2018). Such effect gives rise to higher residence time of freshwater in coastal aquifers leading high chemical alteration such as increase in ionic concentration, metal concentration in groundwater and have been very well documented in various studies (Gantayat and Elumalai 2024; Mthembu et al. 2020; Elumalai et al. 2017).

2.2 *Groundwater and Surface Water Interactions*

The interaction between groundwater and surface water is a critical aspect of hydrogeological systems. These interactions are complex, as groundwater can contribute to surface water bodies (e.g., rivers, lakes, and wetlands) and vice versa, surface water can infiltrate to replenish groundwater. The interactions can affect physical and chemical and biological aspects of groundwater systems (Kalbus et al. 2006). Some of the examples and effect of such interaction might include surface recharging groundwater, groundwater discharging into streams, depletion of groundwater due to surface water withdrawal, pollution of surface water impacting groundwater quality and vice versa (Duque et al. 2023; Kelbe and Germishuyse 2010; Winter 2000). Understanding and managing these interactions are essential for sustainable water resource management, ecosystem health, and addressing challenges like water shortages, contamination, and the effects of climate change.

Groundwater-surface water interaction is influenced by a range of factors including soil permeability, land cover, topography, and seasonal variations in precipitation (Geris et al. 2022). Traditionally, surface water and groundwater have been studied separately, but it has become clear that they form a single interconnected system (Abdelhalim et al. 2020). Integrated hydrological studies seek to address this gap by investigating processes such as: (a) Baseflow Contribution: Groundwater typically provides baseflow to rivers and streams, especially during dry periods. This is important for maintaining stream ecosystems as groundwater acts as a reliable water source during low flow conditions (Cheng et al. 2022) (b) Recharge from Surface Water: Surface water can recharge groundwater through seepage from rivers, lakes, or reservoirs. This is common in regions with abundant surface water and permeable soils (Watson et al. 2021). (c) Wetlands and Riparian Zones: These transitional zones are crucial as they represent areas where groundwater and surface water interact intensely, often acting as filters for pollutants and regulating water flow (Maestas et al. 2023). Several methodologies have been developed to model and monitor groundwater-surface water interactions, combining physical measurements with computational techniques to simulate these processes on various scales. Modelling techniques such as Integrated Hydrological Models (IHMs), which mainly deals with SWAT and MODFLOW integration designed to combine groundwater and surface water into a single framework (Zhang et al. 2022; Chunn et al. 2019) and Coupled Groundwater-Surface Water Models such as MODFLOW-SFR package (Streamflow Routing) integrates streamflow with groundwater models (Diaz et al. 2020; Ou et al. 2013) have been shown promising progress in this field for advancements.

Despite notable advancement in the field, challenges still exist especially with issues such as developing methods to accurately qualify the exchange volume of water within these interconnected systems. Interaction particularly below the water table making it difficult to quantify directly even with traditional techniques like Dracy's law and seepage meters along with stable isotopes tracers, as they pose certain limitation in such studies (Rosenberry et al. 2020; Cook et al. 2018). geological complexity in natural systems pose a bigger challenge understanding such complex interactions and their qualification (Brunner et al. 2017). However advancement in this field not only serves increase in understating of such connected systems but also contribute highly towards artificial recharge research such as MAR (Managed Aquifer Recharge).

2.3 Managed Aquifer Recharge (MAR)

Managed Aquifer Recharge (MAR) is a sustainable water management approach that involves intentionally replenishing groundwater aquifers with water. This practice offers a range of benefits for communities and ecosystems. MAR techniques have successfully helped evading water scarcity or drought situations by reducing dependency on surface water by increasing groundwater storage. This helps improving the quality of the groundwater by filtering contaminants during infiltration to the aquifer especially in regions of high agriculture and industrial waste. They have been helpful controlling seawater intrusion by replenishing freshwater in aquifers while pushing the seawater gradient back. In addition, they have been helpful in prevention of land subsidence in over exploited regions by replenishing the shortage and maintaining required water levels.

Implementation of such technique have been mainly through injection wells, infiltration ponds and riverbank infiltration, however effectiveness is mainly dependant on geological characteristics of aquifers, quality and quantity of available water in the aquifer. Since 1990s, MAR has been implemented in Australia to manage storm water and deemed safe for drinking and water supplies (Vanderzalm et al. 2014). Innovation in water treatment had led to string of water through MAR in multiples aquifers for subsequent use such as in California, USA (Mills 2020). Apart from countries like Spain and Israel with large infiltration systems for agricultural irrigation suggest a diverse applications of MAR techniques with localized requirements proves it to be an efferent tool for sustainable groundwater management. This could be proved by increasing percentage of water reuse with improved quality from MAR in countries including Finland with highest reuse 0f 23%, followed by Netherlands with reuse rate of 13% (Dillon et al. 2019). In last 6 decades, world's MAR capacity has grown from 1 to 10 km^3/year and growing at 4.5% annually with highest installation cantered in India (Dillon et al. 2019). India has been a major beneficiary of MAR techniques and has the highest MAR capacity with 500,000 recharge structures in state of Gujrat alone along with 11 million more structures being planned across the country (Dillon et al. 2019; Dashora et al. 2018). The is followed by the US with 26% of MAR capacity. The total reported volume of implemented MAR structures in 34 countries could be found in Table 1. These structures have proven impact on improvement of water quality, subsidence prevention (IAH-MAR 2018).

Despite significant growth and promising results in MAR utilization, it is considered far from adequate solution to emerging issues like water quality, over extraction and aquifer salinization. Some of the preliminary investigation in Ganges basin, India and Chao Phraya basin Thailand have suggest enormous potential of MAR in controlling water related disasters such as drought and floods (Pavelic et al. 2021). In water scarce countries MAR is expected to expand to 100 km^3/year in recent years to meet at least 10% of the needs especially in Latin America, China, Southern Africa and SE Asia (Bonilla Valverde et al. 2018). Additionally, MAR implementation in mountainous regions has been very limited around the world, probably due to the higher drainage slope and less flat areas for efficient recharge. However, recent studies have explored options in alluvial fans, faults and fractures in Hillslope, mountain front, and road MAR offer fast recharge and better water quality, with extensive modification and research (Meles et al. 2024). Such localized innovation in MAR techniques with existing method have extensive potential in case of hydrogeological research in recent future (Fig. 4).

Table 1 Existing MAR capacity of various country/regions (Dillon et al. 2019)

Country/ region	Average annual MAR volume in the decade centred on date (Mm^3/year)						Annual GW use (Mm^3/year)	MAR as % GW use	MAR as % of global reported capacity	MAR vol growth (%/year)	GW use as % global use
	1965	1975	1985	1995	2005	2015	2010	2015 (%)	2015	2015	2010 (%)
Austria	–	–	–	–	–	56	1,120	5.00	0.60%	–	0.11
Australia	79	144	185	213	257	410	4,960	8.30	4.10%	3.60	0.51
Belgium	–	–	–	–	–	2.5	650	0.40	0.00%	–	0.07
China	20	23	23	24	56	106	112,000	0.10	1.10%	3.60	11.41
Croatia	–	–	42	48	48	46	600	7.70	0.50%	0.30	0.06
Czech Republic	–	–	–	–	–	22	380	5.80	0.20%	–	0.04
Denmark	–	–	–	–	–	0.25	650	0.00	0.00%	–	0.07
Finland	<1	30	35	50	55	65	280	23.20	0.70%	9.30%	0.03
France	20	21	26	30	31	32	5,710	0.60	0.30%	1.00%	0.58
Germany	–	867	766	875	765	870	3,080	28.20	8.70%	0.00%	0.31
Greece	–	–	–	–	–	0.3	3,650	0.00	0.00%	–	0.37
Hungary	–	–	–	–	–	335	370	90.50	3.40%	–	0.04
India (5 states only)	154	430	706	1,020	1,739	3,070	39,800	7.70	30.90%	6.60%	4.05
Israel	87	91	127	132	144	134	1,250	10.70	1.30%	0.90%	0.13
Italy	178	294	301	348	391	461	10,400	4.40	4.60%	2.00%	1.06
Jordan	4	9	9	13	19	20	640	3.10	0.20%	3.50%	0.07
Korea	–	3.7	12	46	91	146	3,800	3.80	1.50%	10.40%	0.39
Latin America	–	–	–	–	–	311	56,660	0.50	3.10%	–	5.77

(continued)

Table 1 (continued)

Country/region	Average annual MAR volume in the decade centred on date (Mm^3/year)						Annual GW use (Mm^3/year)	MAR as % GW use	MAR as % of global reported capacity	MAR vol growth (%/year)	GW use as % global use
	1965	1975	1985	1995	2005	2015	2010	2015 (%)	2015	2015	2010 (%)
Netherlands	181	240	255	241	275	262	1,600	16.40	2.60%	0.80%	0.16
Oman	0	0	0	6	–	84	840	10.00	0.80%	9.90%	0.09
Poland	–	–	–	–	–	143	2,590	5.50	1.40%	–	0.26
Portugal	–	–	–	–	–	9	6,290	0.10	0.10%	–	0.64
Qatar	<	<1	<1	<1	37	44	260	16.90	0.40%	8.40%	0.03
Romania	–	–	–	–	–	7.4	630	1.20	0.10%	–	0.06
Serbia & Montenegro	–	–	–	–	–	9.5	580	1.60	0.10%	–	0.06
Slovakia	–	–	–	–	–	176	360	48.90	1.80%	–	0.04
Slovenia	–	–	–	–	–	9.5	190	5.00	0.10%	–	0.02
Southeast Asia	<1	<1	<1	<1	<1	5	29,270	0.00	0.00%	–	2.98
Southern Africa	1	2	6	6	7	10	4,500	0.20	0.10%	5.10%	0.46
Spain	3	8	12	60	350	380	5,700	6.70	3.80%	10.90%	0.58
Sweden	–	–	–	–	–	44	350	12.60	0.40%	–	0.04
Switzerland	–	–	–	–	–	100	790	12.70	1.00%	–	0.08
UK	0	0	0	5	5	5	2,160	0.20	0.10%	–	0.22
USA	302	494	768	1,218	2,026	2,569	112,000	2.30	25.80%	4.70%	11.41
Total	**1,029**	**2,656**	**3,272**	**4,334**	**6,296**	**9,945**	**414,110**	**2.40**	**100%**	**4.90%**	**42.20**

(continued)

Table 1 (continued)

Country/ region	Average annual MAR volume in the decade centred on date (Mm^3/year)						Annual GW use (Mm^3/year)	MAR as % GW use	MAR as % of global reported capacity	MAR vol growth (%/year)	GW use as % global use
	1965	1975	1985	1995	2005	2015	2010	2015 (%)	2015	2015	2010 (%)
Global gw use	–	–	–	–	–	–	**982,000**	**1.00**	**NA**	**NA**	**100**
Total (km^3/year)	**1**	**2.7**	**3.3**	**4.3**	**6.3**	**9.9**	**414.1**	**2.4**	**100%**	**4.90%**	**42.20**

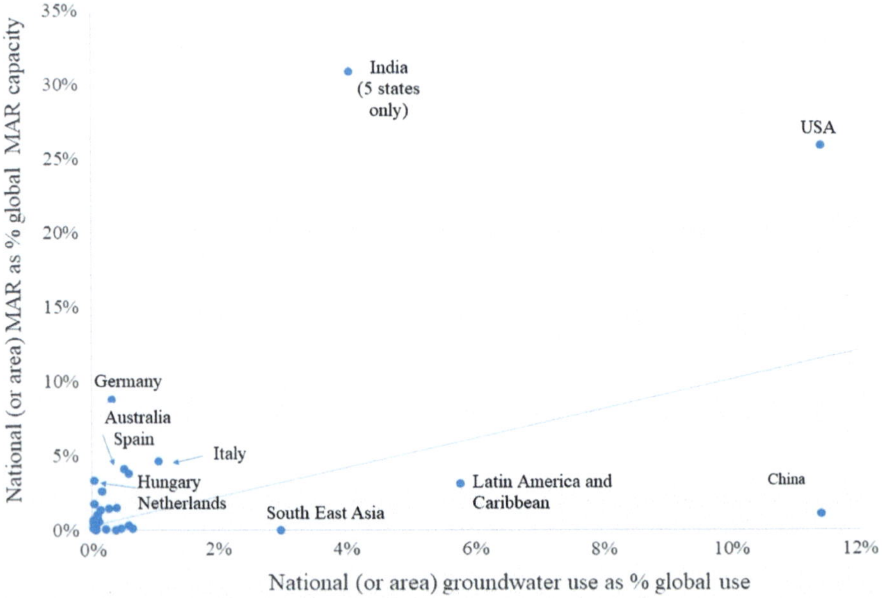

Fig. 4 National or area MAR capacity of country/region (Dillon et al. 2019)

3 Advances in Groundwater Monitoring and Technology

3.1 Remote Sensing in Hydrogeology

Remote sensing technologies, particularly the use of satellite data, have revolutionized hydrogeology by enabling large-scale monitoring of groundwater resources. Satellite-based remote sensing offers an efficient way to monitor groundwater changes over time, especially in areas that are difficult to access.

Satellite data can be utilized for groundwater mapping by observing changes in land surface deformation caused by groundwater extraction or recharge. Various hydrogeological studies frequently utilized satellite images from statistical projection such IDW and Kriging tools (Singh and Verma 2019; Khouni et al. 2021). Another application involves integrating satellite-based precipitation data with hydrological models to improve the accuracy of groundwater recharge estimates (Barbosa et al. 2022; Moges et al. 2022). Remote sensing also facilitates the identification of groundwater potential zones by combining geological, hydrological, and land-use information (e.g. Ponnusamy and Elumalai 2022). This has been particularly useful in arid and semi-arid regions where direct measurement of groundwater resources are challenging.

Recent development of NASA's Gravity Recovery and Climate Experiment (GRACE) satellites have been instrumental in detecting fluctuations in groundwater storage by measuring variations in Earth's gravitational field (Frappart and

Ramillien 2018). These satellites are providing crucial data on the depletion of aquifers and aid in understanding the long-term trends in groundwater availability across large regions. Next-generation satellite missions like GRACE Follow-On will provide more precise measurements of gravitational anomalies linked to groundwater changes along with higher spatial and temporal resolution could open up opportunity of detailed research in hydrogeological field of research. In addition, the development of hyperspectral and thermal infrared remote sensing could aid in detecting subtle changes in surface temperature and vegetation health, which are indicators of groundwater-surface water interactions and recharge zones (Ibrahim et al. 2024). There is also potential for integrating remote sensing with other emerging technologies such as drones and ground-based sensors. These tools can provide more localized data on groundwater recharge areas, helping to refine water management strategies at a regional level.

Innovation in cloud-based platforms such as Amazon Web Services (AWS), Google Earth Engine (GEE) and Microsoft Azure, in the field of remote sensing are supporting planetary based geospatial data from several satellites including Landsat-8 and Sentinel-2 (Pham-Duc et al. 2023). Such access to historical imageries and ability to process them together is allowing researchers to detect changes, map various land-use trends and quantify differences on Earth's surface (Velastegui-Montoya et al. 2023). Among the 3, Microsoft allows utilisation of artificial intelligence and Machine Learning techniques, primarily being utilized in the environmental research including climate change, biodiversity, agriculture and water resources (Mahdavi et al. 2018). However, GEE has been the most popular among the three due to their wide variety of satellite imageries, which includes Landsat series, MODIS, Sentinel-1, -2, -3, and -5P, NOAA, Advanced Land Observing Satellite (ALOS) etc. Additionally, ready to use features like Enhanced Vegetation Index (EVI), Normalized Difference Vegetation Index (NDVI) and popular clustering algorithms such as K-Means, Cobweb or Simple Non-Iterative Clustering (SNIC) are available, which are very helpful in spectral and spatial analyses of historical land use changes is contributing highly to its popularity among researchers (Kumar and Mutanga 2018). Among which NDVI has been 27% of studies since 2010 for draught monitoring and land cover mapping. As a result, one of the highest methodological applications of GEE are water resource assessment and prediction over the last decade (Pérez-Cutillas et al. 2023; Tamiminia et al. 2020). These services are offering wide range of scopes in hydrogeological research including could-based groundwater mapping, modelling, and management of aquifers due to their real time data processing ability. With integration of real-time sensers with cloud-based systems, hydrogeologists could detect anomalies such as groundwater levels, quality, flow rate etc., resulting better decision making. Additionally, integration of GRACE integration with GEE capabilities enables researchers to track groundwater depletion more efficiently.

3.2 Advancement in Stable Isotopes and Tracer Isotopes in Hydrogeological Research

Isotopic methods and tracers have become invaluable in tracing water flow and understanding of groundwater systems, hydrological balance, groundwater origin, water fluxes, controls on water quality etc. (Barbieri 2019; Tweed et al. 2019). Stable isotopes like δ^2H, $\delta^{18}O$, ^{87}Sr, ^{86}Sr along with ^{222}Rn are used commonly in various studies to identify infiltration, percolation, dissolution of minerals, rock-water interaction, ground-surface water interaction (Avery et al. 2018) and seawater-freshwater mixing (González-Trinidad et al. 2017; Boschetti et al. 2018; Tweed et al. 2019). Apart from these, specific tracer isotopes are widely used to study the source of contaminants such as $\delta^{15}N–NO_3^-$ and $\delta^{18}O–NO_3^-$ for nitrate contamination (Meng et al. 2018; Li et al. 2017), $\delta^{34}S$ used for sulfate contamination (Jiménez-Madrid et al. 2017). Another key category of isotope applications involves the decay of radioisotopes such as tritium (^3H), helium-3 ($^3H/^3He$), carbon-14 (^{14}C) (Ortega and Gil 2019). Unlike stable isotopes, which provide insights into the geochemical evolution of aquifers, radioisotopes are primarily used to determine the relative or absolute age of groundwater. This helps revealing the residence time of water in the aquifer after it passes through the vadose zones (Barbieri 2019). However, there are several areas where improvements could make these techniques even more effective. For example, in case of calculation of residence time of groundwater different tracers provide different residence time depending upon their introduction water source and decay duration (Suckow 2014). Another limitation in such studies is inaccuracy, e.g. in few studies ^{14}C based water residence time almost half of original values leading to mass confusion (Vautour et al. 2015). In such cases combining tracers is essential to understanding the distribution of groundwater residence time (Larocque et al. 2009). On the other hand, current isotopic analysis methods are limited by resolution, especially in distinguishing subtle differences in water origins or recharge processes (Scandellari et al. 2024). Increasing the sensitivity of isotopic measurement tools could allow for better detection of small changes in isotopic composition, leading to more accurate interpretations of groundwater movement and age. Instruments that can perform real-time isotopic analysis in the field would greatly enhance data collection efficiency. While isotopes like tritium and carbon-14 are essential for dating groundwater, their analysis often requires specialized equipment and extensive sample preparation. Developing more accessible and cost-effective instruments for radioisotope analysis could expand their use in hydrogeological studies, particularly for smaller research projects and regions with limited resources.

3.3 Hydrogeochemical Modelling Improvements: Induction of Machine Learning and Simulation Techniques

Various numerical and statistical modelling have been a common practice in case hydrogeochemical evaluation such as water quality index, interconnection between parameters for last few decades (Ahmed et al. 2019; Liu et al. 2024). WQI were proposed in 1965 and has been employed ever since for water quality investigation (Mijares et al. 2019; Abbasi and Abbasi 2012). On the other hand, statistical-based water quality models have utilized linear distribution assuming correlation between parameters. Such traditional techniques are not sufficient to address potential impact of various parameters in geochemical studies. Recent introduction of Machine learning (ML) in water research, has driven its applications monitoring, simulation, prediction, pollutant source tracking, pollutant concentration estimation, water resource allocation, and water treatment technology optimization (Zhu et al. 2022). Researchers believe that vast amount of captured data could be processed and optimized faster while discovering interconnected patterns for complex groundwater evacuations and water treatment with the help of ML (Liu et al. 2017). Apart from that, existing models such as Water Quality Index (WQI), Principal Component Analysis (PCA) etc. could be optimized using various algorithms/models such as Support Vector Machine (SVM), Artificial Neural Network (ANN), Linear Regression Model (LRM), Regression Tree (RT), Gaussian Process Regression (GPR), and Boosted Regression Tree (BRT), naive Bayes, k-nearest neighbour (KNN), random forest (RF) (e.g. Khan and Ayaz 2024; Berry et al. 2019). Similarly, in case of risk assessment, implementation of improved statistical receptor models such as EPA PMF, UNIMIX etc. are rapidly being utilized for source tracing due to their ability to integrate data uncertainty in the models and quantifying source contributions (Zanotti et al. 2019). They capable of validation of the results have made these tool more appealing towards researchers. Additionally, recorded source contribution is being integrated with numerical risk assessment with help of Monte Carlo Simulation indices are capable of proving more efficient mitigation strategy (e.g. Eid et al. 2024; Panqing et al. 2023; Huang et al. 2021). As more hydrogeologists integrate ML and simulations into their workflows, these tools will become essential for ensuring sustainable groundwater management in the face of global water challenges.

4 Conclusions

Advancement in hydrogeological research is essential for securing sustainable water supplies in the future. As precipitation patterns shift and droughts become more frequent, groundwater recharge is increasingly threatened, affecting agricultural production, drinking water availability, and ecosystem stability globally. In future scenarios, the persistence of these trends could lead to widespread water shortages, particularly in regions dependent on groundwater for drinking and irrigation. In countries like India, Iran, and parts of Sub-Saharan Africa, where groundwater extraction exceeds recharge, declining aquifers could devastate rural economies and worsen food insecurity. In such case research in widespread application of MAR would be vital around the world. In contrast, regions that invest in adaptive water management—such as artificial recharge, water reuse technologies, and integrated resource planning—may better withstand the impacts of climate change. However, modelling future climate scenarios will also play a critical role in predicting changes to groundwater systems, allowing governments and industries to plan accordingly. Furthermore, advancements in application of Machine learning tools in remote sensing such as in GEE and hydrogeological monitoring technologies will be vital for more accurate data collection and better management of aquifers under foreseeable climate stress.

Without such forward-looking strategies, groundwater depletion and contamination could become one of the most pressing challenges of the twenty-first century. Integrating advanced techniques with hydrogeology is necessary to protect this vital resource, ensure water security, and mitigate future crises related to drought, groundwater contamination, and population growth.

Acknowledgements Authors from the University of Zululand express their gratitude to National Research Foundation (NRF), South Africa (UNIQUE GRANT NO: 151823) for supporting South African Research Chairs Initiative (SARChI—Tier 1) grants. The authors are grateful to UNIZULU and management for providing a research grant for the "Water Crises Project" and Sibusiso Bengu Development Plan (SBDevP) for the 2022/23 to 2026/27 "Sustainable Energy, Water and Agriculture" through Strategic Funds to assist in the successful completion of this research.

References

Abbasi T, Abbasi SA (2012) Water quality indices. Elsevier

Abdelhalim A, Sefelnasr A, Ismail E (2020) Response of the interaction between surface water and groundwater to climate change and proposed megastructure. J Afr Earth Sc 162:103723

Ahmed AN, Othman FB, Afan HA, Ibrahim RK, Fai CM, Hossain MS et al (2019) Machine learning methods for better water quality prediction. J Hydrol 578:124084

Alfarrah N, Walraevens K (2018) Groundwater overexploitation and seawater intrusion in coastal areas of arid and semi-arid regions. Water 10(2):143

Avery E, Bibby R, Visser A, Esser B, Moran J (2018) Quantification of groundwater discharge in a subalpine stream using radon-222. Water 10(2):100

Bajracharya SR, Pradhananga S, Shrestha AB, Thapa R (2023) Future climate and its potential impact on the spatial and temporal hydrological regime in the Koshi Basin, Nepal. J Hydrol: Reg Stud 45:101316

Barbieri M (2019) Isotopes in hydrology and hydrogeology. Water 11(2):291

Barbosa LR, Coelho VHR, Gusmão ACV, Fernandes LA, da Silva BB, Galvão CDO et al (2022) A satellite-based approach to estimating spatially distributed groundwater recharge rates in a tropical wet sedimentary region despite cloudy conditions. J Hydrol 607:127503

Berry MW, Mohamed A, Yap BW (2019) Supervised and unsupervised learning for data science. Springer Nature

Bonilla Valverde JP, Stefan C, Palma Nava A, Bernardo da Silva E, Pivaral Vivar HL (2018) Inventory of managed aquifer recharge schemes in Latin America and the Caribbean. Sustain Water Resour Manag 4:163–178
Boschetti T, Awaleh MO, Barbieri M (2018) Waters from the Djiboutian Afar: a review of strontium isotopic composition and a comparison with Ethiopian waters and red sea brines. Water 10(11):1700
Brauman KA, Richter BD, Postel S, Malsy M, Flörke M (2016) Water depletion: an improved metric for incorporating seasonal and dry-year water scarcity into water risk assessments. Elementa 4:000083
Brunner P, Therrien R, Renard P, Simmons CT, Franssen HJH (2017) Advances in understanding river–groundwater interactions. Rev Geophys 55(3):818–854
Butler SM, Melillo JM, Johnson JE, Mohan J, Steudler PA, Lux H et al (2012) Soil warming alters nitrogen cycling in a New England forest: implications for ecosystem function and structure. Oecologia 168:819–828
Calow PP, Petts GE (2009) The rivers handbook: hydrological and ecological principles. Wiley
Cantelon JA, Guimond JA, Robinson CE, Michael HA, Kurylyk BL (2022) Vertical saltwater intrusion in coastal aquifers driven by episodic flooding: a review. Water Resour Res 58(11):e2022WR032614
Cheng S, Tong X, Illman WA (2022) Evaluation of baseflow separation methods with real and synthetic streamflow data from a watershed. J Hydrol 613:128279
Chunn D, Faramarzi M, Smerdon B, Alessi DS (2019) Application of an integrated SWAT–MODFLOW model to evaluate potential impacts of climate change and water withdrawals on groundwater–surface water interactions in West-Central Alberta. Water 11(1):110
Cook PG, Rodellas V, Stieglitz TC (2018) Quantifying surface water, porewater, and groundwater interactions using tracers: tracer fluxes, water fluxes, and end-member concentrations. Water Resour Res 54(3):2452–2465
Cui Y, Liao Z, Wei Y, Xu X, Song Y, Liu H (2020) The response of groundwater level to climate change and human activities in Baotou City, China. Water 12(4):1078
Dao PU, Heuzard AG, Le TXH, Zhao J, Yin R, Shang C, Fan C (2023) The impacts of climate change on groundwater quality: a review. Sci Total Environ 169241
Dashora Y, Dillon P, Maheshwari B, Soni P, Dashora R, Davande S et al (2018) A simple method using farmers' measurements applied to estimate check dam recharge in Rajasthan, India. Sustain Water Resour Manag 4:301–316
Davamani V, John JE, Poornachandhra C, Gopalakrishnan B, Arulmani S, Parameswari E et al (2024) A critical review of climate change impacts on groundwater resources: a focus on the current status, future possibilities, and role of simulation models. Atmosphere 15(1):122
Diaz M, Sinicyn G, Grodzka-Łukaszewska M (2020) Modelling of groundwater–surface water interaction applying the hyporheic flux model. Water 12(12):3303
Dillon P, Stuyfzand P, Grischek T, Lluria M, Pyne RDG, Jain RC et al (2019) Sixty years of global progress in managed aquifer recharge. Hydrogeol J 27(1):1–30
Dong L, Zhang J, Guo Z, Li M, Wu H (2022) Distributions and interactions of dissolved organic matter and heavy metals in shallow groundwater in Guanzhong basin of China. Environ Res 207:112099
Duque C, Nilsson B, Engesgaard P (2023) Groundwater–surface water interaction in Denmark. Wiley Interdiscip Rev Water 10(5):e1664
Earman S, Dettinger M (2011) Potential impacts of climate change on groundwater resources—A global review. J Water Clim Change 2(4):213–229
Eid MH, Eissa M, Mohamed EA, Ramadan HS, Tamás M, Kovács A, Szűcs P (2024) New approach into human health risk assessment associated with heavy metals in surface water and groundwater using Monte Carlo method. Sci Rep 14(1):1008
Elumalai V, Brindha K, Sithole B, Lakshmanan E (2017) Spatial interpolation methods and geostatistics for mapping groundwater contamination in a coastal area. Environ Sci Pollut Res 24:11601–11617

Faunt CC, Traum JA, Boyce SE, Seymour WA, Jachens ER, Brandt JT et al (2024) Groundwater sustainability and Land Subsidence in California's Central Valley. Water 16(8):1189

Frappart F, Ramillien G (2018) Monitoring groundwater storage changes using the Gravity Recovery and Climate Experiment (GRACE) satellite mission: a review. Remote Sens 10(6):829

Gantayat RR, Elumalai V (2024) Salinity-induced changes in heavy metal behavior and mobility in semi-arid coastal aquifers: a comprehensive review. Water 16(7):1052

Geris J, Comte JC, Franchi F, Petros AK, Tirivarombo S, Selepeng AT, Villholth KG (2022) Surface water-groundwater interactions and local land use control water quality impacts of extreme rainfall and flooding in a vulnerable semi-arid region of Sub-Saharan Africa. J Hydrol 609:127834

González-Trinidad J, Pacheco-Guerrero A, Júnez-Ferreira H, Bautista-Capetillo C, Hernández-Antonio A (2017) Identifying groundwater recharge sites through environmental stable isotopes in an alluvial aquifer. Water 9(8):569

Gurdak JJ, Hanson RT, McMahon PB, Bruce BW, McCray JE, Thyne GD, Reedy RC (2007) Climate variability controls on unsaturated water and chemical movement, High Plains aquifer, USA. Vadose Zone J 6(3):533–547

Huang J, Wu Y, Sun J, Li X, Geng X, Zhao M et al (2021) Health risk assessment of heavy metal (loid) s in park soils of the largest megacity in China by using Monte Carlo simulation coupled with Positive matrix factorization model. J Hazard Mater 415:125629

Hulme M, Doherty R, Ngara T, New M, Lister D (2001) African climate change: 1990–2100. Clim Res 17:145–168

IAH-MAR (2018) International association of hydrogeologists—Commission on managing aquifer recharge. Working group on 60 years history of MAR: national summaries

Ibrahim A, Wayayok A, Shafri HZM, Toridi NM (2024) Remote sensing technologies for unlocking new groundwater insights: a comprehensive review. J Hydrol: X 23:100175

Jehanzaib M, Sattar MN, Lee JH, Kim TW (2020) Investigating effect of climate change on drought propagation from meteorological to hydrological drought using multi-model ensemble projections. Stoch Env Res Risk Assess 34:7–21

Jiménez-Madrid A, Castaño S, Vadillo I, Martinez C, Carrasco F, Soler A (2017) Applications of hydro-chemical and isotopic tools to improve definitions of groundwater catchment zones in a karstic aquifer: a case study. Water 9(8):595

Joubert PD, Hewitson BC (1997) Simulating present and future climates of southern Africa using general circulation models. Progr Phys Geogr 21:51–78

Kalbus E, Reinstorf F, Schirmer M (2006) Measuring methods for groundwater–surface water interactions: a review. Hydrol Earth Syst Sci 10(6):873–887

Kelbe B, Germishuyse T (2010) Groundwater/surface water relationships with specific reference to Maputaland. Water Research Commission Report, 10

Khan I, Ayaz M (2024) Sensitivity analysis-driven machine learning approach for groundwater quality prediction: insights from integrating ENTROPY and CRITIC methods. Groundw Sustain Dev 26:101309

Khouni I, Louhichi G, Ghrabi A (2021) Use of GIS based Inverse Distance Weighted interpolation to assess surface water quality: case of Wadi El Bey, Tunisia. Environ Technol Innov 24:101892

Kumar L, Mutanga O (2018) Google Earth Engine applications since inception: usage, trends, and potential. Remote Sens 10(10):1509

Larocque M, Cook PG, Haaken K, Simmons CT (2009) Estimating flow using tracers and hydraulics in synthetic heterogeneous aquifers. Groundwater 47(6):786–796

Li D, Jiang X, Zheng B (2017) Using $\delta^{15}N$ and $\delta^{18}O$ signatures to evaluate nitrate sources and transformations in four inflowing rivers, north of Taihu Lake. Water 9(5):345

Li N, Lyu H, Xu G, Chi G, Su X (2023) Hydrogeochemical changes during artificial groundwater well recharge. Sci Total Environ 900:165778

Lipczynska-Kochany E (2018) Effect of climate change on humic substances and associated impacts on the quality of surface water and groundwater: a review. Sci Total Environ 640:1548–1565

Liu Y, Zhao T, Ju W, Shi S (2017) Materials discovery and design using machine learning. J Materiomics 3(3):159–177
Liu PW, Famiglietti JS, Purdy AJ, Adams KH, McEvoy AL, Reager JT et al (2022) Groundwater depletion in California's Central Valley accelerates during megadrought. Nat Commun 13(1):7825
Liu Z, Zhou J, Yang X, Zhao Z, Lv Y (2024) Research on water resource modeling based on machine learning technologies. Water 16(3):472
Maestas JD, Wheaton JM, Bouwes N, Swanson SR, Dickard M (2023) Water is life: importance and management of riparian areas for Rangeland Wildlife. Rangeland wildlife ecology and conservation. Springer International Publishing, Cham, pp 177–208
Mahdavi S, Salehi B, Granger J, Amani M, Brisco B, Huang W (2018) Remote sensing for wetland classification: a comprehensive review. Gisci Remote Sens 55(5):623–658
McDonough LK, Santos IR, Andersen MS, O'Carroll DM, Rutlidge H, Meredith K et al (2020) Changes in global groundwater organic carbon driven by climate change and urbanization. Nat Commun 11(1):1279
Meles MB, Bradford S, Casillas-Trasvina A, Chen L, Osterman G, Hatch T et al (2024) Uncovering the gaps in managed aquifer recharge for sustainable groundwater management: a focus on hillslopes and mountains. J Hydrol 131615
Menberg K, Blum P, Kurylyk BL, Bayer P (2014) Observed groundwater temperature response to recent climate change. Hydrol Earth Syst Sci 18(11):4453–4466
Meng Z, Yang Y, Qin Z, Huang L (2018) Evaluating temporal and spatial variation in nitrogen sources along the lower reach of Fenhe River (Shanxi Province, China) using stable isotope and hydrochemical tracers. Water 10(2):231
Mijares V, Gitau M, Johnson DR (2019) A method for assessing and predicting water quality status for improved decision-making and management. Water Resour Manag 33:509–522
Mileham L, Taylor RG, Todd M, Tindimugaya C, Thompson J (2009) The impact of climate change on groundwater recharge and runoff in a humid, equatorial catchment: sensitivity of projections to rainfall intensity. Hydrol Sci J 54(4):727–738
Mills WR (2020) The quest for water through artificial recharge and wastewater recycling. In: Management of aquifer recharge for sustainability. CRC Press, pp 3–10
Milly PCD, Dunne KA, Vecchia AV (2005) Global pattern of trends in streamflow and water availability in a changing climate. Nature 438(7066):347–350
Moges DM, Kmoch A, Uuemaa E (2022) Application of satellite and reanalysis precipitation products for hydrological modeling in the data-scarce Porijõgi catchment, Estonia. J Hydrol: Reg Stud 41:101070
Mthembu PP, Elumalai V, Brindha K, Li P (2020) Hydrogeochemical processes and trace metal contamination in groundwater: impact on human health in the Maputaland coastal aquifer, South Africa. Expos Health 12(3):403–426
Ortega L, Gil L (2019) Isotope hydrology: an overview. IAEA Bull 5
Ou G, Chen X, Kilic A, Bartelt-Hunt S, Li Y, Samal A (2013) Development of a cross-section based streamflow routing package for MODFLOW. Environ Modell Softw 50:132–143
Panqing Y, Abliz A, Xiaoli S, Aisaiduli H (2023) Human health-risk assessment of heavy metal–contaminated soil based on Monte Carlo simulation. Sci Rep 13(1):7033
Pavelic P, Sikka A, Alam MF, Sharma BR, Mutuwatte L, Eriyagama N et al (2021) Case study 15: recharging floodwaters to depleted aquifers for irrigation in the Ganges Basin, India. In: Zheng Y, Ross A, Villholth KG, Dillon P (eds) Managing aquifer recharge: a showcase for resilience and sustainability. UNESCO, Paris, pp 213–222
Pérez-Cutillas P, Pérez-Navarro A, Conesa-García C, Zema DA, Amado-Álvarez JP (2023) What is going on within google earth engine? A systematic review and meta-analysis. Remote Sens Appl: Soc Environ 29:100907
Pham-Duc B, Nguyen H, Phan H, Tran-Anh Q (2023) Trends and applications of google earth engine in remote sensing and earth science research: a bibliometric analysis using scopus database. Earth Sci Inf 16(3):2355–2371

Pitz CF (2016) Predicted impacts of climate change on groundwater resources of Washington State. Environmental Assessment Program, Washington State Department of Ecology

Ponnusamy D, Elumalai V (2022) Determination of potential recharge zones and its validation against groundwater quality parameters through the application of GIS and remote sensing techniques in uMhlathuze catchment, KwaZulu-Natal, South Africa. Chemosphere 307:136121

Pratap S, Markonis Y (2022) The response of the hydrological cycle to temperature changes in recent and distant climatic history. Prog Earth Planet Sci 9(1):30

Rosenberry DO, Duque C, Lee DR (2020) History and evolution of seepage meters for quantifying flow between groundwater and surface water: part 1—Freshwater settings. Earth Sci Rev 204:103167

Scandellari F, Attou T, Barbeta A, Bernhard F, D'Amato C, Dimitrova-Petrova K et al (2024) Using stable isotopes to inform water resource management in forested and agricultural ecosystems. J Environ Manag 365:121381

Sharma E, Molden D, Rahman A, Khatiwada YR, Zhang L, Singh SP et al (2019) Introduction to the hindu kush himalaya assessment. In: The Hindu Kush Himalaya assessment: Mountains, climate change, sustainability and people, pp 1–16

Singh AK (2018) Water security—A reality check. J Indian Soc Soil Sci 66:S179–S192

Singh P, Verma P (2019) A comparative study of spatial interpolation technique (IDW and Kriging) for determining groundwater quality. In: GIS and geostatistical techniques for groundwater science, pp 43–56

Singh KK (2022) Role of nanotechnology and nanomaterials for water treatment and environmental remediation. Int J New Chem 9(3):165–190

Suckow A (2014) The age of groundwater—Definitions, models and why we do not need this term. Appl Geochem 50:222–230

Swain S, Taloor AK, Dhal L, Sahoo S, Al-Ansari N (2022) Impact of climate change on groundwater hydrology: a comprehensive review and current status of the Indian hydrogeology. Appl Water Sci 12(6):120

Tamiminia H, Salehi B, Mahdianpari M, Quackenbush L, Adeli S, Brisco B (2020) Google Earth Engine for geo-big data applications: a meta-analysis and systematic review. ISPRS J Photogramm Remote Sens 164:152–170

Tanner JL, Hughes DA (2015) Understanding and modelling surface water-groundwater interactions (No. 2056/2, p. 14). WRC Report

Telahigue F, Mejri H, Mansouri B, Souid F, Agoubi B, Chahlaoui A, Kharroubi A (2020) Assessing seawater intrusion in arid and semi-arid Mediterranean coastal aquifers using geochemical approaches. Phys Chem Earth Parts A/B/C 115:102811

Tweed S, Leblanc M, Cartwright I, Bass A, Travi Y, Marc V et al (2019) Stable isotopes of water in hydrogeology. In: Encyclopaedia of water. Wiley, Hoboken, NJ, USA, pp 1–10

USGRP (2018) Fourth national climate assessment. US Global Change Research Program: Washington, DC, USA

Van Dijk AI, Beck HE, Crosbie RS, De Jeu RA, Liu YY, Podger GM et al (2013) The Millennium Drought in southeast Australia (2001–2009): natural and human causes and implications for water resources, ecosystems, economy, and society. Water Resour Res 49(2):1040–1057

Vanderzalm J, Page D, Dillon P, Lawson J, Grey N, Sexton D, Williamson D (2014) A risk-based management plan for Mount Gambier Stormwater recharge system: stormwater recharge to the Gambier Limestone aquifer. Goyder Inst Water Res Tech Rep 14(7)

Vautour G, Pinti DL, Méjean P, Saby M, Meyzonnat G, Larocque M et al (2015) 3H/3He, 14C and (U–Th)/he groundwater ages in the st. lawrence lowlands, quebec, eastern canada. Chem Geol 413:94–106

Velastegui-Montoya A, Montalván-Burbano N, Carrión-Mero P, Rivera-Torres H, Sadeck L, Adami M (2023) Google Earth Engine: a global analysis and future trends. Remote Sens 15(14):3675

Watson A, Kralisch S, van Rooyen J, Miller J (2021) Quantifying and understanding the source of recharge for alluvial systems in arid environments through the development of a seepage model. J Hydrol 601:126650

Winter TC (1988) A conceptual framework for assessing cumulative impacts on the hydrology of nontidal wetlands. Environ Manag 12:605–620

Winter TC (2000) Ground water and surface water: a single resource. Diane Publishing

Zanotti C, Rotiroti M, Fumagalli L, Stefania GA, Canonaco F, Stefenelli G et al (2019) Groundwater and surface water quality characterization through positive matrix factorization combined with GIS approach. Water Res 159:122–134

Zhang L, Dai Y, Lin J, Han J, Sun X, Li X et al (2022) Evaluating spatiotemporal variations of groundwater–surface water interaction using an integrated hydrological model in Huashan Basin, China. Sustainability 14(21):14325

Zhu M, Wang J, Yang X, Zhang Y, Zhang L, Ren H et al (2022) A review of the application of machine learning in water quality evaluation. Eco-Environ Health 1(2):107–116

Index

P. Li et al. (eds.), *Sustainable Groundwater and Environment: Challenges and Solutions*, Springer Hydrogeology, https://doi.org/10.1007/978-3-031-82194-3

W

The manufacturer's authorised representative in the EU is Springer Nature Customer Service Centre GmbH, Europaplatz 3, 69115 Heidelberg, Germany. If you have any concerns regarding our products, please contact ProductSafety@springernature.com

Printed and bound by CPI Group (UK) Ltd, Croydon, CR0 4YY
01/10/2026
02225960-0018